# Undergraduate Texts in Mathematics

*Editors*
S. Axler
F. W. Gehring
K.A. Ribet

**Springer**
*New York*
*Berlin*
*Heidelberg*
*Hong Kong*
*London*
*Milan*
*Paris*
*Tokyo*

## Undergraduate Texts in Mathematics

*(continued after index)*

Robin Hartshorne

# Geometry:
# Euclid and Beyond

With 550 Illustrations

 Springer

Robin Hartshorne
Department of Mathematics
University of California
Berkeley, CA 94720
USA
robin@math.berkeley.edu

Mathematics Subject Classification (2000): 51-01

Library of Congress Cataloging-in-Publication Data
Hartshorne, Robin.
    Geometry: Euclid and beyond/Robin Hartshorne.
      p.  cm. — (Undergraduate texts in mathematics)
    Includes bibliographical references and index.
    ISBN 0-387-98650-2 (hc)
    1. Geometry  I. Title.  II. Series.
    QA451 .H37  2000
    516 21—dc21                   99-044789

Printed on acid-free paper.

Production managed by Frank McGuckin; manufacturing supervised by Jerome Basma.
Typeset by Asco Trade Typesetting Ltd., Hong Kong.
Printed and bound by Maple-Vail Book Manufacturing Group, York, PA.
Printed in the United States of America.

9 8 7 6 5 4 3

ISBN 0-387-98650-2          SPIN 10945773

Springer-Verlag  New York Berlin Heidelberg
*A member of BertelsmannSpringer Science+Business Media GmbH*

*For*
*Edie, Ben, and Joemy*

*and*

*In Loving Memory of*
*Jonathan Churchill Hartshorne*
*1972–1992*

I have not found anything in Lobatchevski's work that is new to me, but the development is made in a different way from the way I had started and to be sure masterfully done by Lobatchevski in the pure spirit of geometry.

– letter from Gauss to Schumacher (1846)

# Preface

In recent years, I have been teaching a junior–senior-level course on the classical geometries. This book has grown out of that teaching experience. I assume only high-school geometry and some abstract algebra. The course begins in Chapter 1 with a critical examination of Euclid's *Elements*. Students are expected to read concurrently Books I–IV of Euclid's text, which must be obtained separately. The remainder of the book is an exploration of questions that arise naturally from this reading, together with their modern answers. To shore up the foundations we use Hilbert's axioms. The Cartesian plane over a field provides an analytic model of the theory, and conversely, we see that one can introduce coordinates into an abstract geometry. The theory of area is analyzed by cutting figures into triangles. The algebra of field extensions provides a method for deciding which geometrical constructions are possible. The investigation of the parallel postulate leads to the various non-Euclidean geometries. And in the last chapter we provide what is missing from Euclid's treatment of the five Platonic solids in Book XIII of the *Elements*.

For a one-semester course such as I teach, Chapters 1 and 2 form the core material, which takes six to eight weeks. Then, depending on the taste of the instructor, one can follow a more geometric path by going directly to non-Euclidean geometry in Chapter 7, or a more algebraic one, exploring the relation between geometric constructions and field extensions, by doing Chapters 3, 4, and 6. For me, one of the most interesting topics is the introduction of coordinates into an abstractly given geometry, which is done for a Euclidean plane in Section 21, and for a hyperbolic plane in Section 41.

Throughout this book, I have attempted to choose topics that are accessible to undergraduates and that are interesting in their own right. The exercises are meant to be challenging, to stimulate a sense of curiosity and discovery in the student. I purposely do not indicate their difficulty, which varies widely.

I hope this material will become familiar to every student of mathematics, and in particular to those who will be future teachers.

I owe thanks to Marvin Greenberg for reading and commenting on large portions of the text, to Hendrik Lenstra for always having an answer to my questions, and to Victor Pambuccian for valuable references to the literature. Thanks to Faye Yeager for her patient typing and retyping of the manuscript. And special thanks to my wife, Edie, for her continual loving support.

Of all the works of antiquity which have been transmitted to the present times, none are more universally and deservedly esteemed than the *Elements of Geometry* which go under the name of Euclid. In many other branches of science the moderns have far surpassed their masters; but, after a lapse of more than two thousand years, this performance still maintains its original preeminence, and has even acquired additonal celebrity from the fruitless attempts which have been made to establish a different system.

– from the preface to
Bonnycastle's Euclid
London (1798)

# Contents

# Introduction

 little after the time of Plato, but before Archimedes, in ancient Greece, a man named Euclid wrote the *Elements*, gathering and improving the work of his predecessors Pythagoras, Theaetetus, and Eudoxus into one magnificent edifice. This book soon became the standard for geometry in the classical world. With the decline of the great civilizations of Athens and Rome, it moved eastward to the center of Arabic learning in the court of the caliphs at Baghdad.

In the late Middle Ages it was translated from Arabic into Latin, and since the Renaissance it not only has been the most widely used textbook in the world, but has had an influence as a model of scientific thought that extends way beyond the confines of geometry. As Billingsley said in his preface to the first English translation (1570), "Without the diligent studie of Euclides *Elementes*, it is impossible to attaine unto the perfecte knowledge of Geometrie, and consequently of any of the other Mathematical Sciences." Even today, though few schools use the original text of Euclid, the content of a typical high-school geometry course is the same as what Euclid taught more than two thousand three hundred years ago.

In this book we will take Euclid's *Elements* as the starting point for a study of geometry from a modern mathematical perspective.

To begin, we will become familiar with the content of Euclid's work, at least those parts that deal with geometry (Books I–IV, VI, and XI–XIII). Here we find theorems that should be familiar to anyone who has had a course of high-school

**1**

geometry, such as the fact (I.4) that two triangles are congruent if they have two sides and the included angle equal, or the fact (III.21) that a given arc of a circle subtends the same angle at any point of the circle from which it is seen. (Throughout this book, references such as (I.4) or (III.21) refer to the corresponding Book and Proposition number in Euclid's *Elements*.)

Many of Euclid's propositions pose construction problems, such as (I.1), to construct an equilateral triangle, or (IV.11), to construct a regular pentagon inscribed in a circle. Euclid means to construct the required figure using only the ruler, which can draw a straight line through two points, and the compass, which can draw a circle with given center and given radius. These **ruler and compass constructions** are often taught in high-school geometry. Note that Euclid casts these problems in the form of constructions, whereas a modern mathematician would be more likely to speak of proving the existence of the required figure.

At a second level, we will study the logical structure of Euclid's presentation. Euclid's *Elements* has been regarded for more than two thousand years as the prime example of the **axiomatic method**. Starting from a small number of self-evident truths, called postulates, or common notions, he deduces all the succeeding results by purely logical reasoning. Euclid thus begins with the simplest assumptions, such as Postulate 1, to draw a line through any two given points, or Postulate 3, to draw a circle with given center and radius. He then proceeds step by step to the culmination of the work in Book XIII, where he gives the construction of the five regular solids: the tetrahedron, the cube, the octahedron, the icosahedron, and the dodecahedron.

Upon closer reading, we find that Euclid does not adhere to the strict axiomatic method as closely as one might hope. Certain steps in certain proofs depend on assumptions that, however reasonable or intuitively clear they may seem, cannot be justified on the basis of the stated postulates and common notions. So, for example, the fact that the two circles in the proof of (I.1) will actually meet at some point seems obvious, but is not proved. The **method of superposition** used in the proof of (I.4), which allows one to move the triangle ABC so that it lies on top of the triangle DEF, cannot be justified from the axioms. Also, various assumptions about the relative position of figures in the plane, such as which point lies between the others, or which ray lies in the interior of a given angle, are used without any previous clarification of what such notions should mean.

These lapses in Euclid's logic lead us to the task of disengaging those implicit assumptions that are used in his arguments and providing a new set of axioms from which we can develop geometry according to modern standards of rigor. The logical foundations of geometry were widely studied in the late nineteenth century, which led to a set of axioms proposed by Hilbert in his lectures on the foundations of geometry in 1899. We will examine Hilbert's axioms, and we will see how these axioms can be used to build a solid base from which to develop Euclid's geometry pretty much according to the logical plan that he first laid out.

We will also cultivate an awareness of what additional axioms may be required for certain portions of the theory.

Our third level of reading Euclid's *Elements* involves rather broader investigations than the first two levels mentioned above: We will consider various mathematical questions and subsequent developments that arise naturally from Euclid's presentation.

For example, the modern reader quickly becomes aware that Euclid does not use numbers in his geometry. He speaks of **equality** of line segments, and a notion of one segment being added to another to form a third segment, but he does not mention the **length** of a line segment. When it comes to **area** (I.35 ff.), though Euclid does not say explicitly what he means by equality of area, we can infer from his proofs that he means a notion generated by cutting figures in pieces and adding or subtracting congruent figures. He does not use any number to measure the area of a triangle. So we may note with surprise that the famous Pythagorean theorem (I.47) does not state that the square of the length of the hypotenuse (a number) is equal to the sum of the squares of the lengths of the two sides of a right triangle; rather, it says that the area of a square built on the hypotenuse is equal to the area formed by the union of the two squares built on the sides.

The absence of numbers may seem curious to a student educated in an era in which the real numbers are all-powerful, when an interval is measured by its length (which is a real number), and an area by a certain integral (another real number). In fact some modern educators have gone so far as to build the real numbers into the axioms for geometry with the "ruler postulate," which says that to each interval is assigned a real number, its **length**, and that two intervals are congruent if they have the same length. However, this use of the real numbers at the foundational level of geometry is far from the spirit of Euclid.

So we may ask, what role do numbers play in the development of geometry? As one approach to this question we can take the modern algebraic structure of a field (which could be the real numbers, for example), and show that the **Cartesian plane** formed of ordered pairs of elements of the field forms a geometry satisfying our axioms. But a deeper investigation shows that the notion of number appears intrinsically in our geometry, since we can define purely geometrically an **arithmetic of line segments**. We will show that (up to congruence) one can add two segments to get another segment, and one can multiply two segments (once a unit segment has been chosen) to get another segment. These operations satisfy the usual associative, commutative, and distributive laws, so that we obtain an **ordered field**, whose positive elements are the congruence equivalence classes of line segments.

Thus we establish a connection between the abstract geometry based on axioms and the methods of modern algebra.

I would like to emphasize throughout this course how methods of modern algebra help to understand classical geometry and its associated problems.

For example, in the theory of area, one can formalize Euclid's notion of equality based on adding and subtracting congruent figures. However, we do not know any purely geometric proof that this theory of area is nontrivial, so that, for example, one figure properly contained in another will have a smaller area. Euclid just cites Common Notion 5, "the whole is greater than the part." But unless we are willing to accept this as an axiom, we should give a proof. Such a proof can be provided using algebraic arguments in the field of segment arithmetic.

Concerning ruler and compass constructions, algebraic methods have led to notable results. For example, Gauss made an extraordinary discovery in 1796, when he used roots of unity to show that it is possible to construct a regular 17-sided polygon—the first new polygon construction since Euclid's constructions of the pentagon, hexagon, decagon, and quindecagon. On the other hand, field theory, in particular the Galois theory of finite field extensions of $\mathbb{Q}$, has provided proofs of the impossibility of certain ruler and compass constructions such as the regular 7-sided polygon, the trisection of the angle, or the doubling of the cube. For in the algebraic interpretation, one can construct with ruler and compass only those points whose coordinates lie in successive quadratic extensions of $\mathbb{Q}$, while the three problems just mentioned require the solution of cubic equations. We will see, however, that these three problems can be solved if one allows the use of a **marked ruler**. In fact, constructions using the marked ruler, in addition to ordinary straightedge and compass, correspond exactly to the solution of equations of degrees three and four.

Euclid bases his treatment of similar triangles (Book VI) on a complicated theory of proportion (developed in Book V) where ratios of given quantities are compared by seeing whether arbitrary rational multiples of the one exceed or fall short of the other. This method foreshadows Dedekind's nineteenth-century definition of a real number as a division ("Dedekind cut") of the rational numbers into two subsets, namely those greater than and those less than the given real number. The theory of proportion depends on **Archimedes' axiom**, which states that given any two segments there is an integer multiple of the first that will exceed the second. Using the field of segment arithmetic mentioned above we can give (following Hilbert) an alternative development of the theory of similar triangles that is simpler and does not depend on Archimedes' axiom.

In developing the theory of volume of three-dimensional figures in Books XI and XII, Euclid abandons, remarkably, the finite dissection methods used for the area of plane figures. Instead, he applies the "method of exhaustion" attributed to Eudoxus, which suggests the limiting process used to define the Riemann integral. Gauss (1844) expressed his regret that such an infinite method should be used for something so apparently elementary as the volume of a triangular pyramid (XII.5). Hilbert, in his famous list of problems stated in 1900, asked whether this infinite limiting process was really necessary, and Dehn in the same year provided an answer by showing that a pyramid cannot be dissected into a finite number of pieces and reassembled into a cube. In Dehn's proof, abstract algebra again provides a solution to a geometric problem.

While discussing the foundational and theoretical questions mentioned above, we also have a practical side to this course. We make a point of carrying out many ruler and compass constructions, for example, Euclid's elegant construction of the regular pentagon (IV.11), and carefully counting our steps to heighten awareness of the process. At the same time we will find explicit expressions for various lengths constructed using nested square roots to emphasize the connection with field extensions of $\mathbb{Q}$. When studying area, we will make explicit dissections of figures to show equality, such as for the Pythagorean theorem (I.47) or Dudeney's brilliant dissection in four pieces of an equilateral triangle into a square. When we come to Euclid's construction of the five regular solids, we will make models of these, and we will also explore the thirteen Archimedean solids and the other "face-regular" convex polyhedra made of regular polygons.

Also on the practical side, we will study results that belong to the domain of "Euclidean geometry" although they do not appear in Euclid's *Elements*. Some of these were discovered long ago, such as the fact that the three altitudes of a triangle meet in a point, which was known to Archimedes, while others were found more recently, such as the Euler line and the nine-point circle associated to a triangle. The technique of circular inversion, which became popular in the second quarter of the nineteenth century, provides an example of the modern transformational approach to geometry, and gives a convenient tool for the solution of classical problems such as the problem of Apollonius: to find a circle tangent to three given circles.

Finally, the investigation of the role of the parallel postulate has led to some of the most important developments arising out of Euclid's geometry. Already from the time of Euclid onward, commentators noted that this postulate was less elementary than the others, and they questioned whether it might not be a consequence of the other postulates. Two millennia of efforts to prove the parallel postulate by showing that its negation led to absurd (but not contradictory) results were considered failures until, in the mid-nineteenth century, a brilliant shift of perspective, with lasting consequences for the history of mathematics, admitted that these "absurd" conclusions were merely the first theorems in a new, strange, but otherwise consistent geometry. Thus were born the various non-Euclidean geometries that have been so valuable in the modern theory of topological manifolds, and in the development of Einstein's theory of relativity, to mention just two applications. In this course we will discuss the beginnings of **neutral geometry**, assuming no parallel axiom. We give an analytic model of non-Euclidean geometry over a field due to Poincaré. Then we give an axiomatic treatment of **hyperbolic geometry** based on the axiom of existence of limiting parallel lines. The two approaches are brought together by constructing an abstract field out of the geometry, and showing that any abstract hyperbolic plane is isomorphic to the Poincaré model over its associated field. Once again, algebraic methods help us to understand geometry.

**A note on references:** Propositions in Euclid's *Elements* are given by book and number, e.g., (I.47). Hilbert's axioms are given by initial and number, e.g., (I1)–(I3) are the axioms of incidence. Books and articles are given by author and year, e.g., Hilbert (1971), and listed in the References at the end of the book. Internal references are given by section and number, e.g., (5.9) or (18.4.3). Exercises are labeled, e.g., Exercise 4.5. An exception to this system is that within the exercises, results of the main text are indicated by their full title, e.g., Proposition 20.10.

**A note on diagrams:** Most of the diagrams in this book are drawn by hand, in keeping with the spirit of elementary geometry. I hope you will also draw your own diagrams as you read.

> As lines, so Loves oblique may well
>   Themselves in every Angle greet:
> But ours so truly Parallel,
>   Though infinite can never meet.

– from *The Definition of Love*
by Andrew Marvell (1621–1678)

# 1

**CHAPTER**

# Euclid's Geometry

n this chapter we create a common experience by reading portions of Euclid's *Elements*. We discuss the nature of proof in geometry. We introduce a particular way of recording ruler and compass constructions so that we can measure their complexity. We discuss what are presumably familiar notions from high school geometry as it is taught today. And then we present Euclid's construction of the regular pentagon and discuss its proof.

Throughout this chapter proofs are informal. We do not presuppose any particular knowledge, and yet we assume familiarity with everything in high-school geometry. The purpose of this chapter is to create a common base and language with which to begin our more formal study of geometry in the following chapters.

In the last section of this chapter we present some newer results that do not appear in Euclid's *Elements* but nevertheless belong to the subject of "Euclidean geometry."

**Note:** Reading this chapter should be concurrent with reading Euclid's *Elements* Books I–IV, so as to understand all proofs and constructions. Exercises given here will reinforce this reading.

# 1   A First Look at Euclid's *Elements*

When we first open Euclid's *Elements* to
see what is in this famous book, we find
familiar facts about the geometry of
lines, triangles, and circles in the plane.
I say familiar, because almost every
elementary or high-school curriculum
has some geometry in it, and what has
been taught for thousands of years and
still is commonly taught as "geometry"
is material from Euclid's *Elements*.

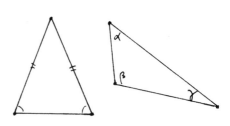

We find, for example, that a triangle is called *isosceles* if two of its sides are
equal, and in that case it follows (I.5) that the two base angles of the triangle are
also equal.

We find the theorem (I.32) that says that the sum of the three angles of a
triangle is 180°: $\alpha + \beta + \gamma = 180°$. However, we note that Euclid does not use degree measure for angles. Instead he says that the sum of the angles of a triangle
is equal to two right angles.

We find the famous "Pythagorean
theorem" (I.47), which says that in a
right triangle the sum of the squares of
the legs is equal to the square of the
hypotenuse:

$$a^2 + b^2 = c^2.$$

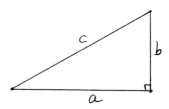

We note, however, that Euclid does not
use algebraic notation to express this
result. Instead, he shows that the area
of the square on the hypotenuse is
equal to the combined area of the
squares on the two sides.

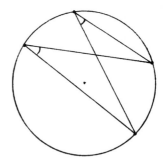

In Book III, which deals with circles,
we find the result (III.21), which I hope
will be familiar to most readers, that an
arc of a circle subtends the same angle
at different points of the circle from
which it is viewed.

Then in Book VI, which deals with
similar triangles and the theory of pro-

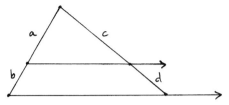

portion, we find the familiar result (VI.2) that a line parallel to the base of a triangle will cut the sides proportionately, namely, $a : b = c : d$.

Euclid's *Elements*, written circa 300 B.C. is a systematic account of the geometry and number theory of his time. What is remarkable is that these same propositions still form the basis of teaching geometry today.

Historically speaking, most of these results were known long before Euclid. Within the realm of Greek mathematics, the theorem on isosceles triangles is attributed to Thales, and the theorem on the sum of the angles of a triangle and the theorem on the sides of a right triangle are attributed to Pythagoras. Both men lived three hundred years before Euclid.

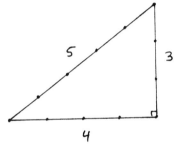

Eves (1953) points out that before the Greeks, the theorem of Pythagoras was known to the ancient Babylonians (1900–1600 B.C.). Also, there are reports that the ancient Egyptians used a rope knotted in twelve equal segments, which could be stretched out to form a triangle with sides 3, 4, 5, to construct right angles for laying out fields.

The great contribution of Euclid, for which he is justly renowned, is that he organized the geometrical knowledge of his time into a coherent logical framework, whereby each result could be deduced from those preceding it, starting with only a small number of "postulates" regarded as self-evident.

To appreciate Euclid's achievement, let us try to put this in perspective.

The most naive approach to geometry is to regard it as a collection of facts, or truths, about the real world. Ancient geometry began as a set of useful rules for measuring fields, laying out cities, building buildings, or constructing altars.

By the time of Euclid, we can detect two important changes in the perception of geometry.

One concerns the nature of geometrical truth. There is a distinction between the real world with all its imperfections, and some kind of abstract or ideal existence that people in this world strive to attain. This point of view is evident in the writings of Plato, who was born about one hundred years before Euclid. Speaking of the geometers, he says (near the end of Book VI of *The Republic*):

> Although they make use of the visible forms and reason about them, they are not thinking of these, but of the ideals which they resemble; not the figures which they draw, but of the absolute square and the absolute diameter, and so on....

Thus geometry is elevated from the status of a practical science to the study of relationships in this ideal existence, with a consequent shift of emphasis to the mathematically exact solution of a problem as opposed to an accurate approximate solution that would be sufficient for practical purposes. Because geometry engages the mind in contemplation of these ideal relationships, Plato also recognizes its value in education, for we find a little later (*Republic*, Book VII) the following exchange:

> The knowledge at which geometry aims is knowledge of the eternal, and not of ought perishing and transient.
>
> That, he replied, may be readily allowed, and is true.
>
> Then, my noble friend, geometry will draw the soul toward truth, and create the spirit of philosophy; and raise up that which is now unhappily allowed to fall down.
>
> Nothing will be more likely to have such an effect.
>
> Then nothing should be more sternly laid down than that the inhabitants of your fair city should by all means learn geometry.

Euclid's geometry is the geometry of this ideal world in the sense of Plato, with its emphasis on exact relationships. In this sense it can be regarded as abstract mathematics. From the point of view of the modern mathematician, however, Euclid's geometry is still tied to the real world because it concerns the unique ideal world of Plato's philosophy of which the real world is a reflection. For example, Euclid does not hesitate to use arguments from time to time (we will look at specific cases later) that seem perfectly acceptable in view of our experience of the real world, yet are not logical consequences of his initial assumptions.

The modern mathematician goes one step further, by trying to make *all* assumptions explicit and create a consistent mathematical structure that no longer derives its validity from the real world. The "truth" of a particular result in the real world is then no longer relevant. The only question is whether that result is consistent with or can be logically deduced from the assumptions of this particular theory. The modern point of view allows for many different equally valid abstract mathematical theories, whereas for Euclid there was only one geometry.

Euclid's *Elements* also differs from the perspective of naive geometry in its emphasis on proofs. It is no longer sufficient to say such-and-such is true, or even to give many instances where its truth is evident. The Greeks since Pythagoras had been concerned with justifying their geometrical results, and Euclid's *Elements* is the ultimate expression of this trend, where all the propositions are proved in one grand logical sequence.

So what exactly is a proof?

The answer to this question depends on the context. Suppose, for example, we are discussing one of those famous hard problems that circulate informally among amateurs, such as the following: Let *ABC* be a triangle. Let *BD* and *CE* be the angle bisectors at *B* and *C*. Suppose that *BD* is equal to *CE*. Then show that the triangle *ABC* is isosceles. This statement is eminently reasonable, but a proof using the usual methods of high-school geometry is surprisingly elusive.

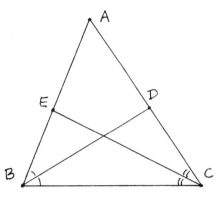

With a hard problem like this, most people would accept as proof any demonstration of its truth based on well-known results that can be found in books, whether the methods used were from geometry, trigonometry, analytic geometry, or even calculus. A purist might increase the difficulty of the problem by insisting on a purely geometric solution. Among experienced mathematicians, there would be little disagreement about what constituted a valid proof, once it was found.

In another context, a proof can be characterized simply as a convincing argument. Suppose you are explaining a result to another person who has a similar general background, but who has not seen this particular result. For example, I wish to inscribe a hexagon with six equal sides in a circle with center *O*.

I choose a point *A* on the circle, and with my compass centered at *A*, and radius *AO*, I mark off a point *B* on the circumference. Then with center *B* and radius *BO* I mark off another point *C* on the circumference. I repeat this process, always with radius equal to the radius of the original circle, to get further points *D*, *E*, and *F*. Then I draw *AB*, *BC*, *CD*, *DE*, *EF*, all of which have the same length, equal to *OA*, by construction. I claim that *FA* also has the same length, so that *ABCDEF* will be an equilateral hexagon inscribed in the circle.

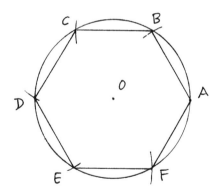

Why does this work? How would you explain this so as to convince another person? To get a real-life answer, I put this question to my seventeen-year-old son, then a high school senior. His first response was, "I have done it myself, so I know it works." "Yes," I said, "from a practical point of view it works. But how do you know this is an exact solution and not just a very good approximation?"

After a few minutes of thought he drew the lines from $O$ to $A, B, C, D, E, F$, and then explained that $OAB$ is an equilateral triangle by construction. Therefore, the angle $\angle AOB$ at the center is 60°. The same is true for the next four triangles $BOC, \ldots, EOF$. Thus we have five 60° angles, so the remaining angle $\angle AOF$ must also be 60°. Then the triangle $AOF$ having two sides the same and the same central angles must be the same as the triangle $AOB$, and so $FA = AB$.

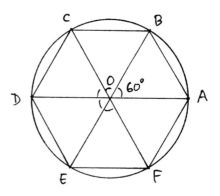

"Fine," I said, "that is very convincing, assuming that your listener knows that the angles of an equilateral triangle are 60°, and the angle of one total revolution is 360°. It seems your listener would have to know the theorem that the sum of the three angles of a triangle is 180°. What if he asked you to explain why that is true?"

I mentioned a proof of the sum of the angles by drawing a line parallel to one side $AB$ of a triangle through the third vertex $C$. Then $\alpha = \alpha'$ because of the parallel lines, and $\beta = \beta'$ because of the parallel lines, so $\alpha + \beta + \gamma = \alpha' + \beta' + \gamma = 180°$ because it is a straight angle. "But then you have to know theorems about the angles formed when a line cuts two parallel lines." There ensued a discussion about proliferation of questions, like the endless "why"s of a three-year-old, and the danger of getting into circular arguments.

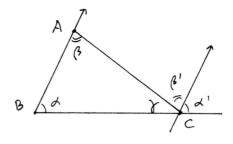

So we see that while the notion of proof as a convincing argument may work well, it depends on who your listener is, and is also subject to the danger of infinite regress if your listener is uncooperative. (At this point you might like to look at (IV.15) to see how Euclid solves this same problem.)

A third and much stricter notion of proof applies to the writer of a mathematical treatise such as Euclid's *Elements*. A proof must deduce the result in question by a series of logical steps based only on those results that have already been proved earlier in the book, and on those definitions, postulates, and common notions that have been set out as self-evident at the beginning and that form the starting point for the logical chain of deductions. Even this notion of proof is not absolute, however, because what constitutes an acceptable proof for a given result will depend on where that result is situated in the logical sequence. So for example, Proclus says that Euclid devised an entirely new proof of the theorem of Pythagoras (I.47). We can infer that he had to do so, because he placed it at the end of Book I and therefore could not use the well-known proof by similar triangles (cf. (20.6)), (which was most likely the method used by Pythagoras), since similar triangles do not appear until Book VI.

It is for this logical structure, perhaps even more than for its mathematical content, that Euclid's *Elements* is famous. The *axiomatic method* of sequential logical deduction, starting from a small number of initial definitions and assumptions, has become the basic structure of all subsequent mathematics. Euclid's *Elements* is the first great example of this method. The importance of the axiomatic method in modern times was emphasized at the turn of the century by David Hilbert, whose work we will study later in this book.

And now, dear reader, it is time for you to open your copy of Euclid and start reading. Abraham Lincoln, speaking of his scanty formal education, says "He studied and nearly mastered the six books of Euclid since he was a member of Congress." You need not go so far as that, but I do urge you to read at least as much as is suggested in the exercises below.

## A Note on the Exercises in This Book

One of my students, in an essay discussing the suitability of Euclid's *Elements* as a text for teaching geometry today, suggested that it would be better to use no text at all, so that students could have the excitement of rediscovering geometry for themselves. If we lived in ancient Athens, when the study of geometry was synonymous with reading Euclid's *Elements*, then I would agree. But we do not live in ancient Athens, and mathematics, including geometry, has developed a great deal since the time of Euclid.

So I propose instead that we take Euclid's *Elements* as a starting point, a touchstone to provoke questions and further investigation, and that we set out to rediscover modern mathematics for ourselves.

My philosophy of mathematics is that you learn by doing. To study mathematics is to do mathematics, not just to learn what other people have done. Many of the results in this book I discovered myself. In almost all cases I learned later that others had discovered them before me, but still I had the pleasure of exploring new territory. As Descartes (1637) says at one point in *La Géométrie*,

> But I will not stop to explain this in more detail, because I would deprive you of the pleasure of learning it yourself, and the utility of cultivating your spirit by the exercise, which in my opinion is the principal benefit one can draw from this science.

Therefore, the exercises in this book are designed (to the best of my ability) to stimulate mathematical activity. There are very few routine exercises. Most require some puzzling, some experimentation. Many offer a challenge of exposition: Once you understand what is happening, how do you explain it clearly in writing? Many allow room for creativity. There may be several ways to give a correct proof or a correct construction. In fact, one of the pleasures of teaching this material has been to see the multitude of imaginative methods with which students have solved the more open-ended problems. I encourage students to work together in groups, to share ideas, and to defend to each other the solutions they have found.

So perhaps the best way to use this book is to treat it as no-text. Go directly to the exercises and start to work, collecting terminology and hints from the main text only as needed!

## Exercises

1.1 See what you can remember from high-school geometry. Make a list of definitions and theorems. Do you remember the "side–angle–side" criterion for congruent triangles? Could you prove it? Can you prove that the three angle bisectors of a triangle meet in a point? Can you prove that the three altitudes of a triangle meet in a point? Do you remember the definition of similar triangles and facts about them?

1.2 Read Euclid's *Elements*, Book I, Propositions 1–34. Be prepared to explain the statements and present the proofs of (I.4), (I.5), (I.8), (I.15), (I.26), (I.27), (I.29), (I.30), and (I.32).

1.3 Discuss the structure of Euclid's proofs.

(a) Proclus describes six parts of a theorem (see Heath (1926), pp. 129 ff.): the *enunciation*, which states what is given and what is sought, the *exposition*, which says again what is given, often in a more specific form; the *specification*, which makes clear what is sought; the *construction*, which adds what is needed; the *proof*, which infers deductively what is sought from what has been previously demonstrated; and the *conclusion*, which confirms what has been proved. Identify these parts (some of which may be missing) in (I.1), (I.4), and (I.5).

(b) Discuss Euclid's habit of presenting only one case of a proposition and leaving the others to the reader. For example, in (I.7) what other cases should we consider, and how would you complete the proof in those other cases?

(c) Discuss the method of *reductio ad absurdum* (arguing to an absurdity) as a method of proof. How does this work in (I.6)? Can you think of a *direct* proof of this result (i.e., without assuming the contrary)?

For the following Exercises 1.4–1.10, present proofs in the style of Euclid, using any results you like from Book I, 1–34 (excluding the theory of area, which starts with (I.35)). Be sure to refer to Euclid's definitions, postulates, common notions, and propositions by number whenever you use one.

## PROP. VII. TH. IV.

Si des extremitez de quelque ligne droite, on mene deux autres lignes droites, se rencontrans à un point; des mêmes extremitez, on n'en pourra pas mener deux autres égales à icélles, chacune à la sienne, & de même part, se rencontrans à un autre point.

Soit la ligne A B, des extremitez de laquelle soient menées deux lignes droites A C & B C se rencontrant à quelconque

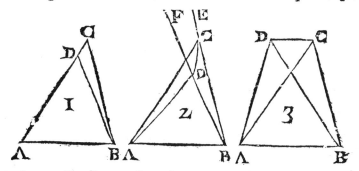

point c. Je dis que des mêmes extremitez A & B, & de la même part que c, on ne peut mener deux autres lignes droites égales à icelles A C & B C chacune à la sienne, qui se rencontrent à un autre point que c; c'est à dire que si de l'extremité A on mene la ligne A D égale à A C, & de l'extremité B la ligne B D égale à B C, il ne peut être que le point de rencontre D, soit autre que le point de rencontre c.

Car si faire se peut, que le point de rencontre D, tombe ailleurs qu'au point C : où iceluy point D tombera sur l'une ou l'autre des lignes AC, BC ; ou dans le triangle ACB; ou hors iceluy.

Premierement, iceluy point de rencontre D, ne peut être sur la ligne AC, comme en la premiere figure : car il faudroit que les deux lignes AD, & AC fussent égales entr'elles, sçavoir est la partie au tout ; ce qui est absurde : partant la rencontre D, ne se fera point sur AC, ny aussi sur B C, à cause de la même absurdité.

Plate I. A page from Henrion's Euclid of (1677) showing three different cases of the proof of (I.7).

**15**

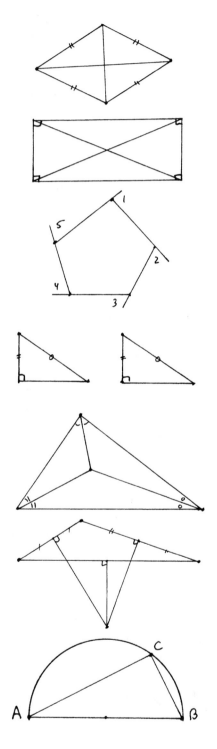

1.4 A *rhombus* is a figure with four equal sides. Show that the diagonals of a rhombus meet at right angles, and that the four small triangles thus formed are congruent to each other.

1.5 A *rectangle* is a four-sided figure with four right angles. Show that the two diagonals of a rectangle are congruent and bisect each other.

1.6 The exterior angles of a pentagon, with sides extended, add up to four right angles.

1.7 If two right triangles have one side and the hypotenuse respectively congruent, then the triangles are congruent. (We call this the right-angle–side–side theorem (RASS). Note in general that "ASS" is false: If two triangles have an angle and two sides equal, they need not be congruent.)

1.8 Show that the three angle bisectors of a triangle meet in a point. Be careful how you make your construction, and in what order you do the steps of your proof. (If you need a hint, look at (IV.4).)

1.9 The three perpendicular bisectors of the sides of a triangle meet in a single point. Be sure to give a reason why they should meet at all. For a hint, look at (IV.5).

1.10 Still using only results from Book I, show that if $AB$ is the diameter of a circle, and $C$ lies on the circle, then the angle $\angle ACB$ is a right angle.

1.11 Read the *Elements*, Book III, Propositions 1–34. Be prepared to present statements and proofs of (III.16), (III.18), (III.20), (III.21), (III.22), (III.31), and (III.32).

For the following exercises, present proofs in the style of Euclid, using any results you like from (I.1)–(I.34) and (III.1)–(III.34) (still excluding the theory of area).

1.12 Let $AB$ and $AC$ be two tangent lines from a point $A$ outside a given circle. Show that $AB \cong AC$.

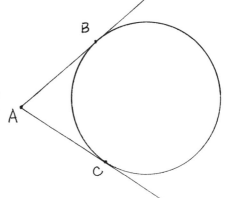

1.13 Let two circles be tangent at a point $A$. Draw two lines through $A$ meeting the circles at further points $B$, $C$, $D$, $E$. Show that $BC$ is parallel to $DE$.

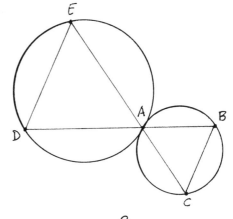

1.14 Given a pentagon $ABCDE$. Assume that all five sides are equal, and that the angles at $A$, $B$, $C$ are equal. Prove that in fact all five angles are equal (so it is a *regular* pentagon).

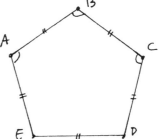

1.15 Let two circles $\gamma$ and $\delta$ meet at a point $P$. Let the tangent to $\gamma$ at $P$ meet $\delta$ again at $B$, and let the tangent to $\delta$ at $P$ meet $\gamma$ again at $A$. Let $\theta$ be the circle through $A$, $B$, $P$. Let the tangent to $\theta$ at $P$ meet $\gamma$ and $\delta$ at $C$, $D$. Prove that $PC \cong PD$. *Hint*: Draw lines joining $P$ and the centers of the three circles, and look for a parallelogram.

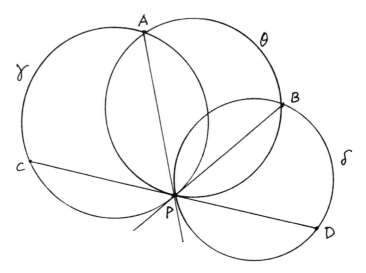

# 2   Ruler and Compass Constructions

One of the notable features of Euclid's *Elements* is his constructive approach to geometry. Many of his propositions are not theorems in the usual sense, that under certain hypotheses a certain result is true. Rather they are construction problems: given certain data, to construct a certain figure. For example, the first proposition of Book I is to construct an equilateral triangle. We could regard these constructions as existence proofs. But they are existence proofs of a very special kind: They are constructive, and the constructions are carried out with specified tools, the ruler (or straightedge) and compass. Almost one-third of the propositions in Book I, and all of the propositions in Book IV, are constructions. The constructive approach is even embedded in the initial assumptions of Euclid's geometry, because Postulate 1 says "to draw a straight line from any point to any point," and Postulate 3 says "to describe a circle with any center and distance." A modern mathematician would be more likely to say that there exists a line through any two points, and replace Postulate 3 by a definition of a circle as the set of points equidistant from a given point.

This constructive approach pervades Euclid's *Elements*. There is no figure in the entire work that cannot be constructed with ruler and compass,[1] and this

---

[1] For the three-dimensional figures of Books XI–XIII we must allow also theoretical tools that can draw a plane through three given points and that can rotate a semicircle about its diameter as axis to construct a sphere.

limits the world of subjects to be discussed to those that are constructible. So for example, in Book IV, where Euclid discusses regular polygons inscribed in a circle, we find the triangle, the square, the pentagon, the hexagon, and the regular 15-sided polygon, all of which can be constructed. But there is no mention of a regular 7-sided polygon, for example, and there are no theorems about regular $n$-gons such as one might find in a modern text. A modern mathematician would never doubt the existence of a regular 7-gon: Just take angles of $2\pi/7$ at the center of the circle, and join corresponding points on the circumference. The question would be rather, is it possible to construct the regular 7-gon with ruler and compass? But for Euclid, it seems that he cannot discuss a figure until he has shown how to construct it. Look, for example, at (I.46), to construct a square on a given line segment. In terms of what is needed for the proof, this result could have been placed immediately after (I.34). Why is it here? Presumably, because in the next proposition, the famous Pythagorean theorem (I.47), he needs to talk about the squares on the three sides of the right triangle, and he does not want to do this until he has shown that a square can be constructed on any given line-segment.

This brings us to the thorny question of what exactly it means to say that a certain mathematical object exists. For some of the structures considered by modern-day mathematicians, this is indeed a difficult question. But for Euclid there was no doubt. I believe we will not be far from the truth if we say simply that in Euclid's geometry only those geometrical figures exist that can be constructed with ruler and compass.

So now let us examine more closely how these ruler and compass constructions work. Look at (I.1), for example: to construct an equilateral triangle on a given line segment. We are given a line segment $AB$. We draw a circle with center $A$ and radius $AB$, and another circle with center $B$ and radius $BA$. These two circles meet at a point $C$ (and also at another point $D$, which we do not need). With the ruler we draw the lines $AC$ and $BC$. Then $ABC$ is the required equilateral triangle.

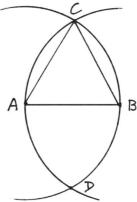

Thus the *construction* consists of a finite number of lines and circles drawn with the ruler and compass, starting from the initial data, obtaining new points as intersections along the way, and ending with the desired figure.

We will distinguish the construction, which is a series of applications of the ruler and compass to create a certain figure, from the *proof* that the figure constructed has the desired properties. The construction can be described, and makes sense, independently of any other constructions or proofs we may have made previously. But the proof that a certain construction gives the desired result depends on its position in the logical sequence of propositions. In the case of (I.1), there are no previous propositions, so Euclid's proof depends only on the definitions, postulates, and common notions set out at the beginning of Book I. His proof says, in substance, that $AC = AB$ because they are both radii of the first circle, and $BA = BC$ because they are radii of the second circle, so $AB = AC = BC$ and hence the triangle is equilateral.

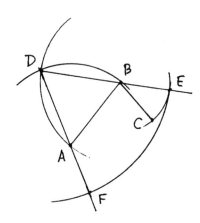

Next, let us look at (I.2). Given a point $A$ and a line segment $BC$, we must construct a line segment $AF$ originating at $A$, equal to $BC$. Euclid's method is as follows: Draw $AB$. Construct the equilateral triangle $ABD$ using the construction of (I.1). Then with center $B$ and radius $BC$ draw a circle to meet $DB$ extended at $E$. With center $D$ and radius $DE$ draw a circle to meet $DA$ extended at $F$. Then $AF$ is the required line segment.

The proof is natural enough: $BC = BE$ by construction; $DE = DF$ by construction; $DB = DA$ by construction, so by subtraction $AF = BE = BC$ as required.

But the question that immediately arises is, why did Euclid go to all this trouble when he could have made a much simpler construction: Set the compass points to the distance $BC$, then draw a circle with center $A$ and radius $BC$, choose $F$ any point on that circle, and join $A$ to $F$? We must infer from the presence of this construction that Euclid allowed himself to use the compass only in its narrow sense to draw a circle with a given center and passing through a given point. It could not be lifted off the paper and used to transport a given distance to another location. So some people call Euclid's compass a *collapsible compass*: when you lift it off the paper the points fall together and do not preserve the radius they were set at. However, the function of this construction (I.2) is to show that with the collapsible compass one can still accomplish the same result, *as if the compass had not been collapsible*, namely, to transport a distance to another point in the plane. So from now on, we will allow ourselves to use the compass in this stronger sense, to draw a circle with given center and radius equal to any given line segment.

### Counting Steps

To increase our awareness of the process of ruler and compass constructions, let us make precise exactly how the tools can be used, and let us set up a way of counting our steps as a measure of the complexity of the construction. The number of steps needed for a construction is not really important of itself, but by counting our steps we become more conscious of the process. This is one of the practical aspects of this course, to have some fun while we are pondering the deeper theoretical questions.

In any construction problem there are usually some points, lines, or circles given at the outset. The ruler may be used to extend a given or previously constructed line in either direction. The ruler may be used to draw a new line through two distinct points either given or constructed earlier. The ruler may not be used to measure distances, and it may not have any markings on it (hence the frequently used term *straightedge* to emphasize that it may be used only to draw straight lines).

The compass may be used to draw a circle with center a given or previously constructed point, and with radius equal to the distance between any two given or previously constructed points.

In addition, at any time one may choose a point at random, or subject to conditions such as that it should lie on a given line or circle, or be on the other side of a line from a given point, etc.

Each time a new line or circle is drawn, those points in which it intersects previously given or constructed lines and circles will be considered to be constructed also.

For counting, we consider each use of the ruler to construct a new line as one step, and each use of the compass to construct a new circle as one step. Extending lines previously given or constructed, choosing points at random, and obtaining new points as intersections do not count as separate steps.

Thus for example, the construction of the equilateral triangle (I.1) above takes four steps:

The line segment $AB$ is given
1. Draw circle with center $A$ and radius $AB$.
2. Draw circle with center $B$ and radius $BA$. Get $C$.
3. Draw $AC$.
4. Draw $BC$.

Then $ABC$ is the required triangle.

When performing more complicated constructions, we will count all of the steps required to perform the entire construction, so that each construction is self-contained and independent of other constructions (though inevitably each construction will contain elements of other constructions). This imposes a different notion of economy of construction from Euclid's. For while Euclid in his sequential development of the propositions finds it most economical to utilize previous constructions, we will find that minimizing the total number of steps will often lead us to different constructions.

Look at (I.9), for example, to bisect a given angle. The angle is given by a point $A$ and two rays $l$, $m$ emanating from $A$. Euclid's method is this: Choose $B$ on $l$ at random. Find $C$ on $m$ such that $AB = AC$ (I.3). Draw $BC$. Construct the equilateral triangle $BCD$ (I.1). Join $AD$. Then $AD$ is the angle bisector.

Euclid's method is economical for him because it makes use of previously described constructions (I.3) and (I.1). If we count the number of steps to carry out this construction, we find seven:

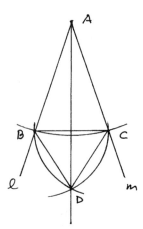

Choose $B$ at random on $l$ (no step)
1. Circle center $A$ radius $AB$, get $C$.
2. Draw $BC$.
3. Circle center $B$ radius $BC$.
4. Circle center $C$ radius $CB$, get $D$.
5. Draw $BD$.
6. Draw $CD$.
7. Draw $AD$, which is the angle bisector.

If we are concerned only with making an independent construction for the angle bisector, there is no need to draw the lines $BC$, $BD$, $CD$. Thus the construction reduces to four steps. In order to prove that this construction works, we might want to draw the lines $BC$, $BD$, $CD$ and argue as Euclid did. The lines then become part of the proof. But they are not part of the construction, so the construction still requires only four steps.

For another example, look at Euclid's construction (I.10) to bisect a given line segment. He first appeals to (I.1) to construct an equilateral triangle, and then to (I.9) to bisect the angle at its vertex. This is an elegant method, making use of what he has done before. But in terms of numbers of steps, it is not efficient. If we add the numbers of steps used in the two previous results, we get 11 steps. If we make use of points already constructed in (I.1) when we do the construction of (I.9), this reduces to 9. But it is possible to give a direct construction of the midpoint of a segment in only three steps (see Exercise 2.2).

### A Note About Accuracy and Exactness of Constructions

When carrying out ruler and compass constructions, we attempt to make our drawings as *accurate* as possible. Using a sharp pencil we draw fine lines and make them pass through given points as closely as possible. Nevertheless, there is always a small error in each step, and those errors will compound throughout a long construction, so that the final figure does not always do just what you want. For example, in constructing the circle circumscribed about a given triangle (Exercise 2.10), you may find that your circle passes nicely through two of the points but misses the third one slightly. This error is inevitable in any drawings we make.

But, to paraphrase the quotation from Plato in Section 1, it is not the line and the circle drawn on the paper that we are thinking of, it is the absolute line and the absolute circle. And in this sense, our construction must be mathematically *exact*. In other words, it must be possible to prove using the reasoning of abstract geometry that this construction in its ideal form gives the exact result we are seeking.

This distinction has caused considerable confusion among amateur mathematicians through the ages, who were trying to make constructions, now known to be impossible, of trisecting the angle or squaring the circle. For many of their constructions are remarkably accurate, while failing to be mathematically exact. (See the interesting book of Dudley (1987), as well as Sections 25, 28 below.)

## Exercises

For each of the following problems, carry out a ruler and compass construction as accurately as you can. Number and label each of your steps as in the text. Feel free to use abbreviations such as "*AB*" for "draw a line *AB*"; "⊙*AB*" to draw a circle with center *A* and radius *AB*; or "⊙c*ArBC*" *to draw a circle with center A and radius BC*. Label each new point as it is constructed and mention it (e.g., "get *F*") in the appropriate step. For the time being, we are not concerned with the proofs. Just do the construction. You should, however, be able to give an informal proof (convincing argument) of why it works, if asked.

After you make your construction, locate the corresponding proposition in Euclid (Book I, III, or IV) and compare. How many steps does his method require? What do you think is the least number of steps possible? I will sometimes give a *par value* for a construction, which is the typical number of steps an experienced constructor would need. By trying harder, you can sometimes succeed with fewer steps.

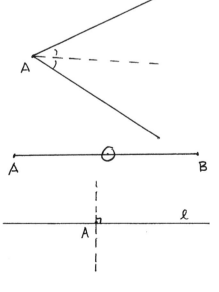

2.1 Given an angle, construct the angle bisector (par = 4).

2.2 Given a line segment, find the midpoint of that segment (par = 3).

2.3 Given a line *l* and a point *A* on *l*, construct a line perpendicular to *l* through *A* (par = 4, possible in 3).

2.4 Given a line $l$ and a point $A$ not on $l$, construct a line perpendicular to $l$ passing through $A$ (par = 4, possible in 3).

2.5 Given an angle at a point $A$, and given a ray emanating from a point $B$, construct an angle at $B$ equal to the angle at $A$ (par = 4).

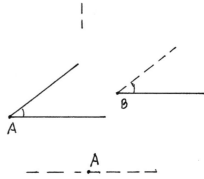

2.6 Given a line $l$ and a point $A$ not on $l$, construct a line parallel to $l$, passing through $A$ (par = 3).

2.7 Given a circumference of a circle, find the center of the circle (par = 5).

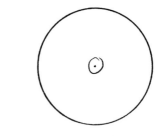

2.8 Given a circle with its center $O$, and given a point $A$ outside the circle, construct a line through $A$ tangent to the circle. (*Warning*: You may not slide the ruler until it seems to be tangent to the circle. You must construct another point on the desired tangent line before drawing the tangent.) (Par = 6.)

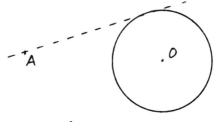

2.9 Construct a circle inscribed in a given triangle $ABC$ (par = 13).

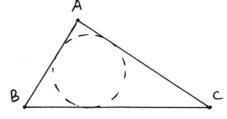

2.10 Construct a circle circumscribed about a given triangle $ABC$ (par $= 7$).

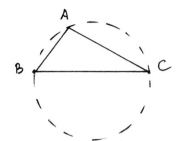

2.11 Given a line $l$, a line segment $d$, and a point $O$, construct a circle with center $O$ that cuts off a segment congruent to $d$ on the line $l$ (par $= 9$).

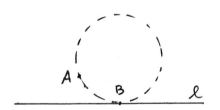

2.12 Given a point $A$, a line $l$, and a point $B$ on $l$, construct a circle that passes through $A$ and is tangent to the line $l$ at $B$ (par $= 8$).

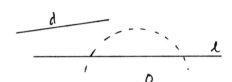

2.13 Construct three circles, each one meeting the other two at right angles. (We say that two circles meet at right angles if the radii of the two circles to a point of intersection make right angles.) (Par $= 10$.)

2.14 Given a line segment $AB$, divide it into three equal pieces (par $= 6$).

2.15 (The one-inch ruler.) Suzie's ruler broke into little pieces, so she can only draw lines one inch long. Fortunately, her compass is still working. She has two points on her paper approximately 3 inches apart. Help her construct the straight line joining those two points.

2.16 (The rusty compass.) Joe's compass has rusted into a fixed position, so it can only draw circles whose radius is one inch. Fortunately, his ruler is still working. Help him construct an equilateral triangle on a segment $AB$ that is approximately $2\frac{1}{2}$ inches long (par $= 6$).

2.17 Using a ruler and rusty compass (cf. Exercise 2.16), construct the perpendicular to a line $l$ at a point $A$ on $l$ (par $= 6$).

2.18 Using a ruler and rusty compass, given a line $l$ and a point $A$ more than 2 inches away from $l$, construct the line through $A$ and perpendicular to $l$ (par $= 12$).

2.19 Using a ruler and rusty compass, given
a segment $AB$ and given a ray $AC$, construct a point $D$ on the ray $AC$ such
that $AB \cong AD$.

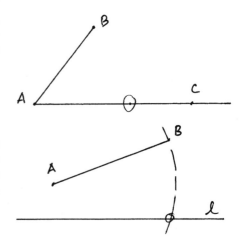

2.20 Using a ruler and rusty compass, given
a line $l$ and given a segment $AB$ more
than one inch long, construct one of the
points $C$ in which the circle of center $A$
and radius $AB$ meets $l$.

2.21 Discussion question: Is it possible with ruler and rusty compass to construct any
figure that can be constructed with ruler and regular compass? What would you
need to know in order to *prove* that this is possible? For starters, can you carry out
all the constructions of Euclid, Book I, with ruler and rusty compass?

2.22 (Back to regular ruler and compass construction.) Given a segment $AB$, given a
circle with center $O$, and given a point
$P$ inside $O$, construct (if possible) a line
through $P$ on which the circle cuts off a
segment congruent to $AB$ (par = 5).

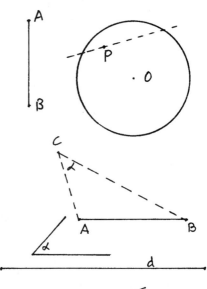

2.23 Given a segment $AB$, given an angle $\alpha$,
and given another segment $d$, construct
a triangle $ABC$ with base equal to $AB$,
angle $\alpha$ at $C$, and such that $AC + BC = d$.

2.24 Given two circles $\Gamma$, $\Gamma'$, with centers
$O$, $O'$, construct a line tangent to both
circles.

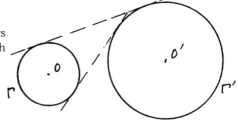

# 3   Euclid's Axiomatic Method

One of the remarkable features of Euclid's *Elements* is its orderly logical structure. Euclid took the great mass of geometrical material that had grown in the previous two or three centuries, and organized it into one coherent logical sequence. This is what we now call the *axiomatic method*: Starting from a small number of definitions and assumptions at the beginning, all the succeeding results are proved by logical deduction from what has gone before. Euclid's text has been a model of mathematical exposition, unchallenged for two thousand years, and only recently (in the last hundred years or so) replaced by newer mathematical systems that we consider more rigorous. As we read Euclid, let us observe how he organizes his material, let us be curious about why he does things the way he does, and let us explore the questions that come to mind when we as modern mathematicians read this ancient text.

### Definitions

Euclid begins with definitions. Some of these definitions are akin to the modern notion of definition in mathematics, in that they give a precise meaning to the term being defined. For example, the tenth definition tells us that if a line segment meets a line so that the angles on either side are equal, then these are called right angles. This tells us the meaning of the term *right angle*, assuming that we already know what is meant by a line, a line segment, an angle, and equality of angles. Similarly, the fifteenth definition, rephrased, defines a circle to be a set of points $C$, such that the line segments $OA$ from a fixed point $O$ to any point $A$ of the circle $C$, are all equal to each other, and the point $O$ is called the center of the circle. This tells us what a circle is, assuming that we already know what a line segment is, and what is meant by equality of line segments.

On the other hand, some of Euclid's other definitions, such as the first, "a point is that which has no part," or the second, "a line is breadthless length," or the third, "a straight line is a line which lies evenly with the points on itself," give us no better understanding of these notions than we had before. It seems that Euclid, instead of giving a precise meaning to these terms, is appealing to our intuition, and alluding to some concept we may already have in our own minds of what a point or a line is. Rather than defining the term, he is appealing to our common understanding of the concept, without saying what that is. This may have been very well in a society where there was just one truth and one geometry and everyone agreed on that. But the modern consciousness sees this as a rather uncertain way to set up the foundations of a rigorous discipline. What if we say now, oh yes, we agree on what points and lines are, and then later it turns out we had something quite different in mind? So the modern approach is to say these notions are *undefined*, that is, they can be anything at all, *provided* that they satisfy whatever postulates or axioms may be imposed on them later. In the algebraic definition of an abstract group, for example, you never say what the elements of the group are, nor what the group operation is. Those are undefined. However, they must satisfy the group axioms that the operation is associative, there exists an identity, and that there exist inverses. The elements of the group can then be anything as long as they satisfy these axioms. They could be integers, or they could be cosets of a subgroup of the integers, or they could be rotations of a geometrical object such as a cube, or anything else. So in our reading of Euclid, perhaps we should regard "point" and "line" as undefined terms.

It may be worth noting some differences of language between Euclid's text and modern usage. By a *line* he means something that may be curved, which we would call a curve. He says *straight line* for what we call line. And then he says a *finite straight line* (as in the statement of (I.1)) for what we would call a line segment. For Euclid, a *plane angle* results where two curves meet, and a *rectilineal plane angle* is formed when two line segments meet. Note that Euclid requires the two sides of an angle not to lie in a straight line. So for Euclid there is no zero angle, and there is no straight angle ($180°$). So we should think of Euclid's concept of angle as meaning an angle of $\alpha$ degrees, with $0 < \alpha < 180°$ (though Euclid makes no mention of the degree measure of an angle).

Euclid's notion of *equality* requires special attention. He never defines equality, so we must read between the lines to see what he means. In Euclid's geometry there are various different kinds of magnitudes, such as line segments, angles, and later areas. Magnitudes of the same kind can be compared: They can be equal, or they can be greater or lesser than one another. Also, they can be added and subtracted (provided that one is greater than the other) as is suggested by the common notions.

Euclid's notion of equality corresponds to what we commonly call *congruence* of geometrical figures. In high-school geometry one has the length of a line segment, as a real number, so one can say that two segments are congruent if they have the same length. However, there are no lengths in Euclid's geometry, so we must regard his equality as an undefined notion. Because of the first common notion, "things which are equal to the same thing are also equal to one another," we may regard equality (which we will call *congruence* to avoid overuse of the word equal) to be an equivalence relation on line segments. Similarly, we will regard congruence of angles as an equivalence relation on angles.

### Postulates and Common Notions

The postulates and common notions are those facts that will be taken for granted and used as the starting point for the logical deduction of theorems. If you think of Euclid's geometry in the classical way as being the one true geometry that describes the real world in its ideal form, then you may regard the postulates and common notions as being self-evident truths for which no proof is required. If you think of Euclid's geometry in the modern way as an abstract mathematical theory, then the postulates and common notions are merely those statements that are arbitrarily selected as the starting point of the theory, and from which other results will be deduced. There is no question of their "truth," because one can begin a mathematical theory from any hypotheses one likes. Later on, however, there may arise a question of relevance, or importance of the mathematical theory constructed. The importance of a mathematical theory is judged by its usefulness in proving theorems that relate to other branches of mathematics or to applications. If you begin a mathematical theory with weird hypotheses as your starting point, you may get a valid logical structure that is of no use. From that point of view the choice of postulates is not so arbitrary. In any case, we can regard Euclid's postulates and common notions collectively as the set of *axioms* on which his geometry is based.

Some commentators say that the postulates (as in Heath's edition) are those statements that have a geometrical content, while the common notions are those statements of a more universal nature, which apply to all the sciences. Other commentators divide them differently, calling "postulates" those statements that allow you to construct something, and calling "axioms" those statements that assert that something is always true. One should also note that some editors give extra axioms not listed in Heath's edition, such as "halves of equals are equal," which is used by Euclid in the proof of (I.37), or "two straight lines cannot contain a space."

We have already noted the constructive nature of Euclid's approach to geometry as expressed in Postulates 1–3. By the way, Euclid makes no explicit statement about the uniqueness of the line mentioned in Postulate 1, though he apparently meant it to be unique, because in the proof of (I.4) he says "otherwise two straight lines will enclose a space: which is impossible."

In the list of Postulates and Common Notions, Postulate 5 stands out as being much more sophisticated than the others. It sounds more like a theorem than an axiom. We will have more to say about this later. For the moment let us just observe that two thousand years of unsuccessful efforts to prove this statement as a consequence of the other axioms have vindicated Euclid's genius in realizing that it was necessary to include Postulate 5 as an axiom.

## Intersections of Circles and Lines

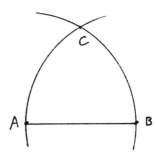

As we read Euclid's *Elements* let us note how well he succeeds in his goal of proving all his propositions by pure logical reasoning from first principles. We will find at times that he relies on "intuition," or something that is obvious from looking at a diagram, but which is not explicitly stated in the axioms. For example, in the construction of the equilateral triangle on a given line segment $AB$ (I.1) how does he know that the two circles actually meet at some point $C$? While the fifth Postulate guarantees that two lines will meet under certain conditions, there is nothing in the definitions, postulates, or common notions that says that two circles will meet. Nor does Euclid offer any reason in his proof that the two circles will meet.

If you carry out the construction with ruler and compass on a piece of paper, you will find that they do meet. Or if you look at the diagram, it seems obvious that they will meet. However, that is not a proof, and we must acknowledge that Euclid is using something that is not explicitly guaranteed by his axioms and yet is essential to the success of his construction.

There are two separate issues here. One is the relative position of the two circles. Two circles need not always meet. If they are far apart from each other, or if one is entirely contained in the other, they will not meet. In the present case, part of one circle is inside the other circle, and part outside, so it appears from the diagram that they must cross each other.

The second issue is, assuming that they are in a position so that they appear to meet, does the intersection point actually exist? Today we will immediately think of continuity and the intermediate value theorem: If $y = f(x)$ is a real-valued continuous function defined on the unit interval $[0, 1]$ of the real numbers, and if $f(0) < 0$ and $f(1) > 0$, then there is some point $a \in [0, 1]$ with $f(a) = 0$. In other words, the graph of the function must intersect the $x$-axis at some point in the interval.

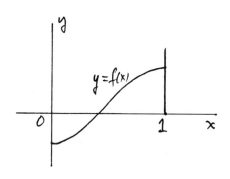

However, we must bear in mind that the concepts of real numbers and continuous functions were not made rigorous until the late nineteenth century, and that this kind of mathematical thinking is foreign to the spirit of Euclid's *Elements*.

To make the same point in a different way, suppose we consider the *Cartesian plane* over the field of rational numbers $\mathbb{Q}$, where points are ordered pairs of rational numbers, and let $AB$ be the unit interval on the $x$-axis. Then the vertex $C$ of the equilateral triangle, which would have to be the point $(\frac{1}{2}, \frac{1}{2}\sqrt{3})$, actually does not exist in this geometry.

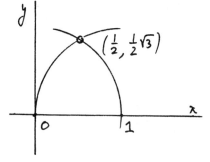

So later on, when we set up a new system of axioms for Euclidean geometry, we will have to include some axiom that guarantees the existence of the intersection points of circles with other circles, or with lines, at least those that arise in the ruler and compass constructions of Euclid's *Elements*. Some modern axiom systems (such as Birkhoff (1932) or the School Mathematics Study Group geometry) build the real numbers into the axioms with a postulate of line measure, or include Dedekind's axiom that essentially guarantees that we are working over the real numbers. In this book, however, we will reject such axioms as not being in the spirit of classical geometry, and we will introduce only those purely geometric axioms that are needed to lay a rigorous foundation for Euclid's *Elements*.

The issue of intersecting circles arises again in (I.22), where Euclid wishes to construct a triangle whose sides should be equal to three given line segments $a$, $b$, $c$. This requires that a circle with radius $a$ at one endpoint of the segment $b$ should meet a circle of radius $c$ at the other end of the segment $b$. Euclid correctly puts the necessary and sufficient condition that this intersection should exist in the statement of the proposition, namely that any two of the line segments should be greater than the third. However, he never alludes to this hypothesis in his proof, so that we do not see in what way this hypothesis implies the existence of the intersection point. While some commentators have criticized Euclid for this, Simson ridicules them, saying "For who is so dull, though only beginning to learn the Elements, as not to perceive ... that these circles must meet one another because $FD$ and $GH$ are together greater than $FG$." Still, Simson has only discussed the position of the circles and has not addressed the second issue of why the intersection point exists. (See Plate V, p. 109)

### The Method of Superposition

Let us look at the proof of (I.4), the side–angle–side criterion for congruence of two triangles (SAS for short). Suppose that $AB = DE$, and $AC = DF$, and the included angle $\angle BAC$ equals $\angle EDF$. We wish to conclude that the triangles are congruent, that is to say, the remaining sides and pairs of angles are congruent to each other, respectively. Euclid's method is to "apply the triangle" $ABC$ to the triangle $DEF$. That

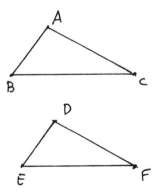

is, he imagines moving the triangle $ABC$ onto the triangle $DEF$, so that the point $A$ lands on the point $D$, and the side $AB$ lands on the side $DE$. Then he goes on to argue that the ray $AC$ must land on the ray $DF$, because the angles are equal, and hence $C$ must land on $F$ because the sides are equal. From here he concludes that the triangles coincide entirely, hence are congruent.

effe. Nam quum de re aliqua fermonem inftituimus: ea nobis tacitè per definitio-
nem fubit in animum : Non enim duos angulos æquales effe cogitabo, nifi quid fit
æquales effe angulos concipiam. Quod refpiciens Euclides, angulorum æqualitatem
proponere, atque eadem opera definire voluit: vt hoc Theorema pro Definitione
haberemus. Nemo enim fignificantius explicabit angulorum æqualitatem, quàm fi
dixerit duos angulos æquales fieri, quum duo latera vnum angulum continentia,
duobus alterum angulum continentibus fiunt æqualia, & bafes quæ latera conne-
ctunt, æquales. Conftat enim angulum tantum effe, quanta eft duarum linearum
ipfum continentium apertio, feu diductio, hanc verò tantam effe, quanta eft bafis,
hoc eft, linea ipfas connectens. Atque vt clarè dicam, tantus eft angulus B A C,
quanta eft remotio lineæ A C ab ipfa A B: tanta verò efficitur remotio, quantam
exhibet linea B C. Hoc autem in Ifofcelibus eft euidentius. Sint enim duo Ifofcelia
A B C & D E F: quorum vnius duo latera A B & A C duobus D E & D F alte-

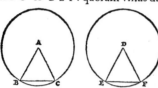

rius fint æqualia: angulusq́; A angulo D. Ac po-
fitis centris in A & D punctis, ducantur duo
Circuli: prior fecundum A B, alter fecundum
D E fpatium. Horum prior manifeftò tranfibit
per B & C: alter verò per E & F puncta: quum
A B & A C, itemq́; D E & E F fint æqualia,
& à centro vtrinq; exeuntia. Atque, ex defini-
tione æqualium angulorum, erunt arcus B C & E F æquales. Angulorum enim
magnitudo defignatur ex arcubus Circulorum qui per extremas lineas quæ angulos
continet, tranfeunt. Ac conuerfo modo, æquales anguli atque æqualibus lineis com-
prehenfi, æquales fubtendunt peripherias. Quum enim æqualia fint fpatia B C &
E F, ea æqualibus rectis lineis claudi oportet: propterea quòd recta linea, eft à pun-
cto ad punctum via breuiffima. Atque haud diffimili iudicio, ex laterum ratione &
bafium, quanta fit angulorum magnitudo æftimabimus. Quur ergo Euclides hoc
inter Theoremata repofuit, non inter Principia præmifit? Nimirùm, quum fpeciem
quodammodo mixtam Principij & Theorematis præ fe ferret: Principij, quòd in
communi animi iudicio confifteret: Theorematis, quòd fpeciatim Triangula Trian-
gulis comparanda proponeret: maluit Euclides inter Theoremata referre: præfer-
tim quum multa haberet capita, Principium verò fimplex ac velut nudum effe de-
beat. Ex hoc præterea Axiomate tanquam ex locupletiffimo Demonftrationum
themate, multæ Propofitiones confequi debebant, eiufdem propè facilitatis & iu-
dicij: quas, quia erant notiffimæ, inter Principia annumerari non conueniebat. Pau-
cis enim Principijs Geometriam contentam effe oportebat: immò multa Principia
confultò fupprimuntur, ne fit onerofa multitudo: vt etiam quæ exprimuntur, tantùm
ad exemplum exprimi videantur. Hûc accedit, quòd primum Theorema facile, per-
fpicuum, ac fenfui obuium effe debebat, pro Geometriæ lege, quæ ex paruis humi-
libusq; initijs, in progreffus mirabiles fefe extollit.

Huius itaque Propofitionis veritatem non aliunde quàm à communi iudicio pe-
temus: cogitabimusq́; Figuras Figuris fuperponere, Mechanicum quippiam effe:
intelligere verò, id demùm effe Mathematicum. Iam verò quum fuerit confeffum
duo Triangula inuicem effe æquilatera, ipfa quoque inter fe æqualia fateri erit ne-
ceffarium. Etenim nulla euidentiori fpecie æqualitas Figurarum dignofcitur, quàm
ex laterum æqualitate: quanquam Circulorum æqualitas ex diametris definitur:
fed non aliam ob caufam, quàm quòd linea obliqua fui copiam adeò apertè non fa-
cit vt recta: Cuius menfuram facilè capimus, ac per eam, obliquarum inter fe com-
parationem facimus.

At fi hæc fuperpofitio aliqua ratione admittenda fit: tolerabilior fane fuerit hoc
qui fequitur modo.

Manente duorum Triangulorum A B C & D E F conditione, continuabo E D

Plate II. The commentary on (I.4) from Peletier's Euclid of (1557). He says the truth of
this proposition belongs among the common notions, because to superimpose one figure
on another is mechanics, not mathematics.

**32**

This is another situation where Euclid is using a method that is not explicitly allowed by his axioms. Nothing in the Postulates or Common Notions says that we may pick up a figure and move it to another position. We call this the *method of superposition*.

Euclid uses this method again in the proof of (I.8), but it appears that he was reluctant to use it more widely, because it does not appear elsewhere. If it were a generally accepted method, for example, then Postulate 4, that all right angles are equal to each other, would be unnecessary, because that would follow easily from superposition.

If we think about the implications of this method, it has far-reaching consequences. It implies that one can move figures from one part of the plane to another without changing their sides or angles. Thus it implies a certain homogeneity of the geometry: The local behavior of figures in one part of the plane is the same as in another part of the plane. If you think of modern theories of cosmology, where the curvature of space changes depending on the presence of large gravitational masses, this is a nontrivial assumption about our geometry.

To state more precisely what assumptions the method of superposition is based on, let us define a *rigid motion* of the plane to be a one-to-one transformation of the points of the plane to itself that preserves straight lines and such that segments and angles are carried into congruent segments and angles. To carry out the method of superposition, we need to assume that there exist sufficiently many rigid motions of our plane that

(a) we can take any point to any other point,
(b) we can rotate around any given point, so that one ray at that point is taken to any other ray at that point, and
(c) we can reflect in any line so as to interchange points on opposite sides of the line.

If we were working in the real Cartesian plane $\mathbb{R}^2$ with coordinates $x, y$, we could easily show the existence of sufficient rigid motions by using *translations, rotations,* and *reflections* defined by suitable formulas in the coordinates.

For example, a translation taking the point $(0, 0)$ to $(a, b)$ is given by

$$\begin{cases} x' = x + a, \\ y' = y + b, \end{cases}$$

and a rotation of angle $\alpha$ around the origin is given by

$$\begin{cases} x' = x \cos \alpha - y \sin \alpha, \\ y' = x \sin \alpha + y \cos \alpha. \end{cases}$$

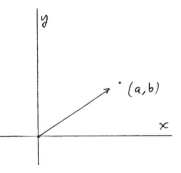

Thus we can easily justify the use of the method of superposition in the real Cartesian plane. However, since there are no coordinates and no real numbers in Euclid's geometry, we must regard his use of the method of superposition as an additional unstated postulate or axiom.

To formalize this, we could postulate the existence of a group of rigid motions acting on the plane and satisfying the conditions (a), (b), (c) mentioned above. Indeed, there is an extensive modern school of thought, exemplified by Felix Klein's *Erlanger Programm* in the late nineteenth century, which bases the study of geometry on the groups of transformations that are allowed to act on the geometry. This point of view has had wide-ranging applications in differential geometry and in the theory of relativity, for example.

We will discuss the rigid motions in Euclidean geometry in greater detail later (Section 17). For the moment let us just note that the proof of the (SAS) criterion for congruence in (I.4) requires something more than what is in Euclid's axiom system. Hilbert's axioms for geometry actually take (SAS) as an axiom in itself. This seems more in keeping with the elementary nature of Euclid's geometry than postulating the existence of a large group of rigid motions.

Finally let us note that Euclid's use of the method of superposition in the proof of (I.4) gives us some more insight into his concepts of "equality" for line segments and angles. In Common Notion 4 he says that things that coincide with one another are equal (congruent) to one another. In the proof of (I.4) he also uses the converse, namely, if things (line segments or angles) are equal to one another (congruent), then they will coincide when one is moved so as to be superimposed on the other. So it appears that Euclid thought of line segments or angles being congruent if and only if they could be moved in position so as to coincide with each other.

**Betweenness**

Questions of betweenness, when one point is between two others on a line, or when a line through a point lies inside an angle at that point, play an important, if unarticulated, role in Euclid's *Elements*. To explain the notion of points on a line lying between each other, one could simply postulate the existence of a linear ordering of the points. Similarly, for angles at a point one could talk of a circular ordering.

But when a hypothesis of relative position of points and lines in one part of a diagram implies a relationship for other parts of the figure far away, it seems clear that something important is happening, and it may be dangerous to rely on intuition.

For example, how do you know that the angle bisector at a vertex $A$ of a triangle $ABC$ meets the opposite side $BC$ between the points $BC$ and not outside? Of course, it is obvious from the picture, but what if you had to explain why without drawing a picture?

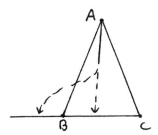

We have already seen that the relative position of two circles may affect whether they meet or not. Let us look at some other instances where betweenness plays an important role in a proof.

Consider (I.7), which is used in the proof of the side–side–side (SSS) criterion for congruence of triangles (I.8). In (I.7) Euclid shows that it is not possible to have two distinct triangles $ABC$ and $ABD$ on the same side of a segment $AB$ and having equal sides $AC = AD$ and $BC = BD$.

The proof goes like this. Since $AC = AD$, the triangle $ACD$ is isosceles, and so the base angles are equal (I.5). In the diagram $\angle 1 = \angle 4$. On the other hand, since $BC = BD$, the triangle $BCD$ is isosceles, so its base angles are equal (I.5) — in our diagram $\angle 2 = \angle 3$. But now $\angle 2$ is less than $\angle 1$, which is equal to $\angle 4$, which is less than $\angle 3$. So $\angle 2$ is much less than $\angle 3$. But they are also equal, and this is impossible.

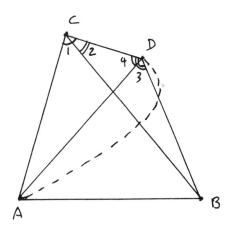

Note that this proof depends in an essential way on the relative position of the lines meeting at $C$ and $D$, which determines the inequalities between the angles. If the line $AD$ should reach the point $D$ outside of the triangle $BCD$, as in our second (impossible) picture, then $\angle 2 < \angle 1$ and $\angle 3 < \angle 4$, and there is no contradiction. Thus the original proof depends on a certain configuration of lines being inside certain angles, which in turn depends on some global properties of the entire two-dimensional figure, and these relationships would be hard to explain convincingly without using a diagram. So as soon as we realize that we are depending on a diagram for part of our proof, a mental red flag should pop up to alert us to the question, What exactly is going on here, and what unstated assumptions are we using?

For another example where similar questions arise, look at the proof of (I.16) to show that an exterior angle of a triangle is greater than the opposite interior angle.

Let $ABC$ be the given triangle. Bisect $AC$ at $E$, draw $BE$, and extend that line to $F$ so that $BE = EF$. Draw $CF$. Then by SAS (I.4), Euclid shows that the triangle $BEA$ is congruent to the triangle $FEC$, and so the angle at $A$ is equal to the angle $\angle ACF$. He then says that the angle $\angle ACF$ is less than the exterior angle $\angle ACD$, which proves the result.

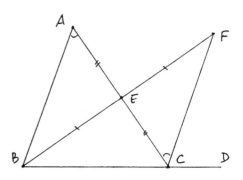

How do we know this relation among the angles? Because the line $CF$ lies inside the angle $ACD$. But why is it inside? Since the line $CF$ was constructed using the point $F$, which in turn was constructed using the point $E$, this is a global property of the whole figure, which is clear from the diagram, but would be hard to explain without a diagram.

To illustrate the danger of relying on diagrams in geometrical proofs, we will present a well-known fallacy due to W.W. Rouse Ball (1940). The following purports to be a proof that every triangle is isosceles. See if you can find the flaw in the argument.

### Example 3.1

Let $ABC$ be any triangle. Let $D$ be the midpoint of $BC$. Let the perpendicular to $BC$ at $D$ meet the angle bisector at $A$ at the point $E$. Drop perpendiculars $EF$ and $EG$ to the sides of the triangle, and draw $BE$, $CE$. The triangles $AEF$ and $AEG$ have the side common and two angles equal, so they are congruent by AAS (I.26). Hence $AF = AG$ and $EF = EG$. The triangles $BDE$ and $CDE$ have $DE$ common, two other sides equal, and the included right angles equal. Hence they are congruent by SAS (I.4). In particular $BE = CE$.

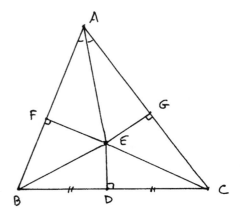

Now, the triangles $BEF$ and $CEG$ are right triangles with two sides equal, so they are congruent (see lemma below), and hence $BF = CG$. Adding equals to equals, we find $AB = AF + FB$ is equal to $AC = AG + GC$. So the triangle $ABC$ is isosceles.

There are several other cases to consider. If the point $E$ lies outside the triangle, one can use this second figure and exactly the same proof to conclude that $AB$ and $AC$ are the differences of equal segments $AF = AG$ and $BF = CG$, hence equal.

If $E$ lands at the point $D$, or if the angle bisector at $A$ is parallel to the perpendicular to $AB$ at $D$, the proof becomes even easier, and we leave it to the reader.

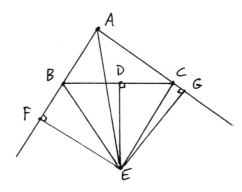

We still need to prove the following lemma.

**Lemma 3.2** (Right-Angle–Side–Side) (RASS)
*If two right triangles have two sides equal, not containing the right angle, they are still congruent.*

*Proof* This result, though not stated by Euclid, is often useful. We give two proofs. The first method is to use (I.47) to conclude that the square on $BC$ is equal to the square on $EF$. Then $BC = EF$, and we can apply (SSS) (I.8).

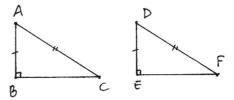

The second proof does not make use of (I.47) and the theory of area. Extend $FE$ to $G$ and make $EG = BC$. Then the triangles $ABC$ and $DEG$ are congruent by SAS (I.4). Therefore, $AC = DG$. It follows that $DF = DG$, so the triangle $DFG$ is isosceles. Therefore, the angles at $F$ and $G$ are equal. Then the triangles $DEG$ and $DEF$ are congruent by AAS (I.26). But $DEG$ is congruent to $ABC$, so the two original triangles are congruent.

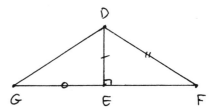

**The Theory of Parallels**

Book I of Euclid's *Elements* can be divided naturally into three parts. The first part, (I.1)–(I.26), deals with triangles and congruence. The second part, (I.27)–

(I.34), deals with parallel lines and their applications, including the well-known (I.32) that the sum of the angles of a triangle is two right angles. The third part, (I.35)–(I.48), deals with the theory of area.

Two lines are *parallel* if they never meet, even if extended indefinitely in both directions (Definition 23). The fifth postulate gives a criterion for two lines to meet under certain conditions, hence to be not parallel, so we often refer to the fifth postulate as the *parallel postulate*. Euclid postponed using this postulate as long as possible so that in fact, the first part of Book I about triangles and congruence does not use the parallel postulate at all. It is first used in (I.29). Let us examine closely Euclid's theory of parallels and his use of the parallel postulate.

The first result about parallel lines, (I.27), says that if a line falling on two other lines makes the alternate interior angles equal, then the lines are parallel. This is proved using (I.16): If not, the lines would meet on one side or the other, and would form a triangle having an exterior angle equal to one of its opposite interior angles, which is impossible.

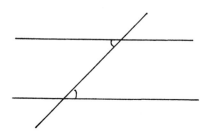

The next result (I.28) is similar, and follows directly from this one using vertical angles (I.15) or supplementary angles (I.13).

The fifth postulate is used to prove the converse of (I.27), which is (I.29): If the lines are parallel, then the alternate interior angles will be equal. For if not, then one would be greater than the other, and so the sum of the interior angles on one side of the transversal would be less than two right angles. In this situation, the fifth postulate applies and forces the lines to meet, which is a contradiction.

As for the existence of parallel lines, Euclid gives a construction in (I.31) for a line through a point $P$, parallel to a given line $l$. Draw any line through $P$, meeting $l$, and then reproduce the angle it makes with $l$ at the point $P$ (I.23). It follows from (I.27) that this line is parallel to $l$.

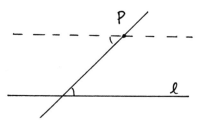

Why does Euclid place this construction after (I.29), even though it does not depend on (I.29) and does not make use of the parallel postulate? Presumably, the answer, although Euclid does not say so, is that using (I.29) one can show that this parallel just constructed is *unique*. If there were any other line parallel to $l$ through $P$, it would make the same angle with the transversal (by (I.29)) and hence would be equal to this one. Thus using the parallel postulate we can prove the following statement:

**P.** For each point $P$ and each line $l$, there exists at most one line through $P$ parallel to $l$.

This statement (P) is often called "Playfair's axiom," after John Playfair (1748–1819), even though it already appears in the commentary of Proclus. Of course, in Euclid's development of geometry, this is not an axiom, but a theorem that can be proved from the axioms. Some authors, however, like to take the statement (P) as an axiom instead of using Euclid's fifth postulate. So I would like to explain in what sense we can say that Euclid's fifth postulate is equivalent to Playfair's axiom (P).

Since the parallel postulate plays such a special role in Euclid's geometry, let us make a special point of being aware when we use this postulate, and which theorems are dependent on its use. Let us call *neutral geometry* the collection of all the postulates and common notions *except* the fifth postulate together with all theorems that can be proved without using the fifth postulate. Thus (I.1)–(I.28) and (I.31) all belong to neutral geometry, while for example, (I.32) and (I.47) do not belong to neutral geometry.

If we take neutral geometry and add back the fifth postulate, then we recover ordinary Euclidean geometry, and we can prove (P) as a theorem as we did above.

But now suppose we take neutral geometry and add (P) as an extra axiom. We will show that in this geometry we can prove Euclid's fifth postulate as a theorem.

Indeed, suppose we are given two lines $l$, $m$ and a transversal $n$ such that the two interior angles 1, 2 on the same side are less than two right angles. Let $P$ be the intersection of the lines $m$ and $n$, and draw a line $l'$ through $P$, making the alternate angle 3 equal to 1. This is possible by (I.23), which belongs to neutral geometry. Then by (I.27), which also belongs to neutral geometry, $l'$ is parallel to $l$.

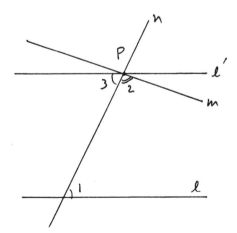

Now, since $1 + 2$ is less than two right angles, it follows that $2 + 3$ is less than two right angles, and hence the line $l'$ is different from $m$ (I.13). Now we can apply (P). Since $l'$ passes through $P$

and is parallel to $l$, it must be the only line through $P$ that is parallel to $l$. In particular, the line $m$, which is different from $l'$, cannot be parallel to $l$, and so by definition it must meet $l$. This proves the fifth postulate.

Thus in the presence of all the results of neutral geometry, we can use Euclid's fifth postulate to prove Playfair's axiom, or we can use Playfair's axiom to prove Euclid's fifth postulate. In this sense we can say that in neutral geometry, Euclid's fifth postulate is equivalent to Playfair's axiom. This means that adding either one of them as an additional assumption to neutral geometry will give the same body of theorems as consequences.

## The Theory of Area

In (I.35), Euclid says that two parallelograms on the same base and in the same parallels (this means their top sides lie on the same line parallel to the base) are equal to each other. In the figure, the parallelogram $ABCD$ is equal to the parallelogram $BCEF$. Clearly, the parallelograms are not congruent.

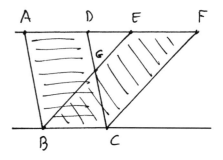

Looking at the proof, which is accomplished by adding and subtracting congruent figures, we conclude that Euclid must be referring to the area of the parallelograms when he says they are equal. But he has not said what the area of a figure is, so we must reflect a bit to see what he means.

Our intuitive understanding of area comes from high-school geometry, where we learn that the area of a rectangle is the product of the lengths of two perpendicular sides, the area of a triangle is one half the product of the lengths of the base and the altitude, etc. The "area" of high-school geometry is a function that attaches to each plane figure a real number; the area of a nonoverlapping union of figures is the sum of the areas, and so forth. Most likely no one ever told you the definition of area, nor did they prove that such an area function exists. Using calculus, you can define the area of a figure in the real Cartesian plane using definite integrals, and in that way it is possible to prove that a suitable area function exists. But in Euclid's geometry there are no real numbers, and we certainly do not want to use calculus to define the concept of area in elementary geometry.

So what did Euclid have in mind? Since he does not define it, we will consider this new equality as an undefined notion, just as the notions of congruence for line segments and angles were undefined. We will call this new notion *equal content*, to avoid confusion with other notions of equality or congruence. We do not want to use the word area, because this notion is quite different from our common understanding of area as a function associating a real number to each figure.

From the way Euclid treats this notion, it is clear that he regards it as an equivalence relation, satisfying the common notions. In particular:

(a) Congruent figures have equal content.
(b) If two figures each have equal content with a third, they have equal content.
(c) If pairs of figures with equal content are *added* in the sense of being joined without overlap to make bigger figures, then these added figures have equal content.
(d) Ditto for subtraction, noting that equality of content of the difference does not depend on *where* the equal pieces were removed.
(e) Halves of figures of equal content have equal content (used in the proof of (I.37)). (Also, doubles of equals are equal, as a consequence of (c) above.)
(f) The whole is greater than the part, which in this case means that if one figure is properly contained in another, then the two figures cannot have equal content (used in the proof of (I.39)).

In terms of the axiomatic development of the subject, at this point Euclid is introducing a new undefined relation, and taking all the properties just listed as new axioms governing this new relation. Later in this book (Section 22), we will discuss Hilbert's reinterpretation of the theory of area where the relationship of having equal content is defined, and all its properties proved, so that it does not require the introduction of new axioms.

Now let us see what Euclid does with this purely geometric notion of equal content of plane figures. In (I.35) he proves that the two parallelograms have equal content (see diagram on previous page) by first showing that the triangle *ABE* is congruent to the triangle *DCF*, so they have equal content. Then by subtracting the triangle *DGE* from each (in different positions!) and adding the triangle *BGC* to each, he obtains the two parallelograms, which therefore have equal content.

In (I.37) he shows that two triangles *ABC* and *DBC* on the same base and in the same parallels have equal content. The method is to double *ABC* to get a parallelogram *EABC*, and to double *DBC* to get a parallelogram *DFBC*.

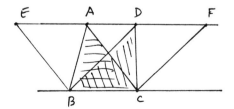

By (I.35) the two parallelograms have equal content, and then he applies the axiom that halves of equals are equal to conclude the triangles have equal content.

This is all that is needed to explain Euclid's beautiful proof of (I.47), the theorem of Pythagoras. The statement of the theorem is that if *ABC* is a right triangle, then the squares on the two legs together have equal content to the square on the hypotenuse. The proof goes like this. The triangle *ABF* is one half of the square *ABFG*. This triangle *ABF* has equal content with the triangle *BFC* by (I.37). The triangle *BFC* is congruent to the triangle *BAD*. And *BAD* has equal content to the triangle *BMD* by (I.37). This latter triangle is equal to one-half of the rectangle *BDLM*. Hence the square *ABFG* has equal content to the rectangle *BDLM*. Doing the same construction on the other side and adding, one has the result.

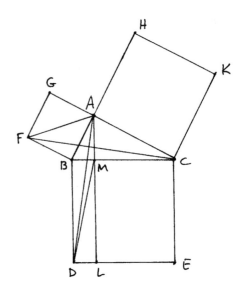

Euclid's statement of (I.47) in terms of equal content of the squares constructed on the sides of the triangle may come as a surprise to the modern student who remembers the formula $a^2 + b^2 = c^2$ (which I suppose in the minds of the general public is rivaled in fame only by Einstein's famous formula $E = mc^2$). We are used to thinking of $a$, $b$, $c$ as the lengths of the sides of the triangle, in which case the theorem becomes an equation among real numbers. How can we reconcile these two points of view?

The modern answer to this question, which we will discuss in more detail later (Section 23), is that after introducing coordinates in our geometry we can prove the existence of an area function. The area of a square of side $a$ will be $a^2$. Furthermore, we will show that having equal content in the sense of Euclid is equivalent to having equal area in the sense of the area function. Then the two formulations of the theorem of Pythagoras become equivalent.

This answer makes sense only when we are able to assign numerical lengths to arbitrary line segments, which the Greeks could not do. Yet there is ample evidence that the Greeks did know special cases of this formula when $a$, $b$, $c$ are integers. The equation $3^2 + 4^2 = 5^2$ was known to the Egyptians, and Proclus in his note on (I.47) mentions two general formulas for generating such "Pythagorean triples" of integers, which he ascribes to Plato and to Pythagoras. So we can presume that the Greeks knew some particular right triangles with integer sides, in which case (I.47) can be represented by the equation among integers $a^2 + b^2 = c^2$. But the geometrical proof given by Euclid is then more general, because it applies to all triangles, and not just those for which one can find integers to fit the sides.

Euclid's theory of area plays an important role in the succeeding books of the *Elements*. It appears not only in results that correspond to our modern notion of area, but also in results, such as the construction of the regular pentagon (IV.11), which at first sight appear to have nothing to do with area. Roughly speaking, Euclid uses arguments involving areas in places where we would expect to see a quadratic equation in analytic geometry. He can add two line segments to get another line segment, but there is no way to multiply line segments so as to get another line segment. Instead, one can regard the rectangle with sides equal to segments *AB*, *CD* as a product of these two segments. The results (I.42)–(I.45) on application of areas and all the results of Book II give a certain flexibility in manipulating and comparing different areas. This creates a sort of "algebra of areas," and one can regard results such as (II.14) as equivalent to the solution of certain quadratic equations. Note also the essential use of area in the proof of (VI.1), which is the cornerstone of Euclid's theory of similar triangles.

# Exercises

3.1 Explain what is wrong with the "proof" in (Example 3.1). (*Hint*: Draw an accurate figure.)

3.2 Read Euclid (I.35)–(I.48), Book II, and (III.35)–(III.37). Be prepared to present proofs of (I.35), (I.41), (I.43), (I.47), (II.6), (II.11), and (III.36).

3.3 Given a triangle *ABC* and given a segment *DE*, construct a rectangle with content equal to the triangle *ABC*, and with one side equal to *DE*.

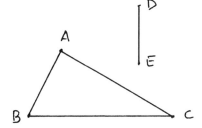

3.4 Given a rectangle, construct a square with the same content.

3.5 Given a line *l* and given two points *A*, *B* not on *l*, construct a circle passing through *A*, *B* and tangent to *l*. (*Hint*: Use (III.36) and/or (III.37).) (Par = 14.)

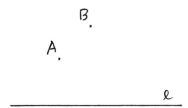

3.6 Given two lines $l$, $m$ and a point $P$ not on either line, construct a circle passing through $P$ and tangent to both $l$ and $m$.

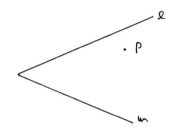

In the following exercises, give proofs based on results of Euclid, Books I–III only.

3.7 Given a triangle $ABC$, let $DE$ be a line parallel to the base $BC$, let $F$ be the midpoint of $DE$, and let $AF$ meet $BC$ in $G$. Prove that $G$ is the midpoint of $BC$. (*Hint*: Draw some extra lines to make parallelograms, and use (I.43).)

3.8 Let $\Gamma$ be a circle with center $O$. Let $AB$ and $AC$ be tangents to $\Gamma$ from a point $A$ outside the circle. Let $BC$ meet $OA$ at $D$. Prove that $OA \times OD = OB^2$ (meaning the rectangle on $OA$ and $OD$ has equal content to the square on $OB$).

3.9 Let $ABC$ be a right triangle, and let $AD$ be the altitude from the right angle $A$ to the hypotenuse $BC$. Prove that $AD^2 = BD \times DC$ (in the sense of content).

3.10 Problem: Given a triangle $ABC$, and given a point $D$ on $BC$, to draw a line through $D$ that will divide the triangle into two pieces of equal content.

Solution (Peletier): Let $E$ be the midpoint of $BC$. Draw $AD$; draw $EF$ parallel to $AD$. Then $DF$ divides the triangle in half.

Prove that the content of the quadrilateral $ABDF$ is equal to the content of the triangle $DFC$.

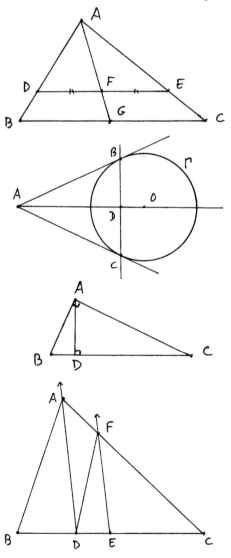

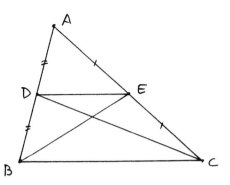

3.11 (Campanus). Use the theory of content to show that the line *DE* joining the midpoint of two sides of a triangle is parallel to the third side. (*Hint*: Draw *BE* and *DC*. Show that the triangles *BDC* and *BEC* have the same content and then apply (I.39).)

# 4   Construction of the Regular Pentagon

One of the most beautiful results in all of Euclid's *Elements* is the construction of a regular pentagon inscribed in a circle (IV.11). The proof of this construction makes use of all the geometry he has developed so far, so that one could say that to understand fully this single result is tantamount to understanding all of the first four books of Euclid's geometry. It also raises questions of exposition that are central to our modern examination of Euclid's methods. For example, why does Euclid use the theory of area in proving a result about the sides of a polygon?

In this section we will present Euclid's construction of the regular pentagon, and begin discussing the issues raised by its proof. Later (see (13.4), Exercise 20.10, (29.1)), we will give other proofs using similar triangles or the complex numbers. Euclid's original geometric proof must be regarded as a tour de force of classical geometry. It depends on the theory of area, which we will discuss in more detail in Section 22. So this section can be regarded as a taste of things to come: a first meeting with one of the deeper topics that is central to Euclid's geometry.

The key point of the construction of the pentagon is the following problem.

**Problem 4.1**
To construct an isosceles triangle whose base angles are equal to twice the vertex angle.

**Construction** ((II.11), (IV.10))
Let *A*, *B* be two points chosen at random.

1. Draw line *AB*.

Next, construct a perpendicular to *AB* at *A*, as follows:

2. Circle *AB*, get *C*.
3. Circle *BC*.
4. Circle *CB*, get *D*.
5. Line *AD*, get *E*.

Next, we bisect *AE* as follows

6. Circle *EA*, get *F*, *G*.
7. Line *FG*, get *H*.

Now comes the unusual part of the construction:

8. Circle *HB*, get *J*.
9. Circle *AJ*, get *K*.
10. Circle center *B*, radius *AK*, get *L*.
11. Line *AL*.
12. Line *BL*.

Then $\triangle ABL$ is the required triangle. The angles at *B* and at *L* will be equal to twice the angle at *A*.

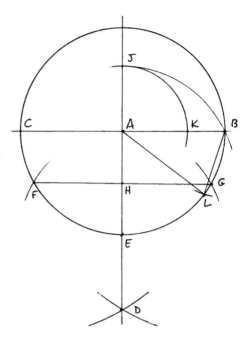

*Proof* From a modern point of view, it would seem that some theory of quadratic equations is essential for the proof. Euclid did not have any algebra available to him, but he was able to deal with quantities essentially equivalent to quadratic expressions via the theory of area. We can think of a rectangle as representing the product of its sides, or a square as the square of its side. These areas, without even assigning a numerical value to them, can be manipulated by cutting up and adding or subtracting congruent pieces. In this way Euclid establishes a "geometrical algebra" for manipulating these quantities (always by geometrical methods), which acts as a substitute for our modern algebraic methods.

Let us then trace the steps by which Euclid proves (IV.10), which is the key point in the construction of the regular pentagon. In Book I, especially (I.35)–(I.47) he discusses the areas of triangles and parallelograms, leading up to the famous Pythagorean theorem (I.47), which is stated in terms of area: The square built on the hypotenuse of a right triangle has area equal to the combined areas of the squares on the two sides. The theorem is proved by cutting these areas into triangles, and proving equality of areas using the cutting and pasting methods just developed. Here area is understood in the sense of content—cf. Section 3.

Book II contains a number of results of geometrical algebra, as described above, all stated and proved geometrically in terms of areas. In particular, (II.5), (II.6), and (II.11) are used in the proof of (IV.10). Note that (II.11), which is sometimes called the division of a segment in extreme and mean ratio, states that the interval $AB$ is divided by a point $K$ (in our notation (4.1) above) such that the rectangle formed by $BK$ and $AB$ has area equal to the square on $AK$. In this way the property of extreme and mean ratio is expressed using area.

From Book III we need (III.36) and its converse (III.37). Proposition (III.36) says that if a point $A$ lies outside a circle, and if $AB$ is tangent to the circle at $B$, and if $ACD$ cuts the circle at $C$ and $D$, then the rectangle formed by $AC$ and $AD$ has area equal to the square on $AB$. This result is proved by several applications of (II.6) and (I.47).

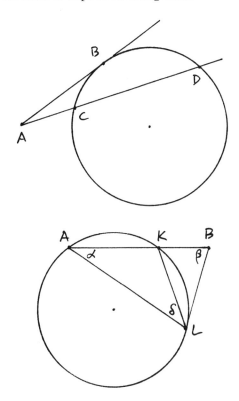

Now Euclid can prove (IV.10) by a brilliant application of (III.37). Let $A$, $K$, $B$, $L$ be as in the construction (4.1) above. Then by (II.11), the rectangle with sides $BK$ and $BA$ has area equal to the square on $AK$. Since $BL$ was constructed equal to $AK$, this is also equal to the square on $BL$.

Now consider the circle passing through the three points $A$, $K$, $L$. Since the rectangle on $BK$ and $BA$ is equal to the square on $BL$, it follows that $BL$ is tangent to this circle (III.37)!

Hence the angle $\angle BLK$ formed by the tangent $BL$ and the line $LK$ is equal to the angle $\alpha$ at $A$, which subtends the same arc (III.32). Let $\angle KLA = \delta$. Then $\angle BKL$ is an exterior angle to the triangle $\triangle AKL$, so $\angle BKL = \alpha + \delta$ (I.32). But $\angle BLK = \alpha$, so $\alpha + \delta = \angle BLA$, and this angle is $\beta$ because $\triangle ABL$ is isosceles. Hence $\angle BKL = \beta$. Now it follows that $\triangle BKL$ is isosceles, so $KL = BL = AK$. Hence $\triangle AKL$ is also isosceles, so $\delta = \alpha$. Now $\beta = \angle BLA = 2\alpha$ as required.

Once we have the isosceles triangle constructed in (4.1), the construction of the pentagon follows naturally. The idea is to inscribe in the circle a triangle equiangular with the given triangle, and then to bisect its two base angles.

### Problem 4.2

Given an isosceles triangle whose base angles are equal to twice its vertex angle, and given a circle with its center, to construct a regular pentagon inscribed in the circle.

**Construction** ((IV.2) and (IV.11))

Let $\triangle ABC$ be the given triangle and let $O$ be the center of the given circle. The first part of the construction is to obtain a tangent line to the circle. Let $D$ be any point on the circle.

1.  Line $OD$.
2.  Circle $DO$, get $E$.
3.  Circle $EO$.
4.  Circle $OE$, get $F$.
5.  Line $DF$.

Then $DF$ will be a tangent line. Next, we reproduce the angle $\beta$ from the base of the isosceles triangle at $D$, on both sides.

6.  Circle $BC$, get $G$.
7.  Circle at $D$ with radius equal to $BC$, get $H$, $I$.
8.  Circle center $H$, radius $CG$, get $K$.
9.  Circle center $I$, radius $CG$, get $L$.
10. Line $DK$, get $M$.
11. Line $DL$, get $N$.
12. Line $MN$.

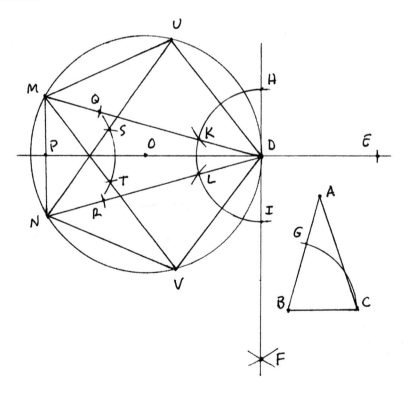

Then $\triangle DMN$ is a triangle inscribed in the circle, equiangular with $\triangle ABC$. Next we bisect the angles at $M$, $N$. Let $P$ be the intersection of $MN$ with $DO$.

13. Circle $MP$, get $Q$.
14. Circle $NP$, get $R$.
15. Circle $PR$.
16. Circle $RP$, get $S$.
17. Circle $QP$, get $T$.
18. Line $NS$, get $U$.
19. Line $MT$, get $V$.

Then $D$, $M$, $N$, $U$, $V$ will be the vertices of the pentagon.

20. Line $DU$.
21. Line $UM$.
22. Line $DV$.
23. Line $VN$.

Then $DUMNV$ is the required pentagon.

*Proof*   We follow the geometric proof given by Euclid. First of all, the line $DF$ is constructed perpendicular to a diameter of the circle, so it is a tangent line to the circle (III.16). Next, the triangles $\triangle DHK$ and $\triangle DLI$ are constructed so that their three sides are equal to the three sides of $\triangle BCG$. Hence by (SSS) = (I.8), it follows that $\angle KDH$ and $\angle LDI$ are both equal to the angle $\beta$ of the triangle $\triangle ABC$ at $B$. From there it follows that the angles of $\triangle DMN$ at $M$ and $N$ are both equal to $\beta$, because they subtend the same arcs cut off by the tangent line and the angles $\beta$ just constructed (III.32). Since the sum of the three angles of a triangle is constant $= 180°$ (I.32), it follows that the triangle $\triangle DMN$ is equiangular with the triangle $\triangle ABC$. In particular, if $\alpha$ is the angle at $D$, then $\beta = 2\alpha$.

The points $U$, $V$ are constructed by taking the angle bisectors of $\triangle DMN$ at $M$ and $N$. Since the angles at $M$ and $N$ are $\beta$, their halves are equal to $\alpha$. Thus the arcs $DU$, $UM$ subtend angles $\alpha$ at $N$; the arc $MN$ subtends an angle $\alpha$ at $D$; and the arcs $DV$, $VN$ subtend angles $\alpha$ at $M$. Hence these five arcs are all equal (III.26), and the line segments on them are also equal. So we have constructed an equilateral pentagon inscribed in the circle. The angle subtended by each side at the center of the circle will be $2\alpha = \beta$. It follows that the angles of the pentagon are also equal, so the pentagon is *regular* in the sense that its sides are all equal and its angles are all equal.

This completes the presentation of Euclid's construction of the pentagon. As usual, his method is adapted to economy of proof, not economy of steps used. The whole construction, as we have presented it here, takes $12 + 23 = 35$ steps. By collapsing separate parts of the construction, in particular, by constructing the triangle of (4.1) on a radius of the given circle, one can make a construction with fewer than half as many steps (cf. (4.3)). Note also that Euclid's construction of the points $U$, $V$ by bisecting the angles at $M$, $N$ makes possible his elegant proof that the five sides of the pentagon are equal. However, in retrospect we see that $MN$ is actually one side of the pentagon, so $U$ and $V$ could have been constructed in a single step by a circle with center $D$ and radius $MN$.

If there is such a thing as beauty in a mathematical proof, I believe that this proof of Euclid's for the construction of the regular pentagon sets the standard for a beautiful proof. In the words of Edna St. Vincent Millay, "Euclid alone has looked on beauty bare."

Now let us use the ideas of Euclid's method to construct a pentagon in as few steps as possible.

### Problem 4.3

Given a circle with center $O$, construct a regular pentagon inscribed in the circle in as few steps as possible.

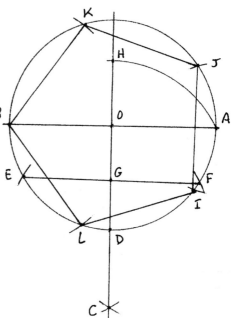

1. Draw any line through $O$. Get $A$, $B$.
2. Circle $AB$.
3. Circle $BA$, get $C$.
4. $OC$, get $D$.
5. Circle $DO$. Get $E$, $F$.
6. $EF$, get $G$.
7. Circle $GA$, get $H$.
8. Circle center $A$, radius $OH$, get $I$, $J$.
9. Circle center $B$, radius $IJ$, get $K$, $L$.
10–14. Draw $BK$, $KJ$, $JI$, $IL$, $LB$.

Then $BKJIL$ is the required pentagon.

## Exercises

4.1 Read Euclid, Book IV.

4.2 Explain why the construction of (Problem 4.3) gives a regular pentagon.

4.3 Given a circle, but not given its center, construct an inscribed equilateral triangle in as few steps as possible (par = 7).

4.4 Construct a square in as few steps as possible (par = 9).

4.5 Given a line segment $AB$, construct a regular pentagon having $AB$ as a side (par = 11).

4.6 Given a circle $\Gamma$ and given its center $O$, construct inside $\Gamma$ three equal circles, each one tangent to $\Gamma$ and to the other two (par = 13).

4.7 Let $ABC$ be an equilateral triangle inscribed in a circle. Let $D$, $E$ be the midpoints of two sides, and extend $DE$ to meet the circle at $F$. Prove that $E$ divides the segment $DF$ in extreme and mean ratio, i.e. the rectangle $EF \times DF$ equals the square $DE^2$. *Hint*: Use (III.35).

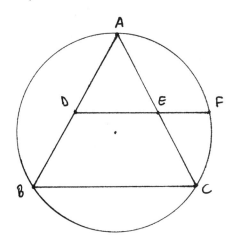

4.8 Take a long thin piece of paper. Tie a simple overhand knot in the paper, and fold the knot flat. Explain why the flat knot makes a regular pentagon.

# 5  Some Newer Results

In this section we mention some results of plane geometry that do not appear in Euclid's *Elements* but that can be proved using the methods developed in Books I–IV. Some of these, such as the three altitudes of a triangle meeting in a point, were known to the Greeks. Others, such as the Euler line and the nine-point circle, were discovered only in the eighteenth and nineteenth centuries.

In some textbooks these results are proved using similar triangles. In Euclid's *Elements*, similar triangles do not appear until Book VI, using the theory of proportion developed in Book V. In modern texts, similar triangles are defined by comparing the lengths of the sides. Since we have not yet discussed either of these techniques, we will use only the pure geometric methods of Books I–IV in this section.

Two theorems taught in modern high-school geometry are that the angle bisectors of a triangle meet in a point (the *incenter* of the triangle), and the perpendicular bisectors of the sides of a triangle meet in a point (the *circumcenter* of the triangle). Although not explicitly stated by Euclid, these two results are implicitly contained in (IV.4) and (IV.5).

On the other hand, the theorems about the three medians and the three altitudes of a triangle do not appear in Euclid, though they were known to Archimedes, so we will start with them.

### Proposition 5.1

*Let ABC be a triangle, and let D, E be the midpoints of AB and AC, respectively. Then the line DE is parallel to the base BC, and equal to one-half of it. In other words, if F is the midpoint of BC, then DE ≅ BF.*

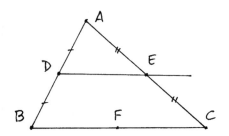

*Proof*  We begin with a slightly different construction. Let $D$ be the midpoint of $AB$, and draw lines through $D$ parallel to $AC$ and $BC$. Let them meet the opposite sides in points $E'$, $F'$. Since $DE'$ is parallel to $BC$, the angles at $B$ and $D$ are congruent ((I.29): Here we use the parallel postulate). Similarly, since $AC$ is parallel to $DF'$, the angles at $A$ and $D$ are equal.

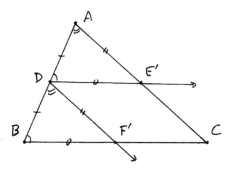

Now $AD \cong DB$, and the angles of the triangle $ADE'$ and $DBF'$ at $A$ and $D$ are equal to those at $D$ and $B$, respectively, so by (ASA) (I.26), the triangles $ADE'$ and $DBF'$ are congruent. We conclude that $AE' \cong DF'$ and $DE' \cong BF'$.

Now look at the parallelogram $DE'F'C$. By (I.34) the opposite sides are equal. So $DF' \cong E'C$ and $DE' \cong F'C$. Thus we see that $E'$ and $F'$ are the midpoints of the sides $AC$ and $BC$. So $E' = E$, the line $DE'$ is equal to the line $DE$, and therefore $DE$ is parallel to $BC$ as claimed. Furthermore, we have seen that $DE' \cong BF'$, and $F'$ is the midpoint of $BC$, so $DE$ is equal to one-half of $BC$.

### Corollary 5.2

*Let ABC be a triangle, and let D, E, F be the midpoints of the three sides. Then the sides of the triangle DEF are parallel to the sides of ABC, and the four small triangles formed are all congruent to each other.*

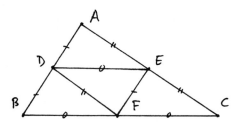

*Proof*  From the proposition it follows that each side of the triangle $DEF$ is parallel to and equal to one-half of a side of the triangle $ABC$. Then by (SSS) (I.8) all four small triangles are congruent.

### Definition

We say a triangle $ABC$ is congruent to the *double* of a triangle $FED$, in symbols $ABC \cong 2FED$, if as in the diagram above, the three sides of $ABC$ are double the sides of $FED$, and the three angles of $ABC$ are equal to the three angles of $FED$.

### Proposition 5.3 (2ASA)

*Let $ABC$ and $A'B'C'$ be two triangles, and assume that the angles at $B$ and $C$ are equal to the angles at $B'$ and $C'$, and that $BC \cong 2B'C'$. Then the triangle $ABC$ is congruent to the double of $A'B'C'$.*

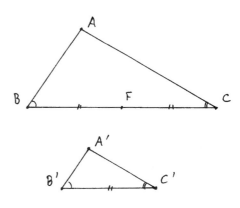

*Proof* Let $D$, $E$, $F$ be the midpoints of the sides of $ABC$, and draw the triangle $DEF$. Then from (5.2) we see that $DE \cong \frac{1}{2}BC \cong B'C'$. Furthermore, because $DE$ is parallel to $BC$, the angles of the triangle $ADE$ at $D$ and $E$ are equal to the angles at $B'$ and $C'$. Now by (ASA), the triangle $ADE$ is congruent to $A'B'C'$. But $ABC$ is a double of $ADE$, so $ABC \cong 2A'B'C'$.

### Remark

One can easily prove other double congruence theorems corresponding to (SAS) and (SSS) (see Exercises 5.1, 5.2). Of course, these are special cases of more general theorems on similar triangles that we will discuss in Section 20.

### Proposition 5.4

*The* medians (*lines from a vertex to the midpoint of the opposite side*) *of a triangle meet in a single point* (*called the* centroid *of the triangle*).

*Proof* Let $ABC$ be the triangle, let $D$, $E$ be the midpoints of $AB$ and $AC$, and draw $DE$. Let the two medians $BE$ and $CD$ meet at a point $G$. Since $DE$ is parallel to $BC$ (5.1), we find that $\angle DEG = \angle CBG$ and $\angle EDG = \angle BCG$. On the other hand, $BC = 2DE$ (5.1). Therefore, we can apply the previous result (5.3) and find that $\triangle BGC \cong 2\triangle EGD$.

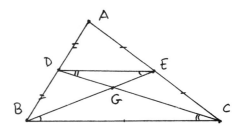

In particular, $BG = 2GE$. Thus $G$ can be described as the point on the median $BE$ that is $\frac{2}{3}$ of the way from $B$ to $E$. Reversing the roles of $A$ and $C$ would therefore show that the third median $AF$ also passes through $G$. Thus all three medians meet in the point $G$.

### Corollary 5.5

*The centroid $G$ lies on each median $\frac{2}{3}$ of the way from the vertex to the midpoint of the opposite side.*

*Proof*  Follows from the proof of (5.4).

### Proposition 5.6

*The three altitudes (lines through a vertex, perpendicular to the opposite side) of a triangle meet in a single point (the orthocenter of the triangle).*

*Proof*  Let $ABC$ be the given triangle. Draw lines through the vertices $A$, $B$, $C$, parallel to the opposite sides, to form a new triangle $A'B'C'$. By (I.34) applied to the parallelograms $BCAC'$ and $BCB'A$ we see that $C'A = BC = AB'$. Thus $A$ is the midpoint of $B'C'$, and similarly for the other two sides of $A'B'C'$.

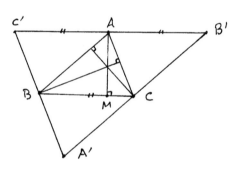

On the other hand, the altitude $AM$ of the triangle $ABC$ is perpendicular to $BC$, and hence also perpendicular to $B'C'$. Thus we see that the altitudes of the triangle $ABC$ are equal to the perpendicular bisectors of the sides of the triangle $A'B'C'$. Hence they meet in a single point ((IV.5), cf. Exercise 1.9).

### Proposition 5.7 (The Euler line)

*In a triangle $ABC$, let $O$ be the circumcenter, let $G$ be the centroid, and let $H$ be the orthocenter. Then $O$, $G$, $H$ lie on a line (called the Euler line of the triangle) and $GH \cong 2OG$.*

*Proof*  For the proof, let $F$ be the midpoint of $BC$, draw the median $AF$, and let the line $OG$ meet the altitude $AM$ in a point $H'$. Note that $OF$ is perpendicular to $BC$, since $O$ is the circumcenter. Hence $OF$ is parallel to $AM$. Therefore, $\angle GAH' \cong \angle GFO$. Also, $\angle AGH' \cong \angle FGO$, since they are vertical angles (I.15). By our previous result on the medians (5.5), $AG \cong 2GF$. Thus we can apply (2ASA) (5.3) to conclude that $\triangle AGH' \cong 2\triangle FGO$. It follows that $GH' = 2OG$.

Thus the point $H'$ is characterized as that point on the ray $OG$ such that $GH' = 2OG$. Now permuting the roles of $A$, $B$, $C$, it follows that $H'$ also lies on the other altitudes of $ABC$, so $H' = H$ is the orthocenter and our conclusions $O$, $G$, $H$ collinear and $GH \cong 2OG$ follow.

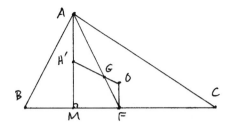

By the way, this argument provides another independent proof of the fact that the three altitudes meet in a point (5.6).

For our next results, we will introduce the very useful method of cyclic quadrilaterals.

## Definition

A *cyclic quadrilateral* is a set of four points $A$, $B$, $C$, $D$ lying in that order on a circle, together with the lines $AB$, $BC$, $CD$, $DA$ joining them. The lines $AC$ and $BD$ are the *diagonals* of the cyclic quadrilateral.

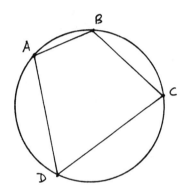

The importance of cyclic quadrilaterals comes from the relationships between the various angles of the figure, which characterize the property of the four points $A$, $B$, $C$, $D$ lying on a circle.

## Proposition 5.8

*Let $A$, $B$, $C$, $D$ be four points in the plane, with $A$, $B$ both on the same side of the line $CD$. Then $A$, $B$, $C$, $D$ lie on a circle if and only if the angles $\angle DAC$ and $\angle DBC$ are equal.*

*Proof*  If $A$, $B$, $C$, $D$ lie on a circle, then Euclid's (III.21) tells us that the angles at $A$ and $B$ are the same, since they both subtend the same arc $DC$.

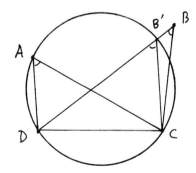

Conversely, suppose the angles at $A$ and $B$ are equal. Draw the circle through $A$, $D$, $C$ (IV.5) and let it meet the line $BD$ at $B'$. (In our figure, $B$ lies outside the circle, but the argument will be similar if $B$ lies inside the circle.)

Arisippus Philosophus Socraticus, naufragio cum ejectus ad Rhodiensium litus animadvertisset Geometrica schemata descripta, exclamavisse ad camites ita dicitur, Bene speremus, Hominum enim vestigia video.
Vitruv. Architect. lib. 6. Præf.

Plate III. The frontispiece to Archimedes (Oxford edition of 1792). When the Socratic philosopher Aristippus was shipwrecked on the shores of Rhodes, he saw geometrical figures in the sand and exclaimed to his comrades: "There is hope: I see traces of men".

Then by (III.21), the angle at $B'$ is also equal to the angles at $A$ and at $B$. If $B \neq B'$, this contradicts (I.16), because the angle $\angle DB'C$ at $B'$ is an exterior angle to the triangle $BCB'$, and so must be greater than the opposite interior angle at $B$. Hence $B = B'$, and all four points lie on the circle.

**Theorem 5.9** (The nine-point circle)
*In any triangle, the midpoints of the three sides, the feet of the three altitudes, and the midpoints of the segments joining the three vertices to the orthocenter all lie on a circle.*

*Proof*  Let $ABC$ be the given triangle. Let $D$, $E$, $F$ be the midpoints of the sides, let $K$, $L$, $M$ be the feet of the altitudes, let $H$ be the orthocenter, and let $P$, $Q$, $R$ be the midpoints of the segments joining the three vertices to $H$. We must show that $D$, $E$, $F$, $K$, $L$, $M$, $P$, $Q$, $R$ all lie on a circle.

We make several uses of (5.1). Applied to the triangle $ABC$, we find that $DE$ is parallel to the base $BC$. Applied to the triangle $BCH$, we find that $RQ$ is parallel to the base $BC$. Hence $DE$ is parallel to $RQ$. Now apply (5.1) to the triangle $ACH$. We find that $EQ$ is parallel to the base $AH$. Similarly, using the triangle $ABH$, $DR$ is parallel to $AH$. Hence $EQ$ and $DR$ are parallel. Furthermore, $EQ$ and $DR$ are perpendicular to $DE$ and $RQ$, since $AH$ is perpendicular to $BC$. Thus $DEQR$ is a rectangle. If $X$ is the center of this rectangle, then $X$ is equidistant from the four corners. Thus $D$, $E$, $Q$, $R$ lie on a circle $\Gamma$ with center $X$. We will show that this circle contains the other required points.

By (5.8), $DLER$ is a cyclic quadrilateral, because the angles at $D$ and $L$ subtending $ER$ are both right angles. Since a circle is determined by three points (III.10), this circle is the same as the circle $\Gamma$; in other words, $L$ also lies on $\Gamma$.

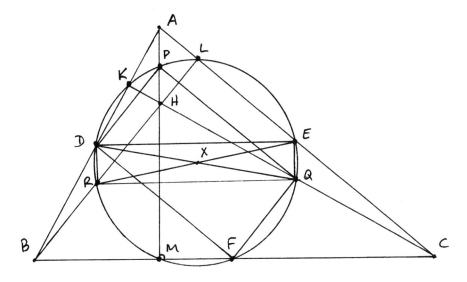

A similar argument shows that $DKEQ$ is a cyclic quadrilateral, and so $K$ lies on $\Gamma$.

Now, shifting perspective so that $AC$ is regarded as the base of the triangle, the same argument shows that $DPQF$ is a rectangle, with the same center $X$, since $X$ is the midpoint of $DQ$. Therefore, $P$ and $F$ also lie on $\Gamma$.

Finally, $MDPF$ is a cyclic quadrilateral for the same reasons as above, so $M$ is also on $\Gamma$.

**Proposition 5.10** (The orthic triangle)
*Let ABC be any acute triangle, and let K, L, M be the feet of the altitudes of ABC. Then the altitudes of ABC are the angle bisectors of the orthic triangle KLM.*

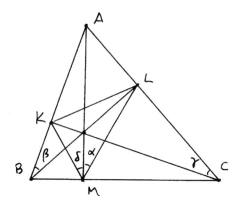

*Proof* We use cyclic quadrilaterals. First, $LMAB$ is a cyclic quadrilateral, because the angles at $L$ and $M$ are right. Hence $\alpha = \angle AML$ is equal to $\beta = \angle ABL$.

Next, $KLCB$ is a cyclic quadrilateral because the angles at $K$ and $L$ are right. Hence $\beta$ is equal to $\gamma = \angle KCL$.

Finally, $MKAC$ is a cyclic quadrilateral because the angles at $M$ and $K$ are right, so $\gamma$ is equal to $\delta = \angle AMK$.

Thus $\alpha = \delta$, so that the altitude $AM$ of $ABC$ is the angle bisector of the angle $\angle KML$ in the orthic triangle. The same argument of course applies to the other two altitudes.

Since the angle bisectors of $\triangle KLM$ meet in a point, this gives another proof that the altitudes of $\triangle ABC$ meet in a point.

We end this section with an ingenious construction given by Pappus in his commentary on the lost book of Apollonius *On Tangencies*.

### Problem 5.11
Given a circle $\Gamma$ and two points $A$, $B$, find a point $C$ on the circle such that if the lines $CA$, $CB$ meet $\Gamma$ in further points $D$, $E$, then $DE$ is parallel to $AB$.

### Construction
Let $\Gamma$ be the given circle, with center $O$, and let $A$, $B$ be the given points.

1. Line $AB$.
2. Circle $AB$, get $F$.
3. Line $AF$, get $G$.

4. Circle $AG$, get $H$.
5. Line $OH$.
6. Circle $O$, any radius.
7. Circle $H$, same radius, get $I, J$.
8. Line $IJ$, get $K$.
9. Circle $KO$, get $D$.
10. Line $AD$, get $C$.
11. Line $CB$, get $E$.
12. Line $DE$ is parallel to $AB$, as required.

The proof of this construction is Exercise 5.11.

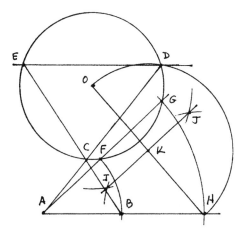

# Exercises

5.1 (2SAS) Suppose we are given two triangles $ABC$ and $A'B'C'$. Assume that $AB \cong 2A'B'$ and $AC \cong 2A'C'$, and the angles at $A$ and $A'$ are equal. Prove that $\triangle ABC \cong 2\triangle A'B'C'$.

5.2 (2SSS) Suppose we are given two triangles $ABC$ and $A'B'C'$ and assume that $AB \cong 2A'B'$, $AC \cong 2A'C'$, and $BC \cong 2B'C'$. Prove that $\triangle ABC \cong 2\triangle A'B'C'$.

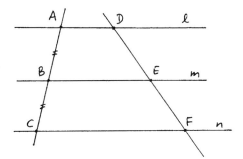

5.3 Let $l$, $m$, $n$ be three parallel lines. Suppose they cut off equal segments $AB \cong BC$ on a transversal line. Show that the segments $DE$, $EF$ cut off by any other transversal line are equal.

5.4 Given three line segments, make a ruler and compass construction of a triangle whose medians are congruent to the three given segments. What condition on the segments is necessary for this to be possible?

5.5 Let $ABCD$ be a quadrilateral. Show that the figure formed by joining the midpoints of the four sides is a parallelogram.

5.6 In any triangle, show that the center $X$ of the nine-point circle lies on the Euler line (Proposition 5.7), and is the midpoint of the segment $OH$ joining the circumcenter $O$ to the orthocenter $H$.

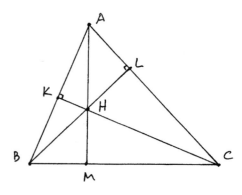

5.7 Use cyclic quadrilaterals to give another proof of Proposition 5.6, as follows. Let *ABC* be the given triangle. Let the altitudes *BL* and *CK* meet at *H*. Let *AH* meet the opposite side at *M*. Then show that *AM* ⊥ *BC*. (This proof is probably the one known to Archimedes.)

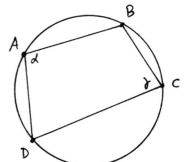

5.8 Show that the opposite angles α, γ of a quadrilateral *ABCD* add to two right angles if and only if *A*, *B*, *C*, *D* lie on a circle.

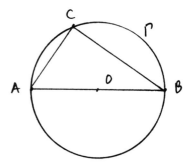

5.9 Let *AB* be the diameter of a circle Γ. Show that a triangle *ABC* has a right angle at *C* if and only if *C* lies on the circle Γ.

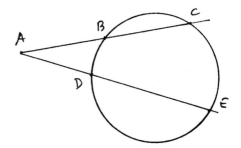

5.10 Let *B*, *C* and *D*, *E* lie on two rays emanating from a point *A*. Show that *B*, *C*, *D*, *E* lie on a circle if and only if *AB* × *AC* = *AD* × *AE* (in the sense of content).

5.11  In the construction to Problem 5.11, prove that *DE* is parallel to *AB*.
      *Hint*: First show that *BCDH* is a cyclic quadrilateral. Then draw *DH*, and compare angles using (III.22) and (III.32).

5.12  In the construction to Problem 5.11, show that the circle through *A*, *B*, *C* is tangent to Γ. Thus this construction solves the problem, "given a circle Γ and given two points *A*, *B*, to find a circle passing through *A*, *B*, and tangent to Γ." This is a special case of the problem of Apollonius (Section 38).

5.13  (The Simson line). Let *ABC* be any triangle. Let *P* be a point on the circumscribed circle of *ABC*. Let *D*, *E*, *F* be the feet of the perpendiculars from *P* to the sides of the triangle (extended as necessary). Then *D*, *E*, *F* lie on a line. (First proved by W. Wallace, 1799.)

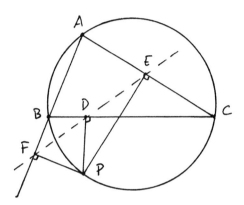

5.14  (The Miquel point). Let *ABC* be a triangle. Let *D*, *E*, *F* be points on the sides of the triangle. Show that the circles through *ADE*, *BDF*, and *CEF* all meet in a common point *G*. *Hint*: Let *G* be the intersection of the first two circles, then show that *CEGF* is a cyclic quadrilateral (due to A. Miquel, 1838).

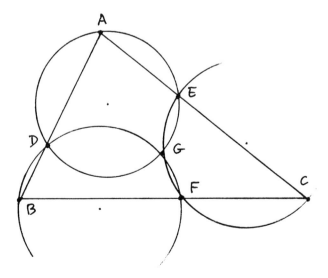

5.15 (Pappus's theorem). Let $A$, $B$, $C$ be points on a line $l$, and let $A'$, $B'$, $C'$ be points on a line $m$. Assume that $AC'\|A'C$ and $B'C\|BC'$. Show that $AB'\|A'B$. *Hint*: Draw a circle through $A, B', C'$ meeting $l$ in $D$. Then use cyclic quadrilaterals (cf. Hilbert, *Foundations*, Section 14).

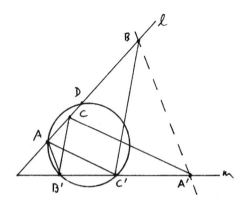

5.16 Construct three circles of different radii, each one tangent to the other two, with noncollinear centers, in as few steps as possible (par = 7).

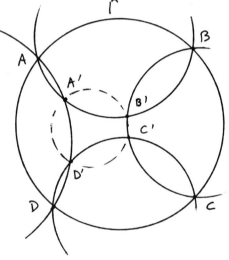

5.17 Let $A$, $B$, $C$, $D$ be four points on a circle $\Gamma$. Let four more circles pass through $AB$, $BC$, $CD$, $DA$, respectively, meeting in further points $A', B', C', D'$. Show that $A'B'C'D'$ is a cyclic quadrilateral.

5.18 (Painting the plane). If the plane has been colored so that each point has one of three colors (red, yellow, blue), prove that for any interval $AB$ there exist two points $C$, $D$ of the same color, with $AB \cong CD$. (It is an unsolved problem whether the same result is true for four colors.)

5.19 Given an angle with vertex $O$ and a point $P$ inside the angle, drop perpendiculars $PA$, $PB$ to the two sides of the angle, draw $AB$, and drop perpendiculars $OC$, $PD$ to the line $AB$. Then show that $AC = BD$.

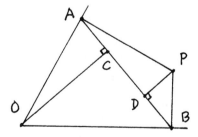

5.20 Given any triangle $ABC$, let $D$, $E$, $F$ be the feet of the altitudes. Show that the six projections $G$, $H$, $I$, $J$, $K$, $L$ of $D$, $E$, $F$ onto the other sides of the triangle lie on a circle.

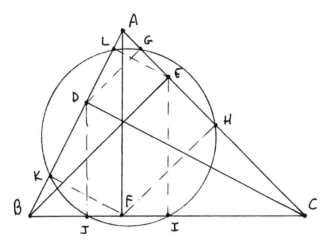

5.21 (Wentworth). Let $ABC$ be a triangle. Construct with ruler and compass a line parallel to $BC$, meeting $AB$ in $D$ and $AC$ in $E$, such that $DE = DB + EC$.

In England the text-book of Geometry consists of the *Elements* of Euclid; for nearly every official programme of instruction or examination explicitly includes some portion of this work. Numerous attempts have been made to find an appropriate substitute for the *Elements* of Euclid; but such attempts, fortunately, have hitherto been made in vain. The advantages attending to a common standard of reference in such an important subject, can hardly be overestimated; and it is extremely improbable, if Euclid were once abandoned, that any agreement would exist as to the author who should replace him.

– from the preface to
Todhunter's Euclid
London (1882)

# 2 Hilbert's Axioms

**CHAPTER**

ur purpose in this chapter is to present (with minor modifications) a set of axioms for geometry proposed by Hilbert in 1899. These axioms are sufficient by modern standards of rigor to supply the foundation for Euclid's geometry. This will mean also axiomatizing those arguments where he used intuition, or said nothing. In particular, the axioms for betweenness, based on the work of Pasch in the 1880s, are the most striking innovation in this set of axioms.

Another choice has been to take the SAS theorem as an axiom, and thus bypass the method of superposition. It is possible to go the other route, and use motions of figures as a basic building block of geometry. This is what Hadamard does in his *Leçons de Géométrie Élémentaire* (1901–06), but the result is a step backward in logical clarity, because he never makes precise exactly what kind of motions he is allowing. See, however, Section 17 for a fuller discussion of rigid motions and SAS.

The first benefit of establishing the new system of axioms is, of course, to vindicate Euclid's *Elements*, and thus establish "Euclidean" geometry as a rigorous mathematical discipline. A second benefit is to pose carefully those problems that have bothered geometers for centuries, such as the question of the independence of the parallel postulate. Unless one has an exact understanding of precisely what is assumed and what is not, one risks going around in circles discussing these questions. In the development of our geometry with the new

**65**

axioms, we will keep the parallel postulate separate and note carefully what depends on it and what does not.

Besides presenting the axioms, this chapter will also contain the first consequences of the axioms, including different proofs of some of Euclid's early propositions, until we have established enough so that Euclid's later results can be deduced without difficulty from the new foundations we have established. In Sections 10, 11, 12, we show how to recover all the results of Euclid, Books I–IV, except for the theory of area, whose proof is postponed until Chapter 5.

# 6   Axioms of Incidence

The axioms of incidence deal with points and lines and their intersections. The points and lines are undefined objects. We simply postulate a set, whose elements are called *points*, together with certain subsets, which we call *lines*. We do not say what the points are, nor which subsets form lines, but we do require that these undefined notions obey certain axioms:

**I1.** For any two distinct points $A, B$, there exists a unique line $l$ containing $A, B$.
**I2.** Every line contains at least two points.
**I3.** There exist three noncollinear points (that is, three points not all contained in a single line).

### Definition
A set whose elements are called points, together with a set of subsets called lines, satisfying the axioms (I1), (I2), (I3), will be called an *incidence geometry*. If a point $P$ belongs to a line $l$, we will say that $P$ lies on $l$, or that $l$ passes through $P$.

From this modest beginning we cannot expect to get very interesting results, but just to illustrate the process, let us see how one can prove theorems based on these axioms.

### Proposition 6.1
*Two distinct lines can have at most one point in common.*

*Proof*   Let $l, m$ be two lines, and suppose they both contain the points $A, B$, with $A \neq B$. According to axiom (I1), there is a unique line containing both $A$ and $B$, so $l$ must be equal to $m$.

Note that this fact, which was used by Euclid in the proof of (I.4) with the rather weak excuse that "two lines cannot enclose a space," follows here from the uniqueness part of axiom (I1). This should indicate the importance of stating explicitly the uniqueness of an object, which was rarely done in Euclid's *Elements*.

Now we have an axiom system, consisting of the undefined sets of points and lines, and the axioms (I1)–(I3). A *model* of that axiom system is a realization of the undefined terms in some particular context, such that the axioms are satisfied. You could also think of the model as an *example* of the *incidence geometry* defined above.

**Example 6.1.1** (The real Cartesian plane).
Here the set of points is the set $\mathbb{R}^2$ of ordered pairs of real numbers. The lines are those subsets of points $P = (x, y)$ that satisfy a linear equation $ax + by + c = 0$ in the variables $x, y$. To verify that the axioms hold, for (I1) think of the "two-point formula" from analytic geometry: Given two points $A = (a_1, a_2)$ and $B = (b_1, b_2)$. They lie on the line

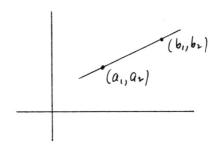

$$y - a_2 = \frac{b_2 - a_2}{b_1 - a_1}(x - a_1)$$

if $a_1 \neq b_1$; if $a_1 = b_1$, they lie on the line $x = a_1$. To verify (I2), take any linear equation involving $y$. Substitute two different values of $x$, and solve for $y$. This gives two points on the line. If the equation did not involve $y$, say $x = c$, take the points $(c, 0)$ and $(c, 1)$. To verify (I3), consider the points $(0, 0), (0, 1),$ $(1, 0)$. One sees easily that there is no linear equation with all three points as solutions.

**Example 6.1.2**
One can also make models out of finite sets. For example, let the set of points be a set of three elements $\{A, B, C\}$, and take for lines the subsets $\{A, B\}, \{A, C\}$, and $\{B, C\}$. We represent this symbolically by the diagram, where the dots represent the elements of the set, and the lines drawn on the page show which subsets are to be taken as lines.

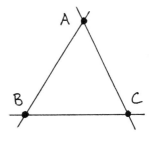

This diagram should be understood as purely symbolic, however, and has nothing to do with a triangle in the ordinary Cartesian plane. The verification of the axioms in this case is trivial.

**Definition**
Two distinct lines are *parallel* if they have no points in common. We also say that any line is parallel to itself.

The parallel postulate, in its equivalent form given by Playfair, can be stated as a further axiom about incidence of lines. However, we do not include this axiom in the definition of incidence geometry. Thus we may speak of an incidence geometry that does or does not satisfy Playfair's axiom.

**P.** (Playfair's axiom, also called the parallel axiom). For each point $A$ and each line $l$, there is at most one line containing $A$ that is parallel to $l$.

Note that the real Cartesian plane (6.1.1) satisfies (P), as you know, and the three-point geometry (6.1.2) satisfies (P) vacuously, because there are no distinct parallel lines at all. Next we give an example of an incidence geometry that does not satisfy (P).

**Example 6.1.3**
Let our set consist of five points $A, B, C, D, E$, and let the lines be all subsets of two points. It is easy to see that this geometry satisfies (I1)–(I3). However, it does not satisfy (P), because, for example, $AB$ and $AC$ are two distinct lines through the point $A$ and parallel to the line $DE$.

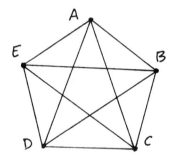

Remember that the word *parallel* simply means that two lines have no points in common or are equal. It does not say anything about being in the same direction, or being equidistant from each other, or anything else.

We say that two models of an axiom system are *isomorphic* if there exists a 1-to-1 correspondence between their sets of points in such a way that a subset of the first set is a line if and only if the corresponding subset of the second set is a line. For short, we say "the correspondence takes lines into lines." So for example, we see that (6.1.1), (6.1.2), and (6.1.3) are nonisomorphic models of incidence geometry, for the simple reason that their sets of points have different cardinality: There are no 1-to-1 correspondences between any of these sets.

On the other hand, we can show that any model of incidence geometry having just three points is isomorphic to the model given in (6.1.2). Indeed, let $\{1, 2, 3\}$ be a geometry of three points. By (I3), there must be three noncollinear points. Since there are only three points here, we conclude that there is no line containing all three. But by (I1), each subset of two points must be contained in a line. Thus $\{1,2\},\{2,3\}$, and $\{1,3\}$ are lines. Now by (I2), every line contains at least two points, so these are all the possible lines. In other words, the lines are just all subsets of two elements. Since (6.1.2) also has this property, any 1-to-1 correspondence between the sets $\{A, B, C\}$ and $\{1, 2, 3\}$ will give an isomorphism.

By the way, this proof shows that the isomorphism just found is not unique. There are six choices. This leads to the notion of automorphism.

**Definition**
An *automorphism* of an incidence geometry is an isomorphism of the geometry with itself, that is, it is a 1-to-1 mapping of the set of points onto itself, preserving lines.

Note that the composition of two automorphisms is an automorphism, and so is the inverse of an automorphism. Thus the set of automorphisms forms a group. In the example above, any 1-to-1 mapping of the set of three elements onto itself gives an automorphism of the geometry, so we see that the group of automorphisms of this geometry is the symmetric group on three letters, $S_3$.

An important question about a set of axioms is whether the axioms are *independent* of each other. That is to say, that no one of them can be proved as a consequence of the others. For if one were a consequence of the others, then we would not need that one as an axiom. To try to prove directly that axiom $A$ is not a consequence of axioms $B, C, D, \ldots$ is usually futile. So instead, we search for a model in which axioms $B, C, D, \ldots$ hold but axiom $A$ does not hold. If such a model exists, then there can be no proof of $A$ as a consequence of $B, C, D, \ldots$, so we conclude that $A$ is independent of the others. This process must be repeated with each individual axiom, to show that each one is independent of all the others. With a long list of axioms this can become tedious and difficult, so we will forgo the process with our full list of axioms. But as an illustration of what is involved, let us show that the axioms (I1), (I2), (I3), and (P) are independent.

**Proposition 6.2**
*The axioms* (I1), (I2), (I3), (P) *are independent of each other.*

*Proof* We have already seen that (6.1.3) is a model satisfying (I1), (I2), (I3), and not (P). Hence (P) is independent of the others.

For a model satisfying (I1), (I2), (P), and not (I3), take a set of two points and the one line containing both of them. Note that (P) is satisfied trivially, because there are no points not on the line $l$.

For a model satisfying (I1), (I3), (P), and not (I2), take a set of three points $A, B, C$, and for lines take the subsets $\{A, B\}, \{A, C\}, \{B, C\}$, and $\{A\}$. The existence of the one-point line $\{A\}$ contradicts (I2). Yet (P) is still fulfilled, because that one-point line is then the unique line through $A$ parallel to $\{B, C\}$.

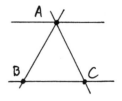

For a model satisfying (I2), (I3), (P) and not (I1), just take a set of three points and no lines at all.

While we are discussing axiom systems, there are a few more concepts we should mention. An axiom system is *consistent* if it will never lead to a contradiction. That is to say, if it is not possible to prove from the axioms a statement $A$ and also to prove its negation not $A$. This is obviously a highly desirable property of a system of axioms. We do not want to waste our time proving theorems from a system of axioms that one day may lead to a contradiction. Unfortunately, however, the logician Kurt Gödel has proved that for any reasonably rich set of axioms, it will be impossible to prove the consistency of that system. So we will have to settle for something less, which is *relative consistency*. As soon as you can find a model for your axiom system within some other mathematical theory $T$, it follows that if $T$ is consistent, then also your system of axioms is consistent. For any contradiction that might follow from your axioms would then also appear in the theory $T$, contradicting its consistency. So for example, if you believe in the consistency of the theory of real numbers, then you must accept the consistency of Hilbert's axiom system for geometry, because all of his axioms will hold in the real Cartesian plane. That is the best we can do about the question of consistency.

Another question about a system of axioms is whether it is *categorical*. This means, does it describe a unique mathematical object? Or in other words, is there a *unique* model (up to isomorphism) for the system of axioms? In fact, it will turn out that if we take the entire list of Hilbert's axioms, including the parallel axiom (P) and Dedekind's axiom (D), the system will be categorical, and the unique model will be the real Cartesian plane. (We will prove this result later (21.3).) Also, if we take all of Hilbert's axioms, together with (D) and the hyperbolic axiom (L) (see Section 40), we will have another categorical system, whose unique model is the non-Euclidean Poincaré model over the real numbers (Exercise 43.2).

However, from the point of view of this book, it is more interesting to have an axiom system that is not categorical, and then to investigate the different possible geometries that can arise. Therefore, we will almost never assume Dedekind's axiom (D), and we will only sometimes assume Archimedes' axiom (A), or the parallel axiom (P).

Finally, one can ask whether the axiom system is *complete*, which means, can every statement that is true in every model of the axiom system be proved as a consequence of the axioms? Again, Gödel has shown that any axiomatic system of reasonable richness cannot be complete. For a fuller discussion of these questions, see Chapter 51 of Kline (1972) on the foundations of mathematics.

# Exercises

6.1 Describe all possible incidence geometries on a set of four points, up to isomorphism. Which ones satisfy (P)?

6.2 The *Cartesian plane over a field F.* Let $F$ be any field (see definition in §14). Take the set $F^2$ of ordered pairs of elements of the field $F$ to be the set of *points*. Define *lines* to be those subsets defined by linear equations, as in Example 6.1.1. Verify that the axioms (I1), (I2), (I3), and (P) hold in this model. (See Section 14 for more about Cartesian planes over fields.)

6.3 A *projective plane* is a set of points and subsets called lines that satisfy the following four axioms:

**P1.** Any two distinct points lie on a unique line.

**P2.** Any two lines meet in at least one point.

**P3.** Every line contains at least three points.

**P4.** There exist three noncollinear points.

Note that these axioms imply (I1)–(I3), so that any projective plane is also an incidence geometry. Show the following:

(a) Every projective plane has at least seven points, and there exists a model of a projective plane having exactly seven points.

(b) The projective plane of seven points is unique up to isomorphism.

(c) The axioms (P1), (P2), (P3), (P4) are independent.

6.4 Let $F$ be a field, and let $V = F^3$ be a three-dimensional vector space over $F$. Let $\Pi$ be the set of 1-dimensional subspaces of $V$. We will call the elements of $\Pi$ "points." So a "point" is a 1-dimensional subspace $P \subseteq V$. If $W \subseteq V$ is a 2-dimensional subspace of $V$, then the set of all "points" contained in $W$ will be called a "line." Show that the set $\Pi$ of "points" and the subsets of "lines" forms a projective plane (Exercise 6.3).

6.5 An *affine plane* is a set of points and subsets called lines satisfying (I1), (I2), (I3), and the following stronger form of Playfair's axiom.

**P′.** For every line $l$, and every point $A$, there exists a unique line $m$ containing $A$ and parallel to $l$.

(a) Show that any two lines in an affine plane have the same number of points (i.e., there exists a 1-to-1 correspondence between the points of the two lines).

(b) If an affine plane has a line with exactly $n$ points, then the total number of points in the plane is $n^2$.

(c) If $F$ is any field, show that the *Cartesian plane* over $F$ (Exercise 6.2) is a model of an affine plane.

(d) Show that there exist affine planes with 4, 9, 16, or 25 points. (The nonexistence of an affine plane with 36 points is a difficult result of Euler.)

6.6 In an incidence geometry, consider the relationship of parallelism, "*l* is parallel to *m*," on the set of lines.

(a) Give an example to show that this need not be an equivalence relation.

(b) If we assume the parallel axiom (P), then parallelism is an equivalence relation.

(c) Conversely, if parallelism is an equivalence relation in a given incidence geometry, then (P) must hold in that geometry.

6.7 Let Π be an affine plane (Exercise 6.5). A *pencil* of parallel lines is the set of all the lines parallel to a given line (including that line itself). We call each pencil of parallel lines an "ideal point," or a "point at infinity," and we say that an ideal point "lies on" each of the lines in the pencil. Now let Π′ be the enlarged set consisting of Π together with all these new ideal points. A *line* of Π′ will be the subset consisting of a line of Π plus its unique ideal point, or a new line, called the "line at infinity," consisting of all the ideal points.

(a) Show that this new set Π′ with subsets of lines as just defined forms a projective plane (Exercise 6.3).

(b) If Π is the Cartesian plane over a field $F$ (Exercise 6.2), show that the associated projective plane Π′ is isomorphic to the projective plane constructed in Exercise 6.4.

6.8 If there are $n+1$ points on one line in a projective plane Π, then the total number of points in Π is $n^2 + n + 1$.

6.9 Kirkman's schoolgirl problem (1850) is as follows: In a certain school there are 15 girls. It is desired to make a seven-day schedule such that each day the girls can walk in the garden in five groups of three, in such a way that each girl will be in the same group with each other girl just once in the week. How should the groups be formed each day?

To make this into a geometry problem, think of the girls as points, think of the groups of three as lines, and think of each day as describing a set of five lines, which we call a pencil. Now consider a *Kirkman geometry*: a set, whose elements we call *points*, together with certain subsets we call *lines*, and certain sets of lines we call *pencils*, satisfying the following axioms:

**K1.** Two distinct points lie on a unique line.

**K2.** All lines contain the same number of points.

**K3.** There exist three noncollinear points.

**K4.** Each line is contained in a unique pencil.

**K5.** Each pencil consists of a set of parallel lines whose union is the whole set of points.

(a) Show that any affine plane gives a Kirkman geometry where we take the pencils to be the set of all lines parallel to a given line. (Hence by Exercise 6.5 there exist Kirkman geometries with 4, 9, 16, 25 points.)

(b) Show that any Kirkman geometry with 15 points gives a solution of the original schoolgirl problem.

(c) Find a solution for the original problem. (There are many inequivalent solutions to this problem.)

6.10 In a finite incidence geometry, the number of lines is greater than or equal to the number of points.

# 7  Axioms of Betweenness

In this section we present axioms to make precise the notions of betweenness (when one point is in between two others), on which is based the notion of sidedness (when a point is on one side of a line or the other), the concepts of inside and outside, and also the concepts of *order*, when one segment or angle is bigger than another. We have seen the importance of these concepts in reading Euclid's geometry, and we have also seen the dangers of using these concepts intuitively, without making their meaning precise. So these axioms form an important part of our new foundations for geometry. At the same time, these axioms and their consequences may seem difficult to understand for many readers, not because the mathematical concepts are technically difficult, but because the notions of order and separation are so deeply ingrained in our daily experience of life that it is difficult to let go of our intuitions and replace them with axioms. It is an exercise in forgetting what we already know from our inner nature, and then reconstituting it with an open mind as an external logical structure.

Throughout this section we presuppose axioms (I1)–(I3) of an incidence geometry. The geometrical notions of betweenness, separation, sidedness, and order will all be based on a single undefined relation, subject to four axioms. We postulate a *relation* between sets of three points $A, B, C$, called "$B$ is between $A$ and $C$." This relation is subject to the following axioms.

**B1.** If $B$ is between $A$ and $C$, (written $A * B * C$), then $A, B, C$ are three distinct points on a line, and also $C * B * A$.

**B2.** For any two distinct points $A, B$, there exists a point $C$ such that $A * B * C$.

**B3.** Given three distinct points on a line, one and only one of them is between the other two.

**B4.** (Pasch). Let $A, B, C$ be three non-collinear points, and let $l$ be a line not containing any of $A, B, C$. If $l$ contains a point $D$ lying between $A$ and $B$, then it must also contain either a point lying between $A$ and $C$ or a point lying between $B$ and $C$, but not both.

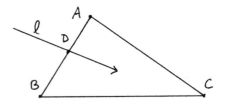

### Definition

If $A$ and $B$ are distinct points, we define the *line segment* $\overline{AB}$ to be the set consisting of the points $A, B$ and all points lying between $A$ and $B$. We define a *triangle* to be the union of the three line segments $\overline{AB}, \overline{BC}$, and $\overline{AC}$ whenever $A, B, C$ are three noncollinear points. The points $A, B, C$ are the *vertices* of the triangle, and the segments $\overline{AB}, \overline{BC}, \overline{AC}$ are the *sides* of the triangle.

**Note:** The segments $\overline{AB}$ and $\overline{BA}$ are the same sets, because of axiom (B1). The *endpoints* $A, B$ of the segment $\overline{AB}$ are uniquely determined by the segment $\overline{AB}$ (Exercise 7.2). The vertices $A, B, C$, and the sides $\overline{AB}, \overline{AC}, \overline{BC}$ of a triangle $ABC$ are uniquely determined by the triangle (Exercise 7.3).

With this terminology, we can rephrase (B4) as follows: If a line $l$ that does not contain any of the vertices $A, B, C$ of a triangle meets one side $\overline{AB}$, then it must meet one of the other sides $\overline{AC}$ or $\overline{BC}$, but not both.

From these axioms together with the axioms of incidence (I1)–(I3) we will deduce results about the separation of the plane by a line, and the separation of a line by a point.

### Proposition 7.1 (Plane separation)

*Let $l$ be any line. Then the set of points not lying on $l$ can be divided into two non-empty subsets $S_1, S_2$ with the following properties:*

(a) *Two points $A, B$ not on $l$ belong to the same set $(S_1$ or $S_2)$ if and only if the segment $\overline{AB}$ does not intersect $l$.*

(b) *Two points $A, C$ not on $l$ belong to the opposite sets (one in $S_1$, the other in $S_2$) if and only if the segment $\overline{AC}$ intersects $l$ in a point.*

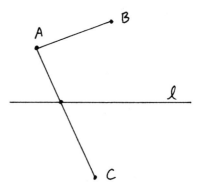

*We will refer to the sets $S_1, S_2$ as the two* sides *of $l$, and we will say "A and B are on the* same *side of $l$," or "A and C are on* opposite *sides of $l$."*

*Proof* We start by defining a relation $\sim$ among points not on $l$. We will say $A \sim B$ if either $A = B$ or if the segment $\overline{AB}$ does not meet $l$. Our first step is to show that $\sim$ is an equivalence relation. Clearly, $A \sim A$ by definition, and $A \sim B$ implies $B \sim A$ because the set $\overline{AB}$ does not depend on the order in which we write $A$ and $B$. The nontrivial step is to show the relation is transitive: If $A \sim B$ and $B \sim C$, we must show $A \sim C$.

*Case 1* Suppose $A, B, C$ are not collinear. Then we consider the triangle $ABC$. Since $A \sim B$, $l$ does not meet $\overline{AB}$. Since $B \sim C$, $l$ does not meet $\overline{BC}$. Now by Pasch's axiom (B4), it follows that $l$ does not meet $\overline{AC}$. Hence $A \sim C$.

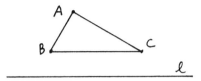

*Case 2* Suppose $A, B, C$ lie on a line $m$. Since $A, B, C$ do not lie on $l$, the line $m$ is different from $l$. Therefore $l$ and $m$ can meet in at most one point (6.1). But by (I2) every line has at least two points. Therefore, there exists a point $D$ on $l$, $D$ not lying on $m$.

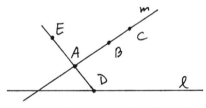

Now apply axiom (B2) to find a point $E$ such that $D * A * E$. Then $D, A, E$ are collinear (B1); hence $E$ is not on $l$, since $A$ is not on $l$, and the line $DAE$ already meets $l$ at the point $D$. Furthermore, the segment $\overline{AE}$ cannot meet $l$. For if it did, the intersection point would be the unique point in which the line $AE$ meets $l$, namely $D$. In that case $D$ would be between $A$ and $E$. But we constructed $E$ so that $D * A * E$, so by (B3), $D$ cannot lie between $A$ and $E$. Thus $\overline{AE} \cap l = \varnothing$, so $A \sim E$. Note also that $E$ does not lie on the line $m$, because if $E$ were on $m$, then the line $AE$ would be equal to $m$, so $D$ would lie on $m$, contrary to our choice of $D$. Therefore, $A, B, E$ are three noncollinear points. Then by Case 1 proved above, from $A \sim E$ and $A \sim B$ we conclude $B \sim E$. By Case 1 again, from $B \sim E$ and $B \sim C$ we conclude $C \sim E$. Applying Case 1 a third time to the three noncollinear points $A, C, E$, from $A \sim E$ and $C \sim E$ we conclude $A \sim C$ as required.

Thus we have proved that $\sim$ is an equivalence relation. An equivalence relation on a set divides that set into a disjoint union of equivalence classes, and these equivalence classes will satisfy property (a) by definition. To complete the proof it will be sufficient to show that there are exactly two equivalence classes $S_1, S_2$ for the relation $\sim$. Then to say that $\overline{AC}$ meets $l$, which is equivalent to $A \not\sim C$, will be the same as saying that $A, C$ belong to the opposite sets.

By (I3) there exists a point not on $l$, so there is at least one equivalence class $S_1$. Given $A \in S_1$, let $D$ be any point on $l$, and choose by (B2) a point $C$ such that

$A * D * C$. Then $A$ and $C$ do not satisfy $\sim$, so there must be at least two equivalence classes $S_1$ and $S_2$.

The last step is to show that there are at most two equivalence classes. To do this, we will show that if $A \not\sim C$ and $B \not\sim C$, then $A \sim B$.

*Case 1*  If $A, B, C$ are not collinear, we consider the triangle $ABC$. From $A \not\sim C$ we conclude that $\overline{AC}$ meets $l$. From $B \not\sim C$ we conclude that $\overline{BC}$ meets $l$. Now by Pasch's axiom (B4) it follows that $\overline{AB}$ does not meet $l$. So $A \sim B$ as required.

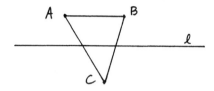

*Case 2*  Suppose $A, B, C$ lie on a line $m$. As in Case 2 of the first part of the proof above, choose a point $D$ on $l$, not on $m$, and use (B2) to get a point $E$ with $D * A * E$. Then $A \sim E$ as we showed above.

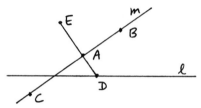

Now, $A \not\sim C$ by hypothesis, and $A \sim E$, so we conclude that $C \not\sim E$, since $\sim$ is an equivalence relation (if $C \sim E$, then $A \sim C$ by transitivity: contradiction). Looking at the three noncollinear points $B, C, E$, from $E \not\sim C$ and $B \not\sim C$ we conclude using Case 1 that $B \sim E$. But also $A \sim E$, so by transitivity, $A \sim B$ as required.

**Proposition 7.2** (Line separation)
*Let $A$ be a point on a line $l$. Then the set of points of $l$ not equal to $A$ can be divided into two nonempty subsets $S_1, S_2$, the two sides of $A$ on $l$, such that*

(a) *$B, C$ are on the same side of $A$ if and only if $A$ is not in the segment $\overline{BC}$;*
(b) *$B, D$ are on opposite sides of $A$ if and only if $A$ belongs to the segment $\overline{BD}$.*

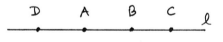

*Proof*  Given the line $l$ and a point $A$ on $l$, we know from (I3) that there exists a point $E$ not on $l$. Let $m$ be the line containing $A$ and $E$. Apply (7.1) to the line $m$. If $m$ has two sides $S'_1, S'_2$, we define $S_1$ and $S_2$ to be the intersections of $S'_1$ and $S'_2$ with $l$. Then properties (a) and (b) follow immediately from the previous proposition.

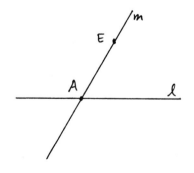

The only mildly nontrivial part is to show that $S_1$ and $S_2$ are nonempty. By (I2), there is a point $B$ on $l$ different from $A$. And by (B2) there exists a point $D$ such that $B * A * D$. Then $D$ will be on the opposite side of $A$ from $B$, and will lie on $l$, so both sides are nonempty.

Now that we have some basic results on betweenness, we can define rays and angles.

### Definition

Given two distinct points $A, B$, the *ray* $\overrightarrow{AB}$ is the set consisting of $A$, plus all points on the line $AB$ that are on the same side of $A$ as $B$. The point $A$ is the *origin*, or *vertex*, of the ray. An *angle* is the union of two rays $\overrightarrow{AB}$ and $\overrightarrow{AC}$ originating at the same point, its *vertex*, and not lying on the same line. (Thus there is no "zero angle," and there is no "straight angle" (180°).) Note that the vertex of a ray or angle is uniquely determined by the ray or angle (proof similar to Exercises 7.2, 7.3).

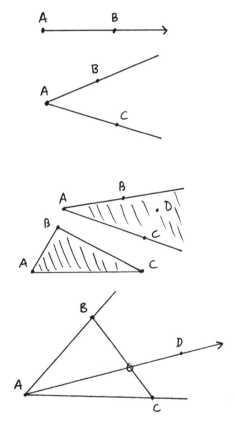

The *inside* (or *interior*) of an angle $\angle BAC$ consists of all points $D$ such that $D$ and $C$ are on the same side of the line $AB$, and $D$ and $B$ are on the same side of the line $AC$. If $ABC$ is a triangle, the *inside* (or *interior*) of the triangle $ABC$ is the set of points that are simultaneously in the insides of the three angles $\angle BAC, \angle ABC, \angle ACB$.

### Proposition 7.3 (Crossbar theorem)

*Let $\angle BAC$ be an angle, and let $D$ be a point in the interior of the angle. Then the ray $\overrightarrow{AD}$ must meet the segment $\overline{BC}$.*

*Proof* This is similar to Pasch's axiom (B4), except that we must consider a line $AD$ that passes through one vertex of the triangle $ABC$. We will prove it with Pasch's axiom and several applications of the plane separation theorem (7.1).

Let us label the lines $AB = l$, $AC = m$, $AD = n$. Let $E$ be a point on $m$ such that $E * A * C$ (B2). We will apply Pasch's axiom (B4) to the triangle $BCE$ and the line $n$. By construction $n$ meets the side $CE$ at $A$. Also, $n$ cannot contain $B$, because it meets the line $l$ at $A$. We will show that $n$ does not meet the segment $\overline{BE}$, so as to conclude by (B4) that it must meet the segment $\overline{BC}$.

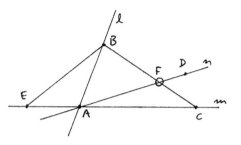

So we consider the segment $\overline{BE}$. This segment meets the line $l$ only at $B$, so all points of the segment, except $B$, are on the same side of $l$. By construction, $C$ is on the opposite side of $l$ from $E$, so by (7.1) all points of $\overline{BE}$, except $B$, are on the opposite side of $l$ from $C$. On the other hand, since $D$ is in the interior of the angle $\angle BAC$, all the points of the ray $\overrightarrow{AD}$, except $A$, are on the same side of $l$ as $C$. Thus the segment $\overline{BE}$ does not meet the ray $\overrightarrow{AD}$.

A similar reasoning using the line $m$ shows that all points of the segment $\overline{BE}$, except $E$, lie on the same side of $m$ as $B$, while the points of the ray of $n$, opposite the ray $\overrightarrow{AD}$, lie on the other side of $m$. Hence the segment $\overline{BE}$ cannot meet the opposite ray to $\overrightarrow{AD}$. Together with the previous step, this shows that the segment $\overline{BE}$ does not meet the line $n$. We conclude by (B4) that $n$ meets the segment $\overline{BC}$ in a point $F$.

It remains only to show that $F$ is on the ray $\overrightarrow{AD}$ of the line $n$. Indeed, $B$ and $F$ are on the same side of $m$, and also $B$ and $D$ are on the same side of $m$, so (7.1) $D$ and $F$ are on the same side of $m$, and so $D$ and $F$ are on the same side of $A$ on the line $n$. In other words, $F$ lies on the ray $\overrightarrow{AD}$.

### Example 7.3.1
We will show that the real Cartesian plane (6.1.1), with the "usual" notion of betweenness, provides a model for the axioms (B1)–(B4).

First, we must make precise what we mean by the usual notion of betweenness. For three distinct real numbers $a, b, c \in \mathbb{R}$, let us define $a * b * c$ if either $a < b < c$ or $c < b < a$. Then it is easy to see that this defines a notion of betweenness on the real line $\mathbb{R}$ that satisfies (B1), (B2), and (B3).

If $A = (a_1, a_2)$, $B = (b_1, b_2)$, and $C = (c_1, c_2)$ are three points in $\mathbb{R}^2$, let us define $A * B * C$ to mean that $A, B, C$ are three distinct points on a line, and that either $a_1 * b_1 * c_1$ or $a_2 * b_2 * c_2$, or both. In fact, if either the $x$- or the $y$-coordinates satisfy this betweenness condition, and if the line is neither horizontal nor vertical, then the other coordinates will also satisfy it, because the points lie on a line, and linear operations (addition, multiplication) of real numbers either preserve or reverse inequalities. Thus linear operations preserve betweenness. So we can verify easily that this notion of betweenness in $\mathbb{R}^2$ satisfies (B1), (B2), and (B3).

For (B4), let $l$ be a line, and let $A, B, C$ be three noncollinear points not on $l$. The line $l$ is defined by some linear equation $ax + by + c = 0$. Let $\varphi : \mathbb{R}^2 \to \mathbb{R}$ be the linear function defined by $\varphi(x, y) = ax + by + c$. Since $\varphi$ is a linear function, $\varphi$ will preserve betweenness. For example, if $l$ meets the segment $\overline{AB}$, then 0 will lie between $\varphi(A)$ and $\varphi(B)$. In other words, one of $\varphi(A), \varphi(B)$ will be positive and the other negative. Suppose $\varphi(A) > 0$ and $\varphi(B) < 0$. Consider $\varphi(C)$. If $\varphi(C) > 0$, then $l$ will meet $\overline{BC}$ but not $\overline{AC}$. If $\varphi(C) < 0$, then $l$ will meet $\overline{AC}$ but not $\overline{BC}$. This proves (B4).

# Exercises

7.1 Using the axioms of incidence and betweenness and the line separation property, show that sets of four points $A, B, C, D$ on a line behave as we expect them to with respect to betweenness. Namely, show that

   (a) $A * B * C$ and $B * C * D$ imply $A * B * D$ and $A * C * D$.

   (b) $A * B * D$ and $B * C * D$ imply $A * B * C$ and $A * C * D$.

7.2 Given a segment $\overline{AB}$, show that there do not exist points $C, D \in \overline{AB}$ such that $C * A * D$. Hence show that the endpoints $A, B$ of the segment are uniquely determined by the segment.

7.3 Given a triangle $ABC$, show that the sides $\overline{AB}$, $\overline{AC}$, and $\overline{BC}$ and the vertices $A, B, C$ are uniquely determined by the triangle. *Hint*: Consider the different ways in which a line can intersect the triangle.

7.4 Using (I1)–(I3) and (B1)–(B4) and their consequences, show that every line has infinitely many distinct points.

7.5 Show that the line separation property (Proposition 7.2) is not a consequence of (B1), (B2), (B3), by constructing a model of betweenness for the set of points on a line, which satisfies (B1), (B2), (B3) but has only finitely many points. (Then by Exercise 7.4, line separation must fail in this model.) For example, in the ring $\{0, 1, 2, 3, 4\}$ of integers (mod 5), define $a * b * c$ if $b = \frac{1}{2}(a + c)$.

7.6 Prove directly from the axioms (I1)–(I3) and (B1)–(B4) that for any two distinct points $A, B$, there exists a point $C$ with $A * C * B$. (*Hint*: Use (B2) and (B4) to construct a line that will be forced to meet the segment $\overline{AB}$ but does not contain $A$ or $B$.)

7.7 Be careful not to assume without proof statements that may appear obvious. For example, prove the following:

   (a) Let $A, B, C$ be three points on a line with $C$ in between $A$ and $B$. Then show that $\overline{AC} \cup \overline{CB} = \overline{AB}$ and $\overline{AC} \cap \overline{CB} = \{C\}$.

   (b) Suppose we are given two distinct points $A, B$ on a line $l$. Show that $\overrightarrow{AB} \cup \overrightarrow{BA} = l$ and $\overrightarrow{AB} \cap \overrightarrow{BA} = \overline{AB}$.

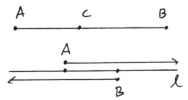

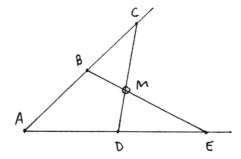

7.8 Assume $A * B * C$ on one line, and $A * D * E$ on another line. Show that the segment $\overline{BE}$ must meet the segment $\overline{CD}$ at a point $M$.

7.9 Show that the interior of a triangle is nonempty.

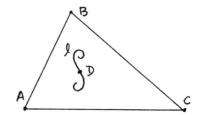

7.10 Suppose that a line $l$ contains a point $D$ that is in the inside of a triangle $ABC$. Then show that the line $l$ must meet (at least) one of the sides of the triangle.

7.11 A set $U$ of points in the plane is a *convex set* if whenever $A, B$ are distinct points in $U$, then the segment $\overline{AB}$ is entirely contained in $U$. Show that the inside of a triangle is a convex set.

7.12 A subset $W$ of the plane is *segment-connected* if given any two points $A, B \in W$, there is a finite sequence of points $A = A_1, A_2, \ldots, A_n = B$ such that for each $i = 1, 2, \ldots,$ $n - 1$, the segment $\overline{A_i A_{i+1}}$ is entirely contained within $W$.

If $ABC$ is a triangle, show that the *exterior* of the triangle, that is, the set of all points of the plane lying neither on the triangle nor in its interior, is a segment-connected set.

7.13 Let $A, B, C, D$ be four points, no three collinear, and assume that the segments $\overline{AB}$, $\overline{BC}$, $\overline{CD}$, $\overline{DA}$ have no intersections except at their endpoints. Then the union of these four segments is a *simple closed quadrilateral*. The segments $\overline{AC}$ and $\overline{BD}$ are the *diagonals* of the quadrilateral. There are two cases to consider.

*Case 1* $\overline{AC}$ and $\overline{BD}$ meet at a point $M$. In this case, show that for each pair of consecutive vertices (e.g., $A, B$), the remaining two vertices $(C, D)$ are on the same side of the line $AB$. Define the *interior* of the quadrilateral to be the set of points $X$ such that for each side (e.g., $\overline{AB}$), $X$ is on the same side of the line $AB$ as the remaining vertices $(C, D)$. Show that the interior is a convex set.

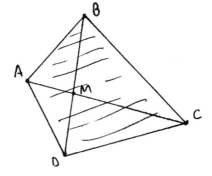

*Case 2* $\overline{AC}$ and $\overline{BD}$ do not meet. In this case, show that one of the diagonals ($\overline{AC}$ in the picture) has the property that the other two vertices $B, D$ are on the same side of the line $AC$, while the other diagonal $\overline{BD}$ has the property that $A$ and $C$ are on the opposite sides of the line $BD$. Define the *interior* of the quadrilateral to be the union of the interiors of the triangles $ABD$ and $CDB$ plus the interior of the segment $\overline{BD}$. Show in this case that the interior is a segment-connected set, but is not convex.

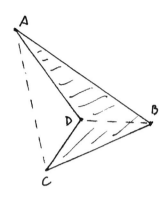

(For a generalization to $n$-sided figures, see Exercise 22.11.)

7.14 (*Linear ordering*) Given a finite set of distinct points on a line, it is possible to label them $A_1, A_2, \ldots, A_n$ in such a way that $A_i * A_j * A_k$ if and only if either $i < j < k$ or $k < j < i$.

7.15 Suppose that lines $a, b, c$ through the vertices $A, B, C$ of a triangle meet at three points inside the triangle. Label them

$$X = a \cdot c,$$
$$Y = a \cdot b,$$
$$Z = b \cdot c.$$

Show that one of the two following arrangements must occur:

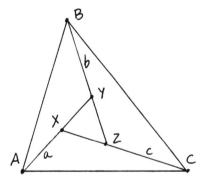

(i) $A * X * Y$ and $B * Y * Z$ and $C * Z * X$ (shown in diagram), or

(ii) $A * Y * X$ and $B * Z * Y$ and $C * X * Z$.

# 8   Axioms of Congruence for Line Segments

To the earlier undefined notions of point, line, and betweenness, and to the earlier axioms (I1)–(I3), (B1)–(B4), we now add an undefined notion of congruence for line segments, and further axioms (C1)–(C3) regarding this notion. This congruence is what Euclid called equality of segments. We postulate an undefined notion of *congruence*, which is a relation between two line segments $\overline{AB}$

and $\overline{CD}$, written $AB \cong CD$. For simplicity we will drop the bars over $AB$ in the notation for a line segment, so long as no confusion can result. This undefined notion is subject to the following three axioms

**C1.** Given a line segment $AB$, and given a ray $r$ originating at a point $C$, there exists a unique point $D$ on the ray $r$ such that $AB \cong CD$.

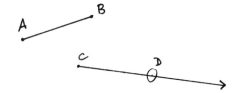

**C2.** If $AB \cong CD$ and $AB \cong EF$, then $CD \cong EF$. Every line segment is congruent to itself.

**C3.** (Addition). Given three points $A, B, C$ on a line satisfying $A * B * C$, and three further points $D, E, F$ on a line satisfying $D * E * F$, if $AB \cong DE$ and $BC \cong EF$, then $AC \cong DF$.

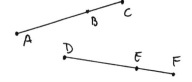

Let us observe how these axioms are similar to Euclid's postulates and how they are different. First of all, while Euclid phrases some of his postulates in terms of constructions ("to draw a line through any two given points," and "to draw a circle with any given center and radius"), Hilbert's axioms are existential. (I1) says for any two distinct points there exists a unique line containing them. And here, in axiom (C1), it is the existence of the point $D$ (corresponding to Euclid's construction (I.3)) that is taken as an axiom. Hilbert does not make use of ruler and compass constructions. In their place he puts the axiom (C1) of the existence of line segments and later (C4) the existence of angles. If you like, you can think of (C1) and (C4) as being tools, a "transporter of segments" and a "transporter of angles," and consider some of Hilbert's theorems as constructions with these tools.

The second congruence axiom (C2) corresponds to Euclid's common notion that "things equal to the same thing are equal to each other." This is one part of the modern notion of an equivalence relation, so to be comfortable in using congruence, let us show that it is indeed an equivalence relation.

### Proposition 8.1
*Congruence is an equivalence relation on the set of line segments.*

*Proof*  To be an equivalence relation, congruence must satisfy three properties.

(1) *Reflexivity*: Every segment is congruent to itself. This is explicitly stated in (C2). And by the way, this corresponds to Euclid's fourth common notion that "things which coincide with each other are equal to each other."

# ΕΥΚΛΕΙΔΟΥ ΣΤΟΙ-
## ΧΕΙΟΝ ΠΡΩΤΟΝ·

## EVCLIDIS ELEMENTORVM GEO-
*metricorum liber primus.*

Est hic liber primus totus ferè elementarius, non tantum ad reliquos ſequentes huius Operis libros, ſed etiam ad aliorum Geometrarum ſcripta intelligenda neceſſarius. Nam in hoc libro communium uocabulorum, quę ſubinde in geometria uerſanti occurrunt, definitiones continentur. Pręceptiones deinde ducendi perpendicularem, quomodo item Trilaterę figurę, ſecundum latera uel angulos diuerſę, & Quadrilaterę, formari debeant. Figura item aliqua propoſita, quomodo illa in alterius formæ figuram permutanda ſit, præceptiones, ut diximus, traduntur. Cum igitur talia doceantur, & plura etiam alia, quàm hoc loco commemorare uoluimus, facile erit cuiuis, non ſolum quàm ſit neceſſarius, ſed etiam ad reliqua perdiſcenda liber iſte quàm utilis, perſpicere.

### ΟΡΟΙ.

Σημεῖόν ἐσιν, οὗ μέρ Θ' οὐθέν. Γραμμὴ δὲ, μῆκος ἀπλαῖις. Γραμμῆς δὲ πέρατα, σημεῖα. Ευθεῖα γραμμή ἐσιν, ἥτις ἐξ ἴσου τοῖς ἐφ' ἱαυτῇ σημείοις κεῖτζ.

### DEFINITIONES.

Punctum eſt, cuius pars nulla. 2. Linea uerò, longitudo latitudinis expers. Lineę autem termini puncta. 3. Recta linea eſt, quę ęqualibiliter inter ſua puncta iacet.

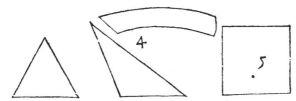

Prima defini.

Επιφάνεια ἐσιν, ὃ μῆκ Θ' καὶ πλατ Θ' μόνον ἔχ. Επιφανείας δὲ πέρατα, γραμμαί. Επίπεδ ς ἐπιφάνεια ἐσιν, ἥτις ἐξ ἴσου ταῖς ἐφ' ἱαυτῇ ινθείαις κεῖται.

**4.** Superficies eſt, quę longitudinem & latitudinem tantum habet.

K  3    Super-

---

Plate IV. The beginning of the *Elements* in the edition of Scheubel (1550), showing the Greek text of the first few definitions.

(2) *Symmetry*: If $AB \cong CD$, then $CD \cong AB$. This is a consequence of (C2): Given $AB \cong CD$, and writing $AB \cong AB$ by reflexivity, we conclude from (C2) that $CD \cong AB$.

(3) *Transitivity*: If $AB \cong CD$ and $CD \cong EF$, then $AB \cong EF$. This follows by first using symmetry to show $CD \cong AB$, and then applying (C2). Notice that Hilbert's formulation of (C2) was a clever way of including symmetry and transitivity in a single statement.

The third axiom (C3) is the counterpart of Euclid's second common notion, that "equals added to equals are equal." Let us amplify this by making a precise definition of the sum of two segments, and then showing that sums of congruent segments are congruent.

**Definition**

Let $AB$ and $CD$ be two given segments. Choose an ordering $A, B$ of the endpoints of $AB$. Let $r$ be the ray on the line $l = AB$ consisting of $B$ and all the points of $l$ on the other side of $B$ from $A$. Let $E$ be the unique point on the ray $r$ (whose existence is given by (C1)) such that $CD \cong BE$.

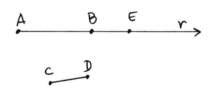

We then define the segment $AE$ to be the *sum* of the segments $AB$ and $CD$, depending on the order $A, B$, and we will write $AE = AB + CD$.

**Proposition 8.2** (Congruence of sums)
*Suppose we are given segments $AB \cong A'B'$ and $CD \cong C'D'$. Then $AB + CD \cong A'B' + C'D'$.*

*Proof*    Let $E'$ be the point on the line $A'B'$ defining the sum $A'E' = A'B' + C'D'$. Then $A * B * E$ by construction of the sum $AB + CD$, because $E$ is on the ray from $B$ opposite $A$. Similarly, $A' * B' * E'$. We have $AB \cong A'B'$ by hypothesis. Furthermore, we have $CD \cong C'D'$ by hypothesis, and $CD \cong BE$ and $C'D' \cong B'E'$ by construction of $E$ and $E'$. From (8.1) we know that congruence is an equivalence relation, so $BE \cong B'E'$. Now by (C3) it follows that $AE \cong A'E'$ as required.

**Note:** Since the segment $AB$ is equal to the segment $BA$, it follows in particular that the sum of two segments is independent of the order $A, B$ chosen, up to congruence. Thus addition is well-defined on congruence equivalence classes of line segments. So we can speak of addition of line segments or congruent segments without any danger (cf. also Exercise 8.1, which shows that addition of line segments is associative and commutative, up to congruence). Later (Section

19) we will also define multiplication of segments and so create a field of segment arithmetic.

Euclid's third common notion is that "equals subtracted from equals are equal." Bearing in mind that subtraction does not always make sense, we can interpret this common notion as follows.

### Proposition 8.3

*Given three points $A, B, C$ on a line such that $A * B * C$, and given points $E, F$ on a ray originating from a point $D$, suppose that $AB \cong DE$ and $AC \cong DF$. Then $E$ will be between $D$ and $F$, and $BC \cong EF$. (We regard $BC$ as the difference of $AC$ and $AB$.)*

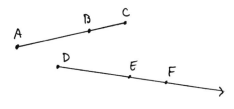

*Proof* Let $F'$ be the unique point on the ray originating at $E$, opposite to $D$, such that $BC \cong EF'$. Then from $AB \cong DE$ and $BC \cong EF'$ we conclude by (C3) that $AC \cong DF'$. But $F$ and $F'$ are on the same ray from $D$ (check!) and also $AC \cong DF$, so by (C2) and the uniqueness part of (C1), we conclude that $F = F'$. It follows that $D * E * F$ and $BC \cong EF$, as required.

Note the role played by the uniqueness part of (C1) in the above proof. We can regard this uniqueness as corresponding to Euclid's fifth common notion, "the whole is greater than the part." Indeed, this statement could be interpreted as meaning, if $A * B * C$, then $AB$ cannot be congruent to $AC$. And indeed, this follows from (C1), because $B$ and $C$ are on the same ray from $A$, and if $AB \cong AC$, then $B$ and $C$ would have to be equal by (C1).

So we see that Euclid's common notions, at least in the case of congruence of line segments, can be deduced as consequences of the new axioms (C1)–(C3). Another notion used by Euclid without definition is the notion of inequality of line segments. Let us see how we can define the notions of greater and lesser also using our axioms.

### Definition

Let $AB$ and $CD$ be given line segments. We will say that $AB$ is *less than $CD$*, written $AB < CD$, if there exists a point $E$ in between $C$ and $D$ such that $AB \cong CE$. In this case we say also that $CD$ is *greater than $AB$*, written $CD > AB$.

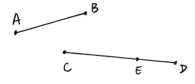

In the next proposition, we will see that this notion of less than is compatible with congruence, and gives an order relation on congruence equivalence classes of line segments.

### Proposition 8.4

(a) *Given line segments $AB \cong A'B'$ and $CD \cong C'D'$, then $AB < CD$ if and only if $A'B' < C'D'$.*

(b) *The relation $<$ gives an order relation on line segments up to congruence, in the following sense:*

(i) *If $AB < CD$, and $CD < EF$, then $AB < EF$.*
(ii) *Given two line segments $AB$, $CD$, one and only one of the three following conditions holds: $AB < CD$, $AB \cong CD$, $AB > CD$.*

*Proof* (a) Given $AB \cong A'B'$ and $CD \cong C'D'$, suppose that $AB < CD$. Then there is a point $E$ such that $AB \cong CE$ and $C*E*D$. Let $E'$ be the unique point on the ray $\overrightarrow{C'D'}$ such that $CE \cong C'E'$. It follows from (8.3) that $C'*E'*D'$. Furthermore, by transitivity of congruence, $A'B' \cong C'E'$, so $A'B' < C'D'$ as required. The "if and only if" statement follows by applying the same argument starting with $A'B' < C'D'$.

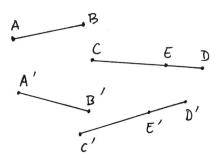

(b) (i) Suppose we are given $AB < CD$ and $CD < EF$. Then by definition, there is a point $X \in CD$ such that $AB \cong CX$, and there is a point $Y \in EF$ such that $CD \cong EY$. Let $Z \in \overrightarrow{EF}$ be such that $CX \cong EZ$. Then by (8.3) we have $E*Z*Y$. It follows that $E*Z*F$ (Exercise 7.1) and that $AB \cong EZ$. Hence $AB < EF$ as required.

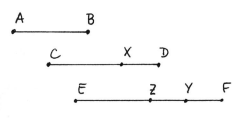

(ii) Given line segments $AB$ and $CD$, let $E$ be the unique point on the ray $\overrightarrow{CD}$ for which $AB \cong CE$. Then either $D = E$ or $C*E*D$ or $C*D*E$. We cannot have $D*C*E$ because $D$ and $E$ are on the same side of $C$. These conditions are equivalent to $AB \cong CD$, or $AB < CD$, or $AB > CD$, respectively, and one and only one of them must hold.

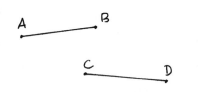

## Example 8.4.1

Let us define congruence for line segments in the real Cartesian plane $\mathbb{R}^2$, so that it becomes a model for the axioms (I1)–(I3), (B1)–(B4), and (C1)–(C3) that we have introduced so far. We have already seen how to define lines and betweenness (7.3.1). Given two points $A = (a_1, a_2)$ and $B = (b_1, b_2)$, we define the *distance* $d(A, B)$ by

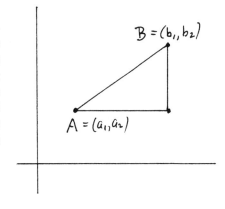

$$d(A, B) = \sqrt{(a_1 - b_1)^2 + (a_2 - b_2)^2}.$$

This is sometimes called the *Euclidean distance* or the *Euclidean metric* on $\mathbb{R}^2$. Note that $d(A, B) \geq 0$, and $d(A, B) = 0$ only if $A = B$.

Now we can give an interpretation of the undefined notion of congruence in this model by *defining* $AB \cong CD$ if $d(A, B) = d(C, D)$. Let us verify that the axioms (C1), (C2), (C3) are satisfied.

For (C1), we suppose that we are given a segment $AB$, and let $d = d(A, B)$. We also suppose that we are given a point $C = (c_1, c_2)$ and a ray emanating from $C$. For simplicity we will assume that the ray has slope $m > 0$ and that it is going in the direction of increasing $x$-coordinate (we leave the other cases to the reader). Then any point $D$ on this ray has coordinates $D = (c_1 + h, c_2 + mh)$ for some $h \geq 0$. The corresponding distance is

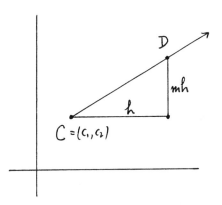

$$d(C, D) = h\sqrt{1 + m^2}.$$

To find a point $D$ with $AB \cong CD$ is then equivalent to solving the equation (in a variable $h > 0$)

$$h\sqrt{1 + m^2} = d,$$

where $m$ and $d > 0$ are given. Clearly, there is a unique solution $h \in \mathbb{R}$, $h > 0$, for given $d, m$. This proves (C1).

The second axiom (C2) is trivial from the definition of congruence using a distance function.

To prove (C3), it will be sufficient to prove that the distance function is additive for points in a line: If $A * B * C$, then

$$d(A, B) + d(B, C) = d(A, C).$$

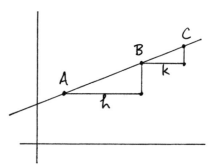

Suppose the line is $y = mx + b$, and $A = (a_1, a_2)$ is the point with smallest $x$-coordinate.

Then there are $h, k > 0$ such that

$$B = (a_1 + h, a_2 + mh),$$
$$C = (a_1 + h + k, a_2 + m(h + k)).$$

In this case

$$d(A, B) = h\sqrt{1 + m^2},$$
$$d(B, C) = k\sqrt{1 + m^2},$$
$$d(A, C) = (h + k)\sqrt{1 + m^2},$$

so the additivity of the distance function follows.

We will sometimes call this model, the real Cartesian plane with congruence of segments defined by the Euclidean distance function, the *standard model* of our axiom system.

## Exercises

The following exercises (unless otherwise specified) take place in a geometry with axioms (I1)–(I3), (B1)–(B4), (C1)–(C3).

8.1 (a) Show that addition of line segments is associative: Given segments $AB, CD, EF$, and taking $A, B$ in order, then $(AB + CD) + EF = AB + (CD + EF)$. (This means that we obtain the same segment as the sum, not just congruent segments.)

(b) Show that addition of line segments is commutative up to congruence: Given segments $AB, CD$, then $AB + CD \cong CD + AB$.

8.2 Show that "halves of equals are equal" in the following sense: if $AB \cong CD$, and if $E$ is a *midpoint* of $AB$ in the sense that $A * E * B$ and $AE \cong EB$, and if $F$ is a midpoint of $CD$, then $AE \cong CF$. (Note that we have not yet said anything about the existence of a midpoint: That will come later (Section 10).) Conclude that a midpoint of $AB$, if it exists, is unique.

8.3 Show that addition preserves inequalities: If $AB < CD$ and if $EF$ is any other segment, then $AB + EF < CD + EF$.

8.4 Let $r$ be a ray originating at a point $A$, and let $s$ be a ray originating at a point $B$. Show that there is a 1-to-1 mapping $\varphi : r \rightarrow s$ of the set $r$ onto the set $s$ that preserves congruence and betweenness. In other words, if for any $X \in r$ we let $X' = \varphi(X) \in s$, then for any $X, Y$, $Z \in r$, $XY \cong X'Y'$, and $X * Y * Z \Leftrightarrow X' * Y' * Z'$.

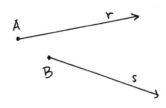

8.5 Given two distinct points $O, A$, we define the *circle* with *center* $O$ and *radius* $OA$ to be the set $\Gamma$ of all points $B$ such that $OA \cong OB$.

(a) Show that any line through $O$ meets the circle in exactly two points.

(b) Show that a circle contains infinitely many points.

(Warning: It is not obvious from this definition whether the center $O$ is uniquely determined by the set of points $\Gamma$ that form the circle. We will prove that later (Proposition 11.1).)

8.6 Consider the *rational Cartesian plane* $\mathbb{Q}^2$ whose points are ordered pairs of rational numbers, where lines are defined by linear equations with rational coefficients and betweenness and congruence are defined as in the standard model (Examples 7.3.1 and 8.4.1). Verify that (I1)–(I3) and (B1)–(B4) are satisfied in this model. Then show that (C2) and (C3) hold in this model, but (C1) fails.

8.7 Consider the real Cartesian plane $\mathbb{R}^2$, with lines and betweenness as before (Example 7.3.1), but define a different notion of congruence of line segments using the distance function given by the sum of the absolute values:

$$d(A, B) = |a_1 - b_1| + |a_2 - b_2|,$$

where $A = (a_1, a_2)$ and $B = (b_1, b_2)$. Some people call this "taxicab geometry" because it is similar to the distance by taxi from one point to anther in a city where all streets run east–west or north–south. Show that the axioms (C1), (C2), (C3) hold, so that this is another model of the axioms introduced so far. What does the circle with center $(0, 0)$ and radius $1$ look like in this model?

8.8 Again consider the real Cartesian plane $\mathbb{R}^2$, and define a third notion of congruence for line segments using the sup of absolute values for the distance function:

$$d(A, B) = \sup\{|a_1 - b_1|, |a_2 - b_2|\}.$$

Show that (C1), (C2), (C3) are also satisfied in this model. What does the circle with center $(0, 0)$ and radius $1$ look like in this case?

8.9 Following our general principles, we say that two models $M, M'$ of our geometry are *isomorphic* if there exists a 1-to-1 mapping $\varphi : M \rightarrow M'$ of the set of points of $M$ onto the set of points of $M'$, written $\varphi(A) = A'$, that sends lines into lines, preserves betweenness, i.e., $A * B * C$ in $M \Leftrightarrow A' * B' * C'$ in $M'$, and preserves congruence of line segments, i.e., $AB \cong CD$ in $M \Leftrightarrow A'B' \cong C'D'$ in $M'$.

Show that the models of Exercise 8.7 and Exercise 8.8 above are isomorphic to each other, but they are not isomorphic to the standard model (Example 8.4.1). Note: To show that the two models of Exercise 8.7 and Exercise 8.8 are isomorphic, you do not need to make the distance functions correspond. It is only the notion of congruence of line segments that must be preserved. To show that two models are not isomorphic, one method is to find some statement that is true in one model but not true in the other model.

8.10 Nothing in our axioms relates the size of a segment on one line to the size of a congruent segment on another line. So we can make a weird model as follows. Take the real Cartesian plane $\mathbb{R}^2$ with the usual notions of lines and betweenness. Using the Euclidean distance function $d(A, B)$, define a new distance function

$$d'(A, B) = \begin{cases} d(A, B) & \text{if the segment } AB \text{ is either horizontal or vertical,} \\ 2d(A, B) & \text{otherwise.} \end{cases}$$

Define congruence of segments $AB \cong CD$ if $d'(A, B) = d'(C, D)$.
Show that (C1), (C2), (C3) are all satisfied in this model. What does a circle with center $(0, 0)$ and radius 1 look like?

8.11 The *triangle inequality* is the statement that if $A, B, C$ are three distinct points, then $AC \leq AB + BC$.

(a) The triangle inequality always holds for collinear points.

(b) The triangle inequality holds for any three points in the standard model (Example 8.4.1) and also in taxicab geometry (Exercise 8.7).

(c) The triangle inequality does not hold in the model of Exercise 8.10. Thus the triangle inequality is not a consequence of the axioms of incidence, betweenness, and congruence of line segments (C1)–(C3). (However, we will see in Section 10 that the triangle inequality, in the form of Euclid (I.20), is a consequence of the full set of axioms of a Hilbert plane.)

# 9   Axioms of congruence for Angles

Recall that we have defined an *angle* to be the union of two rays originating at the same point, and not lying on the same line. We postulate an undefined notion of *congruence* for angles, written $\cong$, that is subject to the following three axioms:

**C4.** Given an angle $\angle BAC$ and given a ray $\overrightarrow{DF}$, there exists a unique ray $\overrightarrow{DE}$, on a given side of the line $DF$, such that $\angle BAC \cong \angle EDF$.

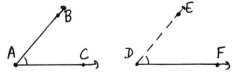

**C5.** For any three angles $\alpha, \beta, \gamma$, if $\alpha \cong \beta$ and $\alpha \cong \gamma$, then $\beta \cong \gamma$. Every angle is congruent to itself.

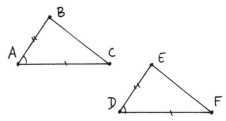

**C6.** (SAS) Given triangles $ABC$ and $DEF$, suppose that $AB \cong DE$ and $AC \cong DF$, and $\angle BAC \cong \angle EDF$. Then the two triangles are congruent, namely, $BC \cong EF$, $\angle ABC \cong \angle DEF$ and $\angle ACB \cong \angle DFE$.

Note that Hilbert takes the existence of an angle congruent to a given one (C4) as an axiom, while Euclid proves this by a ruler and compass construction (I.23). Since Hilbert does not make use of the compass, we may regard this axiom as a tool, the "transporter of angles," that acts as a substitute for the compass.

As with (C2), we can use (C5) to show that congruence is an equivalence relation.

**Proposition 9.1**
*Congruence of angles is an equivalence relation.*

*Proof*   The proof is identical to the proof of (8.1), using (C5) in place of (C2).

As in the case of congruence of line segments, we would like to make sense of Euclid's common notions in the context of congruence of angles. This proposition (9.1) is the analogue of the first common notion, that "things equal to the same thing are equal to each other." The second common notion, that "equals added to equals are equal," becomes problematic in the case of angles, because in general we cannot define the sum of two angles.

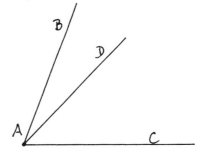

If $\angle BAC$ is an angle, and if a ray $\overrightarrow{AD}$ lies in the interior of the angle $\angle BAC$, then we will say that the angle $\angle BAC$ is the *sum* of the angles $\angle DAC$ and $\angle BAD$.

However, if we start with the two given angles, there may not be an angle that is their sum in this sense. For one thing, they may add up to a straight line, or "two right angles" as Euclid says, but this is not an angle. Or their sum may be greater than $180°$, in which case we get an angle, but the two original angles will not be in the interior of the new angle. So we must be careful how we state results having to do with sums of angles.

Note that we do not have an axiom about congruence of sums of angles analogous to the axiom (C3) about addition of line segments. That is because we can prove the corresponding result for angles. But in order to do so, we will need (C6).

Hilbert's use of (C6) = (SAS) as an axiom is a recognition of the insufficiency of Euclid's proof of that result (I.4) using the method of superposition. To justify the method of superposition by introducing axioms allowing motion of figures in the plane would be foreign to Euclid's approach to geometry, so it seems prudent to take (C6) as an axiom. However, we will show later (17.5) that the (SAS) axiom is essentially equivalent to the existence of a sufficiently large group of rigid motions of the plane. The axiom (C6) is necessary, since it is independent of the other axioms (Exercise 9.3). This axiom is essentially what tells us that our plane is homogeneous: Geometry is the same at different places in the plane.

Now let us show how to deal with sums of angles and inequalities among angles based on these axioms.

### Definition

If $\angle BAC$ is an angle, and if $D$ is a point on the line $AC$ on the other side of $A$ from $C$, then the angles $\angle BAC$ and $\angle BAD$ are *supplementary*.

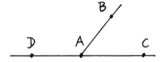

### Proposition 9.2

*If $\angle BAC$ and $\angle BAD$ are supplementary angles, and if $\angle B'A'C'$ and $\angle B'A'D'$ are supplementary angles, and if $\angle BAC \cong \angle B'A'C'$, then also $\angle BAD \cong \angle B'A'D'$.*

*Proof*  Replacing $B', C', D'$ by other points on the same rays, we may assume that $AB \cong A'B'$, $AC \cong A'C'$, and $AD \cong A'D'$. Draw the lines $BC$, $BD$, $B'C'$, and $B'D'$.

First we consider the triangles $ABC$ and $A'B'C'$. By hypothesis we have $AB \cong A'B'$ and $AC \cong A'C'$ and $\angle BAC \cong \angle B'A'C'$. So by (C6) we conclude that the triangles are congruent. In particular, $BC \cong B'C'$ and $\angle BCA \cong \angle B'C'A'$.

Next we consider the triangles $BCD$ and $B'C'D'$. Since $AC \cong A'C'$ and $AD \cong A'D'$, and $C * A * D$ and $C' * A' * D'$, we conclude from (C3) that $CD \cong C'D'$. Using $BC \cong B'C'$ and $\angle BCA \cong \angle B'C'A'$ proved above, we can apply (C6) again to see that the triangles $BCD$ and $B'C'D'$ are congruent. In particular, $BD \cong B'D'$ and $\angle BDA \cong \angle B'D'A'$.

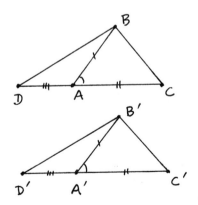

Now we consider the triangles $BDA$ and $B'D'A'$. From the previous step we have $BD \cong B'D'$ and $\angle BDA \cong \angle B'D'A'$. But by hypothesis we have $DA \cong D'A'$. So a third application of (C6) shows that the triangles $BDA$ and $B'D'A'$ are congruent. In particular, $\angle BAD \cong \angle B'A'D'$, which was to be proved.

**Note:** We may think of this result as a replacement for (I.13), which says that the angles made by a ray standing on a line are either right angles or are equal to two right angles. We cannot use Euclid's statement directly, because in our terminology, the sum of two right angles is not an angle. However, in applications, Euclid's (I.13) can be replaced by (9.2). So for example, we have the following corollary.

### Corollary 9.3
*Vertical angles are congruent.*

*Proof*   Recall that *vertical* angles are defined by the opposite rays on the same two lines. The vertical angles $\alpha$ and $\alpha'$ are each supplementary to $\beta$, and $\beta$ is congruent to itself, so by the proposition, $\alpha$ and $\alpha'$ are congruent.

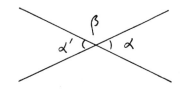

### Proposition 9.4 (Addition of angles)
*Suppose $\angle BAC$ is an angle, and the ray $\overrightarrow{AD}$ is in the interior of the angle $\angle BAC$. Suppose $\angle D'A'C' \cong \angle DAC$, and $\angle B'A'D' \cong \angle BAD$, and the rays $\overrightarrow{A'B'}$ and $\overrightarrow{A'C'}$ are on opposite sides of the line $A'D'$. Then the rays $\overrightarrow{A'B'}$ and $\overrightarrow{A'C'}$ form an angle, and $\angle B'A'C' \cong \angle BAC$, and the ray $\overrightarrow{A'D'}$ is in the interior of the angle $\angle B'A'C'$. For short, we say "sums of congruent angles are congruent."*

*Proof*   Draw the line $BC$. Then the ray $\overrightarrow{AD}$ must meet the segment $\overline{BC}$, by the crossbar theorem (7.3). Replacing the original $D$ by this intersection point, we may assume that $B, D, C$ lie on a line and $B * D * C$. On the other hand, replacing $B', C', D'$ by other points on the same rays, we may assume that $AB \cong A'B'$, and $AC \cong A'C'$, and $AD \cong A'D'$. We also have $\angle BAD \cong \angle B'A'D'$ and $\angle DAC \cong \angle D'A'C'$ by hypothesis.

By (C6) we conclude that the triangles $\triangle BAD$ and $\triangle B'A'D'$ are congruent. In particular, $BD \cong B'D'$ and $\angle BDA \cong B'D'A'$.

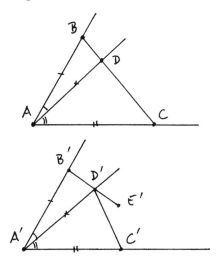

Again by (C6) we conclude that the triangles $\triangle DAC$ and $\triangle D'A'C'$ are congruent. In particular, $DC \cong D'C'$ and $\angle ADC \cong \angle A'D'C'$.

Let $E'$ be a point on the line $B'D'$ with $B' * D' * E'$. Then $\angle A'D'E'$ is supplementary to $\angle A'D'B'$, which is congruent to $\angle ADB$. So by (9.2) and transitivity of congruence, we find that $\angle A'D'E' \cong \angle A'D'C'$. Since these angles are on the same side of the line $A'D'$, we conclude from the uniqueness part of (C4) that they are the same angle. In other words, the three points $B', D',$ and $C'$ lie on a line.

Then from (C3) we conclude that $BC \cong B'C'$. Since $\angle ABD \cong \angle A'B'D'$ by the first congruence of triangles used in the earlier part of the proof, we can apply (C6) once more to the triangles $ABC$ and $A'B'C'$. The congruence of these triangles implies $\angle BAC \cong \angle B'A'C'$ as required. Since $B', D',$ and $C'$ are collinear and $D'A'C'$ is an angle, it follows that $A', B', C'$ are not collinear, so $B'A'C'$ is an angle. Since $B'$ and $C'$ are on opposite sides of the line $A'D'$, it follows that $B' * D' * C'$, and so the ray $\overrightarrow{A'D'}$ is in the interior of the angle $\angle B'A'C'$, as required.

Next, we will define a notion of inequality for angles analogous to the inequality for line segments in Section 8.

**Definition**
Suppose we are given angles $\angle BAC$ and $\angle EDF$. We say that $\angle BAC$ is *less than* $\angle EDF$, written $\angle BAC < \angle EDF$, if there exists a ray $\overrightarrow{DG}$ in the interior of the angle $\angle EDF$ such that $\angle BAC \cong \angle GDF$. In this case we will also say that $\angle EDF$ is *greater than* $\angle BAC$.

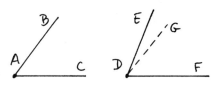

**Proposition 9.5**
    (a) *If $\alpha \cong \alpha'$ and $\beta \cong \beta'$, then $\alpha < \beta \Leftrightarrow \alpha' < \beta'$.*
    (b) *Inequality gives an order relation on angles, up to congruence. In other words:*
(i) *If $\alpha < \beta$ and $\beta < \gamma$, then $\alpha < \gamma$.*
(ii) *For any two angles $\alpha$ and $\beta$, one and only one of the following holds: $\alpha < \beta$;
$\alpha \cong \beta; \alpha > \beta$.*

*Proof*   The proofs of these statements are essentially the same as the corresponding statements for line segments (8.4), so we will leave them to the reader.

**Definition**
A *right angle* is an angle $\alpha$ that is congruent to one of its supplementary angles $\beta$.

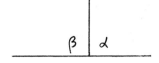

**Note:** In this definition, it does not matter which supplementary angle to $\alpha$ we consider, because the two supplementary angles to $\alpha$ are vertical angles, hence congruent by (9.3). Two lines are *orthogonal* if they meet at a point and one, hence all four, of the angles they make is a right angle.

## Proposition 9.6
*Any two right angles are congruent to each other.*

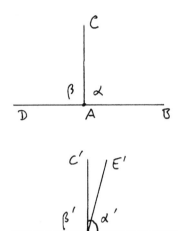

*Proof* Suppose that $\alpha = \angle CAB$ and $\alpha' = \angle C'A'B'$ are right angles. Then they will be congruent to their supplementary angles $\beta, \beta'$, by definition. Suppose $\alpha$ and $\alpha'$ are not congruent. Then by (9.5) either $\alpha < \alpha'$ or $\alpha' < \alpha$. Suppose, for example, $\alpha < \alpha'$. Then by definition of inequality there is a ray $\overrightarrow{A'E'}$ in the interior of angle $\alpha'$ such that $\alpha \cong \angle E'A'B'$.

It follows (check!) that the ray $\overrightarrow{A'C'}$ is in the interior of $\angle E'A'D'$, so that $\beta' < \angle E'A'D'$. But $\angle E'A'D'$ is supplementary to $\angle E'A'B'$, which is congruent to $\alpha$, so by (9.2), $\angle E'A'D' \cong \beta$. Therefore, $\beta' < \beta$. But $\alpha \cong \beta$ and $\alpha' \cong \beta'$, so we conclude that $\alpha' < \alpha$, which is a contradiction.

**Note:** Thus the congruence of all right angles can be proved and does not need to be taken as an axiom as Euclid did (Postulate 4). The idea of this proof already appears in Proclus.

## Example 9.6.1
We will show later that the real Cartesian plane $\mathbb{R}^2$ provides a model of all the axioms listed so far. You are probably willing to believe this, but the precise definition of what we mean by congruence of angles in this model, and the proof that axioms (C4)–(C6) hold, requires some work. We will postpone this work until we make a systematic study of Cartesian planes over arbitrary fields, and then we will show more generally that the Cartesian plane over any ordered field satisfying a certain algebraic condition gives a model of Hilbert's axioms (17.3).

The other most important model of Hilbert's axioms is the non-Euclidean Poincaré model, which we will discuss in Section 39.

# Exercises

9.1 (Difference of angles). Suppose we are given congruent angles $\angle BAC \cong \angle B'A'C'$. Suppose also that we are given a ray $\overrightarrow{AD}$ in the interior of $\angle BAC$. Then there exists a ray $\overrightarrow{A'D'}$ in the interior of $\angle B'A'C'$ such that $\angle DAC \cong \angle D'A'C'$ and $\angle BAD \cong \angle B'A'D'$. This statement corresponds to Euclid's Common Notion 3: "Equals subtracted from equals are equal," where "equal" in this case means congruence of angles.

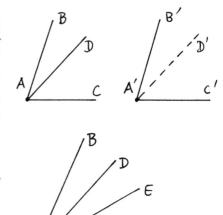

9.2 Suppose the ray $\overrightarrow{AD}$ is in the interior of the angle $\angle BAC$, and the ray $\overrightarrow{AE}$ is in the interior of the angle $\angle DAC$. Show that $\overrightarrow{AE}$ is also in the interior of $\angle BAC$.

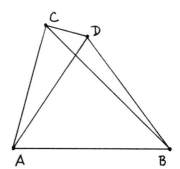

9.3 Consider the real Cartesian plane where congruence of line segments is given by the absolute value distance function (Exercise 8.7). Using the usual congruence of angles that you know from analytic geometry (Section 16), show that (C4) and (C5) hold in this model, but that (C6) fails. (Give a counterexample.)

9.4 Provide the missing betweenness arguments to complete Euclid's proof of (I.7) in the case he considers. Namely, assuming that the ray $\overrightarrow{AD}$ is in the interior of the angle $\angle CAB$, and assuming that $D$ is outside the triangle $ABC$, prove that $\overrightarrow{CB}$ is in the interior of the angle $\angle ACD$ and $\overrightarrow{DA}$ is in the interior of the angle $\angle CDB$.

# 10   Hilbert Planes

We have now introduced the minimum basic notions and axioms on which to found our study of geometry.

## Definition

A *Hilbert plane* is a given set (of *points*) together with certain subsets called *lines*, and undefined notions of *betweenness*, *congruence* for line segments, and *congruence* for angles (as explained in the preceding sections) that satisfy the axioms (I1)–(I3), (B1)–(B4), and (C1)–(C6). (We do not include the parallel axiom (P).)

We could go on immediately and introduce the parallel axiom and axioms of intersection of lines and circles, so as to recover all of Euclid's *Elements*, but it seems worthwhile to pause at this point and see how much of the geometry we can develop with this minimal set of axioms. The main reason for doing this is that the axioms of a Hilbert plane form the basis for non-Euclidean as well as Euclidean geometry. In fact, some people call the Hilbert plane *neutral geometry*, because it neither affirms nor denies the parallel axiom.

In this section we will see how much of Euclid's Book I we can recover in a Hilbert plane. With two notable exceptions, we can recover everything that does not make use of the parallel postulate.

Let us work in a given Hilbert plane. Euclid's definitions, postulates, and common notions have been replaced by the undefined notions, definitions, and axioms that we have discussed so far (excluding Playfair's axiom). We will now discuss the propositions of Euclid, Book I.

The first proposition (I.1) is our first exception! Without some additional axiom, it is not clear that the two circles in Euclid's construction will actually meet. In fact, the existence of an equilateral triangle on a given segment does not follow from the axioms of a Hilbert plane (Exercise 39.31). We will partially fill this gap by showing (10.2) that there do exist isosceles triangles on a given segment.

Euclid's Propositions (I.2) and (I.3) about transporting line segments are effectively replaced by axiom (C1). Proposition (I.4), (SAS), has been replaced by axiom (C6).

Proposition (I.5) and its proof are ok as they stand. In other words, every step of Euclid's proof can be justified in a straightforward manner within the framework of a Hilbert plane. To illustrate this process of reinterpreting one of Euclid's proofs within our new axiom system, let us look at Euclid's proof step by step.

*Proof of (I.5)*   Let $ABC$ be the given isosceles triangle, with $AB \cong AC$ (congruent line segments). We must prove that the base angles $\angle ABC$ and $\angle ACB$ are congruent. "In $BD$ take any point $F$." This is possible by axiom (B2). "On $AE$ cut off $AG$ equal to $AF$." This is possible by (C1). Now $AC \cong AB$ and $AF \cong AG$, and the enclosed angle $\angle BAC$ is the same, so the triangles $\triangle AFC$ and $\triangle AGB$ are congruent by a direct application of (C6). So $FC \cong GB$ and $\angle AFC \cong \angle AGB$ and $\angle ACF \cong \angle ABG$.

Since "equals subtracted from equals are equal," referring in this case to congruence of line segments, we conclude from (8.3) that $BF \cong CG$. Then by another application of (C6), the triangles $\triangle FBC$ and $\triangle GCB$ are congruent. It follows that $\angle CBG \cong \angle BCF$. Now by subtraction of congruent angles (Exercise 9.1), the base angles $\angle ABC$ and $\angle ACB$ are congruent, as required. (We omit the proof of the second assertion, which follows similarly.)

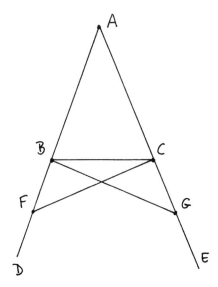

At certain steps in this proof we need to know something about betweenness, which can also be formally proved from our axioms. For example, in order to subtract the line segment $AB$ from $AF$, we need to know that $B$ is between $A$ and $F$. This follows from our choice of $F$. At the last step, subtracting angles, we need to know that the ray $\overrightarrow{BC}$ is in the interior of the angle $\angle ABG$. This follows from the fact that $C$ is between $A$ and $G$.

So in the following, when we say that Euclid's proof is ok as is, we mean that each step can be justified in a natural way, without having to invent additional steps of proof, from Hilbert's axioms and the preliminary results we established in the previous sections.

Looking at (I.6), the converse of (I.5), everything is ok except for one doubtful step at the end. Euclid says, "the triangle $DBC$ is equal to the triangle $ACB$, the less to the greater; which is absurd." It is not clear what this means, since we have not defined a notion of inequality for triangles. However, a very slight change will give a satisfactory proof. Namely, from the congruence of the triangles $\triangle DBC \cong \triangle ACB$, it follows that $\angle DCB \cong \angle ABC$. But also $\angle ABC \cong \angle ACB$ by hypothesis. So $\angle DCB \cong \angle ACB$, "the less to the greater," as Euclid would say. For us, this is a contradiction of the uniqueness part of axiom (C4), since there can be only one angle on the same side of the ray $\overrightarrow{CB}$ congruent to the angle $\angle ACB$. We conclude that the rays $\overrightarrow{CA}$ and $\overrightarrow{CD}$ are equal, so $A = D$, and the triangle is isosceles, as required.

Proposition (I.7), as we have mentioned before, needs some additional justification regarding the relative positions of the lines, which can be supplied from our axioms of betweenness (Exercise 9.4).

For (I.8), (SSS), we will need a new proof, since Euclid's method of superposition cannot be justified from our axioms. The following proof is due to Hilbert.

**Proposition 10.1** (SSS)
*If two triangles ABC and A′B′C′ have their respective sides equal, namely $AB \cong A'B'$, $AC \cong A'C'$, and $BC \cong B'C'$, then the two triangles are congruent.*

*Proof*   Using (C4) and (C1), construct an angle $\angle C'A'B''$ on the other side of the ray $\overrightarrow{A'C'}$ from $B'$ that is congruent to $\angle BAC$, and make $A'B''$ congruent to $AB$. Then $AB \cong A'B''$ by construction, $AC \cong A'C'$ by hypothesis, and $\angle BAC \cong \angle B''A'C'$ by construction, so by (C6), the triangle $\triangle ABC$ is congruent to the triangle $\triangle A'B''C'$. It follows that $BC \cong B''C'$.

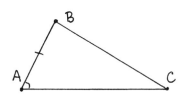

Draw the line $B'B''$. Now $A'B' \cong AB \cong A'B''$, so by transitivity, $A'B' \cong A'B''$. Thus the triangle $A'B'B''$ is isosceles, and so by (I.5) its base angles $\angle A'B'B''$ and $\angle A'B''B'$ are congruent. Similarly, $B'C' \cong B''C'$, so the triangle $C'B'B''$ is isosceles, and its base angles $\angle B''B'C'$ and $\angle B'B''C'$ are congruent. By addition of congruent angles (9.4) it follows that $\angle A'B'C' \cong \angle A'B''C'$.

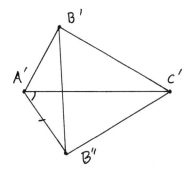

This latter triangle was shown congruent to $\triangle ABC$, so $\angle A'B''C' \cong \angle ABC$. Now by transitivity of congruence, $\angle ABC \cong \angle A'B'C'$, so we can apply (C6) again to conclude that the two triangles are congruent.

**Note:** This proof and the accompanying figure are for the case where $A'$ and $C'$ are on opposite sides of the line $B'B''$. The case where they are on the same side is analogous, and the case where one of $A'$ or $C'$ lies on the line $B'B''$ is easier, and left to the reader.

Starting with the next proposition (I.9) we have a series of constructions with ruler and compass. We cannot carry out these constructions in a Hilbert plane, because we have not yet added axioms to ensure that lines and circles will meet when they ought to (cf. Section 11). However, we can reinterpret these propositions as existence theorems, and these we can prove from Hilbert's axioms. Since we do not have the equilateral triangles that Euclid constructed in (I.1), we will prove the existence of isosceles triangles, and we will use them as a substitute for equilateral triangles in the following existence proofs.

**Proposition 10.2** (Existence of isosceles triangles)
*Given a line segment AB, there exists an isosceles triangle with base AB.*

*Proof* Let $AB$ be the given line segment. Let $C$ be any point not on the line $AB$ (axiom (I3)). Consider the triangle $\triangle ABC$. If the angles at $A$ and $B$ are equal, then $\triangle ABC$ is isosceles (I.6). If not, then one angle is less than the other. Suppose $\angle CAB < \angle CBA$. Then there is a ray $\overrightarrow{BE}$ in the interior of the angle $\angle CBA$ such that $\angle CAB \cong EBA$.

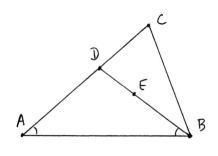

By the crossbar theorem (7.3) this ray must meet the opposite side $AC$ in a point $D$. Now the base angles of the triangle $DAB$ are equal, so by (I.6) it is isosceles.

**Note:** It would not suffice to construct equal angles at the two ends of the interval, because without the parallel axiom, even if the angles are small, there is no guarantee that the two rays would meet.

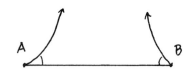

Now let us return to Euclid. We interpret (I.9) as asserting the existence of an angle bisector. We use the same method as Euclid, except that we use (10.2) to give the existence of an isosceles triangle $\triangle DEF$ where Euclid used an equilateral triangle. We may assume that this isosceles triangle is constructed on the opposite side of $DE$ from $A$. Then Euclid's proof, using (SSS), shows that $\angle DAF \cong \angle EAF$. It is not obvious from the construction that the ray $\overrightarrow{AF}$ is in the interior of the angle $\angle DAE$, but it does follow from the conclusion: For if $\overrightarrow{AF}$ were not in the interior of the angle, then $\overrightarrow{AD}$ and $\overrightarrow{AE}$ would be on the same side of $\overrightarrow{AF}$, and in that case the congruence of the angles $\angle DAF \cong \angle EAF$ would contradict the uniqueness in axiom (C4).

For (I.10) to bisect a given line segment, we again use (10.2) to construct an isosceles triangle instead of an equilateral triangle. The rest of Euclid's proof then works to show that a midpoint of the segment exists.

For (I.11) we can also use (10.2) to construct a line perpendicular to a line at a point. By the way, this also proves the existence of right angles, which is not obvious a priori.

For (I.12), to drop a perpendicular from a point $C$ to a line not containing $C$, Euclid's method using the compass does not work in a Hilbert plane. We need a new existence proof (see Exercise 10.4).

Proposition (I.13) has been replaced by the result on congruence of supplementary angles (9.2), and (I.14) is an easy consequence (Exercise 10.7). The congruence of vertical angles (I.15) has already been mentioned above (9.3). The theorem on exterior angles (I.16) is sufficiently important that we will reproduce Euclid's proof here, with the extra justifications necessary to make it work.

**Proposition 10.3** (Exterior angle theorem (I.16))
*In any triangle, the exterior angle is greater than either of the opposite interior angles.*

*Proof* Let $ABC$ be the given triangle. We will show that the exterior angle $\angle ACD$ is greater than the opposite interior angle at $A$. Let $E$ be the midpoint of $AC$ (I.10), and extend $BE$ to $F$ so that $BE \cong EF$ (axiom (C1)). Draw the line $CF$. Now the vertical angles at $E$ are equal (I.15), so by SAS (C6), the triangles $\triangle ABE$ and $\triangle CFE$ are congruent. Hence $\angle A \cong \angle ECF$.

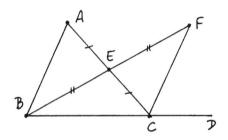

To finish the proof, that is, to show that $\angle ECF$ is less than $\angle ACD$, we need to know that the ray $\overrightarrow{CF}$ is in the interior of the angle $\angle ACD$. This we can prove based on our axioms of betweenness. Since $D$ is on the side $BC$ of the triangle extended, $B$ and $D$ are on opposite sides of the line $AC$. Also, by construction of $F$, we have $B$ and $F$ on opposite sides of $AC$. So from the plane separation property (7.1) it follows that $D$ and $F$ are on the same side of the line $AC$.

Now consider sides of the line $BC$. Since $B * E * F$, it follows that $E$ and $F$ are on the same side of $BC$. Since $A * E * C$, it follows that $A$ and $E$ are on the same side of $AC$. By transitivity (7.1) it follows that $A$ and $F$ are on the same side of the line $BC = CD$. So by definition, $F$ is in the interior of the angle $\angle ACD$, and hence the ray $\overrightarrow{CF}$ is also. Therefore, by definition of inequality for angles, $\angle BAC$ is less than $\angle ACD$, as required.

Propositions (I.17)–(I.21) are all ok as is, except that we should reinterpret the statement of (I.17). Instead of saying "any two angles of a triangle are less than two right angles," which does not make sense in our system, since "two right angles" is not an angle, we simply say; if $\alpha$ and $\beta$ are any two angles of a triangle, then $\alpha$ is less than the supplementary angle of $\beta$.

Proposition (I.22) is our other exception. Without knowing that two circles intersect when they ought to, we cannot prove the existence of the triangle required in this proposition. In fact, we will see later (Exercise 16.11) that there are Hilbert planes in which a triangle with certain given sides satisfying the hypotheses of this proposition does not exist!

The next proposition (I.23), which Euclid proved using (I.22), is replaced by Hilbert's axiom (C4), the "transporter of angles."

The remaining results that Euclid proved without using the parallel postulate are ok as is in the Hilbert plane: (I.24), (I.25), (I.26) = (ASA) and (AAS), (I.27) "alternate interior angles equal implies parallel," and even the existence of parallel lines (I.31).

Summing up, we have the following theorem.

### Theorem 10.4
*All of Euclid's propositions* (I.1) *through* (I.28), *except* (I.1) *and* (I.22), *can be proved in an arbitrary Hilbert plane, as explained above.*

### Constructions with Hilbert's Tools

Euclid used ruler and compass constructions to prove the existence of various objects in his geometry, such as the midpoint of a given line segment. We used Hilbert's axioms to prove corresponding existence results in a Hilbert plane. However, we can reinterpret these existence results as constructions if we imagine tools corresponding to certain of Hilbert's axioms. Thus (I1), the existence of a line through two points, corresponds to the ruler. For axiom (C1), imagine a tool, such as a compass with two sharp points (also called a pair of dividers), that acts as a transporter of segments. For axiom (C4), imagine a new tool, the transporter of angles, that can reproduce a given angle at a new point. It could be made of two rulers joined with a stiff but movable hinge.

We call these three tools, the ruler, the dividers, and the transporter of angles, *Hilbert's tools*. We also allow ourselves to pick points (using (I3) and (B2)) as required.

Now we can regard (10.2) as a construction of an isosceles triangle using Hilbert's tools. Counting steps, with one step for each use of a tool, we have the construction as follows:

Given a line segment $AB$. Pick $C$ not on the line $AB$.

1. Draw line $AC$.
2. Draw line $BC$. Suppose $\angle CAB$ is less than $\angle CBA$.
3. Transport $\angle CAB$ to $\angle ABE$, get point $D$.

Then $ABD$ is the required isosceles triangle.

# Exercises

10.1 Construct with Hilbert's tools the angle bisector of a given angle $(\text{par} = 4)$.

10.2 Construct with Hilbert's tools the midpoint of a given segment $(\text{par} = 4)$.

10.3 Construct with Hilbert's tools a line perpendicular to a given line $l$ at a given point $A \in l \, (\text{par} = 5)$.

10.4 Construct with Hilbert's tools a line perpendicular to a given line $l$ from a point $A$ not on $l \, (\text{par} = 4)$.

10.5 Construct with Hilbert's tools a line parallel to a given line $l$, and passing through a given point $A$ not on $l \, (\text{par} = 2)$.

10.6 Write out a careful proof of Euclid (I.18), justifying every step in the context of a Hilbert plane, and paying especial attention to questions of betweenness and inequalities.

10.7 Rewrite the statement (I.14) so that it makes sense in a Hilbert plane, and then give a careful proof.

10.8 Write a careful proof of (I.20) in a Hilbert plane.

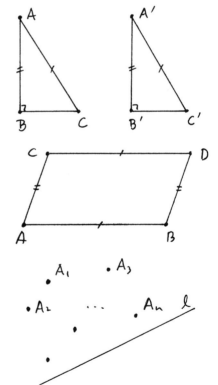

10.9 Show that the right-angle–side–side congruence theorem (RASS) holds in a Hilbert plane: If $ABC$ and $A'B'C'$ are triangles with right angles at $B$ and $B'$, and if $AB \cong A'B'$ and $AC \cong A'C'$, then the triangles are congruent.

10.10 In a Hilbert plane, suppose that we are given a quadrilateral $ABCD$ with $AB = CD$ and $AC = BD$. Prove that $CE$ is parallel to $AB$ (without using the parallel axiom (P)). *Hint*: Join the midpoints of $AB$ and $CD$; then use (I.27).

10.11 Given a finite set of points $A_1, \ldots, A_n$ in a Hilbert plane, prove that there exists a line $l$ for which all the points are on the same side of $l$.

# 11   Intersections of Lines and Circles

In this section we will discuss the intersections of lines and circles in the Hilbert plane, and we will introduce the further axiom (E), which will guarantee that lines and circles will intersect when they "ought" to. With this axiom we can justify Euclid's ruler and compass constructions in Book I and Book III. We work in a Hilbert plane (Section 10) without assuming the parallel axiom (P). Because of (10.4) we can use Euclid's results (I.2)–(I.28) (except (I.22)) in our proofs.

**Definition**
Given distinct points $O$, $A$, the *circle* $\Gamma$ with center $O$ and radius $OA$ is the set of all points $B$ such that $OA \cong OB$. The point $O$ is the *center* of the circle. The segment $OA$ is a *radius*.

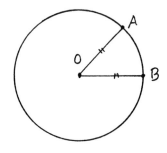

From this definition it is clear that a circle always has points. The point $A$ is on the circle. Moreover, if $l$ is any line through $O$, then by axiom (C1) there will be exactly two points on the line $l$, one on each side of $O$, lying on the circle. However, it is not obvious from the definition that the center is uniquely determined by the set of points of the circle.

**Proposition 11.1**
*Let $\Gamma$ be a circle with center $O$ and radius $OA$, and let $\Gamma'$ be a circle with center $O'$ and radius $O'A'$. Suppose $\Gamma = \Gamma'$ as point sets. Then $O = O'$. In other words the center of a circle is uniquely determined.*

*Proof*   Suppose $O \neq O'$. Then we consider the line $l$ through $O$ and $O'$. Since it passes through the center $O$ of $\Gamma$, it must meet $\Gamma$ in two points $C, D$, satisfy-

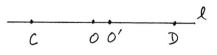

ing $C * O * D$ and $OC \cong OD$.

Since $\Gamma = \Gamma'$, the points $C, D$ are also on $\Gamma'$, so we have $O'C \cong O'D$ and $C * O' * D$. We do not know which of $O$ or $O'$ is closer to $C$, but the two cases are symmetric, so let us assume $C * O * O'$. In this case we must have $O * O' * D$ by the properties of betweenness(!). Then $OC < O'C \cong O'D < OD$, which is impossible, since $OC \cong OD$. Hence $O = O'$.

Now that we know that the center of a circle is uniquely determined, it makes sense to define the inside and the outside of a circle.

**Definition**
Let $\Gamma$ be a circle with center $O$ and radius $OA$. A point $B$ is *inside* $\Gamma$ (or in the *interior* of $\Gamma$) if $B = O$ or if $OB < OA$. A point $C$ is *outside* $\Gamma$ (or *exterior* to $\Gamma$) if $OA < OC$.

**Definition**
We say that a line $l$ is *tangent* to a circle $\Gamma$ if $l$ and $\Gamma$ meet in just one point $A$. We say that a circle $\Gamma$ is *tangent* to another circle $\Delta$ if $\Gamma$ and $\Delta$ have just one point in common.

This definition of tangent circles is a little different from Euclid's: His definition of two circles touching is that they meet in a point but do not cut each other. Since it is not clear what he means by "cut," we prefer the definition above, and we will prove that these notions of tangency have the usual properties.

**Proposition 11.2**
*Let $\Gamma$ be a circle with center $O$ and radius $OA$. The line perpendicular to the radius $OA$ at the point $A$ is tangent to the circle, and (except for the point $A$) lies entirely outside the circle. Conversely, if a line $l$ is tangent to $\Gamma$ at $A$, then it is perpendicular to $OA$. In particular, for any point $A$ of a circle, there exists a unique tangent line to the circle at that point.*

*Proof*   First, let $l$ be the line perpendicular to $OA$ at $A$. Let $B$ be any other point on the line $l$. Then in the triangle $OAB$, the exterior angle at $A$ is a right angle, so the angles at $O$ and at $B$ are less than a right angle (I.16). It follows (I.19) that $OB > OA$, so $B$ is outside the circle. Thus $l$ meets $\Gamma$ only at the point $A$, so it is a tangent line.

Now suppose that $l$ is a line tangent to $\Gamma$ at $A$. We must show that $l$ is perpendicular to $OA$. It cannot be equal to $OA$, because that line meets $\Gamma$ in another point opposite $A$. So consider the line from $O$, perpendicular to $l$, meeting $l$ at $B$. If $B \neq A$, take a point $C$ on the other side of $B$ from $A$, so that $AB \cong BC$

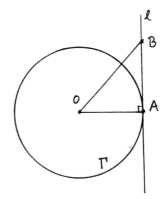

(axiom (C1)). The $\triangle OBA \cong \triangle OBC$ by SAS, so we have $OA \cong OC$, and hence $C$ is also on $\Gamma$. Since $C \neq A$, this is a contradiction. We conclude that $B = A$, and so $l$ is perpendicular to $OA$.

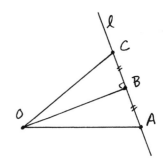

### Corollary 11.3
*If a line l contains a point A of a circle $\Gamma$, but is not tangent to $\Gamma$, then it meets $\Gamma$ in exactly two points.*

*Proof* If $l$ is not tangent to $\Gamma$ at $A$, then it is not perpendicular to $OA$, in which case, as we saw in the previous proof, it meets $\Gamma$ in another point $C$. We must show that $l$ cannot contain any further points of $\Gamma$. For if $D$ were another point of $l$ on $\Gamma$, then $OD \cong OA, OB$ is congruent to itself, so by (RASS) (Exercise 10.9) we would have $\triangle ODB \cong \triangle OAB$. Then $AB \cong BD$, so by axiom (C1) $D$ must be equal to $A$ or $C$.

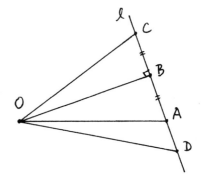

### Proposition 11.4
*Let $O, O', A$ be three distinct collinear points. Then the circle $\Gamma$ with center $O$ and radius $OA$ is tangent to the circle $\Gamma'$ with center $O'$ and radius $O'A$. Conversely, if two circles $\Gamma, \Gamma'$ are tangent at a point $A$, then their centers $O, O'$ are collinear with $A$.*

*Proof* Let $O, O', A$ be collinear. We must show that the circles $\Gamma$ and $\Gamma'$ have no further points in common besides $A$. The argument of (11.1) shows that there is no other point on the line $OO'$ that lies on both $\Gamma$ and $\Gamma'$. So suppose there is a point $B$ not on $OO'$ lying on both $\Gamma$ and $\Gamma'$. We divide into two cases depending on the relative position of $O, O'$, and $A$.

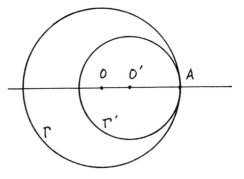

*Case 1* $O * O' * A$. Since $OA = OB$, $\angle OAB \cong \angle OBA$. Also, since $O'A = O'B$, $\angle O'AB \cong \angle O'BA$, using (I.5). It follows that $\angle OBA \cong \angle O'BA$, which contradicts axiom (C4). (This argument also applies if $O' * O * A$.)

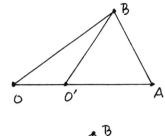

*Case 2* $O * A * O'$. Again using (I.5) we find that $\angle OAB \cong \angle OBA$ and $\angle O'AB \cong \angle O'BA$. But the two angles at $A$ are supplementary, so it follows that the two angles at $B$ are supplementary (9.2). But then $O$, $B$, and $O'$ would be collinear (I.14), which is a contradiction.

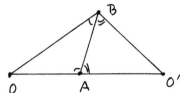

Conversely, suppose that $\Gamma$ and $\Gamma'$ are tangent at $A$, and suppose that $O, O', A$ are not collinear. Then we let $AC$ be perpendicular to the line $OO'$, and choose $B$ on the line $AC$ on the other side of $OO'$ with $AC \cong BC$. It follows by congruent triangles that $OA \cong OB$ and $O'A \cong O'B$, so $B$ also lies on $\Gamma$ and $\Gamma'$, contradicting the hypothesis $\Gamma$ tangent to $\Gamma'$. We conclude that $O, O', A$ are collinear.

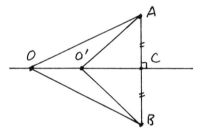

### Corollary 11.5
*If two circles meet at a point $A$ but are not tangent, then they have exactly two points in common.*

*Proof* We have seen above that if they are not tangent, then $O, O', A$ are not collinear, and they meet in an additional point $B$. We must show there are no further intersection points. If $D$ is a third point on $\Gamma$ and $\Gamma'$, then $OD \cong OA$ and $O'D \cong O'A$, so by (I.7), $D$ must be equal to $A$ or $B$.

In the above discussion of lines and circles meeting, we have seen that a line and a circle, or two circles, can be tangent (meeting in just one point), or if they meet but are not tangent, they will meet in exactly two points. There is nothing here to guarantee that a line and a circle, or two circles, will actually meet if they are in a position such that they "ought" to meet according to the usual intuition. For this we need an additional axiom (and we will see later (17.3) that this axiom is independent of the axioms of a Hilbert plane).

**E.** (Circle–circle intersection property). Given two circles $\Gamma, \Delta$, if $\Delta$ contains at least one point inside $\Gamma$, and $\Delta$ contains at least one point outside $\Gamma$, then $\Gamma$ and $\Delta$ will meet. (*Note:* It follows from Exercise 11.3 and (11.5) that they will then meet in exactly two points.)

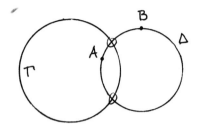

**Proposition 11.6** (Line–circle intersection property LCI)
*In a Hilbert plane with the extra axiom* (E), *if a line $l$ contains a point $A$ inside a circle $\Gamma$, then $l$ will meet $\Gamma$ (necessarily in two points, because of (11.2) and (11.3)).*

*Proof* Suppose we are given the line $L$ with a point $A$ inside the circle $\Gamma$. Our strategy is to construct another circle $\Delta$, show that $\Delta$ meets $\Gamma$, and then show that the intersection point also lies on $l$. Let $OB$ be the perpendicular from $O$ to $l$ (if $O$ is on the line $l$, we already know that $l$ meets $\Gamma$ by (C1)). Find a point $O'$ on the other side of $l$ from $O$, on the line $OB$, with $O'B \cong OB$. Let $\Delta$ be the circle with center $O'$ and radius $r = $ radius of $\Gamma$. (Here we denote by $r$ the congruence equivalence class of any radius of the circle $\Gamma$.)

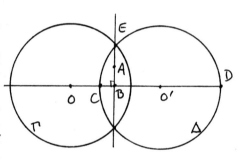

Now the line $OO'$ meets $\Delta$ in two points $C, D$, labeled such that $O, C$ are on the same side of $O'$, and $D$ on the opposite side.

By hypothesis, $A$ is a point on $l$, inside $\Gamma$. Hence $OA < r$. In the right triangle $OAB$, using (I.19) we see that $OB < OA$, so $OB < r$. It follows that $O'B < r = O'C$, so $O'$ and $C$ are on opposite sides of $l$. Hence $O, C$ are on the same side of $l$. We wish to show that $C$ is inside $\Gamma$. There are two cases.

*Case 1*  If $O * C * B$, then $OC < OB < r$, so $C$ is inside $\Gamma$.

*Case 2*  If $C * O * B$, then also $C * O * O'$, so $OC < O'C = r$, and again we see that $C$ is inside $\Gamma$.

On the other hand, the point $D$ satisfies $O * O' * D$, so $OD > O'D = r$, so $D$ is outside $\Gamma$.

# N O T E S.

## PROP. XXII. B. I.

Some Authors blame Euclid becaufe he does not demonftrate that the two circles made ufe of in the conftruction of this Problem fhall cut one another. but this is very plain from the determination he has given, viz. that any two of the ftraight lines DF, FG, GH muft be greater than the third. for who is fo dull, tho' only beginning to learn the Elements, as not to perceive that the circle defcribed from the centre F, at the diftance FD, muft meet FH betwixt F and H, becaufe FD is leffer than FH; and that, for the like reafon, the circle defcribed from the centre G, at the diftance GH or GM muft meet DG betwixt D and G; and that thefe circles muft meet one another, becaufe FD and GH are together greater than FG? and this determination is eafier to be underftood than that which Mr. Thomas Simpfon derives from it, and puts inftead of Euclid's, in the 49. page of his Elements of Geo-

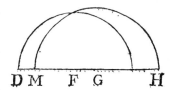

metry, that he may fupply the omiffion he blames Euclid for; which determination is, that any of the three ftraight lines muft be leffer than the fum, but greater than the difference of the other two. from this he fhews the circles muft meet one another, in one cafe; and fays that it may be proved after the fame manner in any other cafe. but the ftraight line GM which he bids take from GF may be greater than it, as in the figure here annexed, in which cafe his demonftration muft be changed into another.

Plate V. Simson's commentary on (I.22) from his English translation of Euclid (1756).

Now we can apply the axiom (E) to conclude that $\Gamma$ meets $\Delta$ at a point $E$. We must show that $E$ lies on $l$. We know that $OE \cong r \cong O'E$ and $OB \cong O'B$ by construction, and $BE$ is equal to itself, so by (SSS) $\Delta OEB \cong \Delta O'EB$. It follows that the angles at $B$ are equal, so they are right angles, so $BE$ is equal to the line $l$, and so $E$ lies on $l$ and $\Gamma$, as required.

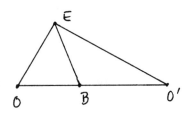

### Remark 11.6.1

We will see later (16.2) that in the Cartesian plane over a field, the circle–circle intersection property is equivalent to the line–circle intersection property. In an arbitrary Hilbert plane, the equivalence of these two statements follows from the classification theorem of Pejas (cf. Section 43), but I do not know any direct proof.

Using the new axiom (E) we can now justify Euclid's first construction (I.1), the equilateral triangle. Given the segment $AB$, let $\Gamma$ be the circle with center $A$ and radius $AB$. Let $\Delta$ be the circle with center $B$ and radius $BA$. Then $A$ is on the circle $\Delta$, and it is inside $\Gamma$ because it is the center of $\Gamma$. The line $AB$ meets $\Delta$ in another point $D$, such that $A * B * D$. Hence $AD > AB$, so $D$ is outside $\Gamma$.

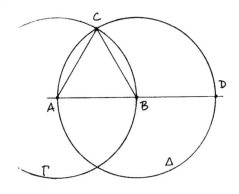

Thus $\Delta$ contains a point inside $\Gamma$ and a point outside $\Gamma$, so it must meet $\Gamma$ in a point $C$. From here, Euclid's proof shows that $\Delta ABC$ is an equilateral triangle.

In a similar way one can justify Euclid's other ruler and compass constructions in Book I. Several of them depend only on using the equilateral triangle constructed in (I.1). For (I.12) and (I.22) see Exercise 11.4 and Exercise 11.5. Thus we have the following theorem.

### Theorem 11.7

*Euclid's constructions* (I.1) *and* (I.22) *are valid in a Hilbert plane with the extra axiom* (E).

We can also justify the results of Euclid, Book III, up through (III.19) (note that (III.20) and beyond need the parallel axiom). The statements (III.10), (III.11), (III.12) about circles meeting and (III.16), (III.18), (III.19) about tangent lines can be replaced by the propositions of this section. (We omit the controversial last phrase of (III.16) about the angle of the semicircle, also called a horned angle or angle of contingency, because in our treatment we consider only angles defined by rays lying on straight lines.) In (III.14) Euclid uses (I.47) to prove (RASS), but that is not necessary: One can prove it with only the axioms of a Hilbert plane (Exercise 10.9). For (III.17), to draw a tangent to a circle from a point outside the circle, we need the line–circle intersection property (11.6) and hence the axiom (E). (Note that the other popular construction of the tangent line using (III.31) requires the parallel axiom!) The other results of Book III, up to (III.19) (except (III.17)), are valid in any Hilbert plane, provided that we assume the existence of the intersection points of lines and circles used in the statement and proofs, and their proofs are ok as is, except as noted.

### Theorem 11.8
*Euclid's propositions* (III.1) *through* (III.19) *are valid in any Hilbert plane, except that for the constructions* (III.1) *and* (III.17) *we need also the additional axiom* (E).

# Exercises

11.1  (a) The interior of a circle $\Gamma$ is a *convex* set: Namely, if $B, C$ are in the interior of $\Gamma$, and if $D$ is a point such that $B * D * C$, then $D$ is also in the interior of $\Gamma$.

(b) Assuming the parallel axiom (P), show that if $B, C$ are two points outside a circle $\Gamma$, then there exists a third point $D$ such that the segments $BD$ and $DC$ are entirely outside $\Gamma$. (This implies that the exterior of $\Gamma$ is a *segment-connected* set. See also Exercise 12.6.)

11.2  Two circles $\Gamma, \Gamma'$ that meet at a point $A$ are tangent if and only if the tangent line to $\Gamma$ at $A$ is equal to the tangent line to $\Gamma'$ at $A$.

11.3  If two circles $\Gamma$ and $\Delta$ are tangent to each other at a point $A$, show that (except for the point $A$) $\Delta$ lies either entirely inside $\Gamma$ or entirely outside $\Gamma$.

11.4  Use the line–circle intersection property (Proposition 11.6) to give a careful justification of Euclid's construction (I.12) of a line from a point perpendicular to a given line.

11.5  Given three line segments such that any two taken together are greater than the third, use (E) to justify Euclid's construction (I.22) of a triangle with sides congruent to the three given segments.

11.6  Show that Euclid's construction of the circle inscribed in a triangle (IV.4) is valid in any Hilbert plane. Be sure to explain why two angle bisectors of a triangle must meet in a point. Conclude that all three angle bisectors of a triangle meet in the same point.

11.7 Using (E), show that Euclid's construction of a hexagon inscribed in a circle (IV.15) makes sense. Without using (P) or results depending on it, which sides can you show are equal to each other?

# 12    Euclidean Planes

Let us look back at this point and see how well Hilbert's axioms have fulfilled their goal of providing a new solid base for developing Euclid's geometry. The major problems we found with Euclid's method have been settled: Questions of relative position of figures have been clarified by the axioms of betweenness; the problematic use of the method of superposition has been replaced by the device of taking SAS as an axiom; the existence of points needed in ruler and compass constructions is guaranteed by the circle–circle intersection property stated as axiom (E). Also, in the process of rewriting the foundations of geometry we have formulated a new notion, the *Hilbert plane*, which provides a minimum context in which to develop the beginnings of a geometry, free from the parallel axiom. Hilbert planes serve as a basis both for Euclidean geometry, and also later, for the non-Euclidean geometries.

In this section we will complete the work of earlier sections by showing how the addition of the parallel axiom allows us to recover almost all of the first four books of Euclid's *Elements*. We will also mention two more axioms, those of Archimedes and of Dedekind, which will be used in some parts of later chapters.

**Definition**
A *Euclidean plane* is a Hilbert plane satisfying the additional axioms (E), the circle–circle intersection property, and (P), Playfair's axiom, also called the parallel axiom. In other words, a Euclidean plane is a set of points with subsets called lines, and undefined notions of betweenness and congruence satisfying the axioms (I1)–(I3), (B1)–(B4), (C1)–(C6), (E), and (P). The Euclidean plane represents our modern formulation of the axiomatic basis for developing the geometry of Euclid's *Elements*.

We have already seen in Section 10 and Section 11 how to recover those results of Euclid's Books I and III that do not depend on the parallel axiom. The first use of the parallel axiom is in (I.29). Since we have replaced Euclid's fifth postulate by Playfair's axiom, we need to modify Euclid's proofs of a few early results in the theory of parallels.

So for example, to prove (I.29) we proceed as follows. Given two parallel lines $l$, $m$, and a transversal line $n$, we must show that the alternate interior angles $\alpha$ and $\beta$ are equal. If not, construct a line $l'$ through $A$ making an angle $\alpha$ with $n$ (axiom (C4)). By (I.27), $l'$ will be parallel to $m$. But then $l$ and $l'$ are two lines through $A$ parallel to $m$, so by (P), we must have $l = l'$, hence $\alpha = \beta$.

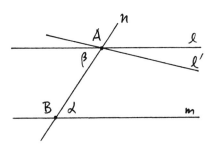

Proposition (I.30) is essentially equivalent to (P). The existence of parallel lines (I.31) follows from (C4) and (I.27) as mentioned before, so now we can reinterpret (I.31) in the stronger form that given a point $A$ not on a line $l$, there exists a unique parallel to $l$ passing through $A$. The remaining propositions using (P), namely (I.32)–(I.34), follow without difficulty. In particular, we have the famous (I.32), that "the sum of the angles of a triangle is equal to two right angles," though if we want to be scrupulous, we would have to say that sum is not defined, and rephrase the theorem by saying that the sum of any two angles of a triangle is supplementary to the third angle.

**Theorem 12.1**

*Euclid's theory of parallels, that is, propositions* (I.29)–(I.34), *hold in any Hilbert plane with* (P), *hence in any Euclidean plane.*

Starting with (I.35), and continuing to the end of Book I and through Book II, is Euclid's theory of area. Since Euclid does not define what he means by this new equality, we must presume that he takes it as another undefined notion, which we call *equal content*, just as the notion of congruence for line segments and angles were taken as undefined notions. Since Euclid freely applies the common notions to this concept, we may say that he has taken the common notions applied to equal content as further axioms, for example, "figures having equal content to a third figure have equal content to each other," or "halves of figures of equal content have equal content."

Hilbert showed that it is not necessary to regard the notion of equal content as an undefined notion subject to further axioms. He shows instead that it is possible to *define* the notion of equal content for figures (by cutting them up, rearranging, and adding and subtracting), and then *prove* the properties suggested by Euclid's common notions. To be more precise, we have the following theorem.

**Theorem 12.2** (Theory of area)

*In a Hilbert plane with* (P) *there is an equivalence relation called* equal content *for rectilineal figures that has the following properties:*

(1) *Congruent figures have equal content.*
(2) *Sums of figures with equal content have equal content.*
(3) *Differences of figures with equal content have equal content.*
(4) *Halves of figures with equal content have equal content.*
(5) *The whole is greater than the part.*
(6) *If two squares have equal content, their sides are congruent.*

We will prove this theorem in Chapter 5, (22.5), (23.1), (23.2). For the present you can either accept this result as something to prove later, or (as Euclid implicitly did) you can regard equal content of figures as another undefined notion, subject to the axioms that it is an equivalence relation and has these properties (1)–(6). For further discussion and more details about the exact meaning of a figure, the notions of sum and difference, etc., see Section 22 and Section 23.

Using this theory of area, the remaining results (I.35)–(I.48) of Book I follow without difficulty. Note in particular the Pythagorean theorem (I.47), which says that the sum of the squares on the legs of a right triangle have equal content with the square on the hypotenuse. Also, the results of Book II, (II.1)–(II.14), phrased as results about equal content, all follow easily. Proposition (II.11), how to cut a line segment in extreme and mean ratio, is used later in the construction of the regular pentagon. Only (II.14), to construct a square with content equal to a given rectilineal figure, uses the axiom (E).

**Theorem 12.3**
*In a Hilbert plane with* (P), *using the theory of area* (12.2), *Euclid's propositions* (I.35)–(I.48) *and* (II.1)–(II.14) *can all be proved as he does, using the extra axiom* (E) *only for* (II.14). *In particular, all these results hold in a Euclidean plane.*

In Book III, the first use of the parallel axiom is in (III.20), that the angle at the center of a circle subtending a given arc is twice the angle on the circumference subtending the same arc. This result uses (I.32), that the exterior angle of a triangle is equal to the sum of the two opposite interior angles, and thus depends on the parallel axiom (P). The following propositions (III.21), (III.22), and then (III.31)–(III.34) follow with no further difficulties. For the propositions (III.23)–(III.30) we need a notion of "equal" segments of circles, a congruence notion that has not been defined by Euclid, though we can infer from the proof of (III.24) that it means being able to place one segment on the other by a rigid motion. Indeed, if we take this as a definition of congruence, then the proofs of these results are all ok (Exercise 17.13). The final propositions (III.35)–(III.37) make use of the theory of area for their statements, and depend on the earlier area results from Books I and II.

**Theorem 12.4**
*In Book III, Euclid's propositions* (III.20)–(III.37) *hold in any Euclidean plane. The last three* (III.35)–(III.37) *make use of the theory of area* (12.2).

Most of the results of Book IV require the parallel axiom (P), some need circle–circle intersection (E), and some, notably (IV.10), (IV.11), require (P), (E), and the theory of area. Thus we may regard the construction of the regular pentagon as the crowning result of the first four books of the *Elements*, making use of all the results developed so far.

**Theorem 12.5**
*All the propositions* (IV.1)–(IV.16) *of Euclid's Book* IV *hold in a Euclidean plane.*

We end this section with a discussion of two further axioms that are not needed for Books I–IV, but will be used later. The first is *Archimedes' axiom*.

**A.** Given line segments $AB$ and $CD$, there is a natural number $n$ such that $n$ copies of $AB$ added together will be greater than $CD$.

This axiom is used implicitly in the theory of proportion developed in Book V, for example in Definition 4, where Euclid says that quantities have a ratio when one can be multiplied to exceed the other. It appears explicitly in (X.1), in a form reminiscent of the $\varepsilon$-arguments of calculus: Given two quantities $AB$ and $CD$, if we remove from $AB$ more than its half, and again from the remainder remove more than its half, and continue in this fashion, then eventually we will have a quantity less than $CD$. In modern texts this would appear as the statement "given any $\varepsilon > 0$, there is an integer $n$ sufficiently large that $1/2^n < \varepsilon$." Euclid applies this "method of exhaustion" to the study of the volume of three-dimensional figures in Book XII. When he cannot compare solids by cutting into a finite number of pieces and reassembling, he uses a limiting process where the solid is represented as a union of a sequence of subsolids so that the remainder can be made as small as you like. See Sections 26, 27 for Euclid's theory of volume.

Archimedes' axiom is independent of all the axioms of a Hilbert plane or a Euclidean plane, so we will see examples of *Archimedean* geometries that satisfy (A) and *non-Archimedean* geometries that do not (Section 18).

The other axiom we would like to consider is *Dedekind's axiom*, based on Dedekind's definition in the late nineteenth century of the real numbers:

**D.** Suppose the points of a line $l$ are divided into two nonempty subsets $S$, $T$ in such a way that no point of $S$ is between two points of $T$, and no point of $T$ is between two points of $S$. Then there exists a unique point $P$ such that for any $A \in S$ and any $B \in T$, either $A = P$ or $B = P$ or the point $P$ is between $A$ and $B$.

This axiom is very strong. It implies (A) and (E), and a Euclidean plane with (D) is forced to be isomorphic to the Cartesian plane over the real numbers. (See Exercise 12.2, Exercise 12.3, (15.5), and (21.3).) So if you want a categorical axiom system, just add (D) to the axioms of a Euclidean plane. From the point of view of this book, however, there are two reasons to avoid using Dedekind's axiom. First of all, it belongs to the modern development of the real numbers and notions of continuity, which is not in the spirit of Euclid's geometry. Second, it is too strong. By essentially introducing the real numbers into our geometry, it masks many of the more subtle distinctions and obscures questions such as constructibility that we will discuss in Chapter 6. So we include this axiom only to acknowledge that it is there, but with no intention of using it.

# Exercises

12.1 Show that in a Hilbert plane with (P), the perpendicular bisectors of the sides of a triangle will meet in a point, and thus justify Euclid's construction of the circumscribed circle of a triangle (IV.5). *Note*: In a non-Euclidean geometry, there may be triangles having no circumscribed circle: cf. Exercise 18.4, Exercise 39.14, and Proposition 41.1.

12.2 Show that in a Hilbert plane Dedekind's axiom (D) implies Archimedes' axiom (A). *Hint*: Given segments $AB$ and $CD$, let $T$ be the set of all points $E$ on the ray $\overrightarrow{CD}$ for which there is no integer $n$ with $n \cdot AB > CE$. Let $S$ be the set of points of the line $CD$ not in $T$, and apply (D).

12.3 Show that in a Hilbert plane (D) implies (E). *Hint*: Follow the discussion in Heath (1926), vol. I, p. 238.

12.4 For the construction and proof of (IV.2), to inscribe a triangle equiangular with a given triangle in a given circle (assume also that you are given the center of the circle), is axiom (E) necessary? Is (P) necessary?

12.5 Same question for (IV.6), to inscribe a square in a given circle.

12.6 In a Hilbert plane with (A), show that the exterior of a circle is a segment-connected set (cf. Exercise 11.1). Without assuming either (P) or (A), this may be false (Exercise 43.17).

> To each book are appended explanatory notes, in which especial care has been taken to guard the student against the common mistake of confounding ideas of number with those of magnitude.
>
> – Preface to Potts' Euclid,
> London (1845)

# 3

**CHAPTER**

# Geometry over Fields

 eginning with the familiar example of the real Cartesian plane, we show how to construct a geometry satisfying Hilbert's axioms over an abstract field. The axioms of incidence are valid over any field (Section 14). For the notion of betweenness we need an ordered field (Section 15). For the axiom (C1) on transferring a line segment to a given ray, we need a property (∗) on the existence of certain square roots in the field $F$. To carry out Euclidean constructions, we need a slightly stronger property (∗∗) — see Section 16.

To prove the (SAS) axiom over a field $F$, we revert to Euclid's method of superposition. In the case of the geometry over a field this can be justified by showing the existence of sufficiently many rigid motions (Section 17).

We end the chapter with some examples of geometries that do not satisfy Archimedes' axiom (Section 18).

We have seen that the geometry developed in Euclid's *Elements* does not make use of numbers to measure lengths or angles or areas. It is purely geometric in that it deals with points, lines, circles, triangles, and the relationships among these.

In the centuries after Euclid, geometers began using numbers more and more. At first number theory (arithmetic) and geometry were kept strictly apart. Number theory dealt with positive whole numbers and their ratios, i.e., rational numbers. Any other magnitude was considered geometrically. Thus $\sqrt{2}$ was not regarded as a number. The fact that $\sqrt{2}$ is irrational was expressed by saying that the diagonal of a square (a geometrical quantity) is not commensurable

**117**

with the side of the square. This means that no integer multiple of the diagonal is equal to any integer multiple of the side. As algebraic notation developed in the Renaissance, the concept of number was enlarged, and geometric quantities were treated more like numbers. A big step was taken by René Descartes (1596–1650), who showed in his book *La Géométrie* how to construct the product, quotient, and square root of line segments, having once fixed a unit line segment. He was thus able to apply algebraic operations to line segments and write algebraic equations relating an unknown line segment to given line segments. Descartes's use of algebra in geometry led to the idea of representing points in the plane by pairs of numbers, and thus to the modern discipline called analytic geometry.

Meanwhile, the concept of number was expanded from rational numbers to include irrational numbers and then transcendental numbers as they were discovered. By the end of the nineteenth century, considerations of limits and continuity made the real numbers $\mathbb{R}$ into the standard to be used in analytic geometry, calculus, and topology. Also at the end of the nineteenth century, the formalization of abstract structures in mathematics led to the concept of a field, so that by analogy with the standard model over $\mathbb{R}$, one could also consider a geometry over any abstract field.

The geometry taught today in high schools and colleges has become a sort of hybrid between the purely geometric methods of Euclid and the algebraic methods of Descartes, with occasional notions of continuity thrown in. One of the purposes of this book is to clarify the blurred distinctions between these different approaches. Therefore, we will pursue two different logical tracks. One is the axiomatic approach of Euclid and Hilbert, starting with geometrical postulates and proving results in logical sequence from them. This theory is built on the platform of the axioms of geometry. The other track is a geometry over a field. In this case the theory is built on a logical platform given by the algebraic definition of a field, or as we may say, the field axioms. The geometrical notions of point, line, betweenness, and congruence are defined in terms of field properties, and all proofs go back to the algebraic foundations. These geometries built from fields will be models of the axiomatic geometries.

In this chapter we start with an informal section on the real Cartesian plane. Then, in the following sections, we develop a rigorous theory of Cartesian planes over an abstract field. In Chapter 4 we will make the two tracks converge by the introduction of coordinates into an abstract geometry (at least in the case where the parallel axiom (P) holds).

# 13   The Real Cartesian Plane

In this section we will make clear what we mean by the *real Cartesian plane*, which is the plane geometry over the real numbers. Our proofs will be informal, using well-known results from high-school geometry and analytic geometry.

We accept as given the field of real numbers $\mathbb{R}$. We call a *point* an ordered pair $(a, b)$ of real numbers, and the set of all such ordered pairs is the Cartesian plane. As usual, we call the set of points $(a, 0)$ the *x-axis*, and the set of points $(0, b)$ the *y-axis*, and their intersection $(0, 0)$ the *origin*.

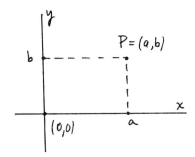

A *line* in this plane is the subset defined by a linear equation $ax + by + c = 0$, with $a, b$ not both zero. Among these are the *vertical* lines, which can be written as $x = a$, and every other line can be written in the form $y = mx + b$. In this case we call $m$ the *slope* of the line, and $b$ its *y-intercept*. For completeness, we will say that a vertical line has slope $\infty$.

Two lines are called *parallel* if they are equal or if they have no points in common. By looking at the equations of two lines, and solving those equations simultaneously, we see that the lines are parallel if and only if they have the same slope. It follows immediately that if $l_1, l_2, l_3$ are three distinct lines, and $l_1 \| l_2$ and $l_2 \| l_3$, then $l_1 \| l_3$. Indeed, all three must have the same slope. In Euclid's *Elements*, this result appears as (I.30) and is proved there using the parallel postulate plus earlier results from Book I. Here in the Cartesian plane, we have a trivial proof just by looking at the equations of the lines.

Let us give another, less trivial, example of how useful the analytic method can be for proving geometric results. We will show that the three altitudes of a triangle meet at a point. (Compare this with the geometric proofs given earlier in Section 5.)

**Proposition 13.1**

*In the real Cartesian plane, the three altitudes of any triangle all meet at a single point.*

*Proof*   Recall that an *altitude* of a triangle is the line through one vertex that is perpendicular to the opposite side. First let us move the triangle so that one edge lies along the x-axis, and the opposite vertex lies on the y-axis.

The we can call the vertices $A = (a, 0)$, $B = (0, b)$, and $C = (c, 0)$. The y-axis is by construction one of the altitudes of the triangle. Our strategy is to find the equations of the other two altitudes, see where they meet the y-axis, and verify that they meet it at the same point.

The line $AB$ has slope $-b/a$, so the altitude through $C$, which is perpendicular to this line, will have slope $a/b$. (Here we use the fact that if two perpendicular lines have slopes $m_1$ and $m_2$, then $m_1 m_2 = -1$.) So the equation of the altitude through $C$, using the point–slope formula, is

$$y = \frac{a}{b}(x - c).$$

To intersect this with the $y$-axis, we set $x = 0$ and obtain $y = -ac/b$.

Now consider the line $BC$. It has slope $-b/c$, so the altitude through $A$ will have slope $c/b$. Its equation becomes

$$y = \frac{c}{b}(x - a).$$

Setting $x = 0$, we obtain $y = -ac/b$. Since this is the same point as the previous calculation, we find that the three altitudes meet at a point.

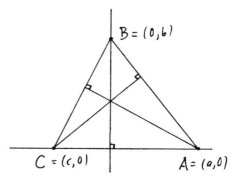

Let us reflect a moment on the significance of this proof.

First of all, the reader may object that we have used some facts without proof, such as the result about the slopes of perpendicular lines, or the possibility of moving the triangle into the special position of the proof. However, I am assuming that anyone who has studied some analytic geometry could fill in those missing arguments satisfactorily.

The more serious question is, how do we respond to someone who says, with a simple analytic proof like that, why bother with geometric proofs from axioms? If you believe that there is only one true geometry, then indeed this proof would be sufficient. But modern mathematics has abandoned the naive position that there is only one truth. Instead it asks, what can be proved within each logical framework, within each separate mathematical theory? This proof shows that the result is true within the logical framework of the real Cartesian plane, using algebra of the real numbers as a logical base. Having found the result to be true in this framework, we certainly expect it to be true in the framework of axiomatic Euclidean geometry. However, this proof gives no hint at all about how to find a proof in the abstract axiomatic geometry. In other words, if an analytic proof shows that a result is true in the geometry of the real Cartesian plane, that does not imply a proof, or even guarantee the existence of a proof, in the abstract axiomatic geometry. For example, think of Archimedes' axiom (18.4.2).

For these reasons we will preserve two separate logical tracks, the abstract axiomatic approach, and the analytic-geometric approach, until such time as we can prove that the two tracks converge again, using abstract ordered fields.

Next we turn to one of the great insights provided by the algebraic perspective, namely Descartes's discovery that the ruler and compass constructions of Euclid's geometry correspond to the solution of quadratic equations in algebra. To be more precise, let us regard a construction problem as giving certain points in the plane, and requiring the construction of certain other points.

# LA
# GEOMETRIE·
## DE
## RENÉ DESCARTES·

# A PARIS,
Chez Charles Angot, ruë faint Iacques,
au Lion d'or.

## M. DC. LXIV.
*AVEC PRIVILEGE DV ROT.*

Plate VI. Title page of *La Géométrie* of Descartes, first separate French edition (1664).

**Theorem 13.2** (Descartes)
*Suppose we are given points $P_1 = (a_1, b_1), \ldots, P_n = (a_n, b_n)$ in the real Cartesian plane. (We also assume that we are given the points $(0,0)$ and $(1,0)$.) Then it is possible to construct a point $Q = (\alpha, \beta)$ with ruler and compass if and only if $\alpha$ and $\beta$ can be obtained from $a_1, \ldots, a_n, b_1, \ldots, b_n$ by field operations $+, -, \cdot, \div$ and the solution of a finite number of successive linear and quadratic equations, involving the square roots of positive real numbers.*

*Proof*   A ruler and compass construction consists of drawing lines through given points, constructing circles with given center and radius, and finding intersections of lines and circles.

Given two points $P_1 = (a_1, b_1)$ and $P_2 = (a_2, b_2)$, the line passing through them has equation

$$y - b_1 = \frac{b_2 - b_1}{a_2 - a_1}(x - a_1).$$

Its coefficients are obtained by field operations from the initial data $a_1, a_2, b_1, b_2$.

A circle with center $(a, b)$ and radius $r$ has equation

$$(x - a)^2 + (y - b)^2 = r^2.$$

This is a quadratic equation whose coefficients depend on $a$, $b$, and $r^2$. Note that $r$ may be determined as the distance between two points $P_1 = (a_1, b_1)$ and $P_2 = (a_2, b_2)$, in which case

$$r^2 = (a_1 - a_2)^2 + (b_1 - b_2)^2.$$

To find the intersection of two lines, we solve two linear equations, which can be done using only field operations.

To intersect a line with a circle, we solve the equations simultaneously, which requires solving a quadratic equation in $x$. Assuming that the line meets the circle, we will need to take square roots of positive numbers only—cf. Exercise 16.6.

To intersect two circles, we first subtract the two equations, which eliminates the $x^2$ and $y^2$ terms. Then we must solve a quadratic with a linear equation, leading to another quadratic equation in $x$.

In other words, to find the coordinates of a point $Q = (\alpha, \beta)$ obtained by a ruler and compass construction from the initial data $P_1, \ldots, P_n$, we must solve a finite number of linear and quadratic equations whose coefficients depend on the coordinates $(a_i, b_i)$ and on quantities constructed in earlier steps.

Conversely, the roots of any linear or quadratic equation can be constructed by ruler and compass, given lengths corresponding to the coefficients of the equations, and given a standard length 1. Indeed, such equations can be solved (using the quadratic formula) by a finite number of applications of field operations $+, -, \cdot, \div$ and extractions of square roots of positive numbers, and each of these five operations can be accomplished using ruler and compass.

For the sum and difference of two line segments, simply lay them out on the same line, end to end for the sum, or overlapping for the difference.

For the product, lay the segment $a$ on the $x$-axis, and the segments $1, b$ on the $y$-axis. Draw the line from 1 to $a$, which will have equation $y = -(1/a)x + 1$. The parallel line through $b$ has equation $y = -(1/a)x + b$. This intersects the $x$-axis in the point $(ab, 0)$, and thus we construct the segment $ab$ out of the segments $1, a, b$.

For the quotient, put 1 on the $x$-axis, and $a, b$ on the $y$-axis. A similar construction gives the point $b/a$ on the $x$-axis.

To construct the square root of a segment $a$, lay out $a$ on the positive $x$-axis, and $-1$ on the negative $x$-axis. Bisect the segment from $-1$ to $a$, and draw the semicircle having that segment as diameter. A brief computation with the equation of the circle shows that it meets the $y$-axis at the point $\sqrt{a}$.

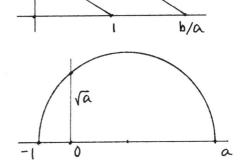

So here we have an algebraic criterion for deciding the possibility of a ruler and compass construction. The method of proof may not lead to an elegant construction, but at least one can determine the possibility of such a construction in a systematic manner. This theorem is a striking example of the insight into geometrical questions given by the algebraic point of view. As Descartes (1637) says:

> One can construct all the problems of ordinary geometry without doing anything more than what little is contained in the four figures which I have explained; which is something I do not believe the ancients had noticed: for otherwise they would not have taken the trouble to write so many fat books, where already the order of their propositions makes it clear that they did not have the true method for finding them all, but merely collected those which they happened to come across.

As a practical application of this result, we will find expressions using nested square roots for some lengths that are constructible with ruler and compass, such as the sides of regular polygons inscribed in a circle. Note that if a particular angle $\alpha$ is constructible, then its trigonometric functions, in particular $\sin \alpha$ and $\cos \alpha$, can be expressed using square roots.

For example, from the right isosceles triangle with sides $1, 1, \sqrt{2}$ we obtain

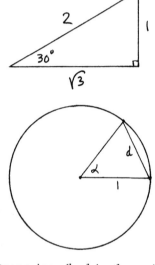

$$\sin 45° = \cos 45° = \frac{1}{2}\sqrt{2}.$$

From the $30°$–$60°$–$90°$ triangle with sides $1, \sqrt{3}$, and $2$ we obtain

$$\cos 60° = \sin 30° = \tfrac{1}{2},$$

$$\sin 60° = \cos 30° = \tfrac{1}{2}\sqrt{3}.$$

### Proposition 13.3
*The length of the chord $d$ of a circle of radius $1$ subtending an angle $\alpha$ at the center of the circle is given by*

$$d = \sqrt{2 - 2\cos\alpha}.$$

*Proof* The law of cosines gives $d^2 = 1^2 + 1^2 - 2\cos\alpha$, from which the result follows immediately.

So for example, the side of the regular octagon inscribed in the unit circle will be

$$d = \sqrt{2 - 2\cos 45°} = \sqrt{2 - \sqrt{2}}.$$

### Proposition 13.4
*In a circle of radius $1$, the length of the side of a regular decagon is $\frac{1}{2}(\sqrt{5} - 1)$.*

*Proof* Let us consider the triangle $ABC$ formed by two radii and one side of the decagon. Then $AB = AC = 1$, and $BC = x$ is the side of the decagon. The angle at $A$ is $2\pi/10$ or $36°$, so the angles at $B$ and $C$ are $72°$ each. Let $BD$ bisect the angle at $B$. Then the two halves are both $36°$ angles. From this it follows that $\triangle ABD$ is an isosceles triangle, and $\triangle BCD$ is an isosceles triangle similar to the original triangle $\triangle ABC$.

Therefore, $BD = x$ and $AD = x$ and $CD = 1 - x$. Writing the ratios of corresponding sides of the similar triangles $\triangle BCD$ and $\triangle ABC$ we have

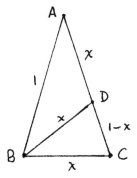

$$\frac{1 - x}{x} = \frac{x}{1}.$$

Hence $x^2 + x - 1 = 0$, and solving with the quadratic formula gives $x = \frac{1}{2}(\sqrt{5} - 1)$, as required.

**Remark 13.4.1**

This result allows us to give an analytic proof of the construction of the regular pentagon (4.3). Indeed, letting the radius $OA$ be 1, then $OG = \frac{1}{2}$, $GA = \frac{1}{2}\sqrt{5}$, and $OH = \frac{1}{2}(\sqrt{5} - 1)$. Thus $A, I, J$ are vertices of the regular decagon, and so $IJ$ is a side of the regular pentagon. For another proof using complex numbers, see (29.1).

**Proposition 13.5**

*In a circle of radius 1, the side of the regular pentagon is $\frac{1}{2}\sqrt{10 - 2\sqrt{5}}$.*

*Proof*  Applying the law of cosines to the triangle $\triangle ABC$ of (13.4), we get

$$1^2 = 1^2 + x^2 - 2x \cos 72°.$$

From this we obtain $\cos 72° = \frac{1}{4}(\sqrt{5} - 1)$. Since a side of the regular pentagon subtends an angle of $72°$ at the center of the circle, from (13.3) we have that the side of the pentagon is

$$d = \sqrt{2 - 2\cos 72°} = \frac{1}{2}\sqrt{10 - 2\sqrt{5}}.$$

# Exercises

13.1  Given $AB = 1$, construct segments of length $\sqrt{2}$, $\sqrt{3}$, $\sqrt{5}$, $\sqrt{6}$, $\sqrt{7}$, $\sqrt{10}$ in 5 steps or fewer each, making the constructions independent of each other.

13.2  Show that any quantity obtainable from the rational numbers by a finite number of operations $+$, $-$, $\cdot$, $\div$, $\sqrt{\phantom{x}}$, can be written in a *standard form* $r \cdot A$, where $r \in \mathbb{Q}$ is a rational number and $A$ is an expression involving only integers, $+$, $-$, $\cdot$, and $\sqrt{\phantom{x}}$. In the following problems, please express your answers always in standard form. (Unfortunately, this standard form is not unique—see Exercises 13.7, 13.12 below.)

13.3  Express $(\sqrt{5} + 1)/\sqrt{10 + 2\sqrt{5}}$ in standard form.

13.4 Find $\sin 22\frac{1}{2}^\circ$ and $\cos 22\frac{1}{2}^\circ$ as expressions involving square roots in standard form. Check your result by finding the decimal equivalent with a calculator.

13.5 Find the side of a regular 16-gon inscribed in the unit circle.

13.6 Find $\sin 11\frac{1}{4}^\circ$ and $\cos 11\frac{1}{4}^\circ$ in standard form.

13.7 Three students working on the same problem came up with the following answers.

   a. $-\sqrt{5} + \sqrt{11 + 6\sqrt{2}}$.

   b. $3 + \sqrt{7 - 2\sqrt{10}}$.

   c. $\sqrt{2} + \sqrt{14 - 6\sqrt{5}}$.

Two answers were correct, and one false. Find which two were correct, and prove that they are equal. Can you express the correct answer in a simpler form? How can you modify the third student's answer so that it becomes correct?

13.8 Find the length of the edge of a regular tetrahedron inscribed in a unit sphere.

13.9 Find the area of the largest equilateral triangle that is contained in a square of side 1.

13.10 If $a, b \in \mathbf{Z}$, and if $a + b\sqrt{2}$ has a square root in $\mathbf{Q}(\sqrt{2})$, then the square root is actually in $\mathbf{Z}[\sqrt{2}]$.

13.11 If $a, b \in \mathbf{Z}$, give a method for deciding whether $\sqrt{a + b\sqrt{2}} \in \mathbf{Z}[\sqrt{2}]$. Are the following squares in $\mathbf{Z}[\sqrt{2}]$? If so, find the square root.

   a. $627 + 442\sqrt{2}$.

   b. $1507 + 1024\sqrt{2}$.

   c. $2107 + 1470\sqrt{2}$.

13.12 Verify

$$\sqrt{5 + 2\sqrt{5}} - \sqrt{5 - 2\sqrt{5}} = \sqrt{10 - 2\sqrt{5}}.$$

Also, show that none of these three nested radicals is in $\mathbf{Q}(\sqrt{5})$. This is another example of nonuniqueness of the standard form.

13.13 Express $\sin 72^\circ$ as nested radicals in standard form. Check by computing decimal equivalents with a calculator.

13.14 Same problem for $\cos 36^\circ$, $\sin 36^\circ$.

13.15 Find $\cos 24^\circ$, $\sin 24^\circ$, $\cos 12^\circ$, $\sin 12^\circ$, and the side of the regular 15-sided polygon inscribed in the unit circle. Express in standard form, and check decimal equivalents with the calculator.

13.16 Find the side of a regular pentagon circumscribed around a unit circle in standard form.

13.17 Given a regular pentagon of side 1, find the distance from the center to a vertex, in standard form.

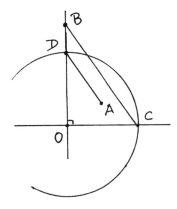

13.18 Given a right angle at $O$, a point $B$ on one arm, and a point $A$, construct with ruler and compass a circle with center $O$, meeting the arms of the right angle at $C$, $D$, such that $AD$ is parallel to $BC$ (par = 9 steps, not counting lines $AD$, $BC$).

13.19 Given segments of lengths $1, a, b$ in the plane, construct with ruler and compass a length $x$ satisfying $x^2 - ax - b = 0$. (If you use the quadratic formula, par = 21; using geometrical ideas from Exercise 13.18, par = 14.)

13.20 Prove Euclid's (XIII.5), which says that the triangle formed of the sides of a pentagon, a hexagon, and a decagon inscribed in the same circle is a right triangle. Conclude that the segment $AH$ in the construction of Problem 4.3 is equal to the side of the pentagon.

13.21 Verify the following construction of a regular pentagon in 13 steps, due to H. Lenstra. The circle and its center $O$ are given.

1. line $OA$.
2. circle $AO$, get $B, C$.
3. line $BC$, get $D$.
4. circle $DO$, get $E$.
5. line $AE$, get $F$.
6. circle $DF$, get $G, H!$
7. circle $FG$, get $I, K$.
8. circle $FH$, get $L, M$.
9–13. ines $FI, IL, LM, MK, KF$.

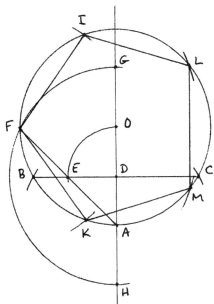

13.22 Find the field extension of $\mathbb{Q}$ obtained by adjoining the coordinates of the point $P = (a, b)$, the center of the inscribed circle of the triangle with vertices $A = (0, 0)$, $B = (-1, 2)$, and $C = (2, 3)$. Answer: $\mathbb{Q}(\sqrt{2}, \sqrt{65})$.

# 14   Abstract Fields and Incidence

In this section we start with the algebraic structure of a field, and based on this field we will obtain a geometry. Thus, using a field, we obtain a *model* of the abstract geometry determined by Hilbert's axioms. Different fields will give different geometries, so we will obtain many different models and many different Euclidean geometries. We will investigate what properties of the field are needed to make each of Hilbert's axioms hold. This will help demonstrate the independence of the axioms.

Hilbert's axiom system is based on the undefined notions of *point*, *line*, *betweenness*, and *congruence* for line segments and for angles. These undefined notions are limited only by having to satisfy all the axioms.

To make a model of the geometry within another mathematical framework, in this case algebra, we must say what the *interpretation* of the undefined notions is to be in our model, and then we must *prove* that the axioms hold in this interpretation.

We start then, with a field, and to fix the ideas we recall the definition of a field.

**Definition**

A *field* is a set $F$, together with two operations, $+$, $\cdot$, i.e., for each $a, b \in F$ there are given $a + b \in F$ and $a \cdot b \in F$, subject to the following conditions:

(1) The set $F$, together with the operation $+$, forms an abelian group, namely,

    (i) $(a + b) + c = a + (b + c)$ for any $a, b, c \in F$,
    (ii) $a + b = b + a$ for any $a, b \in F$,
    (iii) there is an element $0 \in F$ such that $a + 0 = a$ for all $a \in F$,
    (iv) for each $a \in F$ there is an element $-a \in F$ such that $a + (-a) = 0$.

(2) The set $F^* = F - \{0\}$, together with the operation $\cdot$ forms an abelian group, namely,

    (i) $(ab)c = a(bc)$ for all $a, b, c \in F^*$,
    (ii) $ab = ba$ for all $a, b \in F^*$,
    (iii) there is an element $1 \in F^*$ such that $a \cdot 1 = a$ for all $a \in F^*$,
    (iv) for all $a \in F^*$, there is an $a^{-1} \in F^*$ such that $a \cdot a^{-1} = 1$.

(3) The operations $+$ and $\cdot$ are related by the distributive law

$$a(b + c) = ab + ac.$$

Note in particular that in our definition of a field $0 \neq 1$, and multiplication is always commutative. We leave to the reader to verify other elementary properties of a field, such as $0 \cdot a = 0$ for all $a \in F$. The *characteristic* of the field $F$ is the least positive integer $p$ for which $1 + 1 + \cdots + 1$ ($p$ times) is equal to 0, or zero if there is no such integer.

Our first step in making a geometry is to say what we mean by points and lines. Of course, we take our cue from the "standard" model of Euclidean geometry, the real Cartesian plane, given by ordered pairs of real numbers. This is the geometry we might call "high-school geometry," where the axiomatic and the analytic approaches are not clearly distinguished, and we assume that everything is over the real numbers. In that model we suppose that everyone already "knows" what points, lines, angles, betweenness, and congruence mean.

But now, since we are starting with an arbitrary field $F$, which may not be the real numbers, we need to make our definitions precise.

### Definition

The plane $\Pi$ (or $\Pi_F$ if we want to indicate the field), called the *Cartesian plane over the field $F$*, is the set $F^2$ of ordered pairs of elements of the field $F$, which we call the *points* of $\Pi$. A *line* is a subset defined by a linear equation

$$ax + by + c = 0$$

for some $a, b, c \in F$, with $a, b$ not both zero.

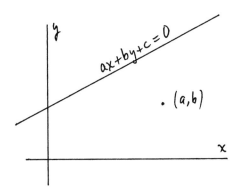

Any line can also be written in either the form $x = c$, in which case we call it *vertical*, or the form $y = mx + b$. In the latter case we say that the line has *slope $m$*, and for the line $x = c$, we say it has slope $\infty$. Here $\infty$ is just a symbol (it is not an element of the field $F$).

### Example 14.0.1

Let $F$ be the field of two elements $F = \{0, 1\}$ with addition and multiplication $(\bmod\, 2)$. Then the plane $\Pi$ over $F$ has exactly four points and six lines, shown schematically in the diagram. Note in particular that the two "diagonal" lines do not meet in this geometry.

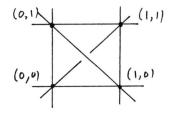

### Proposition 14.1

*If $F$ is any field, the Cartesian plane $\Pi_F$ satisfies Hilbert's incidence axioms* (I1), (I2), (I3), *and the parallel axiom* (P).

*Proof* (I1) says that any two points lie on a line. Since we can perform rational operations $+, -, \cdot, \div$ in the field $F$, the usual "two-point formula" of analytic geometry shows that we can find a linear equation, with coefficients in the field $F$, that determines a line containing the two given points.

(I2) says that every line has at least two points. Since any field $F$ has at least the two distinct elements $0, 1$, by putting $x = 0, 1$ if the line has the form $y = mx + b$, or by putting $y = 0, 1$, if the line is $x = c$, we obtain two points on any line.

(I3) says that there exist three noncollinear points. Indeed, we can always take $(0, 0)$, $(0, 1)$, $(1, 0)$, and we can see easily that these do not lie on any line.

(P) says that there is at most one parallel to a given line $l$ through a given point $P$. In fact, the stronger statement holds that there is *exactly* one line parallel to $l$ through $P$, so that $\Pi$ will be an affine plane, in the terminology of Exercise 6.5. Recall that parallel means that two lines do not meet unless they are equal. In the plane $\Pi_F$, we see immediately that two lines are parallel if and only if they have the same slope. So given a line $l$, let its slope be $m$. Then the familiar "point–slope" formula of analytic geometry shows that there is a unique line of slope $m$ passing through the point $P$. This will be the parallel to $l$.

Before introducing the further notions of betweenness and congruence into our Cartesian plane over a field $F$, there are already some interesting connections between algebraic properties of the field $F$ and incidence properties in the plane $\Pi_F$. To investigate these, it is useful to be able to change coordinates.

**Proposition 14.2**
*In the Cartesian plane* $\Pi$ *over a field* $F$, *it is possible to make a linear change of variables*

$$\begin{cases} x' = ax + by + c, \\ y' = dx + ey + f, \end{cases}$$

*such that the new coordinate axes are any two given intersecting lines, and the new unit points are any given points* $P, Q$ *on them not equal to their intersection point* $E$.

*Proof*  Since a composition of linear changes of variables is again one, we can proceed one step at a time. First, a change of the form

$$\begin{cases} x' = x - a, \\ y' = y - b, \end{cases}$$

will move the origin $(0, 0)$ to the point $E = (a, b)$.

Next, a transformation of the form

$$\begin{cases} x' = ax, \\ y' = by, \end{cases}$$

will move the unit points to any other points on the same axes.

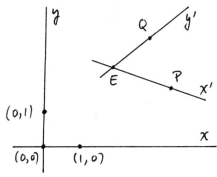

Then, a change of the form

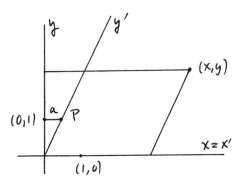

$$\begin{cases} x' = x - ay, \\ y' = y, \end{cases}$$

will keep the $x$-axis fixed, but replace the $y$-axis by another line through the origin. The $x$-axis may be moved by another such transformation, interchanging the roles of $x$ and $y$.

Combining all these gives a transformation that moves the original axes and unit points to any other desired axes and unit points.

**Remark**
Since the change of coordinates is linear, lines in the new coordinate system are still given by linear equations, so it is equivalent to describe the geometry of the plane $\Pi_F$ using either the old or the new coordinates.

Now we give some applications.

**Proposition 14.3**
*There exists a configuration in the plane $\Pi_F$ of four points $A, B, C, D$ such that $AB\|CD$, $AC\|BD$, and $AD\|BC$ if and only if the characteristic of $F$ is 2.*

*Proof* We have already seen the existence (14.0.1), since any field $F$ of characteristic 2 contains the subfield $\{0, 1\}$ of two elements with addition and multiplication (mod 2).

For the converse, suppose that such a configuration exists in $\Pi_F$. Then make a linear change of coordinates such that $C$ becomes the new origin, and $A, D$ are the unit points. Then $B$ will be the point $(1, 1)$; $BC$ will be the line $x = y$, and $AD$ will be the line $x + y = 1$. In this configuration, $AD\|BC$, so the equations $x = y$ and $x + y = 1$ must have no common solution. Solving, we obtain $2x = 1$, which has a solution in $F$ as long as $2 \neq 0$. We conclude that this configuration exists only if $2 = 0$, i.e., the charactistic of $F$ is 2.

**Proposition 14.4** (Pappus's theorem)
*In the Cartesian plane over a field $F$, suppose we are given lines $l, m$ and points $A, B, C \in l$ and $A', B', C' \in m$ such that $AC'\|A'C$ and $BC'\|B'C$. Then also $AB'\|A'B$.*

*Proof*  Suppose that $l, m$ meet at a point
$O$ (we leave the case $l\|m$ as Exercise
14.1). Choose coordinates such that $O$ is
the origin, and $A, B'$ are the unit points.
Let $C$ be the point $(0, a)$ and let $C'$ be
the point $(b, 0)$. Then writing the equations of the lines involved, we find that
$B = (0, ab)$ and $A' = (ab, 0)$. Thus the
line $BA'$ has slope $-1$, hence is parallel
to $AB'$.

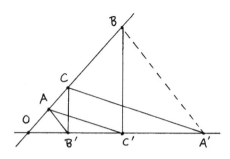

### Remark 14.4.1
It is possible to define a Cartesian plane over a *skew field* $F$ (which is an algebraic structure the same as a field, except that the multiplication need not
be commutative). Then Hilbert (1971) has shown that the skew field $F$ is
commutative if and only if Pappus's theorem holds in the associated plane
$\Pi_F$.

### Example 14.4.2
In the Cartesian plane over the field $F$,
assuming characteristic 0, there is a
configuration such as the one shown
(where all lines that appear parallel are
assumed to be parallel, namely $DE\|BC$,
$DF\|AC$, $EF\|AB$, $GH\|BC$, and $BH\|GE$) if
and only if $\sqrt{2} \in F$.

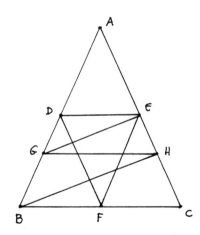

Indeed, to analyze this situation,
take $B$ to be the origin, $BC$ and $BA$ the
axes, and $D, F$ the unit points. Then $A =
(0, 2)$, $E = (1, 1)$, $C = (2, 0)$. Let $G$ have
coordinates $(0, a)$. Then $H = (2 - a, a)$.
The line $BH$ will have slope $a/(2 - a)$,
and the line $GE$ will have slope $1 - a$.
The parallelism $BH\|GE$ then requires
$a/(2 - a) = 1 - a$, or, equivalently, $a^2 -
4a + 2 = 0$. Solving with the quadratic
formula gives $a = 2 \pm \sqrt{2}$.

For this configuration to exist, it is
necessary and sufficient that $a \in F$, and
this is clearly equivalent to $\sqrt{2} \in F$, as
required.

# Exercises

14.1 Show that Pappus's theorem (Proposition 14.4) still holds in $\Pi_F$ in the case that $l\|m$.

14.2 Show that Desargues's theorem holds in the Cartesian plane over a field $F$: Given a configuration as shown, with $AC\|A'C'$ and $AB\|A'B'$, prove that $BC\|B'C'$.

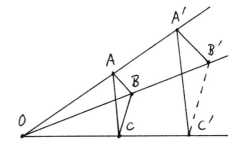

14.3 We define a *skew field* (also called a *division ring*) to be the same as a field, but without assuming property 2(ii), that multiplication is commutative.

(a) Using the same definition of points and lines, show that the Cartesian plane over a skew field $F$ still satisfies the incidence axioms (I1)–(I3) and (P), as in Proposition 14.1.

(b) Show that a skew field is commutative (i.e., is a field) if and only if Pappus's theorem (Proposition 14.4) holds in the Cartesian plane over $F$.

For each of the following problems, assume that you are working in the Cartesian plane $\Pi$ over a field $F$ of characteristic 0. Give necessary and sufficient conditions on the field $F$ for the given configuration to exist. Assume that all lines that appear to be parallel are parallel, and apparent right angles are right angles.

14.4 Ans: $\sqrt{3} \in F$.                    14.5 Ans: $\sqrt{13} \in F$.

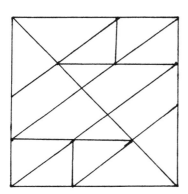 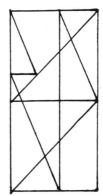

14.6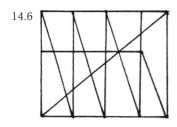

14.7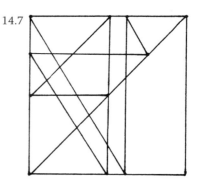

14.8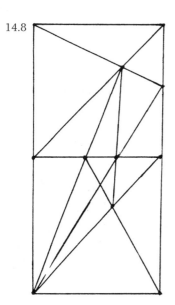

14.9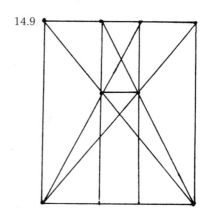

In each of the following four problems, suppose that you are given the triangle $ABC$. Make a ruler and compass construction of the diagram shown. In the first three, $D, E, F$ are the midpoints of the sides. In the last, they are one-third of the way along each side. (Par = 20 to 25 steps each.)

14.10

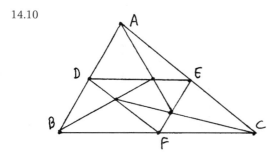

14.11

14.12

14.13

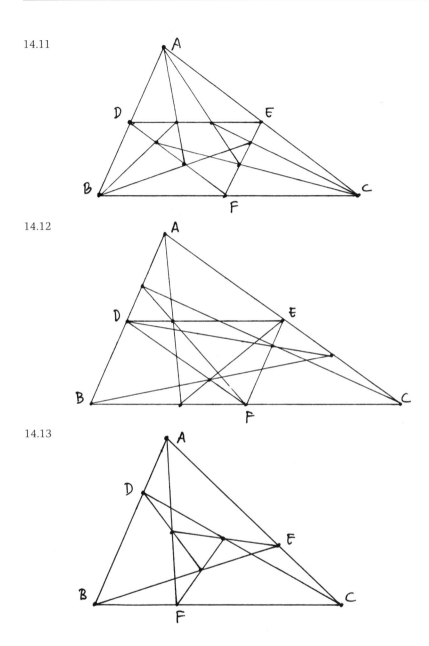

# 15   Ordered Fields and Betweenness

The next undefined notion we need to interpret in the Cartesian plane over a field is *betweenness*. It turns out that this is not possible over an arbitrary field.

We will have to impose some additional structure on the field to make a definition of betweenness possible. To see why this is so, suppose we had a notion of betweenness in our geometry. Then the $x$-axis (whose points are in one-to-one correspondence with the elements of our field $F$) could be divided into subsets consisting of the "positive $x$-axis" meaning all points on the same side of 0 as 1, the origin 0, and the "negative $x$-axis" consisting of all points on the other side of 0 from 1. In this way we can define a notion of "positive" elements of the field $F$, analogous to the usual notion of positive real numbers.

This leads to the concept of an ordered field.

### Definition
An *ordered field* is a field $F$, together with a subset $P$, whose elements are called *positive*, satisfying:

(i)  If $a, b \in P$, then $a + b \in P$ and $ab \in P$.
(ii) For any $a \in F$, one and only one of the following holds: $a \in P$; $a = 0$; $-a \in P$.

Here are a few elementary properties of an ordered field.

### Proposition 15.1
*Let $F, P$ be an ordered field. Then*:

(a) $1 \in P$, *i.e.*, 1 *is a positive element.*
(b) $F$ *has characteristic* 0.
(c) *The smallest subfield of $F$ containing 1 is isomorphic to the rational numbers* $\mathbb{Q}$.
(d) *For any $a \neq 0 \in F$, $a^2 \in P$.*

*Proof*   (a) In any field, $1 \neq 0$, so either $1 \in P$ or $-1 \in P$. If $1 \in P$ we are done. If $-1 \in P$, then by (i), also $(-1) \cdot (-1) = 1 \in P$, which contradicts (ii). Hence $1 \in P$.

(b) Since $1 \in P$, $1 + 1 + 1 + \cdots + 1$ any number of times is also in $P$. In particular, such a sum is never 0, so $F$ has characteristic 0.

(c) The natural map of the positive integers $\mathbb{N}$ to $F$ given by $n$ goes to $1 + 1 + \cdots + 1$ ($n$ times) is injective, by (b), and extends to an injective map of $\mathbb{Q}$ to $F$ whose image is (1) isomorphic to $\mathbb{Q}$ and (2) the smallest subfield of $F$ containing 1. Whenever no confusion can arise, we will identify $\mathbb{Q}$ with its image in $F$. So for example, if $n \in \mathbb{Z}$, then $n$ will also denote the corresponding element of $F$.

(d) If $a \neq 0$, then either $a \in P$ or $-a \in P$. If $a \in P$, then $a^2 \in P$ by (i). If $-a \in P$, then $(-a)(-a) = a^2 \in P$.

### Proposition 15.2
*In an ordered field $F, P$, we define $a > b$ if $a - b \in P$, and $a < b$ if $b - a \in P$. This notion of inequality satisfies the usual properties, namely*:

(i)  *If $a > b$ and $c \in F$, then $a + c > b + c$.*
(ii) *If $a > b$ and $b > c$, then $a > c$.*

(iii)  *If $a > b$ and $c > 0$, then $ac > bc$.*
(iv)  *Given $a, b \in F$, one and only one of the following holds: $a > b$; $a = b$; $a < b$.*

**Examples 15.2.1**
The rational numbers $\mathbb{Q}$ form an ordered field, where we take for $P$ the positive rational numbers, in the usual sense.

**15.2.2** The field of real numbers $\mathbb{R}$ is an ordered field with the usual notion of positive elements.

**15.2.3** The field of complex numbers $\mathbb{C}$ cannot be an ordered field (i.e., there is no subset $P$ of $\mathbb{C}$ satisfying the definition) because $i^2 = -1 < 0$, which contradicts (15.1*d*).

**15.2.4** Since an ordering on a field is extra structure, in general there may be more than one way to make a given abstract field into an ordered field. For example, let $F = \mathbb{Q}(\sqrt{2})$. Then $F$ is a subfield of $\mathbb{R}$, so we can make it into an ordered field by taking $P$ to be the subset of elements of $F$ that are positive in $\mathbb{R}$. But there is another embedding $\varphi : F \to \mathbb{R}$ given by $\varphi(a + b\sqrt{2}) = a - b\sqrt{2}$ for all $a, b \in \mathbb{Q}$, and we can put another ordering on $F$ by taking $P$ to be the set of elements $x \in F$ for which $\varphi(x) > 0$ in $\mathbb{R}$.

**Proposition 15.3**
*If $F$ is a field, and if there is a notion of betweenness in the Cartesian plane $\Pi_F$ satisfying Hilbert's axioms* (B1)–(B4), *then $F$ must be an ordered field. Conversely, if $F, P$ is an ordered field, we can define betweenness in $\Pi_F$ so as to satisfy* (B1)–(B4).

*Proof*  First suppose that $F$ is a field and that there is a notion of betweenness in the plane $\Pi_F$ satisfying (B1)–(B4). We define the subset $P \subseteq F$ to consist of all $a \in F$, $a \neq 0$, such that the point $(a, 0)$ of the *x*-axis is on the same side of 0 as 1. Since addition in the field corresponds to laying out line segments consecutively on the *x*-axis, one can show easily that $a, b \in P \Rightarrow a + b \in P$.

For multiplication, given $a, b \in P$, put $a$ on the *x*-axis, put $1, b$ on the *y*-axis, draw the line from $(0, 1)$ to $(a, 0)$, and draw the line parallel to this one through $(0, b)$. It will meet the *x*-axis in the point $(ab, 0)$. Now clearly, $1, a, b \in P \Rightarrow ab \in P$ (we leave to the reader to see exactly how this follows from (B1)–(B4)!), so $P$ satisfies the first property of the definition of an ordered field. By construction, $F$ is the disjoint union of $P \cup \{0\} \cup -P$, so that $F, P$ is an ordered field.

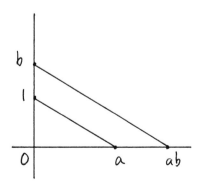

Now suppose conversely that $F, P$ is a given ordered field. We define *betweenness* for points on a line as follows: Let $A = (a_1, a_2)$, $B = (b_1, b_2)$, $C = (c_1, c_2)$ be three distinct points on a line $y = mx + b$. We say that $B$ is *between* $A$ and $C$ $(A * B * C)$ if

$$\text{either } a_1 < b_1 < c_1 \quad \text{or} \quad a_1 > b_1 > c_1.$$

If the line is vertical, we use instead the second coordinates in the same way.

We must verify the axioms (B1)–(B4).

(B1) is obvious from our definition.

(B2) follows from the corresponding fact, true in any ordered field, that given $b > d \in F$, there exist $a, c, e \in F$ such that $a < b < c < d < e$. Indeed, we can always take, for example, $a = b - 1$, $c = \frac{1}{2}(b + d)$, and $c = d + 1$. Note that since $F$ has characteristic 0, by (15.1), $\frac{1}{2} \in F$.

(B3) follows from the fact that in an ordered field $F$, if $a, b, c$ are three distinct elements, then one and only one of the following six possibilities can occur:

$$a < b < c;$$
$$a < c < b;$$
$$b < a < c;$$
$$b < c < a;$$
$$c < a < b;$$
$$c < b < a.$$

(B4). Suppose we are given a triangle $ABC$ and a line $l$ that meets the side $AB$. Assuming $A, B, C \notin l$, we must show that $l$ also meets either $AC$ or $BC$, but not both.

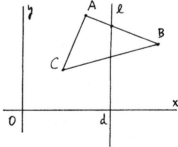

First suppose that the line $l$ is vertical, with equation, say, $x = d$. Let $a, b, c$ be the $x$-coordinates of $A, B, C$. By hypothesis, either $a < d < b$ or $b < d > a$. By symmetry, let us assume $a < d < b$. Then it is clear if $c < d$ (as in the picture), then $l$ will meet $BC$ but not $AC$. If $c > d$, then $l$ will meet $AC$ and not $BC$, as required.

If $l$ is not vertical, we make a change of coordinates (14.2) such that $l$ becomes vertical. Since linear changes of variables either preserve or reverse inequalities, this does not affect the notion of betweenness, and so we are reduced to the previous case.

To complete this section, we will discuss Archimedes' axiom (A) and Dedekind's axiom (D)—cf. Section 12.

## Proposition 15.4

*Let $F, P$ be an ordered field. Then the Cartesian plane $\Pi_F$ will satisfy (A) or (D) if and only if the field $F$ satisfies the corresponding property for a field, namely:*

(A') (Archimedes' axiom for a field). *For any $a > 0$ in F, there is an integer n such that $n > a$.*

(D') (Dedekind's axiom for a field). *Suppose we can write the field F as the disjoint union of two nonempty subsets $F = S \cup T$, and assume that for all $a \in S$ and all $b \in T$ we have $a < b$. Then there exists a unique element $c \in F$ such that for all $a \in S$ and all $b \in T$ we have $a \leq c \leq b$.*

*Proof*  For (A), we can choose coordinates such that the first segment $AB$ is a unit segment. If $C$ and $D$ on the same line correspond to elements $c < d \in F$, then $n$ copies of $AB$ will exceed $CD$ if and only if $n > d - c$.

For (D), choose coordinates such that the line in question is the *x*-axis, and identify its points with elements of $F$. Then the statements are the same.

## Proposition 15.5

*Let F be an ordered field satisfying Archimedes' axiom (A'). Then F is isomorphic, with its ordering, to a subfield of $\mathbb{R}$. Furthermore, in this case, F satisfies Dedekind's axiom (D') if and only if this subfield is equal to $\mathbb{R}$.*

*Proof*  We saw earlier (15.1) that $F$ contains a subfield $F_0$ isomorphic to $\mathbb{Q}$. This gives us a unique isomorphism $\varphi_0 : F_0 \to \mathbb{Q} \subseteq \mathbb{R}$. We will extend $\varphi_0$ to an isomorphism of $F$ into $\mathbb{R}$. Let $\alpha \in F$. Because of Archimedes' axiom, there are integers both smaller and bigger than $\alpha$. So let $a_0$ be the unique integer $n$ such that $n \leq \alpha < n + 1$. Next define $a_1 \in \frac{1}{10}\mathbb{Z}$ to be the unique one-tenth integer such that $a_1 \leq \alpha < a_1 + 1/10$. Similarly, define $a_2 \in (1/100)\mathbb{Z}$ such that $a_2 \leq \alpha < a_2 + 1/100$. Continuing in this way we obtain a sequence $a_0 \leq a_1 \leq a_2 \leq \cdots$ of rational numbers with the property that for each $n$, $a_n \leq \alpha < a_n + 10^{-n}$. In the field of real numbers $\mathbb{R}$, these converge to a certain real number, which we call $\varphi(\alpha)$. This defines a map $\varphi : F \to \mathbb{R}$. It is easy to verify that $\varphi(\alpha + \beta) = \varphi(\alpha) + \varphi(\beta)$ and $\varphi(\alpha\beta) = \varphi(\alpha) \cdot \varphi(\beta)$. So $\varphi$ is a homomorphism of fields, which is necessarily an isomorphism onto its image. One checks also that $\alpha < \beta \Rightarrow \varphi(\alpha) < \varphi(\beta)$, so it is an order isomorphism of $F$ onto $\varphi(F) \subseteq \mathbb{R}$.

Now, condition (D') on $F$ is equivalent to (D') on $\varphi(F)$, since the fields are order-isomorphic. Each real number $r \in \mathbb{R}$ is characterized by the sets $\Sigma_1 = \{a \in \mathbb{R} \mid a \leq r\}$ and $\Sigma_2 = \{a \in \mathbb{R} \mid a > r\}$, so clearly (D') holds in $\varphi(F)$ if and only if $\varphi(F) = \mathbb{R}$.

## Remark 15.5.1

As a converse to this result, note that any subfield $F$ of $\mathbb{R}$ becomes an ordered field if we take for $P \subseteq F$ those elements of $F$ that are positive (in the usual sense) in $\mathbb{R}$. Thus the study of Archimedean ordered fields is equivalent to the study of subfields of $\mathbb{R}$.

See Section 18 for some examples of non-Archimedean ordered fields.

## Exercises

15.1 If $a > 0$ in an ordered field $F$, show that $a^{-1} > 0$ also.

15.2 Let $F$ be an ordered field, and let $a > 0$. Show that if $a$ has a square root in $F$, i.e., an element $b \in F$ such that $b^2 = a$, then $a$ has exactly two square roots in $F$, one of which is positive and the other negative. We use the notation $\sqrt{a}$ to denote the positive square root.

15.3 Let $F$ be an ordered field, let $d > 0$, and suppose that $d$ does not have a square root in $F$. Let $F(\sqrt{d})$ denote the set of all $a + b\sqrt{d}$, with $a, b \in F$, where $\sqrt{d}$ is a square root in some extension field of $F$.

   (a) Show that $F(\sqrt{d})$ is a field.

   (b) Show how to define an ordering on $F(\sqrt{d})$, with $\sqrt{d} > 0$, such that it becomes an ordered field.

15.4 In an ordered field $F$, show that Dedekind's axiom (D$'$) implies Archimedes' axiom (A$'$). *Hint*: If $F$ did not satisfy (A$'$), let $S = \{\alpha \in F \,|\, \exists n \in \mathbb{Z}, \text{ with } \alpha < n\}$, and let $T = F - S$. Then apply (D$'$).

15.5 In the proof of Proposition 15.5, verify that $\varphi(\alpha\beta) = \varphi(\alpha) \cdot \varphi(\beta)$.

15.6 If $F$ is a skew field (Exercise 14.3), together with an ordering defined as in this section that satisfies Archimedes' axiom (A$'$), then in fact $F$ is a field. *Hint*: Show that the proof of Proposition 15.5 still works.

## 16   Congruence of Segments and Angles

Next, we need to define the notion of congruence for line segments and for angles. We assume from now on that we are starting from an *ordered* field $F, P$, so that we have betweenness as studied above. Then we can define the *line segment AB* to be the set of all points on the line $AB$ that are between $A$ and $B$, plus the endpoints $A, B$. We would like to define congruence of line segments using the usual *Euclidean distance* function (motivated by the theorem of Pythagoras (I.47)) for two points $A = (a_1, a_2)$, $B = (b_1, b_2)$, namely,

$$\text{dist}(A, B) = \sqrt{(a_1 - b_1)^2 + (a_2 - b_2)^2}.$$

However, our field $F$ may not have square roots in it, so we will use instead the "distance-squared" function

$$\text{dist}^2(A, B) = (a_1 - b_1)^2 + (a_2 - b_2)^2.$$

This will give the same notion of congruence.

**Definition**
Two line segments $AB$ and $CD$ in the Cartesian plane over an ordered field $F$ are *congruent* if

$$\text{dist}^2(A, B) = \text{dist}^2(C, D).$$

Since congruence is defined using the function $\text{dist}^2$ from line segments to the field, the axiom (C2), transitivity of congruence, will be obvious. Notice that because of the ordering on $F$, if $A, B$ are distinct points, then $\text{dist}^2(A, B) > 0$.

Next we will define congruence for angles, by defining a function $\tan \alpha \in F$ for any angle $\alpha$. This is motivated by the usual tangent function of trigonometry, but since we are working over an abstract field, you should not assume any properties of this function until they have been proved.

Recall that a *ray* is a subset of a line consisting of a point plus all the points of the line on one side of that point. An *angle* is the union of two rays emanating from the same point and not lying on the same line. The *interior* of the angle $\alpha$ consists of all points of the plane that are on the same side of $l$ as $r'$ and on the same side of $l'$ as $r$.

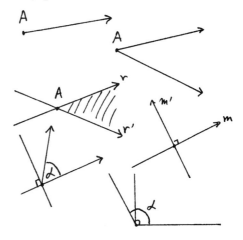

We say that an angle is a *right angle* if the slopes of the lines its rays lie on satisfy $mm' = -1$. Then we say that an angle is *acute* if it is contained in the interior of a right angle; it is *obtuse* if it contains a right angle in its interior.

**Definition**
If $\alpha$ is an angle formed by two rays $r, r'$ lying on lines of slopes $m, m'$, we define the *tangent* of $\alpha$ to be

$$\tan \alpha = \pm \left| \frac{m' - m}{1 + mm'} \right|,$$

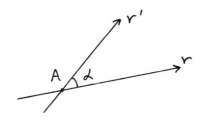

where we take $+$ if the angle is acute and $-$ if the angle is obtuse.

The awkwardness of this definition is due to the fact that the slopes depend only on the lines and do not distinguish the rays on those lines. So any formula using slopes cannot distinguish an angle from its supplement. We cannot use the usual definition of tangent, as side opposite over side adjacent, because that needs square roots (Exercise 16.3).

Note also that by this definition, the tangent of a strictly acute or strictly obtuse angle is an element of the field $F$, while the tangent of a right angle we take to be the symbol $\infty$. In case one of the slopes $m$ or $m'$ in the definition is $\infty$, we can still make sense of the formula by using rules (Exercise 16.2) such as

$$\frac{\infty - m}{1 + m \cdot \infty} = \frac{1}{m}.$$

### Definition

Two angles in the Cartesian plane over an ordered field $F$ are *congruent* if they have the same tangent, considered as an element of the set $F \cup \{\infty\}$.

Because congruence is defined by a function with values in $F \cup \{\infty\}$, axiom (C5), transitivity of congruence, becomes obvious.

### Proposition 16.1

*Let $F$ be an ordered field, and let $\Pi_F$ be the associated Cartesian plane. Then $\Pi_F$ satisfies axioms (C2)–(C5). Furthermore, (C1) holds if and only if $F$ satisfies the condition*

*(\*) For any element $a \in F$, the element $1 + a^2$ has a square root in $F$ (in which case we say that the field $F$ is Pythagorean).*

*Proof* (C2) is transitivity of congruence of segments, which follows immediately from our definition of congruence using the dist$^2$ function.

(C3) is left as an exercise (Exercise 16.1).

(C4) is the axiom about laying off angles. So suppose we are given an angle $\alpha$ and a ray emanating from a point $A$ with slope $m$. We must find a line passing through $A$ with slope $m'$ such that

$$\tan \alpha = \pm \left| \frac{m' - m}{1 + mm'} \right|,$$

where the sign is adjusted according to whether $\alpha$ is acute or obtuse. This gives equations that are linear in $m'$, and so can be solved in $F$. We obtain

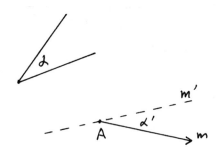

$$m' = \frac{m \pm \tan\alpha}{1 \mp m\tan\alpha}.$$

The two solutions give angles on either side of the given line through $A$, so that we can construct the new angle $\alpha'$ on the desired side of the line.

(C5) is the transitivity of congruence of angles, which is immediate from our definition of congruence using the tangent function.

Now let us consider the axiom (C1) about laying off line segments. This does not hold over an arbitrary field. For example, let $\mathbb{Q}$ be the field of rational numbers. Then the segment from $(0,0)$ to $(1,1)$ cannot be laid off on the $x$-axis, because its length, $\sqrt{2}$, is not in the field.

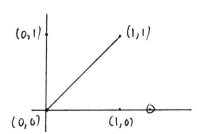

Over an arbitrary field $F$, if $a \in F$ is any element, let us consider the segment from $(0,0)$ to $(a,1)$. There will be a segment congruent to this one, laid off on the $x$-axis starting from $0$, only if there is an element $b \in F$ such that

$$\text{dist}^2((0,0),(a,1)) = \text{dist}^2((0,0),(b,0)).$$

This says that

$$1 + a^2 = b^2.$$

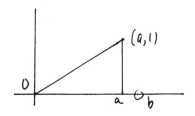

Thus we need $b \in F$ that is a square root of $1 + a^2$. In other words, if (C1) holds in $\text{II}_F$, we must have the condition $(*)$ on the field $F$.

Conversely, suppose that $F$ satisfies $(*)$, namely, for any $c \in F$, we have $\sqrt{1+c^2} \in F$. Then for any $a,b \in F$, with $a \neq 0$, we can write

$$a^2 + b^2 = a^2 \left(1 + \left(\frac{b}{a}\right)^2\right).$$

Now letting $c = b/a$ we see that

$$\sqrt{a^2 + b^2} = |a| \cdot \sqrt{1+c^2}$$

is also in $F$. From this is follows that for any two points $A, B \in \text{II}_F$, the *distance* between $A, B$ is also in $F$, so we have the distance function

$$\text{dist}(A,B) = \sqrt{(a_1 - b_1)^2 + (a_2 - b_2)^2} \in F.$$

Now suppose that we are given a line $y = mx + b$ and a point $A$ on the line, and suppose we wish to lay off a segment of length $d$. We can write $A = (a, ma + b)$, and we are looking for a point $C = (c, mc + b)$ on the same line such that

$$\text{dist}(A, C) = d.$$

This says that

$$\sqrt{(a - c)^2 + (ma + b - (mc + b))^2} = d,$$

which becomes

$$|a - c| \cdot \sqrt{1 + m^2} = d.$$

Since $F$ satisfies $(*)$, the quantity $\sqrt{1 + m^2}$ is in $F$, so we can solve this equation for $c$. Note that there will be two solutions, corresponding to the two directions from $A$ along the line $l$.

### Remark
We defer consideration of (C6), the (SAS) axiom, to the next section, where we discuss rigid motions and Euclid's method of superposition.

To complete this section, we discuss the intersections of lines and circles. Recall from Section 11 the circle–circle intersection property, which we called axiom (E), and the line–circle intersection property (LCI), which was proved in (11.6) as a consequence of (E).

### Proposition 16.2
*Let* $\Pi$ *be the Cartesian plane over an ordered field* $F$. *Then the following conditions are equivalent*:

(i)  $\Pi$ *satisfies the circle–circle intersection property* (E).
(ii) $\Pi$ *satisfies the line–circle intersection property* (LCI).
(iii) *the field* $F$ *satisfies* $(**)$: *For any* $a \in F$, $a > 0$, *there is a square root of* $a$ *in* $F$ *(in which case we say that the field* $F$ *is* Euclidean).

*Proof* (i) $\Rightarrow$ (ii). Let $f = 0$ be the equation of a circle, and let $g = 0$ be the equation of a line. Then $f + g = 0$ is another circle, whose intersections with the first circle are the same as the intersections of the first circle with the line. Thus (E) implies (LCI).

(ii) $\Rightarrow$ (iii). Now we assume (LCI) and we must prove that $F$ has square roots of positive elements. Given an element $a \in F$, $a > 0$, consider the points $O = (0,0)$, $A = (a, 0)$, and $A' = (a + 1, 0)$. Let $\Gamma$ be the circle with center $(\frac{1}{2}(a + 1), 0)$ and radius $\frac{1}{2}(a + 1)$. Consider the vertical line $l$ through the point $A$. Clearly, $A$ is inside the circle, so by (LCI), the line $l$ must meet the circle $\Gamma$ in a point $B$. Solving the equations, we find that $B = (a, \sqrt{a})$, so $\sqrt{a} \in F$.

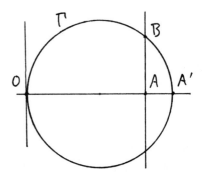

(iii) $\Rightarrow$ (i). This time we assume the existence of square roots of positive elements in $F$, and must prove (E). If $\Gamma$ and $\Gamma'$ are circles in $\Pi$, their equations can be written

$$(x - a)^2 + (y - b)^2 = r^2,$$

$$(x - c)^2 + (y - d)^2 = s^2,$$

where $(a, b)$ and $(c, d)$ are the centers of the two circles, and $r, s$ their radii (which are elements of $F$ because of our hypothesis $(**)$ on existence of square roots).

Normally, solving two quadratic equations simultaneously would lead to a fourth-degree equation, but in this case, the coefficients of $x^2$ and $y^2$ in both equations are 1. Thus we can subtract one equation from the other to get a linear equation. This can be solved simultaneously with one of the quadratic equations using only square roots, and so, using $(**)$, the intersection points of the circles have coordinates in $F$, so they exist in the plane $\Pi$. We leave to the reader the troublesome verification that if $\Gamma$ and $\Gamma'$ satisfy the hypothesis of (E), then the square roots we need will be square roots of *positive* elements of $F$, so will exist by $(**)$; cf. Exercise 16.6.

### Remark 16.2.1
This shows that (E) and (LCI) are equivalent in the Cartesian plane over any ordered field $F$; cf. (11.6.1).

### Proposition 16.3
*Let $\Omega$ be the set of all real numbers that can be expressed starting from the rational numbers and using a finite number of operations $+, -, \cdot, \div$, and $c \mapsto \sqrt{1 + c^2}$. (Note that for any $c \in \mathbb{R}, 1 + c^2 > 0$, so $\sqrt{1 + c^2} \in \mathbb{R}$.) Then $\Omega$ is an ordered Pythagorean field.*

*Proof*  To show that $\Omega$ is a field, let $a, b \in \Omega$. Then each of $a, b$ can be expressed in a finite number of steps using rational numbers and operations $+, -, \cdot, \div$, $c \mapsto \sqrt{1 + c^2}$. Hence the same is true of $a \pm b$, $a \cdot b$, and $a/b$, provided that $b \neq 0$. If $c$ is any element of $\Omega$, then $c$ can be expressed in a finite number of such steps, so $\sqrt{1 + c^2}$ can also, and so $\sqrt{1 + c^2} \in \Omega$. Hence $\Omega$ is Pythagorean. $\Omega$ is an ordered field, because it is a subfield of $\mathbb{R}$, so we can take $P$ to be those elements of $\Omega$ that are positive as real numbers.

### Remark 16.3.1
Clearly, $\Omega$ is the smallest Pythagorean ordered field. We call it *Hilbert's field*, since he studied it in his *Foundations of Geometry* (1971). It is also the smallest field over which all of Hilbert's axioms of betweenness and congruence will hold (17.3).

## Proposition 16.4

*Let K be the set of all real numbers that can be obtained from the rational numbers by a finite number of operations $+, -, \cdot, \div$, and $a > 0 \mapsto \sqrt{a}$. Then K is a Euclidean ordered field.*

*Proof*  Similar to the proof to (16.3). Note that we may take square roots only of positive elements. Since $K$ is also a subfield of $\mathbb{R}$, we get the ordering on $K$ from $\mathbb{R}$ as above.

## Remark 16.4.1

We call this the *constructible field* because it is the smallest field over which we can carry out ruler and compass constructions. Note also that $\Omega \subseteq K$, since $1 + c^2 > 0$ for any $c \in \Omega$. To show $\Omega \neq K$, see Exercise 16.10.

# Exercises

16.1  If $\Pi$ is the plane over an ordered field $F$, show that (C3), congruence of added line segments, holds. Do not assume that $F$ is Pythagorean.

16.2  Make up a set of rules for dealing with $\infty$ so that we can do arithmetic in $F \cup \{\infty\}$ and get the results we want with slopes and tangents of angles.

16.3  Let $ABC$ be a triangle with a right angle at $C$ in the Cartesian plane over an ordered field $F$ satisfying $(*)$. If $\alpha$ is the angle at $A$, show that

$$\tan \alpha = \frac{\text{dist}(B, C)}{\text{dist}(A, C)}.$$

16.4  If $F$ is a Pythagorean ordered field, prove the *triangle inequality* in the corresponding plane $\Pi$, namely, if $A, B, C$ are three points in $\Pi$, then

$$\text{dist}(A, C) \leq \text{dist}(A, B) + \text{dist}(B, C),$$

and equality holds if and only if $A, B, C$ are collinear and $B$ is between $A$ and $C$.

16.5  Using the definition of the tangent of an angle given in the text, verify that for any two acute angles $\alpha, \beta$,

$$\tan(\alpha + \beta) = \frac{\tan \alpha + \tan \beta}{1 - \tan \alpha \tan \beta}.$$

16.6  Let $\Pi$ be the plane over a Euclidean ordered field $F$. Verify that a circle $\Gamma$ meets a line $l$ in two points if and only if $l$ has a point inside $\Gamma$. *Hint*: Compute the shortest distance from the center $O$ of the circle to the line $l$. Show that this is less than the radius of the circle if and only if the square roots needed to solve the equations are square roots of *positive* elements.

16.7 Let $F$ be an ordered field (without assuming Pythagorean or Euclidean).

(a) Show that the associated plane $\Pi$ contains an equilateral triangle if and only if $\sqrt{3} \in F$.

(b) Show that there exists an equilateral (but not necessarily equiangular) pentagon in $\Pi$ if $F = \mathbb{Q}(\sqrt{3})$, $\mathbb{Q}(\sqrt{11})$, or $\mathbb{Q}(\sqrt{15})$, but not if $F = \mathbb{Q}$.

16.8 Let $F$ be an ordered field (without assuming Pythagorean or Euclidean). Let $A, B$ be points of the associated plane $\Pi$. Show that the circle $\Gamma$ with center $A$ and passing through $B$ has infinitely many points on it. *Hint:* First do the case of the circle of radius 1 and center $(0,0)$ over $\mathbb{Q}$.

16.9 (a) Show that $\cos 72°$ and $\sin 72°$ are in Hilbert's field $\Omega$.

(b) Prove that a regular pentagon inscribed in a unit circle exists in the Cartesian plane over the field $\Omega$.

16.10 *Totally real field extensions* (this exercise requires some knowledge of field theory). We consider *algebraic numbers*, which are complex numbers satisfying some polynomial equation with rational coefficients. We denote the set of algebraic numbers by $\bar{\mathbb{Q}}$. An algebraic number $a \in \bar{\mathbb{Q}}$ is *totally real* if it and all its conjugates are real. A subfield $F \subseteq \bar{\mathbb{Q}}$ is *totally real* if all its elements are totally real. We say that $a \in \bar{\mathbb{Q}}$ is *totally positive* if it and all its conjugates are real and positive. Show the following:

(a) If $F \subseteq \bar{\mathbb{Q}}$ is a totally real subfield, and if $a \in F$ is a totally positive element, then the extension field $F' = F(\sqrt{a})$ is totally real.

(b) If $a_1, \ldots, a_r$ are elements of a totally real field $F$, then $\Sigma a_i^2$ is a totally positive element of $F$.

(c) Hilbert's field $\Omega$ (Proposition 16.3) is a totally real field.

(d) The number $a = \sqrt{1 + \sqrt{2}}$ is in the constructible field $K$ (Proposition 16.4) but not in $\Omega$. Thus $\Omega < K$.

16.11 Use ideas from (Exercise 16.10) to give an example of three line segments in the Cartesian plane over $\Omega$, any two exceeding the third, but such that the triangle with sides equal to those segments does not exist. Thus (I.22) fails in this plane.

16.12 The converse of Exercise 16.10b is a theorem of Emil Artin: If $b$ is a totally positive element of a subfield $F \subseteq \bar{\mathbb{Q}}$, then there exist elements $a_1, \ldots, a_n \in F$ such that $b = \Sigma a_i^2$. Fill in the details of the following outline of a proof of this theorem.

(a) Replacing $F$ by $\mathbb{Q}(b)$, we may assume that $F$ is a finite totally real extension of $\mathbb{Q}$.

(b) Let

$$S = \{\Sigma a_i^2 \mid a_i \in F, \text{ not all zero}\}.$$

Show that the set $S$ is closed under addition, multiplication, and multiplicative inverses, and that $0 \notin S$. *Hint:* Write $S^{-1} = (S^{-1})^2 S$.

(c) Now let $b$ be a totally positive element of $F$, and suppose that $b \notin S$. We will show that this leads to a contradiction. Let $S' = S - bS$, and show that $S'$ is closed under $+, \cdot$, inverses, and that $0 \notin S'$.

(d) Let $\mathscr{P}$ be the set of all subsets $P \subseteq F$ such that $0 \notin P$, $S' \subseteq P$, and $P$ is closed under $+, \cdot$, inverses. Use Zorn's lemma to show that $\mathscr{P}$ has a maximal element.

(e) If $P \in \mathscr{P}$ is a maximal element, show that $F, P$ is an ordered field. You only have to check the trichotomy: If $a \neq 0$, and $-a \notin P$, consider $P' = P + aP$ and use maximality to show $P' = P$, so $a \in P$.

(f) Use the fact that $F$ is algebraic over $\mathbb{Q}$ to show that the ordering is Archimedean.

(g) Now use Proposition 15.5 to get an embedding $\varphi : F \hookrightarrow \mathbb{R}$ with $\varphi(b) < 0$, yielding a contradiction.

16.13 Verify the result of Exercise 16.12 directly for the field $\mathbb{Q}(\sqrt{2})$, without using its proof.

16.14 Using Exercise 16.12, show that Hilbert's field $\Omega$ is equal to the set of totally real elements in the constructible field $K$.


# 17   Rigid Motions and SAS

Our first goal in this section is to show that Hilbert's axiom (C6), the "side–angle–side" criterion for congruence of triangles, holds in the geometry over an ordered field $F$. This will complete the proof that all of Hilbert's axioms hold in the Cartesian plane over a field. After that we will study the properties of rigid motions in an arbitrary Hilbert plane.

One could criticize Hilbert for taking a statement as complicated as (SAS) for an axiom, just as one could criticize Euclid for his fifth postulate, which is so much less elementary than his others. The response in both cases is the same: One cannot avoid including a statement as an axiom if one cannot prove it from the other axioms. Now, Euclid did not include (SAS) as an axiom, but "proved" it as (I.4). His proof has been justly criticized, because he used the "method of superposition," which involves moving one triangle and placing it on top of the other. This cannot be justified on the basis of Euclid's postulates and common notions. In fact, if you think about it, the possibility of moving figures around, without distorting their shapes, is a rather strong statement about homogeneity: The geometry is similar in different parts of the space. This is a deep fact not to be taken lightly or assumed without proof.

Curiously enough, in order to show that (SAS) holds in the geometry over a field, we will use Euclid's method of superposition, but only after proving that it makes sense. We will define the notion of *rigid motion* of a plane and show that there are enough of them to make Euclid's method work.

Conversely, we will prove the existence of rigid motions in an arbitrary Hilbert plane. Thus the existence of enough rigid motions is essentially equivalent to the statement (SAS), in the presence of the other axioms. This gives a satisfactory modern understanding of the meaning of Euclid's method of superposition. It also introduces us to the group of rigid motions of the plane and validates Felix Klein's point of view, expressed in his "Erlanger Programm" in the late nineteenth century, that one should classify different geometries according to the groups of motions that act on them.

To start with, we define the notion of a rigid motion.

### Definition

If $\Pi$ is a geometry consisting of the undefined notions of point, line, betweenness, and congruence of line segments and angles, which may or may not satisfy various of Hilbert's axioms, we define a *rigid motion* of $\Pi$ to be a mapping $\varphi : \Pi \to \Pi$ defined on all points, such that:

(1) $\varphi$ is a 1-to-1 mapping of the points of $\Pi$ onto itself.
(2) $\varphi$ sends lines into lines.
(3) $\varphi$ preserves betweenness of collinear points.
(4) For any two points $A, B$, we have $AB \cong \varphi(A)\varphi(B)$.
(5) For any angle $\alpha$, we have $\angle \alpha \cong \angle \varphi(\alpha)$.

In other words, $\varphi$ preserves the structures determined by the undefined notions in our geometry.

### Remark 17.0.1

For example, the *identity* transformation of $\Pi$ to itself, which leaves every point fixed, is a (trivial) rigid motion. It is clear that the set $G$ of all rigid motions forms a *group*, because the composition of any two is another one. We will use functional notation for composition: $\varphi\psi(A) = \varphi(\psi(A))$. However, it is not obvious in general that there are any other rigid motions besides the identity.

Now we can express what is needed to justify Euclid's method of superposition in the following principle

### ERM (Existence of Rigid Motions)

(1) For any two points $A, A' \in \Pi$, there is a rigid motion $\varphi \in G$ such that $\varphi(A) = A'$.
(2) For any three points $O, A, A'$, there is a rigid motion $\varphi \in G$ such that $\varphi(O) = O$ and $\varphi$ sends the ray $\overrightarrow{OA}$ to the ray $\overrightarrow{OA'}$.
(3) For any line $l$, there is a rigid motion $\varphi \in G$ such that $\varphi(P) = P$ for all $P \in l$ and $\varphi$ interchanges the two sides of $l$.

### Proposition 17.1

*In a plane satisfying the incidence and betweenness axioms, and assuming* (C2), (C5), *and the uniqueness portions of* (C1) *and* (C4) *only,* (ERM) *implies* (C6) = (SAS).

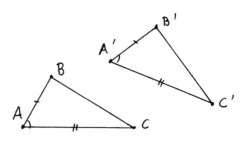

*Proof*  We are now assuming the existence of rigid motions (ERM) and will prove (SAS) by Euclid's method. So suppose we are given two triangles $ABC$ and $A'B'C'$, and suppose $AB \cong A'B'$, $AC \cong A'C'$, and $\angle BAC \cong \angle B'A'C'$. Then we must show that $\triangle ABC \cong \triangle A'B'C'$, namely, that $BC \cong B'C'$ and the angles at $B, C$ are congruent to the angles at $B'$ and $C'$, respectively.

By (ERM) (1), there is a rigid motion $\varphi$ that takes $A$ to $A'$. Let $B'' = \varphi(B)$. Then $AB \cong A'B''$, since $\varphi$ is a rigid motion and $AB \cong A'B'$ by hypothesis, so $A'B' \cong A'B''$ by (C2).

Next, by (ERM) (2), there is a rigid motion $\psi$ that leaves $A'$ fixed and sends the ray $A'B''$ to the ray $A'B'$. Sine $A'B'' \cong A'B'$, and $\psi$ preserves congruence, we conclude from the uniqueness portion of (C1) that $\psi(B'') = B'$. Let $C'' = \psi\varphi(C)$.

Then we consider the line $l = A'B'$, and the two rays $A'C'$ and $A'C''$. If they are on the same side of $l$, we do nothing, but if on opposite sides, then by (ERM) (3) there is a rigid motion $\sigma$ leaving the points of $l$ fixed and interchanging the sides.

Let us denote by $\theta \in G$ the composition $\psi\varphi$, or $\sigma\psi\varphi$ if we used $\sigma$. Then $\theta$ has the following properties: $\theta(A) = A'$, $\theta(B) = B'$, and $C''' = \theta(C)$ is on the same side of $A'B'$ as $C'$.

Since $\theta$ is a rigid motion, $\angle BAC \cong \angle B'A'C''$. But also $\angle BAC \cong \angle B'A'C'$ by hypothesis, so by (C5), $\angle B'A'C' \cong \angle B'A'C'''$. Furthermore, $C'$ and $C'''$ are on the same side of $A'B'$. So by the uniqueness portion of (C4) we conclude that the rays $A'C'$ and $A'C'''$ are equal.

Now, $A'C' \cong AC$ by hypothesis, and $AC \cong A'C'''$, since $\theta$ is a rigid motion, so by (C2) $A'C' \cong A'C'''$. Furthermore, $C'$ and $C'''$ are on the same side of $A'$. So by the uniqueness portion of (C1) we conclude that $C' = C'''$.

Thus $\theta(B) = B'$ and $\theta(C) = C'$. Since $\theta$ is a rigid motion, $BC \cong B'C'$ as required. Similarly, for the angles, $\theta$ takes $\angle ABC$ to $\angle A'B'C'$. So $\theta$ being a rigid motion, we conclude $\angle ABC \cong \angle A'B'C'$. The same method shows $\angle ACB \cong \angle A'B'C'$. This concludes the proof of (SAS).

Next we will show that (ERM) holds in the Cartesian plane over a field.

**Theorem 17.2**

*Let F be an ordered Pythagorean field, and let* $\Pi$ *be the associated Cartesian plane. Then* (ERM) *holds in* $\Pi$.

*Proof*  We think of $\Pi$ as having coordinates $(x, y)$. We will consider certain transformations of $\Pi$ defined by functions of $x$ and $y$, we will show that these are rigid motions, and then we will see that there are enough of them to prove (ERM).

First of all, consider a point $A = (a, b)$ and the transformation $\tau$ (called a *translation*) given by

$$\begin{cases} x' = x + a, \\ y' = y + b. \end{cases}$$

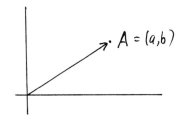

Clearly, $\tau$ is 1-to-1 and onto, because it has an inverse

$$\begin{cases} x = x' - a, \\ y = y' - b. \end{cases}$$

A line $y = mx + k$ under this transformation becomes

$$y' - b = m(x' - a) + k.$$

In particular, its image is a line, so we see that $\tau$ takes lines into lines. Next, we notice that the slope of the new line is the same as the slope of the old line, so $\tau$ preserves angles. Clearly, $\tau$ preserves betweenness, because this reduces to questions of inequalities in the field $F$, which are unchanged by adding constants.

Finally, we must check that $\tau$ preserves the dist$^2$ function to get congruence of segments. This is obvious, since we add the same constant to the coordinates of two points $A, B$, so in computing the dist$^2$ function we get the same value.

Thus the mapping $\tau$ is a rigid motion. Given two points $B, C$, we can take $a, b$ to be the difference of their $x$- and $y$-coordinates, so $\tau(B) = C$, and we have satisfied condition (1) of (ERM).

To prove condition (2) of (ERM) we will consider rotations. A *rotation* of the plane $\Pi$ is a transformation $\rho$ defined by

$$\begin{cases} x' = cx - sy, \\ y' = sx + cy, \end{cases}$$

where $c, s \in F$ and $c^2 + s^2 = 1$. The inverse of this transformation is given by

$$\begin{cases} x = cx' + sy', \\ y = -sx' + cy'. \end{cases}$$

Therefore, $\rho$ is 1-to-1 and onto. Being linear, $\rho$ takes lines to lines, and a brief calculation shows that a line with slope $m$ is transformed to a new line with slope

$$m' = \frac{cm + s}{c - sm}.$$

Since linear transformations either preserve or reverse inequalities, $\theta$ preserves betweenness.

Next, we show that $\rho$ preserves angles. Given two lines with slopes $m_1$ and $m_2$, let $m_1'$ and $m_2'$ be the new slopes. Since congruence of angles is determined by their tangents, it will be enough to show that

$$\frac{m_1' - m_2'}{1 + m_1' m_2'} = \frac{m_1 - m_2}{1 + m_1 m_2}.$$

This is an elementary calculation (left to the reader).

Finally, let us see what happens to the distance function. Let $A$ and $B$ be two points. Then another amusing little calculation, left to the reader, shows that

$$\text{dist}(\rho(A), \rho(B)) = \text{dist}(A, B).$$

Hence $\rho$ is a rigid motion.

Now we can verify condition (2) of (ERM). Given three points $O, A, A'$, we must show that there is a rigid motion leaving $O$ fixed and taking the ray $\overrightarrow{OA}$ to the ray $\overrightarrow{OA'}$. By using a translation, we can reduce to the case $O = $ origin. Let $y = mx$ and $y = m'x$ be the lines containing $A$ and $A'$. Any rotation leaves $O$ fixed, so to send the first line to the second, we have only to find $c, s \in F$ with $c^2 + s^2 = 1$ such that

$$m' = \frac{cm + s}{c - sm}$$

according to the formula above. Solving for $s$ we obtain

$$s = \frac{m' - m}{1 + mm'} c.$$

Let $k$ be the coefficient $(m' - m)/(1 + mm')$. Then we can solve $s = kc$ and $s^2 + c^2 = 1$ by $c = \pm 1/\sqrt{1 + k^2}$, using the Pythagorean property of $F$.

So we have two rotations taking the first line to the second, differing by the rotation $x' = -x$, $y' = -y$. One of these will send the ray $\overrightarrow{OA}$ to the ray $\overrightarrow{OA'}$ as desired.

To complete the proof of (ERM), we must verify condition (3), that for every line $l$, there is a rigid motion (called a *reflection*) leaving $l$ pointwise fixed and interchanging the two sides of $l$. Using a translation from a point of $l$ to the origin $O$, we may assume that $O \in l$. Let $A$ be any other point of $l$, and let $\rho$ be the rotation that sends the positive $x$-axis to the ray $\overrightarrow{OA}$. Let $\sigma$ be the *reflection* in the $x$-axis defined by

$$\begin{cases} x' = x, \\ y' = -y. \end{cases}$$

Clearly, this is a rigid motion that leaves the $x$-axis pointwise fixed and interchanges the two sides. Now $\varphi = \rho\sigma\rho^{-1}$ is the required reflection in the line $l$.

### Theorem 17.3

*If $F$ is any Pythagorean ordered field, then the Cartesian plane $\Pi$ over $F$ is a Hilbert plane satisfying the parallel axiom* (P). *The plane $\Pi$ will be Euclidean if and only if $F$ is Euclidean.*

*Proof* We have previously verified the incidence axioms (I1)–(I3) and (P) in (14.1), the betweenness axioms (B1)–(B4) in (15.3), and the congruence axioms (C1)–(C5) in (16.1). Now, from (17.2) we know that (ERM) holds in $\Pi$, and therefore by (17.1) also (C6) holds. For the plane to be Euclidean, i.e., to satisfy (E), it is necessary and sufficient that the field $F$ be Euclidean (16.2).

Next we will prove a sort of converse to (17.1), namely that (ERM) holds in any Hilbert plane.

### Proposition 17.4

*In any Hilbert plane* (cf. *Section* 10), *there are enough rigid motions*: (ERM) *holds.*

*Proof* First we will show the existence of reflections. Then we will build other rigid motions out of these.

Suppose we are given a line $l$. We will construct a rigid motion $\sigma$, called the *reflection* in $l$, that leaves the points of $l$ fixed and interchanges the two sides of $l$. For any point $P \in l$ we define $\sigma(P) = P$. For any point $A \notin l$, drop the perpendicular $AA_0$ to $l$, and extend it on the far side of $l$ so that $AA_0 \cong A_0A'$. Then we set $\sigma(A) = A'$. Clearly $\sigma^2 = $ id, so $\sigma$ is 1-to-1 and onto.

Let $A, B$ be any two points not on $l$. We will show that $AB \cong A'B'$, where $\sigma(A) = A'$, and $\sigma(B) = B'$. If $A, B$ lie on the same line perpendicular to $l$, this is immediate from subtracting congruent line segments. If $A, B$ are on different perpendiculars, as in the figure, let

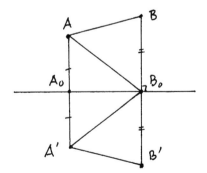

$A_0, B_0$ be the feet of those perpendiculars. Then $\triangle A_0 A B_0$ is congruent to $\triangle A_0 A' B_0$, using the right angles at $A_0$, by (SAS). Therefore, the angles $\angle A B_0 A_0$ and $\angle A' B_0 A_0$ are congruent. Subtracting from the right angles at $B_0$, we find that $\angle A B_0 B \cong \angle A' B_0 B'$. On the other hand, $A B_0 \cong A' B_0$ from the first triangles. Now we can apply (SAS) again to conclude that $\triangle A B_0 B \cong \triangle A' B_0 B'$. In particular, $AB \cong A'B'$ as claimed.

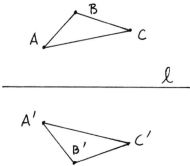

Now suppose $A, B, C$ are three non-collinear points whose images by $\sigma$ are $A', B', C'$. Then by (SSS) we conclude that $\triangle ABC \cong \triangle A'B'C'$, and in particular, $\angle BAC \cong \angle B'A'C'$. Thus $\sigma$ preserves angles. From here it is easy to verify that $\sigma$ preserves lines and betweenness, so in fact, $\sigma$ is a rigid motion (details left to reader).

To verify that (ERM) holds, we have just established property (3) by the existence of reflections. If $A, A'$ are any two points, let $l$ be the perpendicular bisector of the segment $AA'$. Then $\sigma_l$ will send $A$ to $A'$. Thus condition (1) of (ERM) holds. For condition (2), let $O, A, A'$ be three points. Let $l$ be the bisector of the angle $\angle AOA'$. Then the reflection $\sigma_l$ will leave $O$ fixed and send the ray $\overrightarrow{OA}$ to the ray $\overrightarrow{OA'}$. Thus (ERM) holds. Note that for this proof we need the existence of the perpendicular bisector of a line segment (I.10) and (I.11), and the bisector of an angle (I.9), which exist in a Hilbert plane by (10.4).

**Corollary 17.5**
*In the presence of all the axioms of a Hilbert plane except* (C6), *the axiom* (C6) *is equivalent to* (ERM).

*Proof*  Combine (17.1) and (17.4).

**Remark 17.5.1**
One can give the rigid motions an even more prominent position in the foundations of geometry by using them to define congruence, as follows. Suppose we are given a set of points with undefined notions of lines and betweenness satisfying axioms (I1)–(I3) and (B1)–(B4) as before. Suppose also that we are given a group $G$ of transformations of this set, called *motions*, that preserve lines and betweenness and suppose further that $G$ satisfies the following axioms (similar to (ERM)):

(1) Given two rays, and given a side of each line containing one of the rays, there is a *unique* motion $\varphi \in G$ that takes one ray to the other and the given side to the given side.
(2) For any two distinct points $A, B$, there exists a motion of $G$ that interchanges the two points.

(3)  For any two rays emanating from the same point, there exists a motion of $G$ that interchanges the two rays.

Then one can *define* congruence of segments and angles by requiring the existence of a motion in $G$ that sends one to the other, and one can prove that this notion of congruence satisfies the axioms (C1)–(C6) and so makes a Hilbert plane. See Hessenberg–Diller (1967), Sections 37–39. Bachmann (1959) carries this idea a step further, by eliminating points and lines altogether and giving a set of axioms for geometry based on the group $G$ above (cf. discussion in Section 43).

# Exercises

17.1  In the proof of Theorem 17.2 verify that:

(a)  Rotations preserve angles.

(b)  Rotations preserve distances.

17.2  Let $\varphi : \Pi \to \Pi$ be a map of a Hilbert plane into itself. For any point $A$, denote $\varphi(A)$ by $A'$. Assume $AB \cong A'B'$ for any two points $A, B$.

(a)  Prove that $\varphi$ is 1-to-1 and onto.

(b)  Show that in fact, $\varphi$ is a rigid motion.

17.3  In a Hilbert plane $\Pi$, show:

(a)  Any rigid motion with at least three noncollinear fixed points must be the identity.

(b)  Any rigid motion is equal to the product of at most three reflections.

17.4  In a Hilbert plane $\Pi$, define a *rotation* around a point $O$ to be a rigid motion $\rho$ leaving $O$ fixed and such that for any two points $A, B$, the angles $\angle AOA'$ and $\angle BOB'$ are equal, where $\rho(A) = A'$, $\rho(B) = B'$. Show:

(a)  For any two points $A, A'$ with $OA \cong OA'$, there exists a unique rotation around $O$ sending $A$ to $A'$.

(b)  The set of rotations around a fixed point $O$, together with the identity, is an abelian subgroup of the group of all rigid motions.

(c)  Any rotation can be written as the product of two reflections.

(d)  A rigid motion having exactly one fixed point must be a rotation.

17.5  In a Euclidean plane $\Pi$, define a *translation* to be a rigid motion $\tau$ such that for any two points $A, B$, we have $AA' \cong BB'$, where $\tau(A) = A'$, $\tau(B) = B'$. Show:

(a)  For any two points $A, A'$, there exists a unique translation $\tau$ such that $\tau(A) = A'$.

(b) If $\tau$ is a translation, then for any two points $A, B$, we have $AB \| A'B'$ and $AA' \| BB'$.

(c) The set of translations forms an abelian subgroup $T$ of the group of all rigid motions.

(d) Any translation is a product of two reflections.

(e) Is the group $T$ of translations a normal subgroup of the group $G$ of all rigid motions? Prove yes or no.

17.6 In this exercise we establish an algebraic interpretation of the group of rotations around a point in a Cartesian plane. Let $F$ be an ordered field. In the Cartesian plane $\Pi$ over $F$, let $\Gamma$ be the unit circle, and let $E = (-1, 0)$. Let a line $l$ through $E$ meet the circle at a point $A$.

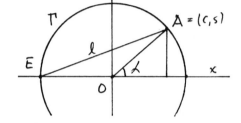

(a) If the line $l$ has slope $t$, show that the coordinates of the point $A$ are $(c, s)$, where

$$c = \frac{1 - t^2}{1 + t^2}, \quad s = \frac{2t}{1 + t^2}.$$

(*Note:* We use this notation because if $F = \mathbb{R}$, and if $\alpha$ is the angle that $OA$ makes with the positive $x$-axis, then by trigonometry, we obtain $t = \tan \frac{1}{2}\alpha$, $c = \cos \alpha$, $s = \sin \alpha$, and these are the usual formulas for expressing $c, s$ in terms of $t$. Do you remember those substitutions used in calculus classes for rationalizing trigonometric integrals?)

(b) Let $\rho_t$ be the corresponding rotation

$$\rho_t : \begin{cases} x' = cx - sy, \\ y' = sx + cy, \end{cases}$$

as defined in the text. Show that $\rho_t$ is also a rotation in the sense of Exercise 17.4, and show that the mapping $t \mapsto \rho_t$ gives a 1-to-1 correspondence between the set $F \cup \{\infty\}$ and the group $R$ of rotations of $\Pi$ with center $O$.

(c) Under the correspondence in (b), show that the group operation in $R$ corresponds to the operation

$$t \circ t' = \frac{t + t'}{1 - tt'}$$

in the set $F \cup \{\infty\}$. We call the set $F \cup \{\infty\}$ with the operation $\circ$ the *circle group* of the field $F$.

(d) If $F = \mathbb{R}$, show that the circle group of $F$ is isomorphic to the abstract group $(\mathbb{R}/\mathbb{Z}, +)$.

17.7 Let $F$ be a field (not necessarily an ordered field) that does not contain a square root of $-1$. In analogy to the situation above, we define the *circle group* of $F$ to be the set $C(F) = F \cup \{\infty\}$ with the operation

$$a \circ b = \frac{a+b}{1-ab} \qquad \text{for} \quad a, b \in C(F).$$

(a) Verify directly that $C(F)$, $\circ$ is an abelian group by verifying the group axioms. Make clear your rules for treating $\infty$ (cf. Exercise 16.2), and point out where you use the hypothesis $\sqrt{-1} \notin F$.

(b) Show that $+1, -1$ are elements of finite order in $C(F)$.

(c) If $F = \mathbb{Q}$, show that $\pm 1, \infty$ are the *only* elements of finite order different from the identity.

(d) If $F = \mathbb{R}$, find explicitly four elements of order 5 in $C(F)$.

17.8 With $F$ a field as in Exercise 17.7, consider an extension field $F(i)$ with $i = \sqrt{-1}$. Elements of $F(i)$ can then be written as $\alpha = a + bi$, $a, b \in F$. Define the *norm* of an element $\alpha$ by $N(\alpha) = a^2 + b^2$.

(a) For any $\alpha, \beta \in F(i)$ verify that $N(\alpha\beta) = N(\alpha)N(\beta)$.

(b) Let $S = \{\alpha \in F(i) \mid N(\alpha) = 1\}$. Then $S$ is a group under multiplication. Show that the map $S \to C(F)$ defined by

$$\alpha = a + bi \mapsto \frac{b}{a+1}$$

is an isomorphism of $(S, \cdot)$ with the circle group $(C(F), \circ)$.

*Note*: This mysterious isomorphism is motivated by the figure in Exercise 17.6, in which $t = \tan \frac{1}{2}\alpha = s/(c+1)$.

17.9 Let $F$ be a finite field of $p$ elements, $p \equiv 3 \pmod 4$.

(a) Show that $-1$ does not have a square root in $F$.

(b) Show that the circle group $C(F)$ is *cyclic* of order $p + 1$.

17.10 Let $ABC$ and $A'B'C'$ be two congruent triangles in a Hilbert plane. Show that there exists a rigid motion $\varphi$ of the plane with $\varphi(A) = A'$, $\varphi(B) = B'$, and $\varphi(C) = C'$.

17.11 In a Euclidean plane, show that the product of two rotations around different points is equal to either a rotation around a third point or a translation. *Hint*: Show that it has at most one fixed point.

17.12 In a Euclidean plane, show that the product of an odd number of reflections cannot be equal to the identity. *Hint*: Use Exercise 17.11 to reduce products of four reflections to products of two reflections, and proceed by induction.

17.13 In a Hilbert plane, let us define one segment of a circle to be *congruent* to another segment of a circle if there exists a rigid motion of the plane that makes the first coincide with the second. Using this notion of congruence of segments, show that Euclid's results (III.23)–(III.30) and their proofs are all ok in a Euclidean plane.

17.14 (Theorem of three reflections).

(a) Given three lines $a, b, c$ through a point $O$, show that there exists a unique fourth line $d$ such that

$$\sigma_c \sigma_b \sigma_a = \sigma_d,$$

where $\sigma$ denotes the reflection in a given line. *Hint:* Let $A$ be a point of $a$, and take $d$ to be the perpendicular bisector of $AC$, where $C = \sigma_c \sigma_b(A)$. (See Proposition 41.2 for an analogous result in hyperbolic geometry.)

(b) Given three lines $a, b, c$ perpendicular to a line $l$, show that there exists a unique fourth line $d$ such that $\sigma_c \sigma_b \sigma_a = \sigma_d$.

# 18    Non-Archimedean Geometry

The Archimedean principle, that given two line segments, some multiple of the first will exceed the second, is so embedded in our experience of the world that it is hard to imagine a geometry in which this would not hold. Even the farthest star has a distance from the earth that can be measured in light years, and even if we take the inch as our standard unit of length, some number of inches, albeit a very large number, will exceed the distance to that farthest star. As long as we retain the notion that geometry somehow represents the real world, we are bound to accept Archimedes' principle as a truth.

In abstract mathematics, on the other hand, a geometry is a theory that satisfies a certain set of axioms. In this chapter we have seen how to construct a geometry over an abstract ordered field. The elements of the field need not be numbers or distances. Any abstract field will do.

We will take advantage of this abstraction to construct some non-Archimedean geometries. These examples will serve two functions. One is to show the independence of Archimedes' axiom (A) and Playfair's axiom (P) from the axioms of a Hilbert plane. The other is to free our minds from the constraints of habit by studying the properties of a logically constructed geometry in which Archimedes' axiom (A) does not hold. Such geometries are called non-Archimedean geometries.

**Proposition 18.1**
*Let* $\mathbb{R}$ *be the field of real numbers, let* $t$ *be an indeterminate, and let* $\mathbb{R}(t)$ *be the field of all rational functions of* $t$, *that is, all quotients* $f(t)/g(t)$ *where* $f$ *and* $g$ *are polynomial functions of* $t$ *with real coefficients and* $g(t)$ *is not identically zero. Then the field* $F = \mathbb{R}(t)$ *has a natural ordering that makes it into a non-Archimedean ordered field.*

*Proof*    If $\varphi \in F$, we think of $\varphi$ as a function $\varphi(t) = f(t)/g(t)$ from $\mathbb{R}$ to $\mathbb{R}$, defined everywhere except at the finite number of points where $g(t) = 0$. We define the set $P$ of "positive" elements of $F$ to be the set of those functions $\varphi$ that are positive for all large enough values:

$$P = \{\varphi \in F \mid \exists a_0 \in \mathbb{R} \text{ such that } \varphi(b) > 0 \text{ for all } b > a_0\}.$$

Note that $\varphi > 0$ if and only if the quotient of the leading coefficients of $f$ and $g$ is positive in $\mathbb{R}$. Now, $F$ is a field, because any sum, difference, product, or quotient of rational functions is again a rational function. The set $P$ is closed under sums and products, because the sum and product of two eventually positive functions is again eventually positive. To show that $(F, P)$ is an ordered field, it remains to show that if $\varphi \in F$, $\varphi \neq 0$, then either $\varphi \in P$ or $-\varphi \in P$, but not both. Indeed, if $\varphi \neq 0$, then it is the quotient of two nonzero polynomials $\varphi = f(t)/g(t)$. Each of these has a finite number of zeros. If we take $a_0 \in \mathbb{R}$ larger than all the zeros of $f(t)$ and $g(t)$, then $\varphi$ is continuous and never 0 for all $b > a_0$. Thus by the intermediate value theorem, $\varphi$ is either always positive for $b > a_0$, or always negative for $b > a_0$. In the first case $\varphi \in P$; in the second case $-\varphi \in P$.

Now consider the element $t \in F$. For any integer $n > 0$, we have $t > n$ as elements of $F$. Indeed, for $b > n$, the function $\varphi(t) = t - n$ is positive. Thus the field $F$ is non-Archimedean. Note that in this field we have

$$0 < 1 < 2 < \cdots < t < t + 1 < t + 2 < \cdots < t^2 < t^3 < \cdots.$$

**Definition**

Let $F$ be a non-Archimedean ordered field. We will say that an element $a \in F$ is *finitely bounded* if there exists a positive integer $n$ for which $-n < a < n$. Otherwise, we say that $a$ is *infinite*. We say that an element $a \in F$ is *infinitesimal* if for every positive integer $n$, we have $-1/n < a < 1/n$. An element of $F$ is *finite* if it is finitely bounded but not infinitesimal.

Next we will construct non-Archimedean fields satisfying the Pythagorean property $(*)$ of $(16.1)$ and the Euclidean property $(**)$ of $(16.2)$.

**Proposition 18.2**
*There is a (non-Archimedean) Pythagorean ordered field $\Omega'$ containing the field $\mathbb{R}(t)$.*

*Proof* We start with the field $\mathbb{R}(t)$ of rational functions in an indeterminate $t$, described above, and we consider $\mathbb{R}(t)$ as a subset of the set $\mathscr{C}$ of all continuous real-valued functions from $\mathbb{R}$ to $\mathbb{R}$, defined at all except a finite number of points, and having only a finite number of zeros (except for the identically 0 function). Beware that $\mathscr{C}$ is not a field(!) because, for example, the functions 2 and $2 + \sin t$ are in $\mathscr{C}$, but their difference $\sin t$ is not in $\mathscr{C}$, because it has infinitely many zeros. Nevertheless, $\mathscr{C}$ has a nice order, because we can define the subset $P_\mathscr{C}$ of positive functions as before: $\varphi(t) \in \mathscr{C}$ is *positive* if $\exists a_0 \in \mathbb{R}$ for which $\varphi(b) > 0$ for all $b > a_0$. Then $P_\mathscr{C}$ satisfies properties (i) and (ii) of the definition of an ordered field, even though $\mathscr{C}$ is not a field. We use the fact that a continuous function on an interval $(a_0, \infty)$ that has no zeros is either always positive or always negative.

Now let $\Omega'$ be the set of all elements of $\mathscr{C}$ that can be obtained from $\mathbb{R}(t)$ by a finite number of operations $+, -, \cdot, \div$, and $c \mapsto \sqrt{1+c^2}$. The hard part is to show that $\Omega'$ is a field. Once we know that $\Omega'$ is a field, the Pythagorean property is easy, because for any $c \in \mathscr{C}$, $1 + c^2$ is a function that is strictly positive whenever it is defined (at all except the finite number of points where $c$ is not defined), so $\sqrt{1+c^2}$ is another such function, hence also in $\mathscr{C}$. Thus if $c \in \Omega'$, $\sqrt{1+c^2} \in \Omega'$ also. We make $\Omega'$ into an ordered field by taking as the positive elements $P' = P_{\mathscr{C}} \cap \Omega'$, and $P'$ satisfies (i) and (ii) because $P_{\mathscr{C}}$ does.

**Lemma 18.3**
*Let $F$ be a subset of $\Omega'$ that is a field, and let $\omega \in F$, $\sqrt{1+\omega^2} \notin F$. Then*

$$F' = \{\alpha + \beta\sqrt{1+\omega^2} \,|\, \alpha, \beta \in F\}$$

*is also a subset of $\Omega'$ that is a field.*

*Proof*   First we show that every element of $F'$ is in $\Omega'$. Since $\alpha, \beta, \omega$ are obtained from $\mathbb{R}(t)$ by a finite number of operations $+, -, \cdot, \div, c \mapsto \sqrt{1+c^2}$, so are the elements of $F'$. The elements of $F'$ are defined except at the finite number of points where $\alpha, \beta, \omega$ may fail to be defined. They are continuous because $\alpha, \beta, \omega$ are. The only problem is to show that $\alpha + \beta\sqrt{1+\omega^2}$ has only finitely many zeros. Any zero $t_0$ of this function satisfies

$$\alpha(t_0) + \beta(t_0)\sqrt{1+\omega(t_0)^2} = 0.$$

Separating the two pieces, squaring, and combining again we obtain

$$\alpha(t_0)^2 - \beta(t_0)^2(1+\omega(t_0)^2) = 0.$$

In other words, $t_0$ is a zero of the function

$$\alpha^2 - \beta^2(1+\omega^2) \in F.$$

Hence there are only finitely many such zeros, since $F \subseteq \Omega'$. Note that $\alpha + \beta\sqrt{1+\omega^2}$ is not identically zero because then $\sqrt{1+\omega^2} \in F$. Thus $\alpha + \beta\sqrt{1+\omega^2}$ has only finitely many zeros, and so $F' \subseteq \Omega'$.

To show that $F'$ is a field is standard. It is clearly closed under $+, -, \cdot$. And to show closure under $\div$ one rationalizes the denominator by multiplying by its conjugate:

$$\frac{a+b\sqrt{f}}{c+d\sqrt{f}} \cdot \frac{c-d\sqrt{f}}{c-d\sqrt{f}} = \frac{(a+b\sqrt{f})(c-d\sqrt{f})}{c^2 - d^2 f}.$$

*Proof of 18.2 (continued)*   To show that $\Omega'$ is a field, suppose $\alpha, \beta \in \Omega'$. We must show that $\alpha \pm \beta$, $\alpha \cdot \beta$, $\alpha/\beta \in \Omega'$ (provided that $\beta \neq 0$). Since $\alpha$ is obtained from

$\mathbb{R}(t)$ by a finite number of operations $+, -, \cdot, \div, \omega \mapsto \sqrt{1 + \omega^2}$, by applying the lemma each time we take a square root, we obtain a subfield $F \subseteq \Omega'$ that contains $\alpha$. Now, starting from $F$, and applying the lemma again each time we use a square root in the description of $\beta$, we obtain a field $F \subseteq G \subseteq \Omega'$, with $\alpha, \beta \in G$. Then clearly, $\alpha \pm \beta, \alpha \cdot \beta, \alpha/\beta \in G \subseteq \Omega'$.

## Proposition 18.4
*There is a (non-Archimedean) Euclidean field $K'$ containing $\mathbb{R}(t)$.*

*Proof* We follow the same plan of proof as for (18.2), except that now we consider the space $\mathscr{C}'$ as follows: $\mathscr{C}'$ consists of continuous real-valued functions defined on some interval $(a_0, \infty)$ of $\mathbb{R}$ that are never 0. Two functions $f$ on $(a_0, \infty)$ and $g$ on $(a_1, \infty)$ are *equivalent* if $\exists a_2 > a_0, a_1$ such that $f = g$ on $(a_2, \infty)$. We say that $f$ is *positive* if for some $a_0$, $f(b) > 0$ for all $b > a_0$. The set $P_{\mathscr{C}'}$ of positive functions clearly satisfies (i) and (ii) of the definition of ordered field. Note again that $\mathscr{C}'$ is not a field. But if $\varphi \in \mathscr{C}'$, $\varphi > 0$, then $\sqrt{\varphi} \in \mathscr{C}'$ also.

Now we take $K$ to be the set of all elements of $\mathscr{C}'$ that can be obtained from $\mathbb{R}(t)$ by a finite number of operations $+, -, \cdot, \div$, and $\varphi > 0 \mapsto \sqrt{\varphi}$.

The proof that $K'$ is a field can be carried out exactly as in the proof of (18.2). Clearly, $K'$ is Euclidean, and taking $P' = K' \cap P_{\mathscr{C}'}$ makes $K'$ into an ordered field.

## Example 18.4.1
Let $\Pi$ be the Cartesian plane over the field $\Omega'$ of (18.2). Then $\Pi$ is a Hilbert plane satisfying (P) but not (A). In particular, this shows that (A) is independent of the axioms of a Hilbert plane.

## Example 18.4.2
Let $\Pi$ be the Cartesian plane over the field $K'$ of (18.4). Then $\Pi$ is a Euclidean plane that does not satisfy (A).

## Example 18.4.3
Let $\Pi$ be the non-Archimedean geometry described in (18.4.2). Let $\Pi_0$ be the subset consisting of all points of $\Pi$ whose distance from the origin is finitely bounded. A *line* of $\Pi_0$ will be the intersection of a line of $\Pi$ with $\Pi_0$, whenever that intersection is nonempty. Take betweenness and congruence to have the same meaning as in $\Pi$. Then $\Pi_0$ is a Hilbert plane satisfying neither (A) nor (P) (Exercise 18.3). In particular, this shows that (P) is independent of the axioms of a Hilbert plane.

To help visualize a non-Archimedean geometry, let us imagine for a moment that we live in a non-Archimedean universe. What we perceive with our telescopes are very large, but still finite, distances; what we observe with our cyclotrons and particle accelerators are very small, but still finite, quantities. And yet out beyond the farthest stars are other parallel universes, and inside each elementary particle are infinitesimal worlds unknown to us. Perhaps they exert some subliminal influence on our lives? How could we determine whether our universe is indeed non-Archimedean when we see only the finite part of it?

# Exercises

18.1 In the ordered field $\mathbb{R}(t)$, arrange the following elements in increasing order: $0, 1, 5, t, 1/t, t+1, 1/(t+1), t-1, \frac{1}{2}t, t^2 - t, t^2 - 1, t + \frac{1}{t}, (t-1)/(t+1)$.

18.2 Show that the field $\Omega'$ of Proposition 18.2 is not Euclidean, by showing that $\sqrt{t} \notin \Omega'$.

18.3 Show that the plane $\Pi_0$ of Example 18.4.3 satisfies the axioms for a Hilbert plane. Pay special attention to (I2), (B2), and (C1).

18.4 Again let $\Pi_0$ be the plane of Example 18.4.3.

   (a) Show that $\Pi_0$ does not satisfy (P).

   (b) Show that $\Pi_0$ does satisfy (I.32): The angle sum of every triangle is two right angles.

   (c) Show that $\Pi_0$ does not satisfy (IV.5), by giving an example of a triangle that has no circumscribed circle.

18.5 Let $\Pi$ be the non-Archimedean plane of Example 18.4.2. Define a subset $\Pi_1$ of $\Pi$ to be all the points of $\Pi$ whose distance from the origin is infinitesimal.

   (a) Show that $\Pi_1$ is a Hilbert plane.

   (b) Show that $\Pi_1$ does not satisfy (P). Thus $\Pi_1$ gives another example of the independence of (P) from the axioms of a Hilbert plane.

18.6 We say that a Hilbert plane is *finitely bounded* if there exists a segment $AB$ such that for every other segment $CD$, there exists an integer $n$, depending on $CD$, for which $CD < n \cdot AB$.

   (a) Any Archimedean Hilbert plane is finitely bounded.

   (b) The plane $\Pi_0$ of Example 18.4.3 is finitely bounded but not Archimedean.

   (c) The plane $\Pi_1$ of Exercise 18.5 is not finitely bounded. In particular, the planes $\Pi_0$ and $\Pi_1$ are not isomorphic Hilbert planes.

18.7 We say that the *rectangle axiom* holds in a Hilbert plane if whenever a quadrilateral has three right angles, then the fourth angle is also a right angle.

   (a) The rectangle axiom holds in any Hilbert plane with (P).

   (b) The rectangle axiom holds in the examples $\Pi_0$ and $\Pi_1$ above. Thus the rectangle axiom does not imply (P).

18.8 Let $F$ be any ordered field. Generalize the proof of Proposition 18.1 to show that the field $F(t)$ of rational functions in an indeterminate $t$ is a non-Archimedean ordered field. (Be careful not to use continuity.)

18.9 Let $F$ be any ordered field. Let $F((t))$ be the set of Laurent series

$$\varphi = \sum_{i \geq n}^{\infty} a_i t^i, \quad a_n \neq 0,$$

where the $a_i \in F$ and $n \in \mathbb{Z}$ can be positive, zero, or negative. Define $\varphi > 0$ if its leading coefficient $a_n > 0$ in $F$.

(a) show that $F((t))$ is a field.

(b) Show that $F((t))$ is a non-Archimedean ordered field.

(c) An element $\varphi \in F((t))$ is a square if and only if its order $n$ is even and its leading coefficient $a_n$ is a square in $F$.

(d) If $F$ is Pythagorean, show that $F((t))$ is also Pythagorean. This gives another method of constructing Pythagorean non-Archimedean ordered fields.

Let man and woman form a circle
From which grows a square;
Around these put a triangle,
Embed them all in a sphere:
Then you will have the philosopher's stone.
If in your mind this does not soon appear,
Geometry, well learned, will make it clear.

– from *Atalanta Fugiens*
by Michael Maier (1618),
Epigramma XXI.

# 4

**CHAPTER**

# Segment
# Arithmetic

egment arithmetic allows us to complete the chain of
logical connections between an abstract geometry satisfying axioms studied in Chapter 2 with the geometries over fields studied in Chapter 3. We will show
how to define addition and multiplication of line segments in a Hilbert plane satisfying the parallel axiom
(P). In this way, the congruence equivalence classes
of line segments become the positive elements of an
ordered field $F$ (Section 19). Using this field $F$ we can
recover the usual theory of similar triangles (Section 20).

To complete the circle, we show that if you start with a Hilbert plane $\Pi$ satisfying (P), and if $F$ is the associated field of segment arithmetic, then $\Pi$ is isomorphic to the Cartesian plane over the field $F$ (Section 21).

## 19  Addition and Multiplication of Line Segments

In studying Euclid's *Elements*, we have noted the absence of numbers in his development. There is no notion of the *length* of a line segment, for example.
There is an undefined notion of congruence of segments, which we can think of
as the segments being the same size. This is in contrast to ordinary high-school

**165**

geometry, where each segment has a length, based on some chosen "unit" segment, which is thought of as a real number, and two segments are congruent if they have the same length.

Similarly, in the case of angles, there is no degree measure attached to an angle, although there is a notion of congruence of angles.

In the study of area, Euclid does not assign a number to a plane figure, in contrast, for example, to high-school geometry, where one takes a triangle and assigns to it the number $\frac{1}{2}bh$ as its area, where $b$ is the length of the base and $h$ is the length of the altitude. Instead, Euclid treats area by adding and subtracting congruent figures.

For the material of Books I–IV of the *Elements*, we have also seen that Euclid succeeds remarkably well in developing a beautiful theory of "pure" geometry without numbers. Hilbert has reinforced this by providing a set of purely geometric axioms on which to base Euclid's geometry in a way that will satisfy modern criteria of rigor.

Just for contrast, you might look at some other twentieth-century proposals for a set of axioms on which to base the study of geometry, where the real numbers are presupposed from the beginning in the axioms. (See, for example, Birkhoff (1932), or the School Mathematics Study Group postulates. Both can be found as appendices to Cederberg (1989).)

For me this is unsatisfactory because it is not purely geometric, and the concept of a real number is a rather sophisticated modern notion, dating from the nineteenth-century, and is not in the elementary spirit of Euclid's geometry. While Euclid was clearly aware of irrational numbers, and studies them extensively in Book X of the *Elements*, I find it difficult to support any argument that Euclid had a concept of the totality of real numbers.

While Euclid was able to develop the material of Books I–IV without any notion of number, it is a different matter when we come to the concept of *similar* triangles as taught in high school. These are triangles whose sides are not equal, but have some common *ratio* to each other. If that ratio is 2, it is not difficult to develop a theory of triangles that are doubles of each other, as we did in Section 5. With a little more effort, one could extend this theory to triangles whose sides are integer multiples of each other, or (with even a little more effort), rational number multiples of each other. But if the ratio is not rational, as for example in comparing an isosceles right triangle to its half formed by drawing an altitude, how can one even express the notion of sides being proportional to each other without having numbers? One would like to say that the ratios of the lengths of the sides are equal, but this is difficult if one has no notion of length as a number and does not have the ability to divide one such number by another.

Euclid handles this difficulty with the theory of *proportion* developed in Book V of the *Elements*. The key concept is in Book V, Definition 5, where he says that magnitudes (which could be line segments, areas, or whatever) are in the *same ratio* (in symbols $a : b = c : d$) if whenever equal integer multiples (say $n$ times) be taken of $a$ and $c$, and whenever equal integer multiples (say $m$ times) be taken of $b$ and $d$, then $na > mb$ or $na = mb$ or $na < mb$ if and only if $nc > md$, or $nc = md$, or $nc < md$, respectively. If $a, b, c, d$ are numbers, this is equivalent to saying that a rational number $m/n$ is less than, equal to, or greater than $a/b$ if and only if that same rational number is less than, equal to, or greater than $c/d$. If furthermore $a, b, c, d$ are real numbers, this is equivalent, as we know, to saying that $a/b$ and $c/d$ are equal as real numbers, since the rational numbers are dense in the set of real numbers. In fact, this is word for word the same notion used by Dedekind in constructing the real numbers by his so-called Dedekind cuts (cf. Dedekind (1872)).

Aha!, you may say, so Euclid did know about the real numbers, and wrote their definition 2000 years before Dedekind! But here is the difference. Euclid used this criterion only to distinguish between ratios that arose naturally in his geometry, such as the ratios of line segments that might be obtained by ruler and compass constructions, and that might be irrational. But I see no evidence that he conceived of the existence of any other real numbers (such as $e$, for example), whereas Dedekind could conceive of the totality of all Dedekind cuts of rational numbers, and take this set to be a new mathematical object called the set of real numbers. It is this process of creating a new mathematical object as a set of all subsets of another set with certain properties that seems very modern to me.

Even in the classical problem of the trisection of the angle it seems that the emphasis was on finding a construction that would produce an angle equal to one-third of a given angle, and there is no evidence that the ancients believed in the existence of such an angle before it was constructed.

Note also that for Euclid's theory of proportion to work, we implicitly need Archimedes' axiom. This is clear from Book V, Definition 4, which says that magnitudes have a ratio to each other if each, when multiplied, is capable of exceeding the other. Without Archimedes' axiom, some quantities would be incomparable. Also, one would fail to distinguish unequal quantities. For example, if $F$ is a non-Archimedean ordered field with an infinite element $t$, then Euclid's test would fail to distinguish between $\sqrt{2}$ and $\sqrt{2} + 1/t$.

Having developed the theory of proportion abstractly in Book V, Euclid proceeds to apply his theory to geometry in Book VI, and develops what we recognize as the familiar theory of similar triangles. The key result here, which forms the basis of the subsequent development, is (VI.2), which says that a line parallel to the base of a triangle, if it cuts the sides, cuts them proportionately, and conversely. Euclid's proof is a tour de force, using the theory of area previously developed in Book I to establish this result.

There are two reasons for us to seek an alternative development of the theory of similar triangles: One is to free ourselves from dependence on Archimedes' axiom, and the other is to avoid Euclid's use of the theory of area, which we have not yet treated satisfactorily (cf. Chapter 5).

So now, after this rather lengthy introduction, we come to the main point of this section, which is to create an arithmetic of line segments. We will define notions of *addition* and *multiplication* for line segments up to congruence, that is, the sum or product of congruent segments will be congruent. Or if you prefer, the operations + and · will be defined on the set $P$ of equivalence classes of line segments modulo congruence. We will show that these operations obey all the usual rules of arithmetic for positive numbers. And then, by a natural construction that introduces an element 0 and negatives of line segments, we will construct an ordered *field* whose positive elements are the congruence classes of line segments. Here is where the concepts of modern abstract algebra play an essential role, because instead of using some preexisting notion of number, such as the rational numbers or the real numbers, we create a set that occurs naturally in our geometry and give this set the structure of an abstract field.

Using this field we will then in the next section be able to define the notion of length of a segment (as an element of this field) and to develop the theory of similar triangles, where ratios are quotients of lengths in the field. Thus we will replace Euclid's theory of proportion as developed in Book V by the use of algebraic relations in the field of segment arithmetic.

We will now define the arithmetic operations on congruence equivalence classes of line segments, following the ideas of Hilbert (1971), with simplifications suggested by material in the supplements to that book, apparently due to Enriques. We will work in a Hilbert plane satisfying the parallel axiom (P).

**Definition**
Given congruence equivalence classes of line segments $a, b$, we define their *sum* as follows. Choose points $A, B$ such that the segment $AB$ represents the class $a$.

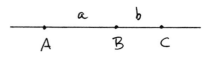

Then on the line $AB$ choose a point $C$ with $A * B * C$, such that the segment $BC$ represents the class $b$. Then we define $a + b$ to be represented by the segment $AC$.

**Proposition 19.1**
*In any Hilbert plane, addition of line segment classes has the following properties:*

(1) $a + b$ *is well-defined, i.e., different choices of $A, B, C$ in the definition will give rise to congruent segments.*
(2) $a + b = b + a$, *i.e., the corresponding line segments are congruent.*
(3) $(a + b) + c = a + (b + c)$.

(4) *Given any two classes a, b, one and only one of the following holds*:
    (i) $a = b$.
    (ii) *There is a class c such that $a + c = b$*.
    (iii) *There is a class d such that $a = b + d$*.

*Proof* (1) If we choose a different representative $A'B'$ of the class $a$, and lay off $C'$ on the line $A'B'$ such that $B'C'$ represents $b$, then $AC \cong A'C'$ by axiom (C3); cf. (8.2).

(2) Let $AB$ represent $a$, and choose $C$ such that $A * B * C$ and $BC$ represents $b$, as in the definition. Then $AC$ represents $a + b$. Now take $DE$ to represent $b$, and lay off $F$ such that $D * E * F$ and $EF$ represents $a$. Then $DF$ represents $b + a$. But $AB \cong FE$ and $BC \cong ED$, so $AC \cong FD$ by (C3). This shows that $a + b = b + a$.

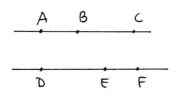

(3) To get $(a + b) + c$ we first choose $AB \in a$, then find $C$ such that $A * B * C$ and $BC \in b$, then find $D$ such that $A * C * D$ and $CD \in c$. Then $AD$ represents $(a + b) + c$.

On the other hand, let $EF \in b$ and choose $G$ such that $FG \in c$. Then $EG$ represents $b + c$. To get $a + (b + c)$ we need to find a point $H$ with $A * B * H$ and $BH \cong EG$. But $BD \cong EG$ by (C3), so $H = D$ by the uniqueness part of (C1). Therefore, $(a + b) + c = a + (b + c)$.

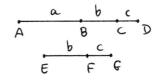

(4) Given two classes $a, b$ on a ray from a point $A$, lay off points $B, C$ such that $AB \in a$ and $AC \in b$. If $B = C$, then $a = b$. If $A * B * C$, then $a + [BC] = b$. If $A * C * B$, then $a = b + [CB]$. By (B3) these are the only possibilities, and this proves (4).

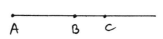

Before we define multiplication, we need a standard unit segment. So choose arbitrarily, and then fix once and for all, a segment class we call the *unit segment*, and denote it by 1. We also need the parallel axiom (P), even for the definition of the product (Exercise 19.1).

### Definition

Given two segment classes $a, b$, we define their *product* $ab$ as follows. First make a right triangle $ABC$ with $AB \in 1$ and $BC \in a$, where the right angle is at $B$. Let $\alpha$ be the angle $\angle BAC$. Now make a new right triangle $DEF$ with $DE \in b$ and having the same angle $\alpha$ at $D$. Then we define $ab$ to be the class of side $EF$ of this new triangle.

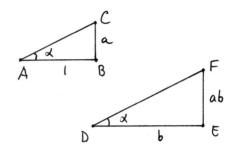

### Proposition 19.2

*In any Hilbert plane with* (P), *multiplication of segment classes has the following properties:*

(1) $ab$ *is well-defined.*
(2) $a \cdot 1 = a$ *for all* $a$.
(3) $ab = ba$ *for all* $a, b$.
(4) $a(bc) = (ab)c$ *for all* $a, b, c$.
(5) *For any* $a$, *there is a unique* $b$ *such that* $ab = 1$.
(6) $a(b + c) = ab + ac$ *for all* $a, b, c$.

*Proof* (1) The product is well-defined. If $A'B'C'$ is another right triangle with sides $1, a$, then it is congruent to $ABC$ by (SAS). Hence we get a congruent angle $\alpha$. If $D'E'F'$ is another right triangle with angle $\alpha$ and side $b$, then it is congruent to $DEF$ by (ASA). So we get a congruent segment $E'F'$.

(2) To compute $a \cdot 1$, we take the triangle $DEF$ to have side $b = 1$ and angle $\alpha$. Then $DEF \cong ABC$ by (ASA), so $a \cdot 1 = a$.

(3) Given $a, b$, first make a right triangle $ABC$ with sides $1, a$. This determines the angle $\alpha = \angle BAC$. Now extend $CB$ on the other side of $AB$ to $D$, so that $BD \in b$, and draw a line through $D$ making an angle $\alpha$ with $BD$, on the far side of $BD$ from $A$. Let this line meet $AB$ extended to $E$. Then $DBE$ is a right triangle with side $b$ and angle $\alpha$, so the segment $BE$ represents $ab$ by definition.

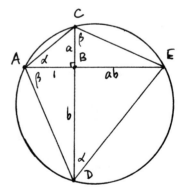

Now consider the four points $ACDE$. We will use the method of cyclic quadrilaterals developed in Section 5. Because the angles $\angle CAE$ and $\angle CDE$ are

de ces deux, comme l'vnité est a l'autre, ce qui est le mefme que la Diuifion ; ou enfin trouuer vne, ou deux, ou
plufieurs moyennes proportionnelles entre l'vnité, &
quelque autre ligne ; ce qui est le mefme que tirer la racine quarrée, ou cubique, &c. Et ie ne craindray pas d'in-
troduire ces termes d'Arithmetique en la Geometrie,
afin de me rendre plus intelligible.

La Multi-
plication.

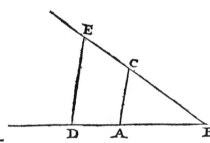

Soit par exemple A B
l'vnité, & qu'il faille
multiplier B D par B C,
ie n'ay qu'a ioindre les
poins A & C, puis tirer
DE parallele a CA, &
B E est le produit de
cette Multiplication.

La Divi-
fion.

Oubien s'il faut diuifer
BE par BD, ayant ioint les poins E & D, ie tire A C parallele a DE, & BC est le produit de cette diuifion.

L'Extra-
ction de la
racine
quarrée.

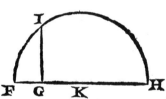

Ou s'il faut tirer la racine
quarrée de G H, ie luy ad-
ioufte en ligne droite F G,
qui est l'vnité, & diuifant FH
en deux parties égales au
point K, du centre K ie tire
le cercle FIH, puis efleuant du point G vne ligne droite
iufques à I, à angles droits fur F H, c'est G I la racine
cherchée. Ie ne dis rien icy de la racine cubique, ny des
autres, à caufe que i'en parleray plus commodement cy-
aprés.

Comment
on peut

Mais fouuent on n'a pas befoin de tracer ainfi ces li-

Plate VII. A page from *La Géométrie* of Descartes (1664), showing how he multiplies two
line segments to get another, and how he finds the square root of a line segment.

both equal to $\alpha$, they satisfy the hypotheses of (5.8), so the points $ACDE$ form a cyclic quadrilateral. Applying (5.8) to the same points in a different order, it follows that the angles $\angle DAE$ and $\angle DCE$ are equal; call this class $\beta$. To compute the product $ba$ we first use the triangle $ABD$, obtaining the angle $\beta$, and then use the triangle $CBE$, which has angle $\beta$ and side $a$. This shows that $BE$ represents the product $ba$. Thus $ab = ba$.

(4) For the associative law, we proceed as follows. Make right triangles with $1, a$ to define the angle $\alpha$, and with $1, c$ to define the angle $\gamma$. Make a right triangle $ABC$ with angle $\alpha$ and side $b$ to determine $ab$.

Extend $CB$ on the other side of $AB$ to meet a line from $A$ making an angle $\gamma$ with $AB$. Then $BD$ represents $cb$. Now make a line at $D$ with angle $\alpha$ to meet $AB$ extended at $E$. Then $BE$ will represent $a(cb)$.

As in the previous proof, the angles $\alpha$ at $A$ and $D$ show that $ACDE$ is a cyclic quadrilateral. Then by (5.8) again we conclude that $\angle BCE = \gamma$. It follows that $BE$ also represents the segment $c(ab)$. Thus $a(cb) = c(ab)$. Then using the commutative law already proved, we get $a(bc) = (ab)c$.

(5) Given $a$, make a right triangle with sides $1, a$ to define $\alpha$, and let $\beta$ be the other acute angle in that triangle. Then make a right triangle with angle $\beta$ and side $1$ to determine a new segment $b$. Since the other angle in this triangle is $\alpha$ (I.32), this second triangle shows that $ab = 1$.

(6) Given $a, b, c$, let $\alpha$ be determined by the right triangle with sides $1, a$. Make a right triangle $ABC$ with side $b$ and angle $\alpha$ to determine $BC \in ab$. Choose $D$ on the line $AB$ such that $A * B * D$ and $BD \in c$. Draw $CE$ parallel to $AB$, and $DEF$ perpendicular to $AB$. Then $\angle ECF = \alpha$, and $CE \in c$, so $EF$ represents $ac$.

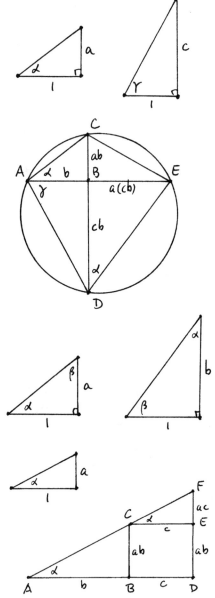

Because *BCDE* is a rectangle, $DE \in ab$. Now by definition of sum, *AD* represents $b + c$ and *DF* represents $ab + ac$. On the other hand, the triangle *ADF* has side $b + c$ and angle $\alpha$, so *DF* also represents $a(b + c)$. Hence $a(b + c) = ab + ac$.

## Remark 19.2.1

Let us examine carefully the hypotheses needed for the validity of these two results, (19.1) and (19.2). The first, concerning addition of line segments, is valid in any Hilbert plane (Exercise 8.1). On the other hand, even for the definition of the product, we need (P), or its equivalent, Euclid's fifth postulate, to guarantee that the point *F* exists. For the proof of (19.2) you might feel more comfortable assuming the hypotheses of a Euclidean plane, in which case we have justified the needed results from Book III (12.4). But if you look closely, we do not need the Euclidean axiom (E) on intersection of circles for (5.8): See Exercise 19.2. We also need (I.32), which uses (P) but not (E). Thus these two results hold in a Hilbert plane with (P). We do not need (E), nor did we ever use Archimedes' axiom (A).

## Proposition 19.3

*Given a Hilbert plane satisfying* (P), *and a unit segment* 1 *having been chosen, there is a unique (up to isomorphism) ordered field F whose set of positive elements P is the set of congruence equivalence classes of line segments with operations* $+, \cdot$ *defined above.*

*Proof* This is a consequence of the purely algebraic lemma that follows.

## Lemma 19.4

*Let P be a set, with two operations* $+, \cdot$ *defined on it that satisfy the properties listed in* (19.1) *and* (19.2). *Then there is a unique ordered field F whose positive elements form the set P.*

*Proof* One is tempted to define *F* to be the set $P \cup \{0\} \cup -P$, for intuitively, this is what is happening. *F* will consist of the original set *P*, plus a 0 element, plus another set of "negative" elements that is in 1-to-1 correspondence with the set of positive elements. However, I believe that we can obtain a cleaner proof by imitating the definition of the quotient field of an integral domain using ordered pairs, except that this time our ordered pairs will represent differences of elements of *P*.

So here is the formal construction. Let *F* be the set of equivalence classes $(a, b)$ of ordered pairs (think of $(a, b)$ as being $a - b$) of elements of *P*, where

$$(a, b) \sim (a', b') \quad \text{if } a + b' = a' + b.$$

Define addition by

$$(a, b) + (c, d) = (a + c, b + d)$$

and multiplication by

$$(a, b)(c, d) = (ac + bd, ad + bc).$$

We must verify that these operations are well-defined, i.e., if we replace an ordered pair by an equivalent ordered pair, the result is equivalent (!). (The symbol (!) means a trivial verification left to the reader. There will be lots of these, and all will result by using properties of $+$ and $\cdot$ in the set $P$.)

Then we let 0 denote the equivalence class of $(a, a)$ for any $a \in P$, and note that 0 acts as an additive identity (!). Also note that addition is commutative (!) and associative (!). For any pair $(a, b)$ note that $(b, a)$ acts as an additive inverse (!). Thus the set $F$ together with the operation $+$ is an abelian group.

Next verify that multiplication is associative (!), commutative (!), and distributive over addition (!). Let 1 be the class of $(1 + a, a)$ for any $a \in P$. Thus 1 acts as a multiplicative identity (!), and there exist multiplicative inverses (!). Hence $F$ together with $+, \cdot$ is a field.

We define a mapping $\varphi : P \to F$ by $a \in P$ goes to $(a + b, b)$ for any $b \in P$. This mapping is 1-to-1 onto its image (!), which we therefore identify with $P$. Also, $\varphi$ preserves $+, \cdot$ (!), so that $P$ has already two of the three properties required for $P$ to be the set of positive elements of an ordered field. It remains to verify the trichotomy, namely, for any $x = (a, b)$ in $F$, either $x \in P$ or $x = 0$ or $-x \in P$. If $a = b$, then $x = 0$. We will use property (4) of (19.1). If there exists a $c$ such that $a + c = b$, then $x = (a, b) = (a, a + c)$, and the negative of this element satisfies $-x = (a + c, a) \in P$. If on the other hand there is a $d$ such that $a = b + d$, then $x = (a, b) = (b + d, b) \in P$.

This concludes the proof modulo a million tedious verifications (!) left to the reader!

**Remark 19.4.1**
We will see in the next section (20.7) that $F$ is necessarily Pythagorean.

# Exercises

19.1  Explain where and how (P) is needed in the definition of the product.

19.2  Show that the result (Proposition 5.8) about cyclic quadrilaterals holds in any Hilbert plane with (P).

19.3  Supply the missing verifications in the proof of (Lemma 19.4).

19.4  If we start with the Cartesian plane over a field $F_0$, show that the field $F$ of segment arithmetic constructed in Proposition 19.3 is naturally isomorphic to the original field $F_0$.

19.5 Is Lemma 19.4 still true if we omit property (4) of Proposition 19.1, but keep all the other properties of Proposition 19.1 and Proposition 19.2? Give a proof or counterexample.

19.6 In Proposition 19.3, show that if we choose a different unit segment $1'$, the resulting field $F'$ is isomorphic to $F$.

# 20 Similar Triangles

We continue to work in a Hilbert plane satisfying (P).

Now that we have defined the arithmetic of line segments and have constructed a field $F$ whose positive elements correspond to congruence classes of line segments, we can establish a theory of proportion and similar triangles. The results are the same as Euclid's in Book VI, but our methods are different.

For any line segment $AB$, its congruence equivalence class $a$ is an element of the field $F$. We will call $a$ the *length* of $AB$, to conform with the usual terminology. If $AB$ and $CD$ are two segments with lengths $a, b$, we can speak of their *ratio* as the quotient $a/b \in F$. We say that four segments with lengths $a, b, c, d$ are *proportional* if $a/b = c/d$ as elements of the field $F$.

**Definition** (VI, Definition 1)
Two triangles $ABC$ and $A'B'C'$ are *similar* if the three angles of one are respectively equal to the three angles of the other, and the corresponding sides are proportional, i.e.,

$$a/a' = b/b' = c/c'.$$

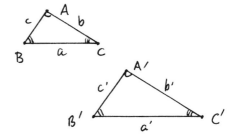

**Proposition 20.1** (Sim AAA) (VI.4)
*If two triangles $ABC$ and $DEF$ have their three angles respectively equal, then the two triangles are similar.*

*Proof* Our definition of multiplication in the field of segment arithmetic was based on a special case of the notion of similar triangles, namely, comparing the legs of equal-angled right triangles. So we will prove this result, following Hilbert, by reducing to this case.

In the first triangle, draw the angle bisectors of the three angles, and let them meet at the point $I$ (cf. Exercise 1.8 or (IV.4)). Recall from the proof of (IV.4) that $I$ is equidistant from the three sides of the triangle: If we drop perpendiculars from $I$ to the three sides, we obtain three congruent segments $h$. Also, in the course of the proof we obtained congruent triangles about each vertex: $AFI \cong AEI$, etc. Thus we get congruent segments $AE \cong AF$, which we call $x$, $BD \cong BF = y$, and $CD \cong DE = z$.

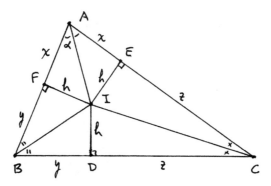

Make a similar construction in the second triangle $A'B'C'$, obtaining points $D', E', F', I'$ and segments $x', y', z', h'$.

Let $\alpha$ be one-half of the angle at $A$, draw a right triangle with one leg equal to 1, and let $r$ be the other leg. Then by the definition of segment multiplication, $h = rx$. In the second triangle, the angle at $A'$ is equal to the angle at $A$ by hypothesis, so one-half of it is also $\alpha$, so we find similarly that $h' = rx'$. Dividing one equation by the other, we find that $x/x' = h/h'$.

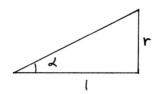

In the same way, working from the other two vertices of the triangle, we obtain $y/y' = h/h'$ and $z/z' = h/h'$. If we let $h/h' = k$, then we can write these results as

$$x = kx',$$

$$y = ky',$$

$$z = kz'.$$

The sides of the original triangle are formed of sums of these. Thus $a = y + z$ and $a' = y' + z'$. It follows from the distributive law that

$$a = ka',$$

and by the same reasoning also

$$b = kb',$$

$$c = kc'.$$

Then $a/a' = b/b' = c/c'$, so the two triangles are similar.

While this proof has an entirely different basis from Euclid's, the other results on similar triangles in Book VI will now follow easily, but in a different order.

**Proposition 20.2** (VI.2)
*In any triangle ABC, let B'C' be drawn parallel to BC. Then the sides AB and AC are proportional to AB' and AC'. Conversely, if the sides are divided by points B', D such that AB, AC are proportional to AB', AD, then B'D is parallel to BC.*

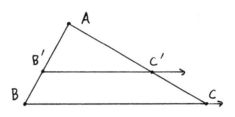

*Proof* Since $B'C'$ is parallel to $BC$, the angles at $B', C'$ are equal to the angles at $B, C$, respectively (I.29). Since the angle at $A$ is common, the triangles $ABC$ and $AB'C'$ have their three angles equal, and so they are similar (20.1). It follows that the sides are proportional.

Conversely, suppose we are given $B', D$ such that $AB, AC$ are proportional to $AB', AD$. Draw $B'C'$ parallel to $BC$. Then also $AB, AC$ are proportional to $AB', AC'$. Since we are working in a field $F$, the fourth proportional to three given quantities is uniquely determined. Hence $AD \cong AC'$. Since the points $D, C'$ lie on the same ray from $A$, the points $D, C'$ are equal (axiom (C1)). Hence $B'D$ is parallel to $BC$.

**Proposition 20.3** (Sim SSS) (VI.5)
*Suppose two triangles ABC and A'B'C' have their three sides respectively proportional to each other. Then the two triangles are similar.*

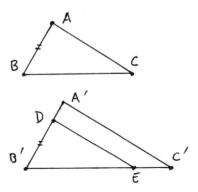

*Proof* Supposing the sides of the second triangle to be larger, find a point $D$ on the segment $B'A'$ such that $B'D \cong BA$. Then draw a line through $D$ parallel to $A'C'$. It follows (20.2) that the triangles $A'B'C'$ and $DB'E$ are similar, and in particular, their sides are proportional. But the sides of $ABC$ are also proportional to the sides of $A'B'C'$, so it follows (from field arithmetic) that the sides of $ABC$ are proportional to the sides of $DB'E$.

On the other hand, $B'D$ was chosen congruent to $BA$. So the proportionality factor is 1, and it follows that all three sides of $ABC$ are congruent to all three sides of $DB'E$. Then by the congruence criterion (SSS) = (I.8), the triangle $ABC$ is congruent to $DB'E$. In particular, the three angles of $ABC$ are equal to the three angles of $DB'E$, which in turn are equal to the three angles of $A'B'C'$, since the latter two are similar. Thus we have proved that the angles of $ABC$ are equal to the angles of $A'B'C'$, and so the two triangles are similar.

**Proposition 20.4** (Sim SAS) (VI.6)
*Suppose that two triangles $ABC$ and $A'B'C'$ have the angles at $A$ and $A'$ equal, and the two sides $AB, AC$ are proportional to the two sides $A'B', A'C'$. Then the two triangles are similar.*

*Proof* (Exercise 20.1).

**Theorem 20.5**
*In a Hilbert plane with* (P), *the results of Euclid's theory of similar triangles* (VI.2)–(VI.13) *all hold.*

*Proof* The propositions (VI.2)–(VI.6) appear as results in this section, or Exercises 20.1, 20.2. Also, (VI.8) is covered in the proof of (20.6) below. The remaining results follow easily, replacing Euclid's references to Book V by algebraic reasoning in the field of segment arithmetic.

**Remark 20.5.1**
Proposition (VI.1) and most of the latter part of Book VI, namely Propositions (VI.14)–(VI.31), deal with the connection between proportionality of figures and their area, so we postpone discussion of these until Chapter 5 (Exercise 23.7).

Next, using our segment arithmetic and the theory of similar triangles, we can prove some analogues of theorems that Euclid stated in terms of area, but that we will state as equations in the field $F$.

**Proposition 20.6**
*If $ABC$ is a right triangle with legs $a$, $b$ and hypotenuse $c$, then*

$$a^2 + b^2 = c^2$$

*in the field $F$ of segment arithmetic.*

*Proof* This, of course, is another version of the Pythagorean theorem (I.47), which Euclid proved in terms of the areas of the squares built on the sides of the triangle. The present statement in terms of segment arithmetic is of a totally different nature, and neither implies, nor is implied by, the previous statement, until we have made some connection between area and segment arithmetic (cf. Chapter 5).

To prove the current statement, drop a perpendicular $CD$ from the vertex with the right angle to the hypotenuse. Then we find that the original triangle $ABC$ has the same angles as the two new triangles $ACD$ and $CBD$. Hence all three triangles are similar, by (20.1). (This statement is (VI.8) in Euclid.) Then corresponding sides are proportional, and we obtain

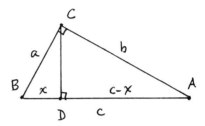

$$\frac{x}{a} = \frac{a}{c}$$

using $CBD$ similar to $ABC$, and we obtain

$$\frac{c-x}{b} = \frac{b}{c}$$

using $ACD$ similar to $ABC$. Cross multiplying, we obtain

$$cx = a^2,$$

$$c^2 - cx = b^2,$$

from which by substituting we obtain

$$a^2 + b^2 = c^2.$$

**Corollary 20.7**
*In a Hilbert plane satisfying* (P), *the field of segment arithmetic* (19.3) *is Pythagorean.*

*Proof* We must show for any $a \in F$ that $\sqrt{1 + a^2} \in F$. If $a = 0$, this is trivial; if $a$ is negative, we can replace $a$ by $-a$, so we may assume that $a$ is positive. Then $a$ is the length of a certain segment. If we construct a right triangle with legs 1 and $a$, then by (20.6) the hypotenuse will be a segment whose class in $F$ is $\sqrt{1 + a^2}$. Thus $F$ is Pythragorean.

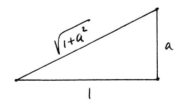

**Proposition 20.8** (cf. (III.35))
*If two chords of a circle meet, cutting each
other in segments of lengths $a, b, c, d$, then*

$$ab = cd$$

*in the field $F$ of segment arithmetic.*

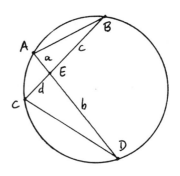

*Proof* Draw the lines $AB$ and $CD$. Then
by (III.21) the angles at $B$ and $D$ are
equal, and the angles at $A$ and $C$ are
equal. Hence the triangles $ABE$ and
$CDE$ have the same angles, so are similar (20.1). It follows that corresponding
sides are proportional: $a/c = d/b$. Cross
multiplying, we obtain $ab = cd$.

**Proposition 20.9** (cf. (III.36))
*Let $A$ be a point outside a circle, let the line
$AB$ be tangent to the circle at $B$, and let the
line $ACD$ cut the circle at $C$ and $D$. Then,
in the field of segment arithmetic,*

$$(AB)^2 = (AC) \cdot (AD).$$

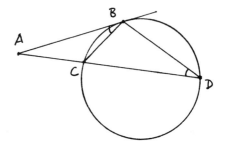

*Proof* Draw the lines $BC$ and $BD$. Then $\angle ABC = \angle ADB$ by (III.32). Since the
angle at $A$ is common, the triangles $ABC$ and $ADB$ have two (and hence three)
angles equal, so they are similar (20.1). It follows that corresponding sides are
proportional, namely,

$$\frac{AB}{AC} = \frac{AD}{AB}.$$

Cross multiplying gives

$$(AB)^2 = (AC) \cdot (AD).$$

As applications of similar triangles, we give some other well-known theorems here and in the exercises.

**Proposition 20.10** (Menelaus's theorem)
*Let $ABC$ be any triangle, and let a line $l$ cut the sides of the triangle (extended if nec-
essary) in points $D, E, F$. Then*

$$\frac{AD}{BD} \cdot \frac{BF}{CF} \cdot \frac{CE}{AE} = 1.$$

*Proof* Draw a line through $A$ parallel to $BC$, and let it meet $l$ at $G$. Then the triangle $ADG$ is similar to $BDF$, and the triangle $AEG$ is similar to $CEF$. From this we obtain

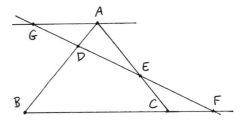

$$\frac{AD}{AG} = \frac{BD}{BF} \quad \text{and} \quad \frac{AE}{AG} = \frac{CE}{CF}.$$

Eliminating $AG$ from these equations and rearranging gives the result.

# Exercises

These exercises take place in a Hilbert plane with (P).

20.1 Prove (Sim SAS) (Proposition 20.4).

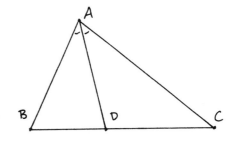

20.2 (VI.3) Let $ABC$ be any triangle, and let $AD$ be the angle bisector at $A$. Prove that $AB$ and $AC$ are proportional to $BD$ and $DC$.

20.3 Let $A$ be a point outside a circle, and draw any two lines through $A$ cutting the circle at $B, C$ and $D, E$. Then show that

$$(AB) \cdot (AC) = (AD) \cdot (AE).$$

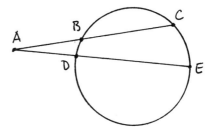

The product $(AB) \cdot (AC)$, which thus depends only on $A$, is called the *power* of the point $A$ with respect to the circle. If $A$ is inside the circle, we use signed lengths, so that the power of $A$ will be positive if $A$ is outside the circle, and negative if $A$ is inside the circle.

20.4 If two circles intersect in two points, the line through those points is called the *radical axis* of the two circles. Show that the radical axis is equal to the set of those points $A$ in the plane for which the power of $A$ with respect to the first circle is equal to the power of $A$ with respect to the second circle. (Even when the two circles do not intersect, this latter property defines a straight line that is taken to be the radical axis in that case.)

20.5 If three circles each meet the other two in two points, and their centers are not collinear, show that the three radical axes of the circles, taken two at a time, meet in a single point. (We will see later (Exercise 39.20) that this result also holds in the Poincaré model of non-Euclidean geometry. So we can ask, is it true in any Hilbert plane?)

20.6 In a Hilbert plane with (P), given two circles by their centers and one point each, but without being given their intersection points, show that the following construction (which can be done with Hilbert's tools) gives the radical axis of the two circles.

Let the two circles be defined by their centers $O_1, O_2$ and their points $A_1, A_2$. Let $B$ be the midpoint of $A_1A_2$. Drop a perpendicular from $A_1$ to $O_1B$, and a perpendicular from $A_2$ to $O_2B$, and let these two lines meet at $P$. Then the perpendicular from $P$ to $O_1O_2$ is the required radical axis of the two circles.

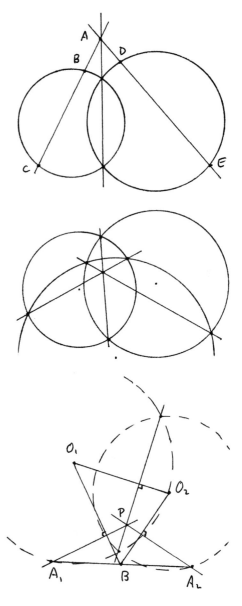

20.7 (Ceva's theorem). Let $ABC$ be any triangle, and let $P$ be any point inside the triangle. Draw lines from the vertices through $P$ meeting the opposite sides at $D, E, F$. Then show that

$$\frac{AD}{BD} \cdot \frac{BF}{CF} \cdot \frac{CE}{AE} = 1.$$

*Hint*: Imitate the proof of Proposition 20.10.

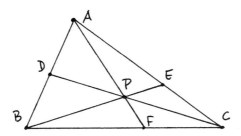

20.8 (Desargues's theorem). Let $ABC$ and $A'B'C'$ be two triangles. Assume that $AA', BB', CC'$ all pass through a single point $O$ (we can say that $ABC$ and $A'B'C'$ are *perspective* from $O$). Assume further that $AB$ is parallel to $A'B'$, and $BC$ is parallel to $B'C'$. Prove that $AC$ is parallel to $A'C'$. (Compare Exercise 14.2, where the same result is proved in a different situation.)

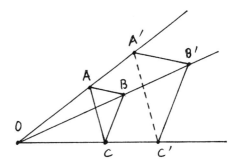

20.9 Prove the field analogue of (II.11), as follows. Let $AB$ be a given line segment. Construct $AC \cong AB$ and perpendicular to it. Let $D$ be the midpoint of $AC$. Then find $E$ on $AC$ such that $DE \cong DB$. Find $F$ on $AB$ such that $AE \cong AF$. Prove that

$$(AF)^2 = (AB)(BF)$$

in the field of segment arithmetic. *Hint*: Use Proposition 20.6. We say that $AB$ has been divided in *extreme and mean ratio*.

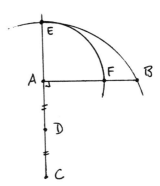

20.10 Give a new proof of (IV.10) as follows. Let the segment $AB$ be divided in extreme and mean ratio as in Exercise 20.9 above: $(AF)^2 = (AB) \cdot (BF)$. Construct a triangle $ABC$ such that $AC \cong AB$ and $BC \cong AF$. Prove that the base angles of the isosceles triangle $ABC$ are each equal to twice the vertex angle at $A$. *Hint*: Use (Sim SAS) (Proposition 20.4) to obtain similar triangles.

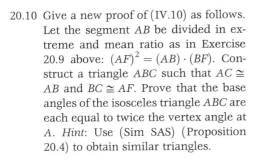
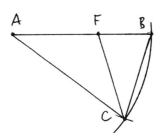

20.11 Let $OA$ and $OB$ be two perpendicular
radii of a circle. Let $C$ be the midpoint
of $OB$. Let $CD$ be the angle bisector of
$\angle ACO$. Let $DE$ be perpendicular to
$OA$. Then prove that $AE$ is a side of
the regular pentagon inscribed in the
circle. *Hint*: Use Exercise 20.2.

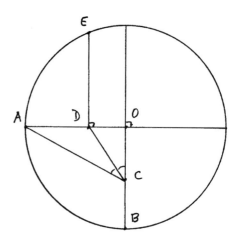

How many steps would it take to
construct the pentagon by this method
(given the circle and its center)?

20.12 Prove that in a Hilbert plane with (P) (without assuming (E)), there exists an equilateral triangle with given side $AB$. *Hint*: First show that the field $F$ of segment
arithmetic contains an element $\frac{1}{2}\sqrt{3}$.

20.13 Given a triangle $ABC$ with acute
angles at $B$ and $C$, make a ruler and
compass construction for a square
with one edge along the side $BC$, and
the other two vertices on the sides
$AB, AC$. We call this an *inscribed
square* (par $= 17$).

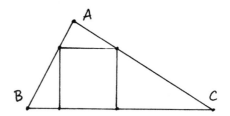

20.14 Match wits with the great nineteenth-century geometer Jakob Steiner: This is one
of his many theorems published without any indication of proof (Werke (1881)
vol. I, p. 128). Suppose you are given four lines in the plane, no two parallel, and
no three concurrent. Taken three at a time, they make four triangles. Show that the
orthocenters (intersection of the altitudes) of these four triangles are collinear.

20.15 (Trigonometry). In a Hilbert plane
with (P), suppose that you are given a
right triangle $ABC$ with sides $a, b, c$
and angle $\alpha$ at $A$. Define

$$\sin \alpha = \frac{a}{c}, \quad \cos \alpha = \frac{b}{c}, \quad \tan \alpha = \frac{a}{b}$$

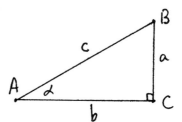

as elements of the field of segment
arithmetic $F$.

(a) Show that the functions $\sin \alpha$, $\cos \alpha$, $\tan \alpha$ depend only on the angle $\alpha$, and not
on the particular triangle chosen.

(b) Prove the identity

$$\sin^2 \alpha + \cos^2 \alpha = 1.$$

20.16 (Law of cosines). Let $ABC$ be any triangle, with sides $a, b, c$, and angle $\alpha$ at $A$. Using the cosine function defined in Exercise 20.15, prove the law of cosines

$$a^2 = b^2 + c^2 - 2bc \cos \alpha.$$

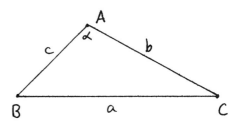

*Hint*: Draw an altitude to make two right triangles, and use Proposition 20.6. How does this result relate to Euclid (II.13)?

20.17 (Law of sines). With the same notation as in Exercise 20.16, prove the law of sines:

$$\frac{\sin \alpha}{a} = \frac{\sin \beta}{b} = \frac{\sin \gamma}{c}.$$

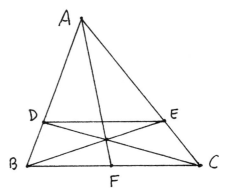

20.18 Let $ABC$ be a triangle, and let $D, E, F$ be points on the sides such that $AF, BE$, and $CD$ are concurrent. Show that $DE$ is parallel to $BC$ if and only if $F$ is the midpoint of $BC$. *Hint*: Use Ceva's theorem (Exercise 20.7).

20.19 (a) Given a line segment $BC$ and its midpoint $F$, construct with ruler alone a line through a given point $D$ parallel to the line $BC$ (par = 6).

(b) Given a segment $BC$ and given a line $m$ parallel to the line $BC$ and distinct from it, construct with ruler alone the midpoint of $BC$ (par = 5).

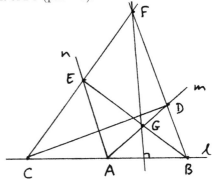

20.20 Verify the following construction due to Hilbert. Given a line $l$, to construct a line perpendicular to $l$ (at an unspecified point) using only ruler and dividers (cf. Section 10). Take any two points $A$, $B$ on $l$ and any two rays $m$, $n$ emanating from $A$. Lay off segments $AC, AD, AE$ equal to $AB$. Let $BD$ meet $CE$ at $F$; let $CD$ meet $BE$ at $G$. Then the line $FG$ is perpendicular to $l$ (10 steps).

20.21 In a Hilbert plane with (P), given an
angle $\alpha$, given a ray $r$, and given a side
of $r$, construct with ruler and dividers
only an angle equal to $\alpha$ on the given
side of the ray $r$ (par = 30). *Hint*: Embed $\alpha$ in a right triangle and transport
its legs, using Exercises 20.19, 20.20.

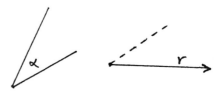

It follows that any construction possible with Hilbert's tools (Section 10) is
actually possible using only the ruler and the dividers: The present construction
makes the transporter of angles superfluous.

# 21   Introduction of Coordinates

In this section we will complete a logical circle by showing that if $\Pi$ is an
abstract geometry satisfying the axioms of a Hilbert plane plus (P) and if $F$ is the
field of segment arithmetic for $\Pi$ (19.3), then $\Pi$ is isomorphic to the Cartesian
plane over the field $F$.

Let me explain this in greater detail. We started our study of geometry from
two different perspectives. On the one hand we considered a purely geometric
development, where points, lines, congruence, etc., were undefined notions
subject to certain axioms, from which we prove theorems. This is Euclid's
approach, improved by Hilbert, who gave us a set of axioms including Euclid's
unstated assumptions, so that we could develop his geometry on a rigorous basis.

On the other hand, we constructed examples, or *models*, of this abstract geometry, based on the logical foundations of modern algebra, by starting with an
ordered field $F$ (for example the real numbers), and making a geometry whose
points are ordered pairs of elements of the field $F$. This is the Cartesian
approach (cf. Section 13). In this model we defined lines and congruence, using
linear equations and a distance function, and then proved, by algebraic methods, that the axioms of abstract geometry are true.

For any particular field $F$, it may happen that certain things are true that do
not hold in every geometry: For the plane $F^2$ is just one of many possible models
of an abstract geometry. For example, if $F = \mathbb{R}$, then Dedekind's axiom (D)
holds, but it does not hold in the field of *constructible* numbers (16.4).

Perhaps more interesting is that we can prove certain results in the geometry over any field $F$, though we do not know how to prove the corresponding
statement in abstract geometry. For example, over any field $F$, the line–circle
intersection property (LCI) is equivalent to the circle–circle intersection property (E), because we have shown that both of these are equivalent to the Euclidean condition on the field $F$, (16.2). We do not know any purely geometric
proof of this equivalence.

Of course, it might be that the geometries constructed over fields were only some possible geometries, and that there were other abstract geometries, not corresponding to any field, with properties different from the geometries over fields. If we drop the parallel axiom, this is indeed the case, as we will see with the non-Euclidean geometries (Chapter 7). However, we will show that any abstract geometry with (P) is isomorphic to a geometry over a field.

To understand this, we need to be clear what we mean by an *isomorphism* of geometries. The Greek roots $iso + morph$ mean "the same form." Intuitively, both geometries behave the same way. Their outer structures may be different, but there is no way that they can be distinguished internally.

The formal definition of isomorphism of geometries is as follows.

**Definition**
Let $\Pi$ and $\Pi'$ be two Hilbert planes. An *isomorphism* between $\Pi$ and $\Pi'$ is a one-to-one mapping $\varphi : \Pi \to \Pi'$ of $\Pi$ onto $\Pi'$ that is compatible with the undefined notions. This means:

(1) A subset $L \subseteq \Pi$ is a line if and only if $\varphi(L) \subseteq \Pi'$ is a line.
(2) Three points $A, B, C \in \Pi$ satisfy the betweenness property $A * B * C$ if and only if $\varphi(A) * \varphi(B) * \varphi(C)$ in $\Pi'$.
(3) Given four points $A, B, C, D \in \Pi$, the line segments $AB$ and $CD$ are congruent if and only if the line segments $\varphi(A)\varphi(B)$ and $\varphi(C)\varphi(D)$ are congruent in $\Pi'$.
(4) If $\alpha$ is an angle formed by the rays $AB$ and $AC$ in $\Pi$, we denote by $\varphi(\alpha)$ the angle formed by the rays $\varphi(A)\varphi(B)$ and $\varphi(A)\varphi(C)$ in $\Pi'$. If $\alpha$ and $\beta$ are two angles in $\Pi$, then $\alpha$ and $\beta$ are congruent if and only if $\varphi(\alpha)$ and $\varphi(\beta)$ are congruent in $\Pi'$.

**Theorem 21.1** (Introduction of coordinates)
*Let $\Pi$ be a Hilbert plane satisfying the parallel axiom* (P). *Let F be the ordered field of segment arithmetic in $\Pi$ (19.3). Then F is Pythagorean (20.7), and $\Pi$ is isomorphic to the Cartesian plane $F^2$ over the field F.*

*Proof* We start by fixing two perpendicular lines in the plane $\Pi$, which we call the *x-axis* and the *y-axis*. We call their intersection point $O$ the *origin*. On each axis choose a point $1_x$ and $1_y$ such that the segments $O1_x$ and $O1_y$ both represent 1 in the field $F$. These then define the *positive rays* on the $x$-axis and the $y$-axis.

Now for any point $P$ in the plane, we drop perpendicular $PA$ to the $x$-axis and $PB$ to the $y$-axis. Let the segment $OA$ represent $a \in F$ and let $OB$ represent $b \in F$.

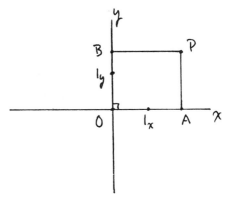

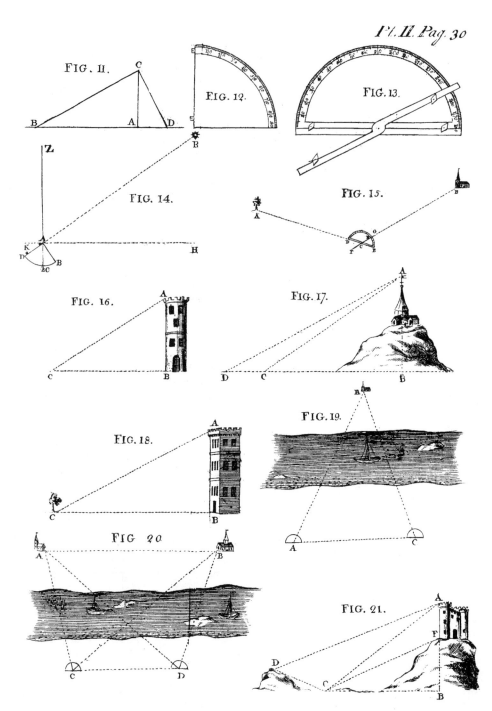

Plate VIII. Figures from Gregory's *Treatise of Practical Geometry* (1751).

Now we can define a mapping $\varphi : \Pi \rightarrow F^2$ by $\varphi(P) = (\pm a, \pm b)$, where we choose the $+$ sign if $A$ (resp. $B$) is on the positive $x$-axis (resp. $y$-axis) and the $-$ sign if not. Clearly, this construction gives a bijective correspondence between the set of points of $\Pi$ and the set of ordered pairs of the field $F$; so $\varphi$ is 1-to-1 and onto.

We must verify that $\varphi$ is compatible with the notions of line, betweenness, congruence of segments, and congruence of angles. And remember that in $\Pi$ these are undefined notions, whose properties are known only through the axioms and propositions of the geometry, while in $F^2$ they were defined in terms of algebraic conditions (Chapter 3).

*Step 1* Let $l$ be a line in $\Pi$. (For simplicity we will consider a general line, and let the reader check the special cases of horizontal and vertical lines(!).) Let $l$ meet the $x$-axis at $A$. Measure off $AB \in 1$, let $BC$ be a perpendicular, and let $m \in F$ be the class of $BC$. We call $m$ the *slope* of the line.

Let $l$ meet the $y$-axis at $D$, and let $b \in F$ represent that point (i.e., $b = OD$ if $D$ is on the positive $y$-axis; otherwise, $b = -OD$).

Now consider an arbitrary point $P = (x, y)$ in the plane. Make a triangle $DPE$ using horizontal and vertical lines. Then $DE = x$ and $PE = y - b$ (in the case shown; otherwise, adjust signs $\pm$ as needed (!)). This point $P$ will lie on the line $l$ if and only if the angle $PDE = \alpha$. Because of the definition of our segment arithmetic, this condition is equivalent to saying $y - b = mx$. In other words, $P = (x, y)$ lies on the line $l$ if and only if $y = mx + b$. Since lines in $F^2$ were defined by linear equations, this establishes the first property of an isomorphism: $L \subseteq \Pi$ is a line $\Leftrightarrow \varphi(L) \subseteq \Pi'$ is a line.

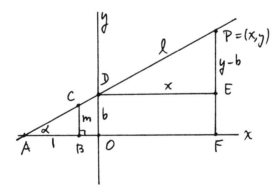

*Step 2* Let $A, B, C$ be three collinear points in $\Pi$ (which by Step 1 will guarantee that their images in $\Pi'$ are collinear). Let $A', B', C'$ be their projections on the $x$-axis (and again for simplicity we will treat the special case that $A, B, C$ are in the first quadrant, leaving other cases to the reader (!)).

Since the lines $AA', BB', CC'$ are parallel, $A$ and $C$ will be on opposite sides of the line $BB'$ if and only if $A'$ and $C'$ are on opposite sides of the line $BB'$ (7.2), so $A * B * C$ if and only if $A' * B' * C'$. Let the segments $OA', OB', OC'$ represent $a, b, c, \in F$. Then $A' * B' * C'$ means that either the segments are related $OA' \leq OB' \leq OC'$ or vice versa $OC' \leq OB' \leq OA'$. This is equivalent to saying $a < b < c$ or $c < b < a$, by the way we defined inequality in the field $F$ of segment arithmetic. And this, in turn, is equivalent to saying $\varphi(A) * \varphi(B) * \varphi(C)$ because of the definition of betweenness in $F^2$ (Section 15).

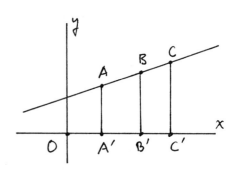

*Step 3*   Let $A, B$ be two points in $\Pi$, and let the segment $AB$ represent $d \in F$. On the other hand, let $\varphi(A) = (a_1, a_2)$ and $\varphi(B) = (b_1, b_2)$. Then if we draw the right triangle $ABC$ with legs parallel to the axes, we find that $AC = b_1 - a_1$ and $BC = b_2 - a_2$ by construction. We use the field version of the Pythagorean theorem (20.6) to conclude that

$$d^2 = (b_1 - a_1)^2 + (b_2 - a_2)^2$$

in $F$.

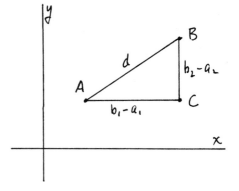

Now let $A'B'$ be another segment, with length $d' \in F$. Then similarly, if $\varphi(A') = (a_1', a_2')$ and $\varphi(B') = (b_1', b_2')$, we have

$$d'^2 = (b_1' - a_1')^2 + (b_2' - a_2').$$

Now $AB \cong A'B'$ if and only if $d = d'$, because $F$ was constructed from the set $P$ of congruence equivalence classes of line segments. On the other hand, $d = d'$ if and only if $d^2 = d'^2$, because both are positive elements of $F$. But the equations above show that $d^2$ and $d'^2$ are equal to the "distance squared" function that we used to define congruence of segments in $F^2$ (Section 16). Thus $AB \cong A'B'$ if and only if $\varphi(A)\varphi(B) \cong \varphi(A')\varphi(B')$.

*Step 4*   We show that two angles $\alpha, \alpha'$ in $\Pi$ are congruent if and only if $\varphi(\alpha)$ and $\varphi(\alpha')$ are congruent in $\Pi'$. For economy of exposition, we give an indirect proof, though a direct proof is also possible (see Exercise 21.2).

Suppose we are given angles $\alpha$ and $\alpha'$ in $\Pi$. Let the vertices be $A$ and $A'$, and choose any two points $B, C$ on the two rays of $\alpha$. Then find $B', C'$ on the rays of $\alpha'$ such that $AB \cong A'B'$ and $AC \cong A'C'$. Draw the lines $BC$ and $B'C'$ to make triangles.

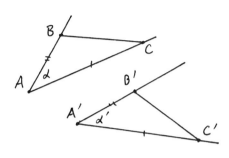

If $\alpha \cong \alpha'$, then by (SAS) it follows that the triangles $ABC$ and $A'B'C'$ are congruent, and in particular, $BC \cong B'C'$. Conversely, if $BC \cong B'C'$, then by (SSS) the two triangles are congruent, and so $\alpha \cong \alpha'$. Hence $\alpha \cong \alpha' \Leftrightarrow BC \cong B'C'$.

Apply $\varphi$ to the six points $A, B, C, A', B', C'$. Then $\varphi(A)\varphi(B) \cong \varphi(A')\varphi(B')$ and $\varphi(A)\varphi(C) \cong \varphi(A')\varphi(C')$ by Step 3. Furthermore, we have shown that the geometry $F^2$ satisfies Hilbert's axioms (Section 17), and in particular, (SAS) and (SSS) hold also in $F^2$. So by the same argument in $F^2$ we see that $\varphi(\alpha) \cong \varphi(\alpha')$ if and only if $\varphi(B)\varphi(C) \cong \varphi(B')\varphi(C')$.

Combining this result with Step 3 for the segments $BC$ and $B'C'$, we see that $\alpha \cong \alpha' \Leftrightarrow BC \cong B'C' \Leftrightarrow \varphi(B)\varphi(C) \cong \varphi(B')\varphi(C') \Leftrightarrow \varphi(\alpha) \cong \varphi(\alpha')$.

**Corollary 21.2**
*In any Hilbert plane $\Pi$ satisfying* (P), (LCI) *is equivalent to* (E), *and both are equivalent to saying that the field $F$ of segment arithmetic is Euclidean.*

*Proof*   Indeed, we have shown that this is true over a field (16.2), so by the theorem it is true in $\Pi$ also.

**Corollary 21.3**
*A Hilbert plane $\Pi$ satisfying* (P) *and* (D) *is isomorphic to the real Cartesian plane.*

*Proof*   By (21.1), $\Pi$ is isomorphic to the Cartesian plane over a Pythagorean ordered field $F$. By (15.4), the plane $\Pi$ satisfies (D) if and only if the field $F$ satisfies $(D')$. And then by (15.5), $F \cong \mathbb{R}$.

# Exercises

21.1 Given two adjacent nonoverlapping angles $\alpha, \beta$ at a point $A$, use the diagram shown, plus similar triangles and Proposition 20.6, to show that

$$\tan(\alpha + \beta) = \frac{\tan \alpha + \tan \beta}{1 - \tan \alpha \tan \beta}$$

in $F$ (cf. Exercise 20.15 for the definition of $\tan \alpha$).

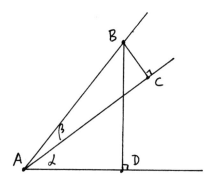

21.2 Use Exercise 21.1 above to give a direct proof of step (4) of Theorem 21.1, namely, that two angles $\alpha, \alpha'$ in $\Pi$ are congruent if and only if $\varphi(\alpha)$ and $\varphi(\alpha')$ are congruent in $F^2$, using the definition of congruence we gave for angles in $F^2$ (Section 16).

21.3 Give another proof that (LCI) is equivalent to (E) in a Hilbert plane with (P) by using the construction of Exercise 20.6.

21.4 In this and the following exercises we consider, in a Hilbert plane with (P), constructions with a ruler alone, but we are given a fixed circle $\Gamma$ and its center $O$, and we are allowed to intersect lines with this circle. The key observation is that any line through $O$ cuts off a diameter, with $O$ as its midpoint, and this allows us to draw parallel lines (Exercise 20.19).

Given $\Gamma$ and $O$, construct with ruler alone the midpoint of a given segment ($par = 15$). The diagram is given as a hint of one possible construction.

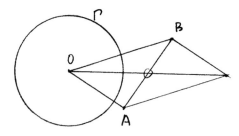

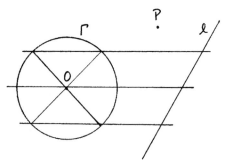

21.5 Given $\Gamma$ and $O$, construct with ruler alone a line parallel to a given line $l$ and passing through a given point $P$ ($par = 16$). *Hint*: First construct a bisected segment on $l$, as shown.

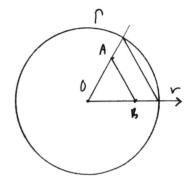

21.6 Given $\Gamma$ and $O$, and given a segment $OA$ and a ray $r$ originating at $O$, construct with ruler alone a point $B$ on $r$ with $OA \cong OB$ (par = 17).

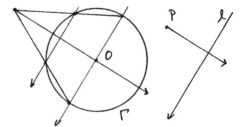

21.7 Given $\Gamma$ and $O$, and given a point $P$ and a line $l$, construct a perpendicular to $l$ through $P$ (par = 33).

21.8 Given $\Gamma$ and $O$, and given a circle $\Delta$ defined by its center $A$ and a point $B$, and given a line $l$, construct with ruler alone an intersection point of $\Delta$ and $l$ (par = 54).

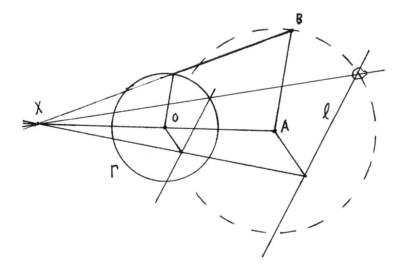

21.9 Now prove the theorem of Poncelet–Steiner, that any ruler and compass construction can be accomplished with ruler alone if we are given a single circle $\Gamma$ and its center $O$. *Hint*: Use the construction of Exercise 20.6 to reduce the problem of intersecting two circles to intersecting a circle and a line. Or else proceed algebraically and show first that the operations $+, -, \cdot, \div, \sqrt{}$ can be carried out on line segments; then use Theorem 13.2.

21.10 (Extra credit) Given a circle and its center, construct with ruler alone an inscribed regular pentagon (par = about 50).

But when the sceptre devolved to Almamon, the seventh of the Abbassides, he completed the designs of his grandfather, and invited the Muses from their ancient seats. His ambassadors at Constantinople, his agents in Armenia, Syria, and Egypt, collected the volumes of Grecian science: at his command they were translated by the most skillful interpreters into the Arabic language: his subjects were exhorted assiduously to peruse these instructive writings; and the successor of Mahomet assisted with pleasure and modesty at the assemblies and disputations of the learned ...

The sages of Greece were translated and illustrated in the Arabic language, and some treatises, now lost in the original, have been recovered in the version of the East, which possessed and studied the writings of Aristotle and Plato, of Euclid and Apollonius, of Ptolemy, Hippocrates, and Galen.

– from *The History of the Decline and Fall of the Roman Empire*
by Edward Gibbon, vol V, ch 52
Bigelow, NY (1845)

# 5 | Area

**CHAPTER**

ooking at Euclid's theory of area in Books I–IV, Hilbert saw how to give it a solid logical foundation. We define the notion of equal content by saying that two figures have equal content if we can transform one figure into the other by adding and subtracting congruent triangles (Section 22). We can prove all the properties of area that Euclid uses, except that "the whole is greater than the part." This is established only when we relate the geometrical notion of equal content to the notion of a measure of area function (Section 23).

In an Archimedean Euclidean plane, we prove the theorem of Bolyai and Gerwien, that figures of equal area are equivalent by dissection (Section 24). We also investigate the practical problem of dissecting one figure into another.

We briefly discuss the classical problem of squaring the circle (Section 25) and its influence.

In comparing the volumes of three-dimensional figures, Euclid uses a limiting process, the "method of exhaustion" (Section 26). We give Dehn's solution of Hilbert's third problem, that solid figures of equal volume are not necessarily equivalent by dissection (Section 27), thus vindicating Euclid's use of an infinite limiting process in the study of volume.

# 22   Area in Euclid's Geometry

Starting with (I.35), Euclid introduces a new notion of equality between figures, which corresponds to what we would call "equal area." The area we are familiar with from high school attaches a number to each figure. So the area of a rectangle with sides $a$ and $b$ is the number $ab$; the area of a triangle with base $b$ and height $h$ is $\frac{1}{2}bh$. In Euclid's geometry, there are no numbers, so we cannot explain his concept of area this way.

Euclid does not define this new notion of equality, but we can infer from his proofs that he considers it another undefined notion, like congruence of segments or angles, that satisfies certain properties similar to the common notions. In particular, he assumes that:

1. Congruent figures are "equal."
2. Sums of "equal" figures are "equal."
3. Differences of "equal" figures are "equal."
4. Halves of "equal" figures are "equal."
5. The whole is greater than the part.
6. If squares are "equal," then their sides are equal.

Properties 1, 2, and 3 are used in the proof of (I.35). Property 4 appears in the proof of (I.37), and property 5 appears in the proof of (I.39). Property 6, which is actually a consequence of 5, is used in the proof of (I.48).

We could accept this notion of "equality" between figures as another undefined notion, with these properties as additional axioms. However, one is reluctant to encumber the foundations of geometry with unnecessary undefined notions and axioms. So instead, following Hilbert, we will show that one can define a suitable notion of "equal area" and prove its properties, thus providing a new foundation for the theory of area. To avoid overuse of the word "equal," we introduce a new terminology and will say that certain figures have "equal content."

To begin with, let us be precise about our terminology. We presuppose the axioms of a Hilbert plane. When we speak of a *triangle ABC* in this chapter, we mean that subset of the plane consisting of the three line segments $AB$, $AC$, $BC$, the *sides* of the triangle, plus all the points in the interior of the triangle.

Recall (Section 7) that the *interior* of a triangle $ABC$ is the set of points that are on the same side of the line $AB$ as $C$, on the same side of $AC$ as $B$, and on the same side of $BC$ as $A$. Two triangles are *nonoverlapping* if they have no interior points in common. They may have common vertices or parts of edges.

**Definition**

A *rectilineal figure* (or *figure* for short) is a subset of the plane that can be expressed as a finite nonoverlapping union of triangles. A point $D$ is in the *interior* of a figure $P$ if there is a triangle $ABC$ entirely contained in $P$ such that $D$ is in the interior of the triangle $ABC$. Two figures are nonoverlapping if they have no interior points in common. Note that our definition of a figure includes its edges and all its interior points.

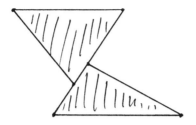

## Proposition 22.1

*The intersection of any two figures is a figure. The union of any two figures is a figure. The complement of one figure inside another figure (plus the line segments that form its sides) is a figure. In particular, any finite union of triangles is a figure.*

*Proof*   The basic idea is to deal with one triangle at a time. For example, if a triangle $ABC$ is cut by a line $l$, then that portion of the triangle that lies on one side of the line is a figure. One side $BDE$ in this example is a triangle. The other side is a union of two triangles, after we draw the line $DC$. We leave details to the reader (Exercises 22.1, 22.2, 22.3).

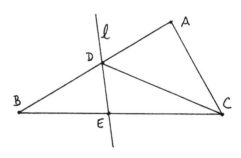

## Definition

Two figures $P, P'$ are *equidecomposable* if it is possible to write them as non-overlapping unions of triangles

$$P = T_1 \cup \cdots \cup T_n,$$

$$P' = T_1' \cup \cdots \cup T_n',$$

where for each $i$, the triangle $T_i$ is congruent to the triangle $T_i'$.

Two figures $P, P'$ have *equal content* if there are other figures $Q, Q'$ such that:

(1) $P$ and $Q$ are nonoverlapping.
(2) $P'$ and $Q'$ are nonoverlapping.
(3) $Q$ and $Q'$ are equidecomposable.
(4) $P \cup Q$ and $P' \cup Q'$ are equidecomposable.

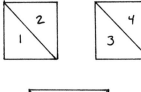

### Example 22.1.1

If $P$ is the union of two congruent squares in a Euclidean plane, and $P'$ is a square built on the diagonal of one of the squares of $P$, then $P$ and $P'$ are equidecomposable. Indeed, we cut $P$ and $P'$ into four congruent triangles each, as shown in the diagram.

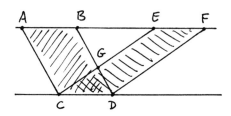

### Example 22.1.2

In a Euclidean plane, let $ABCD$ and $CDEF$ be two parallelograms on the same base $CD$ and lying in the same parallels. The Euclid's proof of (I.35) shows that $ABCD$ and $CDEF$ have equal content.

Indeed, if we let $P = ABCD$ and $P' = CDEF$, take $Q = Q' = $ triangle $BGE$. Then $P \cup Q$ and $P' \cup Q'$ are the unions of the congruent triangles $ACE$ and $BDF$ and the equal triangles $CDG$ and $CDG$.

### Example 22.1.3

If two figures $P$ and $P'$ are equidecomposable, then they have equal content, but the converse is not necessarily true. For example, consider the Cartesian plane over a non-Archimedean field $F$ (Section 18). Let $t$ be an infinite element of the field $F$. Consider the unit square $ABCD$ and the parallelogram with base $CD$ and top side $EF$, where $E = (t, 1)$ and $F = (t + 1, 1)$.

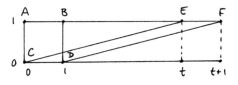

Then according to (I.35), $ABCD$ and $CDEF$ have equal content. However, they are not equidecomposable. Indeed, any triangle contained in the unit square has sides of length less than or equal to $\sqrt{2}$. Any finite number of these sides, placed end to end, still has finite length in the field $F$. But the side $CE$ of the parallelogram has length $\sqrt{t^2 + 1} > t$, which is infinite. Thus no finite number of triangles contained in $ABCD$ can ever fill up the parallelogram $CDEF$.

**Remark 22.1.4**
This example suggests the following question: If we assume Archimedes' axiom (A) in addition, are the notions of equidecomposable and equal content equivalent? We will see that the answer is yes in any Hilbert plane with (P) and (A), by using a measure of area function with values in the field of segment arithmetic (24.7.3). I do not know any purely geometric proof of this fact. In the non-Euclidean case, we obtain the same result using the defect of a triangle as a measure of area function (36.7.1).

**Proposition 22.2**
*In a Hilbert plane, the relation of two figures being equidecomposable is an equivalence relation. Nonoverlapping unions of equidecomposable figures are equidecomposable.*

*Proof*   The relation is obviously reflexive ("$P$ is equidecomposable with $P$") and symmetric ("if $P$ and $P'$ are equidecomposable, then $P'$ and $P$ are equidecomposable"). The nontrivial part is transitivity. So suppose that $P$ and $P'$ are equidecomposable, and $P'$ and $P''$ are equidecomposable. Let

$$P = T_1 \cup \cdots \cup T_n,$$

$$P' = T_1' \cup \cdots \cup T_n',$$

where $T_i$ and $T_i'$ are congruent triangles, for each $i$. Also let

$$P' = S_1' \cup \cdots \cup S_m',$$

$$P'' = S_1'' \cup \cdots \cup S_m'',$$

where $S_j'$ and $S_j''$ are congruent triangles, for each $j$. We must show that $P$ and $P''$ are equidecomposable.

To do this, we will refine the decompositions of $P$ and $P''$ in order to express them both as unions of congruent triangles. For each $i, j$ consider the intersection $T_i' \cap S_j'$ in $P'$. It may be empty, or may consist of points or line segments only. We ignore those. When the intersection has a nonempty interior, it will be a figure (Exercise 22.1) that can be written as a union of triangles

$$T_i' \cap S_j' = \bigcup_{k=1}^{l} U_{ijk}'.$$

Now let $\varphi_i : T_i \to T_i'$ be a rigid motion (Exercise 17.10) taking the triangle $T_i$ to the congruent triangle $T_i'$. We use $\varphi_i$ to transport the triangles $U_{ijk}'$ to new triangles $U_{ijk} = \varphi_i^{-1}(U_{ijk}')$ contained in $T_i$. Then

$$T_i = \bigcup_{j,k} U_{ijk},$$

and each $U_{ijk}$ is congruent to $U_{ijk}'$.

Similarly, for each $j$, let $\psi_j$ be a rigid motion taking $S'_j$ to the congruent triangle $S''_j$. Let $U''_{ijk} = \psi_j(U'_{ijk})$. Then

$$S''_j = \bigcup_{i,k} U''_{ijk},$$

and each $U''_{ijk}$ is congruent to $U'_{ijk}$.

By construction, the $U_{ijk}$ and the $U''_{ijk}$ are nonoverlapping triangles, and we can write

$$P = \bigcup_{i,j,k} U_{ijk},$$

$$P'' = \bigcup_{i,j,k} U''_{ijk},$$

where $U_{ijk}$ is congruent to $U''_{ijk}$ for each $i, j, k$. Thus $P$ and $P''$ are equidecomposable.

If $P$ and $P'$ are equidecomposable, and $Q$ and $Q'$ are equidecomposable, and if $P$ does not overlap $Q$ and $P'$ does not overlap $Q'$, then it is obvious that $P \cup Q$ and $P' \cup Q'$ are equidecomposable.

**Proposition 22.3**
*In a Hilbert plane, the relation of two figures having equal content has the following properties:*
(a) *Equal content is an equivalence relation.*
(b) *Equidecomposable figures have equal content.*
(c) *Nonoverlapping unions of figures of equal content have equal content.*
(d) *If $Q \subseteq P$ and $Q' \subseteq P'$, and if $Q$ and $Q'$ have equal content, and $P$ and $P'$ have equal content, then $P - Q$ and $P' - Q'$ have equal content.*

**Lemma 22.4**
*Suppose $P$ and $P'$ are equidecomposable figures, and suppose $P$ is expressed as a nonoverlapping union of subfigures $P = P_1 \cup P_2$. Then there are subfigures $P'_1, P'_2$ of $P'$ such that $P'$ is the nonoverlapping union of $P'_1$ and $P'_2$, and $P_i$ and $P'_i$ are equidecomposable for $i = 1, 2$.*

*Proof*  Suppose

$$P = T_1 \cup \cdots \cup T_n,$$

$$P' = T'_1 \cup \cdots \cup T'_n,$$

where $T_i$ and $T'_i$ are congruent triangles for each $i$. As in the proof of (22.2) we will refine decompositions appropriately.

For each $i$, consider the intersections $T_i \cap P_1$ and $T_i \cap P_2$. We can write each as unions of triangles (22.1)

$$T_i \cap P_1 = \bigcup_j S_{ij1},$$

$$T_i \cap P_2 = \bigcup_j S_{ij2}.$$

Use rigid motions $\varphi_i : T_i \to T_i'$ to transport these triangles and define

$$S_{ijk}' = \varphi_i(S_{ijk})$$

for each $i, j, k$. Let

$$P_1' = \bigcup S_{ij1}',$$
$$P_2' = \bigcup S_{ij2}'.$$

Then $P_1', P_2'$ satisfy the requirements of the lemma.

*Proof of 22.3* (a) The relation of equal content is obviously reflexive and symmetric. The nontrivial part is to show that it is transitive. So suppose figures $P$ and $P'$ have equal content, and $P'$ and $P''$ have equal content. Then there are equidecomposable figures $Q$ and $Q'$ such that $P \cup Q$ and $P' \cup Q'$ are equidecomposable, and there are further equidecomposable figures $R'$ and $R''$ such that $P' \cup R'$ and $P'' \cup R''$ are equidecomposable.

The difficulty is that while the unions mentioned above are all nonoverlapping, it may happen that $Q'$ and $R'$ overlap. To avoid this situation, we apply the lemma to the equidecomposable figures $P' \cup R'$ and $P'' \cup R''$ and the given decomposition of the first of these. Thus we may assume that the triangulation of $P' \cup R'$ arises from separate triangulations of $P'$ and $R'$. Once this is so, we can move $R'$ to some other position in the plane $R^*$ and still have $P' \cup R^*$ equidecomposable with $P'' \cup R''$. In particular, we may choose $R^*$ in such a way that $Q'$ and $R^*$ do not overlap (Exercise 22.4).

Now let $R$ be a figure congruent to $R^*$ that does not overlap $P$ or $Q$, and let $Q''$ be a figure congruent to $Q'$ that does not overlap $P''$ or $R''$. Then, by additivity of equidecomposability, we find $P \cup Q \cup R$ equidecomposable with $P'' \cup Q'' \cup R''$, and $Q \cup R$ equidecomposable with $Q'' \cup R''$, so by definition, $P$ and $P''$ have equal content. This completes the proof of (a).

Statement (b) is trivial.

Statements (c) and (d) are not difficult to prove after using the lemma to avoid overlaps (Exercise 22.5).

Now we have defined a notion of equal content for rectilineal figures in the plane, and we have established enough properties to recover most of Euclid's results on area. In particular, this notion of equal content satisfies the properties, 1, 2, 3 listed at the beginning of this section. However, our theory does not seem to be strong enough to establish properties 4, 5, 6, so we formulate what is missing as follows.

**Z. (de Zolt's axiom).** If Q is a figure contained in another figure P, and if $P - Q$ has a nonempty interior, then P and Q do not have equal content.

We can think of (Z) as a precise formulation of Euclid's Common Notion 5, "the whole is greater than the part," for the notion of content. However, we avoid the use of the words "greater" and "lesser," because these imply the existence of an order relation among figures, which we have not yet established. In fact, the existence of an order relation for content depends on (Z) (Exercise 22.7). We will also see in the exercises that (Z) implies the other two properties 4 and 6 listed at the beginning of this section (Exercises 22.6, 22.8).

I do not know of any purely geometric proof of (Z) from the definition of content we have given. We will see in the next section, however, that (Z) holds whenever there is a measure of area function in the geometry. In particular, (Z) will hold in any Hilbert plane with (P) (Section 23), and also in the non-Euclidean geometries (Section 36).

### Corollary 22.5
*In a Hilbert plane, the relation equal content has properties 1, 2, 3 at the beginning of this section. In a Euclidean plane with (Z), 4, 5, 6 also hold.*

*Proof*   1, 2, 3 are contained in (22.3), and 4, 5, 6 are in Exercises 22.6, 22.7, 22.8.

Now let us review Euclid's results about area and their proofs, substituting everywhere his "equality" of figures by the notion of equal content developed in this section. We work in a Euclidean plane (Section 12), i.e., a Hilbert plane with (P) and (E).

We have already seen (22.1.2) that Euclid's proof of (I.35) shows that the two parallelograms have equal content. The next result (I.36) follows using transitivity of equal content. But in the proof of (I.37) Euclid uses the property that "halves of equals are equal," which depends on (Z) (Exercise 22.8). So we will give another proof, which does not depend on (Z).

### Proposition 22.6 (I.37)
*In a Euclidean plane, triangles on the same base, whose top vertices are on the same line parallel to the base, have equal content.*

*Proof*  Let *ABC* and *DBC* be the given triangles, lying in the parallel lines *l, m*. Let *E* be the midpoint of *AB*, and draw a line *n* through *E*, parallel to *l*. Let this line meet *DC* at *F*. Then, from (5.1) (cf. Exercise 5.3) it follows that *F* is the midpoint of *DC*. Draw a line through *B*, parallel to *AC*, to meet *n* at *K*, and draw a line through *C*, parallel to *BD*, to meet *n* at *L*.

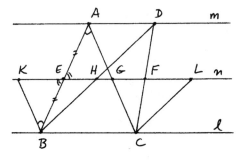

Then because of parallel lines, we get equal angles, which shows that $\triangle AEG \cong \triangle BEK$ using (ASA). So the triangle *ABC* has equal content with the parallelogram $BCGK$. Similarly, $\triangle DHF \cong \triangle CLF$, so that $\triangle BCD$ has equal content to the parallelogram $BCHL$.

Now by (I.35), these two parallelograms have equal content. So by transitivity it follows that the triangles $ABC$ and $DBC$ have equal content.

Continuing with our examination of Euclid's results, (I.38) follows by transitivity. Its converse (I.39) uses "the whole is greater than the part" in its proof, and so depends on (Z). Generally speaking, all of the results in which Euclid shows that two figures are equal will be valid for the notion of equal content. However, when a hypothesis of equal content is used to conclude something involving congruence of segments or angles in the figure, then (Z) will be necessary. So (I.40) also depends on (Z). In (I.48) Euclid says that if squares have equal content, then their sides are equal, so this result also depends on (Z) (Exercise 22.6). The remaining results of Book I, namely (I.41)–(I.45) and (I.47) are all valid for content. So for example, (I.47), the Pythagorean theorem, says that the square on the hypotenuse of a right triangle has equal content to the union of the squares on the two legs of the triangle.

In Book II, all of the results make statements about certain figures having equal content to certain others, and all of these are valid in our framework. Note that the line–circle intersection property (11.6) is used in (II.11) to divide a line such that the square on the larger piece has equal content to the rectangle on the whole and the smaller piece. It is also used in (II.14) to construct a square with equal content to any given rectilineal figure.

In Book III, Propositions (III.35) and (III.36) hold for equal content. The converse (III.37) of (III.36) requires property 6 above—if squares are equal, their sides are equal—and so depends on (Z).

In Book IV the only result needing Euclid's theory of area is (IV.10), to construct an isosceles triangle whose base angles are twice the angle at the vertex. The proof uses (III.37) and so depends on (Z). In particular, Euclid's proof of the construction of the regular pentagon (Section 4) depends on (Z). So we see again how the construction of the pentagon involves all the subtleties of Euclid's geometry!

At this point we could take (Z) as an additional axiom, and then we would have a satisfactory basis for Euclid's theory of area. However, we will see in the next section that (Z) holds in the Cartesian plane over a field, and hence, using the theorem of introduction of coordinates (21.1), it holds in any Hilbert plane with (P), and hence in any Euclidean plane. See (23.6) for a summary.

# Exercises

22.1 Show that the intersection of any two figures is a figure. *Hint*: First do the intersection of two triangles.

22.2 Show that the complement of one figure contained in another figure is a figure. *Hint*: First do the case where the smaller figure is a single triangle.

22.3 Show that the union of two figures is a figure.

22.4 In a Hilbert plane, given two figures $P, Q$, show that there is a rigid motion $\varphi$ of the plane (cf. Section 17) such that $P$ and $\varphi(Q)$ do not overlap.

22.5 Prove parts (c) and (d) of Proposition 22.3.

22.6 In a Euclidean plane, assuming (Z), suppose that you are given segments $AB$ and $CD$ such that the squares on $AB$ and $CD$ have equal content. Show that $AB$ and $CD$ are congruent.

22.7 In a Euclidean plane with (Z), show that there is a total ordering on the set of figures with the property that $P \le Q$ whenever there exists a figure $P'$ with the same content as $P$, and $P'$ is contained in $Q$. *Hint*: Use (I.44) to show that any figure $P$ has equal content with a rectangle $P'$ of given fixed side $AB$.

22.8 In a Euclidean plane with (Z) show that "halves of equals are equal," in the following sense: If $P$ and $Q$ are figures with equal content, and if $P = P_1 \cup P_2$ is a non-overlapping union, where $P_1$ and $P_2$ have equal content, and similarly $Q = Q_1 \cup Q_2$ with $Q_1$ and $Q_2$ having equal content, then $P_1$ and $Q_1$ have equal content.

22.9 In a Euclidean plane satisfying (A), suppose that (Z) fails in the sense that there exists a triangle $ABC$ and a point $D$ between $B$ and $C$, such that $ABC$ has equal content with the smaller triangle $ADC$. Then show that for any figure $P$, there exists a figure $Q$ containing $P$, and $Q$ has equal content with the empty set. (If you can find a contradiction resulting from this, you will have discovered a proof of (Z)!)

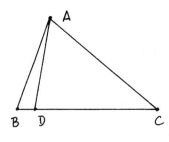

22.10 In a Euclidean plane with Archimedes' axiom (A), give a direct proof of the analogue of (I.35) for equidecomposability: Two parallelograms on the same base and within the same parallels are equidecomposable.

22.11 *Simple closed polygons.* A *simple closed polygon* is a finite union of line segments $A_1A_2, A_2A_3, \ldots, A_iA_{i+1}, \ldots, A_nA_1$, where $A_1, \ldots, A_n$ are distinct points in the plane, and the line segments have no other points in common except their endpoints, each of which lies on two segments.

(a) Show that a simple closed polygon divides the plane into two segment-

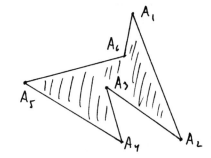

connected (cf. Exercise 7.12) subsets, its *interior* and its *exterior*. (This is the much easier polygonal analogue of the famous Jordan curve theorem for simple closed curves in $\mathbb{R}^2$.)

(b) Show that the simple closed polygon, together with its interior, is a figure $P$ (i.e., a finite union of triangles). Is it always possible to write an $n$-sided polygon $P$ as a union of $n - 2$ triangles?

# 23  Measure of Area Functions

In this section we will relate Euclid's theory of area discussed in the last section with the familiar notion of area as a number. In high-school geometry, most likely you learned how to compute the area of various figures, but never saw a definition of area or a proof that it exists. So we will first define an area function by the properties we want it to have. Then we discuss the question of existence and uniqueness. We will see that an area function exists in a Hilbert plane with (P). We will also see that the existence of an area function implies de Zolt's axiom (Z) discussed in the last section. These two results then put Euclid's theory of area on a firm basis.

**Definition**
An *ordered abelian group* is an abelian group $G$, together with a subset $P$, whose elements are called *positive*, satisfying:

(i) If $a, b \in P$, then $a + b \in P$.
(ii) For any $a \in G$, one and only one of the following holds: $a \in P$; $a = 0$; $-a \in P$.

As in the case of an ordered field (cf. Section 15) we define $a > b$ if $a - b \in P$. Then the relation $>$ has all the usual properties (15.2).

**Definition**
A *measure of area function* on a Hilbert plane is a function $\alpha$, defined on the set $\mathscr{P}$ of all *figures* (see definition in Section 22), with values in an ordered abelian group $G$, such that:
(1) For any triangle $T$, we have $\alpha(T) > 0$ in $G$.
(2) If $T$ and $T'$ are congruent triangles, then $\alpha(T) = \alpha(T')$.
(3) If two figures $P$ and $Q$ do not overlap, then

$$\alpha(P \cup Q) = \alpha(P) + \alpha(Q).$$

We call $\alpha(P)$ the *area* of the figure $P$, with respect to the given measure of area function.

**Proposition 23.1**
*Suppose α is a measure of area function on a Hilbert plane.*
   (a) *If P is any figure with nonempty interior, then* $\alpha(P) > 0$.
   (b) *If P and P′ are equidecomposable figures, then* $\alpha(P) = \alpha(P')$.
   (c) *If P and P′ are figures with equal content, then* $\alpha(P) = \alpha(P')$.
   (d) *If a figure Q is contained in a figure P, and* $P - Q$ *has nonempty interior, then* $\alpha(Q) < \alpha(P)$. *In particular, P and Q cannot have equal content, so* (Z) *holds* (*cf.* Section 22).

*Proof* (a) Writing the figure $P$ as a union of triangles $T_i$, it follows from the definition of a measure of area function that $\alpha(P) = \sum \alpha(T_i)$, and each $\alpha(T_i) > 0$, so $\alpha(P) > 0$.
   (b) This follows from the property that congruent triangles have equal area function.
   (c) If $P$ and $P'$ have equal content, then there are figures $Q$ and $Q'$ that are equidecomposable and do not overlap with $P$ and $P'$, respectively, such that $P \cup Q$ and $P' \cup Q'$ are equidecomposable. Hence $\alpha(Q) = \alpha(Q')$ and $\alpha(P \cup Q) = \alpha(P' \cup Q')$. Using the additivity property (3) and subtracting in the group $G$, we find that $\alpha(P) = \alpha(P')$.
   (d) Write $P = Q \cup (P - Q)$. Since $P - Q$ has nonempty interior, $\alpha(P - Q) > 0$. Hence by additivity, $\alpha(Q) < \alpha(P)$. It follows from (c) that $P$ and $Q$ cannot have equal content. In other words, de Zolt's axiom (Z) stated in Section 22 holds.

Now that we know what a measure of area function is and have seen some of its properties, the main work of this section is to prove the existence of such a function in a Hilbert plane with (P). See Section 36 for the existence of measure of area functions in non-Euclidean geometry.

**Theorem 23.2**
*In a Hilbert plane with* (P), *there is an area function α, with values in the additive group of the field of segment arithmetic F* (19.3), *that satisfies and is uniquely determined by the following additional condition: For any triangle ABC, whenever we choose one side AB to be the* base *and let it have length* $b \in F$, *and let h be the length of an altitude perpendicular to the base, then* $\alpha(ABC) = \frac{1}{2}bh$.

*Proof* In this theorem, the uniqueness is obvious, because the additional condition tells us the value of $\alpha$ for any triangle, and any figure is a finite union of triangles.

For the existence of $\alpha$, there is no choice: For any figure $P \in \mathscr{P}$, write $P$ as a union of triangles $P = T_1 \cup \cdots \cup T_n$, for each triangle $T_i$ choose one side to be the base $b_i$, let $h_i$ be the correspond-

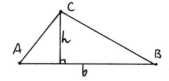

ing altitude, and define

$$\alpha(P) = \sum \frac{1}{2} b_i h_i.$$

The problem is to show that this notion is well-defined: We must show that $\alpha$ of a triangle is independent of which side we choose as base, and then we must show that $\alpha(P)$ is independent of the triangulation chosen. After that we must verify that $\alpha$ satisfies the properties of an area function. We will deal with some of these questions as separate lemmas.

**Lemma 23.3**

*In a Hilbert plane with* (P)*, let* $ABC$ *be any triangle. Let* $b$ *be one choice of base, with corresponding altitude* $h$*, and let* $b'$ *be another choice of base, with altitude* $h'$*. Then* $\frac{1}{2}bh = \frac{1}{2}b'h'$ *in the field of segment arithmetic.*

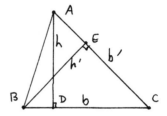

*Proof* Let $ABC$ be the triangle, with $b = BC$, $h = AD$, $b' = AC$, $h' = BE$. The two right triangles $ADC$ and $BEC$ have the angle at $C$ in common; hence all three angles equal (I.32). So by (Sim AAA) (20.1) they are similar triangles. It follows that the ratios of corresponding sides are equal, so

$$\frac{h}{b'} = \frac{h'}{b}.$$

Cross multiplying, we obtain $bh = b'h'$, and so $\frac{1}{2}bh = \frac{1}{2}b'h'$, as required.

Thus the function $\alpha$ is well-defined for triangles. The key point in studying arbitrary triangulations is to see what happens when a triangle is divided into smaller triangles.

**Lemma 23.4**

*If a triangle* $T$ *be subdivided into smaller triangles* $T_i$ *in any way whatsoever (but still a finite number), then the measure of area of the big triangle is equal to the sum of the measures of area of the small ones:* $\alpha(T) = \sum \alpha(T_i)$.

*Proof Step 1* We consider the special case where a triangle $ABC$ is divided into two triangles by a single *transversal*, namely a line that goes from one vertex (say $C$) to a point $D$ on the opposite side $AB$.

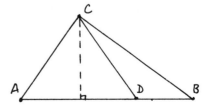

Then choosing side $AB$ as a base for the big triangle, and sides $AD$ and $DB$ as bases for the two smaller triangles, all three triangles have the same height, and the bases $AD, DB$ add up to the base $AB$, so clearly $\alpha(ABC) = \alpha(ACD) + \alpha(BCD)$.

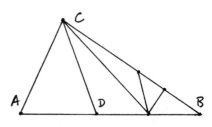

*Step 2* Next we do a slightly harder case. Suppose a triangle $ABC$ is subdivided into triangles $T_i$ such that there are no new vertices in the *interior* of the triangle, and at least one edge ($AC$ in the picture) is free from new vertices on the sides of the triangle. Then $\alpha(ABC) = \sum \alpha(T_i)$.

We prove this result by induction on the number of the smaller triangles $T_i$. If there are just two of them, then we are in the situation of Step 1 above. So suppose there are more than two. The free side (say $AC$) must belong to one of the small triangles, say $T_1$, and the third vertex $D$ of $T_1$ must be on one of the sides $AB$ or $BC$ (suppose it is $AB$). Then by Step 1,

$$\alpha(ABC) = \alpha(T_1) + \alpha(BCD).$$

Notice that $BCD$ has one fewer triangle in its subdivision than $ABC$. Furthermore, $BCD$ satisfies the hypotheses of Step 2, because it has no interior vertices (since $ABC$ didn't) and the side $CD$, being interior to $ABC$, has no vertices on it. Thus by the induction hypothesis, $\alpha(BCD) = \sum_{i=2}^{n} \alpha(T_i)$, and we are done.

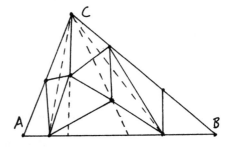

*Step 3* The general case. Let the triangle $ABC$ be divided into subtriangles $T_i$. Choose one vertex of $ABC$, say $C$, and draw lines (dotted in the diagram) from $C$ to each of the vertices of the triangulation, including those on the opposite side $AB$, and extend these lines down to the side $AB$.

Then we have another subdivision of $ABC$, this time by triangles $S_j$. Note that the subdivision $S_j$ satisfies the hypotheses of Step 2, so

$$\alpha(ABC) = \sum_j \alpha(S_j). \qquad (1)$$

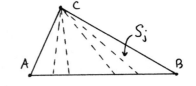

Taking all the $T_i$ and the $S_j$ together gives another subdivision $ABC = \bigcup_{i,j} T_i \cap S_j$. These figures $T_i \cap S_j$ may be triangles or may have four sides. Add extra lines so that they become unions of triangles: $T_i \cap S_j = \bigcup_k U_{ijk}$. Thus we have a third triangulation of $ABC$ into the triangles $U_{ijk}$.

Next, observe that each $S_j$ is the union of triangles $U_{ijk}$ as $i$ and $k$ vary, and this triangulation of $S_j$ satisfies the conditions of Step 2! It has no interior vertices, because the lines forming the triangles $S_j$ went through all the vertices of the original triangulation. Furthermore, the side of $S_j$ along the base $AB$ is free of vertices, for the same reason. Thus by Step 2, for each $j$ we have

$$\alpha(S_j) = \sum_{i,k} \alpha(U_{ijk}). \tag{2}$$

Combining this with the earlier result (1) about $ABC$ as the union of the $S_j$, we obtain

$$\alpha(ABC) = \sum_{i,j,k} \alpha(U_{ijk}). \tag{3}$$

It remains to discuss the division of each $T_i$ into the smaller triangles $U_{ijk}$. Typically, one of the lines from $C$ through the three vertices $XYZ$ of $T_i$ will cut $T_i$ in halves, say $T_i', T_i''$. (In the drawing the line through $Z$ cuts $T_i$ in half.) Then by Step 1, $\alpha(T_i) = \alpha(T_i') + \alpha(T_i'')$.

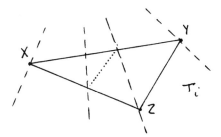

Each of these halves is further subdivided by lines through $C$ and additional lines we have added to cut quadrilaterals into two triangles. These triangulations of $T_i'$ and $T_i''$ satisfy the conditions of Step 2: There are no interior vertices, and the side through $Z$, which we used to separate $T_i$ into $T_i'$ and $T_i''$, contains no vertices. Hence by Step 2 again, each of $T_i'$ and $T_i''$ has measure equal to the sum of the measures of the $U_{ijk}$ of which it is composed, and hence

$$\alpha(T_i) = \sum_{j,k} \alpha(U_{ijk}). \tag{4}$$

Now, finally, combining this with equation (3) we obtain

$$\alpha(ABC) = \sum_i \alpha(T_i) \tag{5}$$

as required.

**Lemma 23.5**

*The measure of area of a rectilineal figure is independent of the triangulation used to define it.*

*Proof*   If a figure $P$ has two triangulations

$$P = T_1 \cup \cdots \cup T_n$$

and

$$P = T'_1 \cup \cdots \cup T'_m,$$

then the intersections of the $T_i$ and $T'_j$ can be further subdivided into triangles $U_{ijk}$, as in the proof of (22.2). Now applying Lemma 23.4 to each $T_i = \bigcup_{j,k} U_{ijk}$ and to each $T'_j = \bigcup_{i,k} U_{ijk}$, we find that

$$\sum_i \alpha(T_i) = \sum_{i,j,k} \alpha(U_{ijk}) = \sum_j \alpha(T'_j),$$

so that $\alpha(P)$ comes out the same either way.

*Proof of 23.2, continued*   From the lemmas (23.3), (23.4), and (23.5) we know that the function $\alpha$ is well-defined. We need only verify that it has the properties required of an area function. Since segments give positive elements of $F$, $\alpha(T) > 0$ for any triangle. Congruent triangles have congruent sides and congruent altitudes, so $\alpha(T) = \alpha(T')$ if $T$ and $T'$ are congruent triangles.

   If $P$ and $Q$ are two figures with nonoverlapping interior, and if we write $P = T_1 \cup \cdots \cup T_n$ and $Q = T'_1 \cup \cdots \cup T'_m$, we can use all the $T_i$ and $T'_i$ to triangulate $P \cup Q$. In that case the additivity is obvious.

**Corollary 23.6**

*In a Euclidean plane, all of Euclid's theory of area, namely (I.35)–(I.48), (II.1)– (II.14), (III.35)–(III.37), and (IV.10) hold, where we interpret "equality" of figures to mean equal content in the sense of Section 22.*

*Proof*   We have already seen in Section 22 that all these results follow from the definition of equal content plus the statement (Z). In this section we saw that (Z) is a consequence of the existence of an area function (23.1) and that an area function exists in a Hilbert plane with (P) (23.2).

**Remark 23.6.1**

This proof is analytic in that it makes use of the field of segment arithmetic and similar triangles. We do not know any purely geometric proof, for example of (I.39), that triangles on the same base with equal content have the same altitude.

**Remark 23.6.2**
It may be worthwhile to point out that we are not using circular reasoning here. To be sure, the proof of (Z) we have just given uses the field of segment arithmetic (19.3), whose proof uses some results from Euclid, Book III (19.2.1). But the results used were only those that did not depend on the theory of area (cf. 12.4), so that we may now use them to go back and validate earlier results from Book I on that need the theory of area.

**Proposition 23.7**
*In a Hilbert plane with* (P), *let α be the measure of area function of* (23.2). *Then two figures P, Q have equal content if and only if* $α(P) = α(Q)$.

*Proof*   If $P$ and $Q$ have equal content, then $α(P) = α(Q)$ by (23.1). Conversely, suppose $α(P) = α(Q)$. By (I.44) we can find rectangles $P'$ and $Q'$ with content equal to $P, Q$, respectively, and furthermore, we may assume that one side of these rectangles is the unit 1 in the field of segment arithmetic. Let $a, b$ be the other sides of these two rectangles. Then cutting each into two triangles, we see that $α(P') = 1 \cdot a = a$, and $α(Q') = 1 \cdot b = b$. On the other hand, figures of equal content have equal $α$, so $α(P) = a$ and $α(Q) = b$. Our hypothesis now implies $a = b$. Thus the two rectangles are congruent, so $P'$ and $Q'$ have the same content. By transitivity, $P$ and $Q$ have the same content. For this proof, note that (I.44) does not use (E), and so is valid in any Hilbert plane with (P).

**Remark 23.7.1**
Thus in a Hilbert plane with (P), the theory of content is essentially equivalent to the theory of area given by the area function of (23.2), so we can also restate Euclid's theory of area in terms of the area function.

# Exercises

23.1  In Theorem 23.2, the uniqueness of the area function was established by requiring it to have the expected value on every triangle. Suppose instead, in the Cartesian plane over a field $F$, we consider measure of area functions, with values in the additive group of the field, with the weaker requirement that $α$ of the unit square should be equal to 1.

(a) If the field is Archimedean, show that $α$ is uniquely determined by the above condition.

(b) If the field $F$ is non-Archimedean, show that there can be more than one area function having value 1 on the unit square.

23.2  Use the measure of area function of Theorem 23.2 to show that if two squares have equal area, then their sides are congruent.

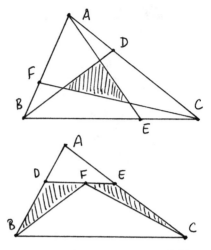

23.3 Let *ABC* be any triangle. Let *D*, *E*, *F* divide the sides in thirds. Draw *AE*, *BD*, *CF*. Show that the small triangle formed in the inside has area equal to $\frac{1}{7}$ the area of the whole triangle. *Hint*: Review the proof of Proposition 5.4 and use similar triangles.

23.4 Let *ABC* be any triangle, let *DE* be a line parallel to the base, and let *F* be any point on *DE*. Show that the area of the union of the two triangles *DBF* and *ECF* is less than or equal to one-fourth the area of the whole triangle, with equality if and only if *D* and *E* are the midpoints of *AB* and *AC*.

23.5 To get a feeling for ordered abelian groups, try this one. Let $\mathbb{Q}(\sqrt{2}) = \{a + b\sqrt{2} \mid a, b \in \mathbb{Q}\}$, and similarly $\mathbb{Q}(\sqrt{3})$.

(a) Show that $\mathbb{Q}(\sqrt{2})$ and $\mathbb{Q}(\sqrt{3})$ are isomorphic as abelian groups (under addition).

(b) Show that $\mathbb{Q}(\sqrt{2})$ and $\mathbb{Q}(\sqrt{3})$ are not isomorphic as fields.

(c) Show that $\mathbb{Q}(\sqrt{2})$ and $\mathbb{Q}(\sqrt{3})$ are isomorphic as ordered sets, with the orderings induced by the natural ordering on $\mathbb{R}$. (Here an *ordered set* is a set *S*, together with a relation $a < b$, having the two properties (i) $a < b$ and $b < c$ implies $a < c$; (ii) if $a, b \in S$, then one and only one of the following holds: $a < b$, $a = b$, $b < a$.)

(d) $\mathbb{Q}(\sqrt{2})$ and $\mathbb{Q}(\sqrt{3})$ are not isomorphic as ordered abelian groups (again taking addition as the group operation).

23.6 Prove Ptolemy's theorem: The rectangle formed of the two diagonals of a cyclic quadrilateral is equal in content to the sum of the two rectangles made by opposite sides of the quadrilateral.

23.7 In a Euclidean plane, show that the results (VI.1) and (VI.14)–(VI.31) hold using equal content for equality of figures.

## 24   Dissection

As we have seen in the previous sections, Euclid bases his theory of area on adding and subtracting congruent figures. This notion is formalized by Hilbert's definition of equal content (Section 22), which can also be interpreted using a measure of area function (Section 23). While the notion of content gives a good general theory, one could complain that it is not practical, in the sense that if two figures have equal content, one does not know how much must be added in order to make them equidecomposable. In Example 22.1.3, the two figures being compared both have area 1, but to make them equidecomposable, one must add a triangle whose area is infinite in the non-Archimedean field *F*.

In this section we will investigate the stricter notion of when two figures are equidecomposable (Section 22). This leads to the practical problem of *dissection*: given two figures, to find, if possible, an efficient decomposition of the first as a nonoverlapping union of smaller figures, not necessarily triangles, that can be reassembled into the second. A dissection exists if and only if the two figures are equidecomposable. In this case we will also say that one figure can be dissected into the other, or that they are *equivalent by dissection*.

We work in a Hilbert plane with (P), and in some cases use also (A) or (E). We will see that certain of the results, such as the Pythagorean theorem (I.47), which Euclid proved for content, are also true in the stronger sense of dissection. We will prove the theorem, due to Bolyai and Gerwien, that in an Archimedean plane, any two figures of equal area (Section 23) are equivalent by dissection.

The practical problem of finding dissections of one figure into another has received a certain amount of attention among recreational mathematicians, but at this point it seems to be more of an art than a science. Amateurs have discovered a number of clever dissections, but proofs that certain dissections are minimal, or effective bounds on the number of pieces required, seem to be lacking.

We begin this section with some general results on existence of dissections.

**Proposition 24.1**

*In a Hilbert plane with* (P), *any triangle can be dissected into a parallelogram.*

*Proof* (cf. 22.6) Let *ABC* be the triangle. Let *D* be the midpoint of *AC*. Draw lines through *D* parallel to *BC* and through *C* parallel to *AB*, meeting at *F*. Then $\triangle ADE \cong \triangle CDF$ by (ASA). So $\triangle ABC$ can be dissected into the parallelogram *BCEF*.

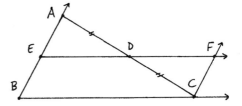

**Lemma 24.2**

*Let ABC be a triangle, and suppose that the foot D of the altitude from A to side BC lands outside the interval BC. Then one of the angles of the triangle, at B or C, is obtuse.*

*Proof* Suppose *B* is between *D* and *C*. Then the angle $\angle ABC$ is an exterior angle of the right triangle *ADB*, so $\angle ABC$ is greater than a right angle, by (I.16).

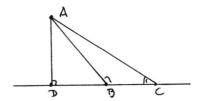

### Proposition 24.3

*Any parallelogram can be dissected into a rectangle.*

*Proof* Let ABCD be the given parallelogram. Drop the altitudes from C and D
to the side AB, with feet E, F. Suppose,
for example, that E is inside the interval
AB. Then the triangle △ACE is congruent to the triangle BDF. So the parallelogram is dissected into the rectangle
CDEF.

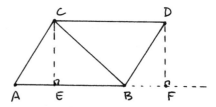

Now let us see that we can always apply this construction. In any parallelogram, the opposite angles are equal (I.34) and the sum of the angles is four right
angles, so two of the opposite angles are acute. (If not, all four are right angles
and there is noting to prove.) So we may assume that the angle at A is acute.

Now, if the altitude from C lands
outside AB, then by (24.2) the angle
ABC must be obtuse. This forces the
angle ACB to be acute (I.32), and in that
case, exchanging the roles of B and C,
the altitude from B to AC will land
inside the segment AC (24.2), so we can
apply the construction above.

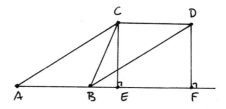

### Proposition 24.4

*Given a rectangle ABCD, and given a segment AE such that AB < AE < 2AB, the
rectangle ABCD can be dissected into a rectangle having one side equal to AE.*

*Proof* Suppose we are given the rectangle ABCD and the point E. Let C and
E be joined, meeting BD at F. Let G be
chosen on AC such that CG ≅ BF. Construct the rectangle AEGH, and let K, L
be as shown. Because of parallel lines,
the angles at C and F are equal. And CG
was equal to BF by construction, so by
(ASA), △CGK ≅ △FBE. It follows that
GK ≅ BE, and so by subtraction, CD ≅
AB ≅ KH. Now by (ASA) again, △CDF
≅ △KHE. This gives a dissection of
the rectangle ABCD into the rectangle
AEGH, as required.

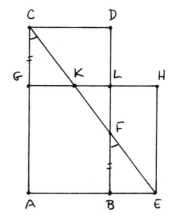

Note that in order for this dissection to work, we need to know that $F$ is below the midpoint of $BD$, so that $G$ is above the midpoint of $AC$, and so $F$ is below $L$. This follows from the hypothesis $AB < AE < 2AB$, because the line from $C$ to the midpoint of $BD$ would meet the line $AB$ at a point $M$ with $AM = 2AB$.

## Proposition 24.5
*Assume Archimedes' axiom* (A). *Given any rectangle ABCD and given any segment EF, there is a rectangle EFGH equivalent by dissection to ABCD.*

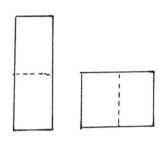

*Proof* Given any rectangle, by cutting it in half and reassembling the two halves along the other sides, we can dissect the original rectangle into a new rectangle with one-half the height and twice the base of the original one. Now by Archimedes' axiom, after doubling or halving $AB$ a finite number of times, we may assume $AB \le EF < 2AB$. Then we apply (24.4) to get a rectangle $EFGH$ as required.

## Corollary 24.6
*Assuming* (A) *and given a segment EF, any rectilineal figure* (Section 22) *can be dissected into a rectangle with one side EF.*

*Proof* Divide the figure $P$ into a finite number of triangles $T_1, T_2, \ldots, T_n$. For each $i$ first dissect the triangle $T_i$ into a parallelogram (24.1), then to a rectangle (24.3), then to a rectangle $R_i$ with base equal to $EF$ (24.5). Now stacking the rectangles $R_1, R_2, \ldots, R_n$, all of which have base equal to $EF$, on top of each other, we obtain one big rectangle with base $EF$, as required.

## Theorem 24.7 (Bolyai, Gerwien)
*In a Hilbert plane with* (P) *and* (A), *let $\alpha$ be the area function of* (23.2). *Then two figures P and Q are equivalent by dissection if and only if they have the same area.*

*Proof* One direction follows from (23.1). The new part, which requires the use of Archimedes' axiom, is that figures of equal area are equivalent by dissection. The proof parallels (23.7), where the analogous result was proved for content.

So suppose figures $P$ and $Q$ have equal area: $\alpha(P) = \alpha(Q)$. By (24.6), they can each be dissected into rectangles with sides $1, a$ and $1, b$. Then $\alpha(P) = a$, $\alpha(Q) = b$, so $a = b$, and the rectangles are equal. Thus by transitivity of equidecomposability (22.2), $P$ can be dissected into $Q$.

**Remark 24.7.1**
This result is false without (A), as is shown by Example 22.1.3.

**Remark 24.7.2**
This proof is effective, since given figures $P$ and $Q$ of equal area, and given a decomposition of each into triangles, the method of proof of the results of this section will lead to a dissection of one to the other. Of course, the dissection found in this way may not be efficient in terms of the number of pieces required.

**Remark 24.7.3**
Combining this result with (23.7) we see that in a Hilbert plane with (P) and (A), two figures have equal content if and only if they are equidecomposable (cf. 22.1.4).

**Proposition 24.8**
*Assume* (E). *Let ABCD be a rectangle with sides $a = AB$ and $b = AC$ satisfying $a \leq b \leq 4a$. Then there is a segment $c = AE$ such that the rectangle ABCD can be dissected into a square with side AE.*

*Proof*   Form a segment of length $a + b$, let $O$ be its midpoint, draw a circle with center $O$ passing through the segment's endpoints, and let $c$ be the segment cut off on a perpendicular dividing the segment into $a + b$, from the segment to where it meets the circle. (Here we use (E) for the existence of the intersection point.)

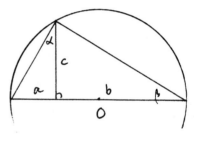

Now let $ABCD$ be the given rectangle, let $AE$ be congruent to the segment $c$, and let $AEFG$ be the square on side $AE$. Since the angle in a semicircle is a right angle (III.31), the angles $\alpha, \beta$ in the circle diagram are equal. These are the same as the angles $\alpha, \beta$ of the second diagram, by congruent triangles (SAS). Hence the two diagonal lines $FB$ and $CE$ are parallel (I.28). It follows that $CF \cong MB$ and $FK \cong BE$ (I.33). From here the same argument as in the proof of (24.4) shows that $ABCD$ can be dissected into the square $AEFG$.

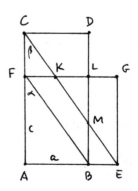

In order for this construction to work, we need to know that $AB \leq AE < 2AB$. We assumed that $a \leq b$, so the angle $\alpha = \beta$ is less than one-half a right angle, and so $a \leq c \leq b$. On the other hand, if $c \geq 2a$, then by similar triangles it follows that $b \geq 2c$. This implies $b \geq 4a$, contrary to our hypothesis.

**Corollary 24.9**
*In a Euclidean plane with* (A), *any rectilineal figure can be dissected into a square.*

*Proof*  Choose a segment $AB$. By (24.6) the original figure can be dissected into a rectangle $ABCD$ with side $AB$. Cutting the rectangle in half and reassembling a finite number of times (again using (A)), as in the proof of (24.5), we may assume that the sides $a = AB$ and $b = AC$ satisfy $a \leq b \leq 4a$. Then by (24.8) this rectangle can be dissected into a square.

Next, let us look at the Pythagorean theorem (I.47). Euclid's proof shows that the union of the squares on the legs of the right triangle has equal *content* to the square on the hypotenuse. Using Archimedes' axiom (A) one can improve Euclid's results about parallelograms (I.35) and triangles (I.37) to hold also for dissection. Then Euclid's proof will show that (I.47) holds for dissection assuming (A). However, even better than this are several direct proofs of (I.47) by dissection that do not need Archimedes' axiom.

Let us first consider the proof attributed to Thabit b. Qurra (826–901 A.D.).

**Proposition 24.10** (I.47)
*In a Hilbert plane with* (P), *the union of the squares on the legs of a right triangle can be dissected into the square on the hypotenuse.*

*Proof*  Let $ABC$ be the original triangle; let $ABDE$ be the square on $AB$; let $ACFG$ be the square on $AC$; fill in the square $GHEK$ and the line $LM$ so that $DM \cong BC$. Then the three triangles $ADF, FHG, CKG$ are congruent to the original triangle $ABC$, and $GHEK$ is congruent to the square on $BC$.

Now the square on the hypotenuse is dissected into the five pieces $1, 2, 3, 4, 5$. The first three are congruent to $1', 2', 3'$ by construction. The square on $AB$ is dissected into $1', 3', 5$, while the square on $BC$ is dissected into $2', 4$. So the result is proved.

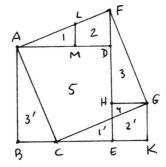

**Remark 24.10.1**

Here is another proof of (I.47) by dissection. Let $ABC$ be the original right triangle. Draw the squares on the three sides as shown. Let $O$ be the center of the square $ACFG$, and draw lines through $O$ parallel to the sides of the square $BCRS$.

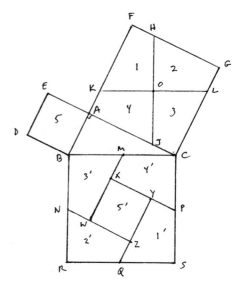

Now let $M, N, P, Q$ be the midpoints of the sides of the square $BCRS$, and draw lines through them parallel to $AB$ and $AC$, to form the figure shown.

We claim that $1', 2', 3', 4', 5'$ are respectively congruent to $1, 2, 3, 4, 5$, which will complete the proof by dissection.

Since $O$ is the center of the square $ACFG$, it follows that $O$ is also the midpoint of the segments $KL$ and $HJ$. These two segments are also equal to each other, and from the parallelogram $BCKL$, they are equal to $BC$. Thus $OK$, $OL$, $OH$, $OJ$ are all equal to one-half of $BC$; hence are congruent to the segments $BM$, $MC$, $DP$, $PS$, etc. Since the lines drawn in the big square are parallel to the sides of the square $ACFG$, it follows that $1, 2, 3, 4$ are congruent to $1', 2', 3', 4'$.

It remains to show that $5 \cong 5'$. Note that $MW \cong LC$, since they are corresponding sides of 3 and $3'$. From the parallelogram $BCKL$ we see that $LC \cong BK$. But also $BK \cong MX$. So by subtraction $XW \cong AB$. The figure is symmetrical, so one can also show that $AJ \cong LC$ and $JC \cong KA$; then a similar argument shows that $XY \cong AB$. Thus $5'$ is a square congruent to 5.

Note that a proof by dissection requires

(1) a diagram to show the pieces and how they correspond,
(2) a careful construction stating how and in what order the diagram was drawn, and
(3) a proof, based on the construction (2), showing that corresponding pieces are congruent.

There may be many ways of making the construction, and it is worth contemplating which will make the proof easier.

# Exercises

The following exercises take place in a Hilbert plane with (P). Assume that it is a Euclidean plane, or the Cartesian plane over a field, as needed.

24.1 Using Archimedes' axiom (A) give a direct proof of (I.37) by dissection.

24.2 Dissect a square into three equal smaller squares.

24.3 Use the accompanying diagram to provide another proof of the Pythagorean theorem (I.47) by dissection. You must supply a construction of the diagram, and then a proof based on your construction, that corresponding pieces are congruent.

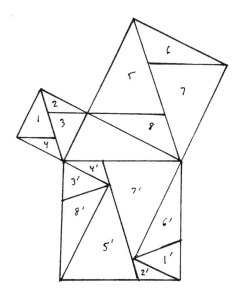

24.4 Find a dissection of an equilateral triangle into a square. After writing your construction and proof, make a model with construction paper and numbered pieces to illustrate your dissection.

24.5 Prove or disprove the existence of a dissection of a square of side 21 into a rectangle of sides 13, 34 according to the diagram below.

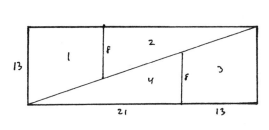

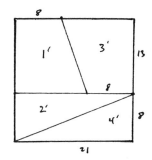

24.6 Let $ABC$ be any triangle. Let $D$ be a point on $AC$. Prove that there is a dissection of $ABC$ into a trapezoid $PQRS$ as shown. What condition is necessary on the point $D$ for this to work?

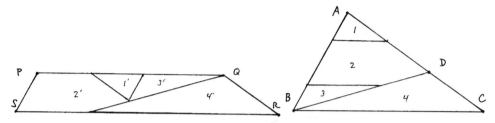

24.7 Use the previous exercise to find a dissection of two equilateral triangles of side 1 into a single equilateral triangle of side $\sqrt{2}$.

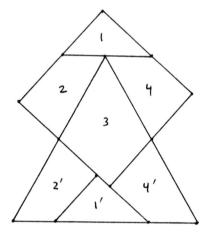

24.8 Prove the existence of Dudeney's (1929) dissection of an equilateral triangle into a square using four pieces, as suggested in the accompanying diagram.

24.9 Dissect a regular pentagon into a square.

24.10 Prove or disprove: For any integer $n \geq 1$, the minimum number of triangles required to dissect a square of side 1 into a rectangle of sides $n$ and $1/n$ is $2n$.

24.11 Dissect a square into eight strictly acute triangles (all angles less than 90°).

24.12 If a square is dissected into $n$ triangles of equal area, then $n$ must be an even number.

24.13 Consider a restricted form of dissection problem, where you can cut the figure in pieces and move the pieces around in the plane, *using translations only* (i.e., no rotations, and no turning pieces over). Show that you can dissect the unit square, with translations only, into the unit square with any other given orientation.

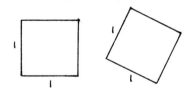

24.14 Divide the unit square by its diagonal into two (congruent) triangles $T_1$, $T_2$. Show that it is impossible to dissect $T_1$ into $T_2$ using translations only.

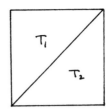

24.15 If $P$ and $P'$ are any two figures with the same area, then assuming Archimedes' axiom, show that it is possible to dissect $P$ into $P'$ using translations and 180° rotations only.

24.16 Let $T$ be a triangle whose smallest angle is greater than or equal to 45°. Then $T$ can be dissected into a square using 5 or fewer pieces.

24.17 Let $T$ be a triangle having an angle $\alpha$ with $\tan \alpha < 1/a$ for some $a \in F$, $a > 1$. Then any dissection of $T$ into a square requires $\geq \frac{1}{2}\sqrt{a}$ pieces. In particular, there exist triangles requiring arbitrarily many pieces to dissect into a square.

24.18 Let $T$ be a triangle whose smallest angle $\alpha$ satisfies $\tan \alpha > 1/a$, for some $a \in F$, $a \geq 1$. Then $T$ can be dissected into a square using no more than $3\sqrt{a} + 4$ pieces.

# 25   Quadratura Circuli

No discussion of area would be complete without mention of the classical problem of squaring the circle. Through the ages this problem has had an influence way beyond the confines of mathematics as an archetype of the insoluble problem. In one of his religious meditations, John Donne (early seventeenth century) mentions squaring the circle as something only God can achieve. Perhaps the expression "to put a round peg in a square hole" is a modern vestige of this old problem. And yet what a natural problem it is! The circle is the perfect round form, while the square is complete with its four equal sides and equal angles. What could be more natural than to transform the one into the other? But the solution has been as elusive as the alchemist's search for a way to transform lead into gold.

The problem is to construct a square with area equal to a given circle. What does this mean, exactly? By "construction," of course, is meant a geometrical construction with ruler and compass. But what is the area of a circle? In the theory of area presented so far, which corresponds to Euclid's usage in Books I–VI of the *Elements*, we have discussed the area of rectilineal figures only. So there is another problem implicit in the problem of squaring the circle, namely to give a good definition of the area of a circle. In the language of Section 23, we might ask for a measure of area function defined on all plane regions that can be bounded by straight line segments and arcs of circles. A modern approach to this (in the real Cartesian plane) is to use a definite integral to define the area.

Euclid never defines the area of a circle, and wisely avoids the problem of squaring the circle, but in Book XII, he does have a proposition saying that the ratio of the area of one circle to another is equal to the ratio of the squares of their radii. His proof makes use of the "method of exhaustion," which we would call a limiting process. The key result, which is remarkably similar to the treatment of the Riemann integral in modern calculus classes, is this.

**Proposition 25.1**
*In a Euclidean plane with* (A), *given a circle* $\Gamma$, *one can find an inscribed polygon P and a circumscribed polygon P' such that the difference in area* $\alpha(P') - \alpha(P)$ *is less than any preassigned quantity (as an element of the field F of segment arithmetic).*

*Proof*  Start with any inscribed polygon $P$ and any circumscribed polygon $P'$, where the vertices of $P$ are the points of tangency of $P'$. If we double the number of sides of $P$ and $P'$ by bisecting all the central angles of $P$, then we get new polygons $P_1$ and $P_1'$. We compare the difference of $P_1$ and $P_1'$ to the difference of $P$ and $P'$. Let us look at the triangle $ABC$, which is one piece of $P' - P$. The new difference $P_1' - P_1$ intersected with $ABC$ is represented by the shaded area.

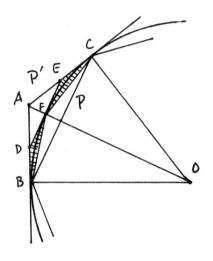

Euclid shows that the shaded area is less than one-half of the area of the triangle $ABC$. In fact, its area is less than or equal to $\frac{1}{4}$ of the area of that triangle (Exercise 23.4). So every time we double the number of sides, the new difference in areas of outside and inside polygons is less than half of the previous difference. Hence, using Archimedes' axiom (A), it can be made less than any preassigned quantity.

Approximating the circle by its inscribed and circumscribed polygons, Euclid could show that the ratio of the area of a circle to the square of its radius was a constant independent of the size of the circle. This constant, of course, is what we now call $\pi$. Archimedes, who lived shortly after Euclid, used the method of approximating by polygons to obtain the estimate $3\frac{10}{71} < \pi < 3\frac{1}{7}$, or in fractions, $\frac{223}{71} < \pi < \frac{22}{7}$. Thus began one facet of the work inspired by the problem of squaring the circle, namely to find increasingly accurate numerical approximations for $\pi$. Energetic human calculators worked hard on this problem, so that by

1600, the value of $\pi$ was known to 15 decimal places, by 1700 to 71 places, and by 1873 to more than 500 places. Even today, the urge to compute is still strong, and with the help of electronic calculators, the value of $\pi$ has been found to more than one billion decimal places.

Coming back to the original problem of squaring the circle, it has always been regarded as very difficult, if not impossible. Since Lindemann proved in the 1880s that $\pi$ is transcendental we have known that the problem as originally stated is mathematically impossible (cf. 13.2). Nevertheless, the problem has exerted a great fascination on the inquiring mind, and over the years, many people have come forward with solutions. The study of these solutions forms a colorful chapter in the history of mathematics.

The solutions can be divided into three general categories.

I. Honest ruler and compass constructions that, however, give only an approximate solution to the problem. One of the best of these depends on the coincidental proximity of the number $\frac{1}{5}(3\sqrt{5}+9) = 3.14164\ldots$ to $\pi$.

II. Constructions using additional tools that give an exact solution to the problem. One such known from antiquity uses an auxiliary curve, the *quadratrix*.

Take a quadrant $AOB$ of the unit circle. Let $OA_1$ bisect the right angle, and let $B_1$ bisect the interval $OB$. Let the radius $OA_1$ meet the horizontal line through $B_1$ at $C_1$. Bisect the two 45° angles at $O$ with $OA_2$ and $OA_3$, and bisect the intervals $OB_1$ and $B_1B$ with $B_2, B_3$. Let the radii meet the corresponding horizontal lines in $C_2$ and $C_3$. Continuing in this manner, bisecting angles and segments, one can construct as many points as one likes $C_1, C_2, C_3, \ldots$. The curve passing through these points is called the quadratrix.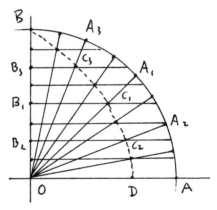

With a little help from modern trigonometry we can express the curve parametrically as

$$\begin{cases} x = \dfrac{2\theta}{\pi \tan\theta}, \\[2ex] y = \dfrac{2\theta}{\pi}, \end{cases}$$

where $\theta$ is the angle of the radius to the $x$-axis. As $\theta$ tends to 0 we obtain the point $D$ where the curve meets the $x$-axis. Its $x$-coordinate is $2/\pi$ (Exercise 25.1). So clearly, if we are given a copy of thc quadratrix curve, we can obtain the quantity $\pi$, and hence can construct a square with area equal to a given circle. Note, however, that this is not an honest ruler and compass construction:

Though we can obtain as many of the $C_i$ as we like by ruler and compass, the point $D$ is obtained only as a limit of the $C_i$ as they approach the $x$-axis.

III. Solutions by people I would like to call fringe mathematicians, though they have been called by worse names, who have little mathematical training but are attracted to famous problems like the moth to the flame. The hint that a problem may be insoluble inspires them to work even harder, as they have a conviction that they will succeed where others have failed. Their solutions are a jumble of facts and fallacies lacking in logical coherence. Pointing out their errors and quoting the results of established mathematicians only confirms their belief that they have discovered the real truth that all the scholars were too blind to see. Sometimes their solutions come from divine inspiration, which leaves little room for discussion, since one cannot argue with God.

A case in point is Joseph Ignati Carl von Leistner, a knight in the service of Francis I (1708–1765), duke of Lorraine, and later Holy Roman emperor. Herr Leistner discovered that the exact value of $\pi$ was 3844/1225. When his discovery did not occasion the acclaim he expected, he used his influence with the duke to appoint a imperial commission to report on his work. This commission naturally pointed out his errors, upon which he published a lengthy rebuttal attacking his critics and defending his work. Apparently, Herr Leistner believed that the true value of $\pi$ would be a ratio of integers, but that no one had yet been clever enough to find the right ones, as he did. One of his arguments goes like this: "Archimedes proposed the values 22/7 and 223/71, but even he acknowledged that they were not correct, while mine is the correct value." Herr Leistner failed to notice that his value fell outside of the bounds proved by Archimedes. On the occasion of his lord's marriage to Maria Theresa of Austria in 1736, Herr Leistner published a treatise on the wonderful coincidence of the numbers 3844 and 1225 with various dates and other numbers from the life of the duke and his bride.

# Exercises

25.1 (a) Derive the parametric equations for the quadratrix given in the text.

   (b) Use a little first-year calculus to show that the point $D$ has $x$-coordinate $2/\pi$.

25.2 (a) Let $\Gamma$ be the circle of radius 1. Find the area of the inscribed polygons of 6, 12, 24, 48, 96, 192 sides, and express them in standard form (Exercise 13.2) as a rational number times an expression involving integers and square roots. Then verify that the area of the inscribed 192-gon is greater than 223/71, thus confirming Archimedes' estimate $223/71 < \pi$. (Check: To four decimals, area of inscribed 48-gon is 3.1326, 96-gon is 3.1394, 192-gon is 3.1410.)

   (b) Show that the area of the circumscribed 96-gon is less than 22/7, confirming the inequality $\pi < 22/7$.

# Unwiderrufflicher, Wohlgegründter

## Und

# Ohnendlicher Beweiß

### Der Wahren

# QUADRATUR

## des Circuls,

### Oder

## Des Durchmessers zu seinem Umcreyß,

Wie 1225. zu 3844. oder 3844. zu 1225.

Zu

## Abermaligen und noch mehreren Ehren-Ruf der

gantzen Hoch-Edlen Teutschen Nation und Aufnahm der
edlen Mathematischen Wissenschafften, auch zum Besten des gemeinen
Weesens und Nutzens der studirenden Jugend,
Besonders aber

## Zur Erkandtnuß der goldenen Wahrheit

Wider die sich anmassende und ungegründte Widerlegung
Hn. J. J. MARINONII.
Verabfasset,
Und zu einem unpartheyischen Urtheil der gantzen Welt in
offentlichen Druck befördert
Von

# JOSEPHO IGNATIO CAROLO

## DE LEISTNER,

Der Römisch-Kayserlichen Majestät Rittmeistern,
Als

## INVENTORE QUADRATURÆ CIRCULI.

*Cum Gratia & Privilegio S.C.R.C. Majestatis, & Licentia Superiorum.*

Wienn, gedruckt und zu finden bey Johann Ignatz Heyinger, Universitäts-
Buchdruckern, 1737.

Plate IX. Title page of Leistner's treatise on the squaring of the circle (1737), in which he
claims that the exact value of $\pi$ is 3844/1225.

# 26  Euclid's Theory of Volume

Euclid discusses three-dimensional geometry in Book XI, and beginning with (XI.28) he treats the volume of solid figures. As with area in Book I, there is no definition of volume, and no acknowledgment that we are dealing with a new sort of equality different from congruence. From his proofs it becomes apparent that he is using "equality" of solid figures to mean equality of volume. Exactly what his notion of volume is becomes more clear as we progress; to begin with, we can think of a notion similar to the notion of *content* for plane figures, achieved by adding and subtracting congruent pieces of a dissection of the figure.

To simplify our discussion of Euclid's text, let us suppose that we are working over a field $F$ and are given a *volume function* $v$ that to each solid figure $P$ associates a nonnegative element $v(P) \in F$, that assigns the same volume to congruent figures, and that is additive for unions with nonoverlapping interiors. Each of Euclid's results can be interpreted as saying that certain figures have equal volume. Our attention will be focused on *how* Euclid proves this equality.

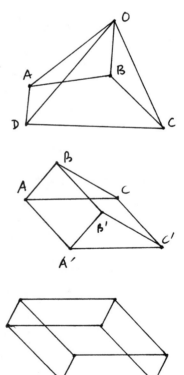

First we fix our terminology. A *pyramid* is a solid figure formed by taking a plane figure, say $ABCD$, and joining its vertices to a point $O$ outside the plane of $ABCD$. If the base is a triangle, we speak of a *triangular pyramid*, if the base is a square, a *square pyramid*, etc.

A *prism* is a solid figure formed by taking two congruent figures in parallel planes, with parallel edges, and joining their corresponding vertices. Its faces consist of the two original congruent figures, plus parallelograms joining the corresponding sides. If the base is a triangle, it is a *triangular prism* (as shown).

A *parallelepiped* is a prism whose base is a parallelogram. It is formed by three pairs of parallelograms in parallel planes.

Euclid's first results concerning volume are proved by methods entirely analogous to the methods of Book I. So for example, in (XI.28) he shows that a parallelepiped is bisected by the plane through two opposite edges. Note that Euclid proves that the two halves have congruent faces and angles. But they cannot be superimposed on each other in three dimensional space, because they are mirror images. However, Legendre gives a proof that they are equivalent in the sense of dissection, and we will see later (Exercise 26.1) that any solid figure is equivalent by dissection to its mirror image.

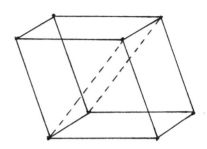

Using techniques exactly analogous to those of Book I, Euclid shows (XI.29) and (XI.30) that parallelepipeds with the same base and of the same height have the same volume. Then (XI.31) he shows, using the method of (I.44), that parallelepipeds having equal bases (in the sense of same content) and of the same height have equal volume. Using these results, one could show as a corollary that any parallelepiped has the same volume as a rectangular parallelepiped with sides $1, 1$, and $a$, for some $a \in F$. In fact, assuming the field to be Archimedean, one can accomplish this by *dissection*, that is, by cutting the solid into pieces and rearranging.

As an interesting application of these methods, Euclid shows (XI.39) that if we are given two triangular prisms, one lying on its side, and the other standing up, and if the parallelogram base of the one is equal to twice (in the sense of content) the triangular base of the other, and if they have the same height, then they have equal volume.

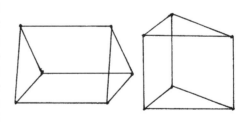

The method is to double each to get parallelepipeds, which have equal volume by the earlier results, and then appeal to the principle that "halves of equals are equal." When we think of volume as a function with values in a field, this principle poses no problem. However, if we sought to develop a purely geometric theory of volume by dissection or by content, as we did for area, it would require justification.

So we see that the theory of volume of parallelepipeds and prisms can be handled by methods familiar from the theory of area, and this involves nothing really new beyond translating ideas from the plane into three-dimensional space.

When we come to the study of the volume of pyramids in Book XII, however, the situation is quite different. The key result here is (XII.5), which, in a special case, is the content of the following proposition.

**Proposition 26.1** (XII.5)
*Triangular pyramids having equal bases (in the sense of content) and equal height have equal volume.*

*Proof*  For the proof, Euclid uses the *method of exhaustion*, attributed to Eudoxus. The idea is to write both figures as infinite unions of subfigures of equal volume, each one equal to more than one-half of what was left over after removing the previous subfigures. In this way the subfigures "exhaust" the entire figure, and from the equality of volume of the subfigures one concludes the equality of volume of the whole figures. This last step requires an explicit use of Archimedes' axiom, because as Euclid says in the proof of (XII.5), if the two pyramids were different, let this exhaustion process be done until the remainder left over is less than the difference of the two pyramids. This is possible because each remainder is less than one-half of the previous remainder, and so by repeating this process can be made smaller than any preassigned quantity.

In modern language, we would say that the volume of the whole pyramid is the *limit* of the volumes of the subfigures, and the definition of limit using $\varepsilon$'s and $\delta$'s is nothing but a modern rewording of the argument Euclid uses here.

Now let us examine the particulars of the proof. Let $ABCD$ be one of the given triangular pyramids. Let $E$, $F$, $G$, $H$, $J$, $K$ be the midpoints of the sides. Then the pyramid $P$ is decomposed into four pieces: two smaller pyramids $P_1 = AEFG$ and $P_2 = FBHK$, which are congruent to each other and have edges equal to $\frac{1}{2}$ the edges of $P$; and two triangular prisms $T_1 = FHKGJD$ and $T_2 = EFGCHJ$. Of these, $T_1$ is lying on its side, while $T_2$ is standing up, and the base of $T_1$ is a parallelogram $HJKD$ equal to twice the triangle $CHJ$ (in fact, it is the union of two triangles congruent to this one, if one draws the line $JK$).

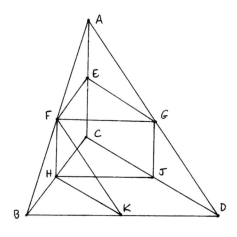

So by (XI.39), $T_1$ and $T_2$ have the same volume. Furthermore, since $T_2$ and $P_1$ have the same base $EFG$, and the same height, and one is a prism while the other is a pyramid, it is clear that $T_2$ is greater than $P_1$ in volume. (In fact, the pyramid $ECHJ$ is congruent to $P_1$ and is contained in $T_2$.) So we conclude that $T_1 + T_2$ is more than one-half of $P$ in volume.

Now suppose we divide the second pyramid $P'$ similarly into two pyramids $P'_1, P'_2$ and two prisms $T'_1, T'_2$. Since the base triangles of $P$ and $P'$ are equal in content, it follows that parallelepipeds on the bases of $T_1$ and $T'_1$ of the same height will have equal volume (XI.39), and hence their halves $T_1$ and $T'_1$ will have equal volume. So $T_1 + T_2$ and $T'_1 + T'_2$ have equal volumes, which are respectively more than one-half the volumes of $P$ and $P'$. Moreover, the remainders $P_1 + P_2$ and $P'_1 + P'_2$ are unions of pyramids of equal height and having bases of equal content.

Thus, inductively, we can repeat this process and write each of $P_1, P_2, P'_1, P'_2$ as unions of four pieces, two pyramids, and two prisms, and continue in this manner.

Since the prisms constructed at each step have equal volume, and these exhaust the pyramids as explained above, in the limit, we find that $P$ and $P'$ have the same volume.

### Corollary 26.2 (XII.7)
*A triangular pyramid has volume equal to one-third of a triangular prism on the same base and of the same height.*

*Proof* Let $ABCDEF$ be a triangular prism. We can regard the prism $T$ as the union of three pyramids $P_1 = CDEF$, $P_2 = ACDE$, and $P_3 = ABCE$.

Think of $P_1$ and $P_2$ as pyramids with vertex $E$ and bases the triangles $CDF$ and $ACD$. These two triangles are halves of the parallelogram $ACDF$, hence have equal area (or content). So by (26.1) $P_1$ and $P_2$ have equal volume.

The two pyramids $P_2$ and $P_3$ can be regarded as pyramids with vertex $C$ and bases the triangles $ADE$ and $ABE$. These triangles are halves of the parallelogram $ABDE$, hence have equal content. The two pyramids have the same height, so by (26.1) again, $P_2$ and $P_3$ have the same volume.

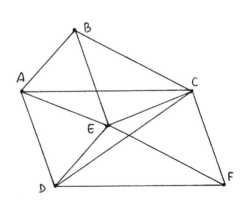

Thus $P_1, P_2, P_3$ all have the same volume, so this volume is equal to one-third the volume of the prism.

Euclid's use of the method of exhaustion here requires some comment. First of all, it is clear that a theory of volume based on this result will be considerably more complicated than the notion of content for plane figures. If we presuppose the existence of a volume function and work over an Archimedean field, this is a perfectly good proof. However, if we wish to establish a purely geometric theory of volume, we will have to allow limiting processes such as this method of exhaustion in our definition, along with the earlier notions of dissection and complementation.

Secondly, we can wonder whether this method is really necessary. Note that (XII.5) applies to the problem of mirror images mentioned earlier. Any figure bounded by planes can be cut up into a finite number of triangular pyramids, so it is enough to show that a triangular pyramid $P$ has the same volume as its mirror image $P'$. Since $P$ and $P'$ have congruent bases (hence same content) and the same height, (XII.5) applies to show that $P$ and $P'$ have the same volume.

Gauss, in a letter to Gerling in 1844, says that it is too bad that the equality of volume of two symmetrical but not congruent figures can be proved only using the method of exhaustion. Gerling in his reply gives a direct proof that any triangular pyramid can be dissected into twelve pieces that are congruent to those in a dissection of its mirror image (Exercise 26.1). Gauss responds yes, that is nice, but it is still unfortunate that the proof of (XII.5) seems to require the method of exhaustion.

These reflections, contrasted with the theorem of Bolyai and Gerwien that plane figures of equal area are equivalent by dissection (24.7), led Hilbert to include this question in his famous list of the 23 most important problems facing mathematics in the twentieth century, which he presented at the International Congress of Mathematicians in 1900. The third problem was to show that the method of exhaustion really is necessary, by exhibiting two solid figures of the same volume that cannot be subdivided in any way into a finite union of congruent pieces. This was done in the same year by Max Dehn, who showed that indeed, the method of exhaustion is necessary. In particular, it is not possible to dissect a regular tetrahedron into a finite number of pieces that can be reassembled into a cube (Section 27).

# Exercises

26.1 Gerling's proof that a triangular pyramid and its mirror image are equal by dissection.

(a) Let $ABCD$ be a given triangular pyramid. Show that it can be inscribed in a sphere, i.e., there is a sphere containing the points $A, B, D, D$. Then by drawing planes through the center $O$ of the sphere and the edges of the original pyramid $P$, show that $P$ can be dissected into four triangular pyramids $P_1, P_2, P_3, P_4$ with bases $ABC$, etc., and such that the vertex $O$ is equidistant from the three vertices of the base of each $P_i$.

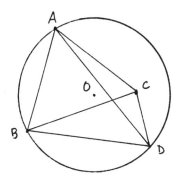

(b) Now let $P_1 = OABC$ be a triangular pyramid with $O$ equidistant from $A, B, C$. Drop a perpendicular $OD$ to the plane of $ABC$, and show that $D$ is equidistant from $A, B, C$. Then by drawing planes through $OD$ and $A, B, C$ respectively, show that $P_1$ is subdivided into three *right isosceles triangular pyramids* $P_{11}, P_{12}, P_{13}$. So for example, $P_{11} = OABD$, and $OA \cong OB$, $DA \cong DB$, and $OD \perp DA$ and $OD \perp DB$.

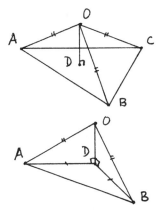

(c) Show that a right isosceles triangular pyramid $P_{11}$ as just described is congruent to its mirror image, by a motion in 3-space.

(d) If $P$ is the original pyramid, and $P'$ its mirror image, show that $P = \sum P_{ij}$ and $P' = \sum P'_{ij}$ are two decompositions into twelve pieces each, and $P_{ij} \cong P'_{ij}$.

26.2 Show that a regular tetrahedron of edge 1 can be dissected into four regular tetrahedra of edge $\frac{1}{2}$ plus a regular octahedron of edge $\frac{1}{2}$.

# 27  Hilbert's Third Problem

As mentioned in the last section, Hilbert's third problem was to show that the method of exhaustion really is necessary in Euclid's proof of (XII.5). More precisely, the problem is to find two solid figures of equal volume that are not equivalent by dissection, even after possibly adding on other figures that are equivalent by dissection. This problem was solved in 1900 by Max Dehn, and here we give a modern algebraic version of his solution.

This is another example of the use of methods of modern algebra to solve a purely geometric problem. To show that a certain dissection is possible, it is enough to exhibit the dissection and prove that the parts are congruent. This we have done with many areas (Section 24), and it is a purely geometric process. However, to show that a certain dissection is not possible requires a different approach.

In contemporary mathematics, a proof that certain geometric objects are not equivalent in some way is usually accomplished by defining an invariant that is the same for equivalent figures, and then showing that the invariant of the figures in question are different. This is the philosophy behind the study of homotopy and homology groups in topology, for example. Modern algebra renders a great service in providing the tools for defining these invariants and calculating them.

In the present case, we will define a certain abelian group $G$, and for each polyhedral figure $P$ we will define an element $\delta(P) \in G$, called the *Dehn invariant* of $P$. We will show that $\delta$ of congruent figures is the same, and that $\delta$ is additive, in the sense that $\delta(P_1 \cup P_2) = \delta(P_1) + \delta(P_2)$ for figures $P_1, P_2$ with nonoverlapping interiors. It follows that figures that are equivalent by dissection or by complementation (the three-dimensional analogue of equal content—cf. Section 22) have the same invariant $\delta$. We will compute $\delta$ of a tetrahedron, which will be nonzero, and $\delta$ of a cube, which will be zero. This will show that a tetrahedron is not equivalent to any cube by dissection or complementation.

We start with the definition of the group $G$.

**Definition**

Let $G$ be the set of all expressions

$$(a_1, \alpha_1) + (a_2, \alpha_2) + \cdots + (a_n, \alpha_n),$$

where $a_i$ are real numbers ($a_i \in \mathbb{R}$), and the $\alpha_i$ are real numbers modulo $\pi$ ($\alpha_i \in \mathbb{R}/\pi\mathbb{Z}$), modulo the equivalence relation generated by the two following types of operations:

$$(a, \alpha) + (a, \beta) = (a, \alpha + \beta),$$

$$(a, \alpha) + (b, \alpha) = (a + b, \alpha).$$

We define addition by adding one expression to another to make a longer expression, and the order of the terms does not matter. Thus addition is associative and commutative.

**Lemma 27.1**

*G is an abelian group (called the* tensor product $\mathbb{R} \otimes_{\mathbb{Z}} \mathbb{R}/\pi\mathbb{Z}$).

*Proof* We must show the existence of an additive identity ($O$) and additive inverses. First we note that for any $a \in \mathbb{R}$,

$$(a,0) = (a,0) + (0,0)$$
$$= (a,0) + (a,0) + (-a,0)$$
$$= (a,0) + (-a,0)$$
$$= (0,0)$$

using the rules of operation above. Similarly, one can show that $(0,\alpha) = (0,0)$ for any $\alpha \in \mathbb{R}/\pi\mathbb{Z}$.

Now let $O = (0,0)$. Then for any $(a,\alpha)$,

$$O + (a,\alpha) = (0,0) + (a,\alpha)$$
$$= (a,0) + (a,\alpha)$$
$$= (a,\alpha),$$

so $O$ is an additive identity.

Given any $(a,\alpha)$, if we add $(-a,\alpha)$ we get $(0,\alpha) = O$, so $(-a,\alpha)$ is an additive inverse. Thus $G$ is an abelian group. For simplicity we will henceforth denote $O$ by 0.

**Definition**

For any polyhedral solid $P$ in Euclidean 3-space over the real numbers, we define its *Dehn invariant* $\delta(P) \in G$ as follows. For each edge of $P$, let the length of the edge be $a$, and let $\alpha$ be the *dihedral angle* (measured in a plane perpendicular to the edge) in the interior of the solid between the two planes meeting along the edge.

We take $\alpha$ in radians to be a positive number and reduce $(\bmod \pi)$. Then we define

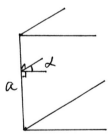

$$\delta(P) = \sum(a_i, \alpha_i),$$

where the sum is taken over all the edges of $P$.

**Example 27.1.1**

Let $P$ be a cube whose edge has length $a$. The dihedral angle between any two faces is a right angle, so

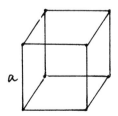

$$\delta(P) = 12(a, \pi/2),$$

since there are 12 edges of length $a$. Now, in the group $G$,

$$(a, \pi/2) = (a/2, \pi/2) + (a/2, \pi/2)$$

$$= (a/2, \pi) = 0.$$

Hence $\delta(P) = 0$.

In fact, the same method shows that $\delta(P) = 0$ for any rectangular parallelepiped.

### Example 27.1.2

Let $P$ be a right triangular prism with base edges $a, b, c$, base angles $\alpha, \beta, \gamma$, and height $h$. Then

$$\delta(P) = 2(a, \pi/2) + 2(b, \pi/2) + 2(c, \pi/2)$$

$$+ (h, \alpha) + (h, \beta) + (h, \gamma).$$

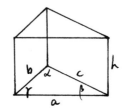

As above, $(a, \pi/2) = 0$, and similarly for $b$ and $c$. Now, in $G$ we have

$$(h, \alpha) + (h, \beta) + (h, \gamma) = (h, 2\pi) = 0.$$

So $\delta(P) = 0$.

### Example 27.1.3

Let $P$ be a regular tetrahedron of edge $a$. Then $P$ has six edges of length $a$, all having the same dihedral angle $\alpha$, so

$$\delta(P) = 6(a, \alpha).$$

We can compute the angle $\alpha$ by drawing the altitudes of two of the faces. We get a triangle $AHC$, with $AC = a$, and with $AH = HC = \sqrt{3}/2a$, the altitude of an equilateral triangle. The perpendicular from $A$ to the face $BCD$ of the tetrahedron meets it at a point $K$ equidistant from $A, B, C$. So $K$ is the centroid of $ABC$, and $HK = \frac{1}{3}HC$. From this we conclude that

$$\cos \alpha = \frac{1}{3},$$

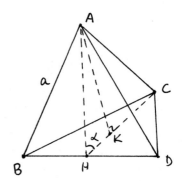

and this determines the angle $\alpha$.

We will show later that $\delta(P) \neq 0$ in $G$.

## Proposition 27.2

*The Dehn invariant $\delta$ has the following properties.*

(a) *If $P$ and $P'$ are congruent, then $\delta(P) = \delta(P')$.*

(b) *If $P_1$ and $P_2$ have nonoverlapping interiors, then*

$$\delta(P_1 \cup P_2) = \delta(P_1) + \delta(P_2).$$

*Proof*  Statement (a) is obvious, because congruent figures have congruent edges and congruent dihedral angles, so their lengths $a_i$ and measure of angles $\alpha_i$ are equal.

For (b), in comparing the union $P_1 \cup P_2$ to the two pieces $P_1$ and $P_2$, there are three ways in which the edges of $P_1 \cup P_2$ can be different than the aggregate of the edges of $P_1$ and $P_2$. In each case we will show that the contribution to $\delta$ is the same.

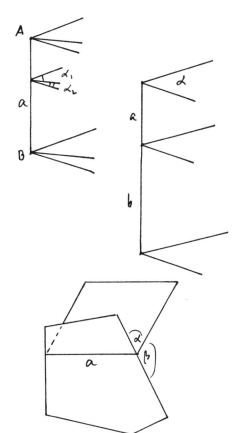

*Case 1*  An edge of $P_1$ and an edge of $P_2$ can be glued to form a single edge of $P_1 \cup P_2$. In this case the angle $\alpha$ of that edge in $P_1 \cup P_2$ will be the sum $\alpha_1 + \alpha_2$ of the angles in $P_1$ and $P_2$.

Since $(a, \alpha_1 + \alpha_2) = (a, \alpha_1) + (a, \alpha_2)$ in $G$, the contributions to $\delta$ are the same.

*Case 2*  Two edges of length $a$ in $P_1$ and of length $b$ in $P_2$ having the same angle $\alpha$ can be attached end to end to form a single edge of $P_1 \cup P_2$. Since $(a + b, \alpha) = (a, \alpha) + (b, \alpha)$ in $G$, the contributions to $\delta$ are the same.

*Case 3*  Two edges of $P_1$ and $P_2$ can be glued so as to make a single face. In this case there is no corresponding edge of $P_1 \cup P_2$. But the dihedral angles $\alpha, \beta$ of that edge in $P_1$ and $P_2$ will add up to $\pi$. Since

$$(a, \alpha) + (a, \beta) = (a, \pi) = 0 \quad \text{in } G,$$

the contribution to $\delta$ is unchanged.

There are some further possibilities, essentially equivalent to those above, which we leave to the reader. This shows that $\delta$ is additive, as claimed.

**Proposition 27.3** (Dehn's solution to Hilbert's third problem)
*In Euclidean 3-space over the real numbers, a tetrahedron cannot be dissected into a cube.*

*Proof* Because of the properties (27.2), any two figures equivalent by dissection must have the same Dehn invariant. We have seen that the invariant of a cube of any size is 0, so it is sufficient to show that the invariant of a tetrahedron is nonzero. This will be a consequence of the following two lemmas:

**Lemma 27.4**
*An element of the form $(a, \alpha) \in G$ is zero if and only if $a = 0$ or $\alpha$ is a rational multiple of $\pi(\alpha \in \pi\mathbb{Q})$.*

*Proof* First suppose $\alpha \in \pi\mathbb{Q}$, so $\alpha = (r/s)\pi$, with $r, s \in \mathbb{Z}$. Then in $G$ we can write

$$(a, \alpha) = s\left(\frac{1}{s}a, \alpha\right) = \left(\frac{1}{s}a, s\alpha\right) = \left(\frac{1}{s}, r\pi\right) = 0.$$

Conversely, suppose $a \neq 0$. We will define a group homomorphism $\varphi : G \to \mathbb{R}/\pi\mathbb{Q}$ as follows. Think of $\mathbb{R}$ as a vector space over $\mathbb{Q}$. Since $a \neq 0, a\mathbb{Q}$ is a 1-dimensional subvector space. Choose a complementary subspace $V$, so that every element $b \in \mathbb{R}$ can be written uniquely as

$$b = ra + v$$

with $r \in \mathbb{Q}$ and $v \in V$. For any element $g = \sum(b_i, \beta_i) \in G$ write each $b_i = r_i a + v_i$ with $r_i \in \mathbb{Q}, v_i \in V$, and define

$$\varphi(g) = \sum r_i \beta_i \in \mathbb{R}/\pi\mathbb{Q}.$$

We must check that $\varphi$ is well-defined. First, note that if $\beta_i \in \pi\mathbb{Z}$, then $r_i \beta_i \in \pi\mathbb{Q}$, so it is well-defined on $\beta_i \pmod{\pi}$. Second, we must see whether $\varphi$ respects the equivalence relation used to define $G$. If

$$(b, \beta) = (b_1, \beta) + (b_2, \beta)$$

and

$$b_1 = r_1 a + v_1,$$
$$b_2 = r_2 a + v_2,$$

then

$$b = (r_1 + r_2)a + (v_1 + v_2),$$

so

$$\varphi(b,\beta) = (r_1 + r_2)\beta = \varphi(b_1,\beta) + \varphi(b_2,\beta).$$

On the other hand, if

$$(b,\beta) = (b,\beta_1) + (b,\beta_2)$$

and $b = ra + v$, then

$$\varphi(b,\beta) = r(\beta_1 + \beta_2) = \varphi(b,\beta_1) + \varphi(b,\beta_2).$$

Thus $\varphi$ is well-defined.

Now observe that $a = 1 \cdot a + 0$, so $\varphi(a,\alpha) = \alpha \in \mathbb{R}/\pi\mathbb{Q}$. So if $(a,\alpha) = 0$ in $G$, it follows that $\varphi(a,\alpha) = 0$ in $\mathbb{R}/\pi\mathbb{Q}$, and hence $\alpha \in \pi\mathbb{Q}$, which is what we wanted to prove.

### Lemma 27.5
*If $\alpha$ is an angle with $\cos\alpha = \frac{1}{3}$, then $\alpha$ is not a rational multiple of $\pi$.*

*Proof*  We will offer two proofs of this fact. The first is "elementary" in the sense of using nothing more than trigonometry, but does not give much insight into why the result is true. The second is more conceptual, but uses results from the Galois theory of cyclotomic extensions of $\mathbb{Q}$.

*1st Proof*  From a small right triangle we see that $\tan\alpha = 2\sqrt{2}$. Our proof will consist in showing that for every positive integer $n$, $\tan n\alpha \neq 0, \infty$. This is sufficient to show that $a \notin \pi\mathbb{Q}$, because if $\alpha$ were a rational multiple of $\pi$, then some $n\alpha$ would be an integral multiple of $\pi$, in which case $\tan n\alpha = 0$ or $\infty$.

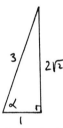

We compute $\tan n\alpha$ inductively using the angle sum formula for tangents:

$$\tan(n+1)\alpha = \frac{\tan\alpha + \tan n\alpha}{1 - \tan\alpha\tan n\alpha}.$$

In this way we find that

$$\tan\alpha = 2\sqrt{2},$$
$$\tan 2\alpha = -\frac{4}{7}\sqrt{2},$$
$$\tan 3\alpha = \frac{10}{23}\sqrt{2},$$

and so on. More generally, if $\tan n\alpha = a/b\sqrt{2}$, with $a, b \in \mathbb{Z}$, then

$$\tan(n+1)\alpha = \frac{a+2b}{b-4a}\sqrt{2}.$$

We conclude, inductively, that $\tan n\alpha$ is a rational multiple of $\sqrt{2}$ for all $n \geq 1$. While the rational numbers that occur seem more and more complicated, matters are simpler if we look at the numerator and the denominator of this fraction (mod 3). To be precise, consider the transformation that takes an ordered pair $(a, b)$ and sends it to $(a + 2b, b - 4a)$, and regard $a, b$ as elements of $\mathbb{Z}/3\mathbb{Z}$. Starting with $a = 2, b = 1$, we obtain

$$(2, 1) \rightarrow (1, 2) \rightarrow (2, 1).$$

After two steps it repeats. We conclude that for any $n \geq 1$, $\tan n\alpha = (a/b)\sqrt{2}$, with $(a, b) \equiv (2, 1)$ or $(1, 2)$ (mod 3). In particular, neither of $a, b$ can be 0, so $\tan n\alpha \neq 0, \infty$, as required.

*2nd Proof* If $\cos\alpha = \frac{1}{3}$, then $\sin\alpha = \frac{2}{3}\sqrt{2}$. Consider the complex number

$$z = \cos\alpha + i\sin\alpha = \tfrac{1}{3} + \tfrac{2}{3}\sqrt{-2}.$$

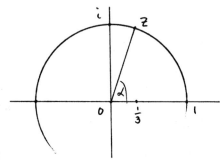

Obviously, $z$ is a root of a quadratic equation over $\mathbb{Q}$ (in fact, $3z^2 + 2z + 3 = 0$), so $z$ generates a field extension $\mathbb{Q}(z)$ of degree 2 over $\mathbb{Q}$.

Now, if $\alpha$ is a rational multiple of $\pi$, then we can write $\alpha = \dfrac{p}{q} \cdot 2\pi$ with $p$, $q \in \mathbb{Z}$, $(p, q) = 1$. In this case $z$ will be a primitive $q$th root of unity. Its minimal polynomial over $\mathbb{Q}$ will be the *cyclotomic polynomial* $\Phi_q(z)$, which has degree $\varphi(q)$, where $\varphi$ is the Euler $\varphi$ function (32.7).

In our case $z$ has degree 2 over $\mathbb{Q}$, so we conclude that $\varphi(q) = 2$. If $q = p_1^{e_1} \ldots p_r^{e_r}$ is the prime factorization of $q$, then

$$\varphi(q) = \prod p_i^{e_i - 1}(p_i - 1).$$

So the only values of $q$ that give $\varphi(q) = 2$ are $q = 3, 4, 6$. The corresponding field extensions are $\mathbb{Q}(\sqrt{-3})$ and $\mathbb{Q}(i)$. Neither of these is equal to $\mathbb{Q}(z) = \mathbb{Q}(\sqrt{-2})$, so we have a contradiction. This shows that $\alpha$ is not a rational multiple of $\pi$.

# Exercises

27.1 Compute the Dehn invariant of a regular octahedron of side $a$. Show that this element is nonzero in $G$, so the octahedron cannot be dissected into a cube. Show furthermore that an octahedron cannot be dissected into a regular tetrahedron.

27.2 Show that two tetrahedra of edge 1 cannot be dissected into a single tetrahedron of edge $\sqrt[3]{2}$. Is it possible to dissect 8 tetrahedra of edge 1 into a single tetrahedron of side 2?

27.3 Is there any triangular pyramid with Dehn invariant 0?

27.4 Compute the Dehn invariant of a *right isosceles triangular pyramid* (Exercise 26.1) *OABC* with $AB = AC = a$, base angle $\angle BAC = \theta$, with *OA* perpendicular to the plane of *ABC*, and dihedral angle $\alpha$ along *BC*. Use this to determine the image of the map $\delta$, i.e., the subgroup of $G$ consisting of elements that are the Dehn invariant of some polyhedron.

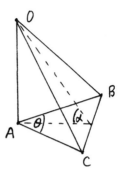

27.5 Imitate the first proof of Lemma 27.5 to show that if $\tan\alpha \in \mathbb{Q}$, and $\alpha$ is a rational multiple of $\pi$, then $\tan\alpha = 0, \pm 1, \infty$. *Hint*: Let $\tan\alpha = (r/s)$ with $r, s \in \mathbb{Z}, (r, s) = 1$. If $r, s$ are not both $\pm 1$, take an odd prime divisor $p$ of $r^2 + s^2$, and make a calculation $(\bmod\, p)$ similar to the one $(\bmod\, 3)$ in (Lemma 27.5).

27.6 Use the second proof of (Lemma 27.5) to show that if $\cos\alpha \in \mathbb{Q}$ and if $\alpha$ is a rational multiple of $\pi$, then $\cos\alpha = 0, \pm\frac{1}{2}, \pm 1$.

Hobbes, on being asked why he did not read more? answered, if I read as much as other men, I should know as little; his library consisted of Homer, Thucydides, Euclid, and Virgil.

– from the introduction to the Retrospective Review, Vol. I (1820)

# 6

**CHAPTER**

# Construction Problems and Field Extensions

uring the earlier parts of this book, we started always from Euclid's geometry, developing and expanding it using our modern mathematical awareness. Because of the construction of the field of segment arithmetic, one could even argue that the use of fields in Chapter 4 arises naturally from the geometry. In this chapter, however, we will make use of modern algebra, the theory of equations and field extensions, and in particular the Galois theory, as it developed in the late nineteenth and early twentieth centuries.

We use algebra as a lens to examine and interpret geometrical questions in a new way. It was this algebraic interpretation that first provided the tools for rigorously proving the unsolvability of classical construction problems. If these proofs seem easy today, just think of all the modern algebra we take for granted that did not exist even two hundred years ago.

We begin by examining three classical construction problems, two of which we can prove to be impossible using field theory. (The third depends on the transcendence of $\pi$, whose proof would carry us too far afield.)

Next we consider the problem of constructing regular polygons, and give Gauss's proof of the constructibility of the regular 17-sided polygon.

In Section 30 we consider constructions made with an additional tool, the marked ruler. Using the marked ruler we show how to extract cube roots and trisect angles. Then in Section 31 we study the real roots of cubic and quartic equations, and show that the constructions with compass and marked ruler are equivalent to solutions of quadratic, cubic, and quartic equations.

**241**

In an appendix Section 32 we collect, mostly without proof, the results from field theory that we use in the earlier sections of this chapter.

In this chapter we work always over a subfield of the real numbers $\mathbb{R}$.

# 28   Three Famous Problems

In this section we will discuss three classical construction problems via field theory. Recall (13.2) that a point with coordinates $(\alpha, \beta)$ in the real Cartesian plane can be constructed with ruler and compass if and only if the coordinates $\alpha, \beta$ can be obtained from the initial data by a finite number of field operations and extractions of square roots of positive quantities. If we suppose that the initial data are given by rational numbers, then this is equivalent to saying that $\alpha$ and $\beta$ are elements of the constructible field $K$ (16.4.1), which is just the set of those real numbers that can be obtained by a finite number of field operations and square roots from $\mathbb{Q}$. We say that a real number $\alpha$ is *constructible* if it is an element of $K$. So a point $(\alpha, \beta)$ in the real Cartesian plane is constructible by ruler and compass from the rational numbers if and only if $\alpha$ and $\beta$ are constructible numbers.

For our applications we need some criteria to decide when a number is constructible. The first is merely a restatement of what we have just said.

**Proposition 28.1**
*A real number $\alpha \in \mathbb{R}$ is constructible if and only if there is a tower of subfields of $\mathbb{R}$,*

$$\mathbb{Q} = F_0 \subseteq F_1 \subseteq \cdots \subseteq F_k \subseteq \mathbb{R}$$

*such that $\alpha \in F_k$ and each $F_i = F_{i-1}(\sqrt{a_i})$ for some $a_i \in F_{i-1}$.*

*Proof*   In writing $\alpha$ as obtained by a finite number of field operations and square roots, each time there is a square root, let $F_i$ be the field generated by that square root over the previously constructed field. Conversely, if $\alpha \in F_k$, then by construction it is written using field operations and square roots.

**Corollary 28.2**
*If $\alpha \in \mathbb{R}$ is constructible, then $\deg \mathbb{Q}(\alpha)/\mathbb{Q}$ is a power of 2.*

*Proof*   Here we use the notion of the degree of a field extension (32.1). Adjoining a square root of an element that is not already a square creates a field extension of degree 2. So, assuming that the extensions in (28.1) are all nontrivial, we find that $\deg F_k/\mathbb{Q} = 2^k$. Since $\mathbb{Q} \subseteq \mathbb{Q}(\alpha) \subseteq F_k$, by multiplicativity of degrees of field extensions (32.1) we see that $\deg \mathbb{Q}(\alpha)/\mathbb{Q}$ is a power of 2.

## Remark 28.2.1

This simple statement is already enough to dispose of two of the three classical construction problems. Beware, however! The converse of this statement is false (28.7.1).

## Duplication of the Cube

The problem is, given a cube, to construct a new cube whose volume is twice the volume of the old cube. We will interpret the phrase "given a cube" to mean that we are given its edge, as a line segment in the plane, and the problem is then to construct the edge of the new cube. If the first cube has edge $a$, then we are looking for a length $b$ such that $b^3 = 2a^3$. This means that $b = \sqrt[3]{2}a$. If $b$ is constructible, then $\sqrt[3]{2}$ is constructible, and conversely.

## Theorem 28.3

$\sqrt[3]{2}$ *is not in* $K$. *Hence the duplication of the cube is not possible by ruler and compass.*

*Proof*  Consider $\alpha = \sqrt[3]{2}$. It is a root of the equation $x^3 - 2 = 0$. The polynomial $x^3 - 2$ is irreducible over $\mathbf{Q}$. Indeed if it factored, then at least one factor would be linear, so it would have a root. Let this root be $a/b$ with $a, b \in \mathbf{Z}$, $a, b$ relatively prime. Then

$$a^3 = 2b^3.$$

It follows that $a$ is even. Then $2^3 | a^3$, so $2^2 | b$, so $b$ is also even. This contradicts the hypothesis that $a$ and $b$ were relatively prime.

Now, since $x^3 - 2$ is irreducible, the field $\mathbf{Q}(\sqrt[3]{2})$ is an extension of degree 3 of $\mathbf{Q}$. Since 3 is not a power of 2, we conclude $\sqrt[3]{2} \notin K$, by (28.2).

## Trisection of the Angle

The problem is, given an angle in the plane, to construct by ruler and compass a new angle equal to one-third of the given angle. This problem is a bit more complicated because some angles can be trisected. For example, if the given angle is a right angle, we can construct an angle equal to one-third of it (30°) by first constructing an equilateral triangle (I.1) and then bisecting one of its angles. See also Exercise 28.4.

We will show that the problem of trisecting the angle is not always possible by exhibiting one angle that exists in the Cartesian plane II over the constructible field $K$, namely an angle of 60°, whose third, an angle of 20°, does not exist in II. If an angle $\alpha$ exists in II, then the trigonometric functions $\sin \alpha$, $\cos \alpha$, $\tan \alpha$ will be elements of the field $K$ (Exercise 20.15).

## Theorem 28.4

*The real number* $\alpha = \cos 20°$ *is not in* $K$. *Hence an angle of 60° cannot be trisected by ruler and compass.*

*Proof*   Using the sum angle formulas for sin and cos, we find for any angle $\theta$ that

$$\cos 3\theta = 4\cos^3 \theta - 3\cos\theta.$$

Since $\cos 60° = \frac{1}{2}$, the real number $\alpha = \cos 20°$ satisfies

$$4\alpha^3 - 3\alpha = \frac{1}{2}.$$

In other words, $\alpha$ is a root of the equation

$$4x^3 - 3x - \frac{1}{2} = 0.$$

We will show that this equation has no roots in $K$. Making the substitution $y = 2x$, it is sufficient to show that the equation

$$y^3 - 3y - 1 = 0$$

has no roots in $K$.

First we show that it has no roots in $\mathbf{Q}$. If it did, let the root be $a/b$, with $a, b \in \mathbf{Z}$, relatively prime. Then

$$a^3 - 3ab^2 - b^3 = 0.$$

Consequently any prime factor $p$ of $a$ also divides $b$, and conversely, any prime factor of $b$ divides $a$. Since $a, b$ are relatively prime, we conclude that $a, b = \pm 1$, so $a/b = \pm 1$. But by inspection we see that neither $+1$ nor $-1$ is a root of the equation.

Hence $y^3 - 3y - 1$ is an irreducible polynomial over $\mathbf{Q}$, since it is of degree 3 with no rational roots. We conclude that $\mathbf{Q}(\alpha)$ is an extension of degree 3 of $\mathbf{Q}$. But if $\alpha$ were constructible, then $\mathbf{Q}(\alpha)$ would have degree a power of 2 over $\mathbf{Q}$ (28.2). We conclude that $\alpha \notin K$, so the angle of 20° is not constructible.

### Squaring of the Circle

The problem is, given a circle, to construct by ruler and compass a square of area equal to the area enclosed by the circle. If the circle has radius $r$, its area is $\pi r^2$, so we need a square of side $a$ with $a^2 = \pi r^2$. Thus $a = \sqrt{\pi} \cdot r$. If $a$ were constructible from $r$, then $\sqrt{\pi}$, and hence $\pi$ would be constructible.

### Theorem 28.5
$\pi \notin K$, so the problem of squaring the circle cannot be solved by ruler and compass.

The fact that $\pi \notin K$ is a consequence of the stronger fact that $\pi$ is *transcendental* over $\mathbf{Q}$: There is no polynomial with rational coefficients having $\pi$ as a root. But every element of $K$ is algebraic over $\mathbb{Q}$. The proof that $\pi$ is transcendental is analytic, so we will not give it here. See, for example, Stewart (1989), Chapter 6. See also Section 25 for further discussion of this problem and its history. And see (42.4.1) for a surprise: In the hyperbolic plane, some circles can be squared!

Up to here we have used only the necessary condition (28.2) for a number to be constructible. Using a little more field theory, we will now derive a necessary and sufficient condition for a real number to be constructible. It depends on a remarkable coincidence of three separate notions: adjoining square roots, solving quadratic equations, and making field extensions of degree 2.

### Proposition 28.6

*Let $F \subseteq E$ be an extension of fields of characteristic not equal to 2. Then the following conditions are equivalent:*

(i) $E = F(\sqrt{a})$ *for some $a \in F$, $\sqrt{a} \notin F$.*
(ii) $\deg E/F = 2$.
(iii) $E = F(\alpha)$, *where $\alpha$ is a root of an irreducible quadratic polynomial over $F$.*

*Proof* (i) $\Rightarrow$ (ii). If $E = F(\sqrt{a})$, then every element of $E$ can be written uniquely in the form $c_1 + c_2\sqrt{a}$, $c_1, c_2 \in F$, so $\deg E/F = 2$.

(ii) $\Rightarrow$ (iii). If $\deg E/F = 2$, and if $\alpha \in E$ is any element not in $F$, then $1, \alpha, \alpha^2$ are linearly dependent, so $\alpha$ satisfies an irreducible quadratic equation with coefficients in $F$.

(iii) $\Rightarrow$ (i). Follows from the quadratic formula: The roots of the equation $ax^2 + bx + c = 0$ are given by $\alpha = (1/2a)(-b \pm \sqrt{b^2 - 4ac})$.

### Remark 28.6.1

This is indeed an elementary result, but we thought it worth stating explicitly because of the contrast with what happens in higher degrees—cf. Exercise 28.5 and Section 31.

### Proposition 28.7

*A real number $\alpha$ is constructible if and only if the Galois group of its minimal polynomial is a group of order $2^n$ for some $n$.*

*Proof* Suppose $\alpha$ is constructible. Then there is a tower of fields

$$\mathbb{Q} = F_0 \subseteq F_1 \subseteq \cdots \subseteq F_k \subseteq \mathbb{R}$$

as in (28.1). For any isomorphism $\sigma : F_k \to F'$ to a subfield $F'$ of $\mathbb{C}$, $F'$ will also be the top of a similar tower

$$\mathbb{Q} = F_0 \subseteq \sigma(F_1) \subseteq \sigma(F_2) \subseteq \cdots \subseteq \sigma(F_k) = F'.$$

Hence every element of $F'$ can be obtained by field operations and successive square roots. The subfield $\bar{F}$ of $\mathbb{C}$ generated by the fields $\sigma(F_k)$ for all possible such $\sigma$ is then a normal extension of $\mathbb{Q}$, whose degree over $\mathbb{Q}$ is a power of 2.

Now, if $\alpha$ is an element of $F_k$ with minimal polynomial $f(x)$, then all the conjugates of $\alpha$ lie in the various fields $\sigma(F_k)$, so the splitting field $E$ of $f(x)$ will be a subfield of $\bar{F}$. Hence also the degree of $E/\mathbb{Q}$, which is equal to the order of the Galois group $G$ of $f(x)$, is a power of 2.

Conversely, let $\alpha \in \mathbb{R}$, and assume that the Galois group $G$ of its minimal polynomial has order a power of 2. An elementary result in group theory states that the center of a $p$-group is nontrivial (see, e.g., Herstein (1975), p. 86). Hence $G$ has a normal subgroup $G_1$ of order 2. Applying the same argument to $G/G_1$ and continuing, we find that there is a chain of subgroups

$$\{e\} = G_0 \subseteq G_1 \subseteq G_2 \subseteq \cdots \subseteq G_n = G$$

where each $G_i$ is normal in $G$, and $G_i/G_{i-1}$ has order 2.

By the fundamental theorem of Galois theory (32.4), if we denote by $E$ the splitting field of the minimal polynomial $f(x)$ of $\alpha$, there is a chain of fields

$$\mathbb{Q} = E_0 \subseteq E_1 \subseteq \cdots \subseteq E_n = E \subseteq \mathbb{C}$$

with each $E_i/E_{i-1}$ of degree 2, and each $E_i$ is a normal extension of $\mathbb{Q}$. Let $F_i = E_i \cap \mathbb{R}$. Then we have

$$\mathbb{Q} = F_0 \subseteq F_1 \subseteq F_2 \subseteq \cdots \subseteq F_n = E \cap \mathbb{R}.$$

Now, each $E_i$ is normal over $\mathbb{Q}$, so complex conjugation $\tau$ acts on $E_i$, with fixed field $F_i$. Hence $\deg E_i/F_i = 1$ or 2 for each $i$. More precisely, if $k$ is the largest index for which $E_k \subseteq \mathbb{R}$, then $F_i = E_i$ for $i \leq k$; $F_k = F_{k+1}$, and $\deg E_i/F_i = 2$ for all $i > k$. In particular, $\deg F_i/F_{i-1}$ is equal to 2 for all $i$ except $i = k+1$, in which case it is 1. By (28.6), $F_i = F_{i-1}(\sqrt{a_i})$ for some $a_i \in F_{i-1}$. Furthermore $\alpha \in F_n = E \cap \mathbb{R}$, so $\alpha$ is constructible.

### Example 28.7.1

For $\alpha \in \mathbb{R}$ to be constructible, it is not sufficient that $\deg \mathbb{Q}(\alpha)/\mathbb{Q}$ be a power of 2. Take, for example, the polynomial $f(x) = x^4 - x^3 - 5x^2 + 1$. A little curve-sketching from first-year calculus shows that it has four distinct real roots. Let $\alpha$ be one of them. It is shown in (32.5.1) that $f(x)$ is irreducible, and that its Galois group has order 12 or 24. Hence $\deg \mathbb{Q}(\alpha)/\mathbb{Q} = 4$, but by (28.7), $\alpha$ is not constructible.

# Exercises

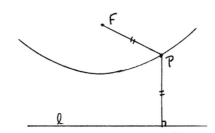

28.1 A *parabola* Γ can be defined as the locus of points $P$ equidistant from a fixed point $F$, the *focus*, and a fixed line $l$, the *directrix*.

(a) If the focus is $(0,1)$ and the directrix is $y = -1$, find the equation of the parabola.

(b) If $y = mx + b$ is a line with $m, b \in K$ (the constructible field) and $b > 0$, show that the two intersection points of the line with Γ have coordinates in $K$.

(c) Now give a ruler and compass construction for the intersection points of a line $m$ with the parabola having focus $F$ and directrix $l$. (Par $= 9$ to get one point.) Thus, even though we cannot "draw" the parabola, yet we can find its intersection points with any line, as if it were drawn.

28.2 In the real Cartesian plane:

(a) Show that the circle with center $(a,1)$ passing through the origin intersects the parabola Γ given by $y = \frac{1}{2}x^2$ at the point $(2\sqrt[3]{a}, 2\sqrt[3]{a^2})$.

(b) Show that if we are given a single parabola drawn in the plane, then the problem of duplication of the cube becomes possible.

(c) If a parabola is given by its focus and directrix, conclude that its intersection points with a circle are not always constructible.

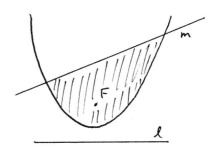

28.3 *Squaring the parabola.* Let a parabola be given by its focus $F$ and directrix $l$ (as in Exercise 28.1), and let another line $m$ be given in the plane. Prove that it is possible to construct with ruler and compass a square whose area is equal to the area bounded by the line and the parabola. (You need not find the construction: Just prove that it is possible.)

28.4 In the real Cartesian plane, if we are given an angle $\alpha$ of radian measure $2\pi a/b$ with $a, b \in \mathbb{Z}$, and $b$ not divisible by 3, show that it is possible to construct an angle $\frac{1}{3}\alpha$ starting from the given angle.

28.5 (a) Sketch the graph of $y = x^3 - 3x - 1$ in the real Cartesian plane, and use the intermediate value theorem to show that it has three real roots.

(b) If $\alpha$ is one of the roots of $x^3 - 3x - 1$, show that the field $\mathbb{Q}(\alpha)$ is not equal to a field of the form $\mathbb{Q}(\sqrt[3]{d})$ for any $d \in \mathbb{Q}$. Thus a field extension of degree 3 need not be generated by cube roots, in contrast to the situation in degree 2 (Proposition 28.6).

(c) However, one can express roots of this equation using square roots and cube roots of complex numbers: Verify that

$$\alpha = \sqrt[3]{\frac{1}{2} + \frac{1}{2}\sqrt{-3}} + \sqrt[3]{\frac{1}{2} - \frac{1}{2}\sqrt{-3}}$$

is a root of the above equation. *Hint*: Of course you could substitute directly in the equation and multiply out, but using a little geometry you should be able to show without any messy calculation that $\alpha$ is a root of this equation.

28.6 Let $\alpha = \cos 20°$. Show that $\alpha$ is totally real and $\alpha + 1$ is totally positive in the sense of Exercise 16.10. Find an explicit representation of $\alpha + 1$ as a sum of squares in the field $\mathbb{Q}(\alpha)$ — cf. Artin's theorem (Exercise 16.12).

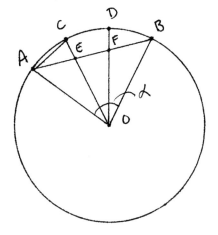

28.7 Instead of using trigonometry, as in the text, one can also use geometry to obtain a cubic equation expressing the trisection of an angle. Let an angle $\alpha = AOB$ be given in a circle of radius 1, and let the length of $AB$ be $a$. Suppose $\alpha$ can be trisected by $OC$, $OD$. Let $x = $ length $AC$. Show that $\triangle ACE$ and $\triangle OEF$ are both similar to $\triangle OAC$. From this derive a cubic equation for $x$ in terms of $a$. If $\alpha = 60°$, what equation do you get?

28.8 Prove a theorem analogous to (Theorem 13.2) for constructions with Hilbert's tools (Section 10): Given a number of points $P_i$, including $(0,0)$ and $(1,0)$ in the real Cartesian plane, a new point $Q = (\alpha, \beta)$ can be constructed with Hilbert's tools from the initial points if and only if $\alpha$ and $\beta$ can be obtained from the coordinates of the initial points by a finite number of field operations $+, -, \cdot, \div$, and operations $a \mapsto \sqrt{1 + a^2}$.

28.9 Conclude from Exercise 28.8 that a point $Q = (\alpha, \beta)$ is constructible with Hilbert's tools from the rational numbers if and only if its coordinates $\alpha, \beta$ are in Hilbert's field $\Omega$ (Proposition 16.3). Using Exercise 16.14 this is also equivalent to saying that $\alpha, \beta$ are constructible and totally real.

28.10 (a) Show that $\cos 72°$ and $\sin 72°$ are totally real, and conclude that it is possible to construct a regular pentagon with Hilbert's tools.

(b) Given a segment *OA* in the plane, construct with Hilbert's tools a regular pentagon inscribed in the circle with center *O* and radius *OA* (but without drawing the circle!).

28.11 Show that it is not possible with Hilbert's tools to construct a square with area equal to a given equilateral triangle. (Of course, it is possible with ruler and compass by (II.14).)

28.12 (Origami) In the traditional art of paper folding, you start with a square piece of paper. Consider the corners and the edges to be given. You get new points and lines as images of previously constructed points and lines by the following three operations (the "restricted" rules of origami):

(1) Make a fold passing through two given points.
(2) Make a fold that places one known point on another known point.
(3) Make a fold that places a known line on another known line.

If we consider the original paper to be the unit square $[0,1] \times [0,1]$, show that the points obtainable by these rules are the same as those constructed by Hilbert's tools, namely points in the unit square with coordinates in Hilbert's field $\Omega$ (Proposition 16.3).

28.13 In practice, most origami constructions make use of a fourth rule (the "general" rules of origami), namely:

(4) Make a fold through a given point that places another given point on a given line.

Show that with the general rules, we obtain all points with constructible coordinates (Proposition 16.4).

28.14 Fold a piece of origami paper into an equilateral triangle as follows, and explain why it works.

1. Fold *A* to *B*, get *C*.
2. Fold *B* to *C*, get line *d*.
3. Fold *A* to *d*, passing through *C*, get *E*.
4. Fold *CB* to *CE*, get *F*.
5. Fold *EF*.

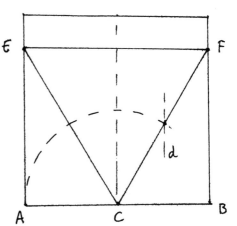

28.15 Fold a piece of origami paper into a regular pentagon, and explain why it works. *Hint*: The side of an inscribed pentagon, with one side parallel to a side of the paper will be $\frac{1}{2}(\sqrt{5}-1)$. (About 10 folds.)

28.16 For a real challenge, suppose that we are given two segments $a, b$ on opposite edges of the paper. Fold the triangle with sides $1, a, b$. (About 10 folds.)

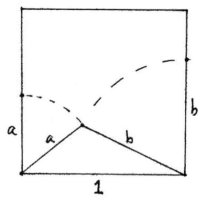

28.17 The problem of Apollonius is, given three circles $\Gamma_1, \Gamma_2, \Gamma_3$, to construct a circle $\Delta$ that is tangent to all three. Let $\Gamma_i$ have center $(a_i, b_i)$ and radius $r_i$. If we assume that $\Delta$ is in between the $\Gamma_i$, as in the figure, and has unknown center $(x, y)$ and radius $r$, we obtain equations

$$\text{dist}((x, y), (a_i, b_i)) = r + r_i$$

for $i = 1, 2, 3$. Show that these equations can be solved for $x, y, r$ using field operations and square roots. Thus the problem is solvable by ruler and compass. (For an actual construction, see Section 38).

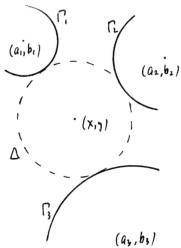

# 29    The Regular 17-Sided Polygon

A *regular* polygon is a polygon with all sides equal and all angles equal. Euclid knew how to construct regular polygons with $n$ sides for $n = 3$ (I.1), $n = 4$ (I.46), $n = 5$ (IV.11), $n = 6$ (IV.15), and $n = 15$ (IV.16). By bisecting the central angles, one can construct a regular polygon of $2n$ sides, given one of $n$ sides. Thus Euclid could construct regular polygons of $n$ sides for $n = 2^k$, $2^k \cdot 3$, $2^k \cdot 5$, $2^k \cdot 3 \cdot 5$. For two thousand years these were the only known constructible regular polygons until Carl Friedrich Gauss in 1796, at the age of 19, made the remarkable discovery that the regular polygon of 17 sides is constructible with ruler and compass. He was so proud of this result that he requested that a regular 17-gon be inscribed on his tombstone.

In this section we will explain Gauss's method using the algebra of complex numbers. We start with a proof of the construction of the regular pentagon, and finish with a criterion for constructibility of regular $n$-gons.

Our first step is to represent points of the real Cartesian plane by complex numbers. To the point $(a, b)$ we associate the complex number $z = a + bi$. If $\zeta$ is a point on the unit circle making an angle $\theta$ with the positive real axis, then we can write

$$\zeta = \cos\theta + i\sin\theta.$$

If $\theta = 2\pi k/n$ for some integers $k, n$, then according to the laws of multiplication of complex numbers, $\zeta^n = 1$. In other words, $\zeta$ is a complex root of the equation $x^n - 1 = 0$. Letting $k = 0, 1, \ldots, n-1$, we obtain $n$ distinct roots of this equation, so these are all the roots. They are called the $n$th *roots of unity*.

To construct a regular polygon of $n$ sides, it is sufficient to construct the quantity $\cos 2\pi/n$, and hence also $\sin 2\pi/n$. Since the complex number $\zeta = \cos 2\pi/n + i\sin 2\pi/n$ is then an $n$th root of unity, the constructibility of the regular $n$-gon in the real Cartesian plane is related to the solutions of the equation $x^n - 1 = 0$ in the complex numbers.

Let us illustrate this situation by considering the case $n = 5$. Take $\theta = 2\pi/5$, and let $\zeta = \cos 2\pi/5 + i\sin 2\pi/5$ be the complex number that corresponds to the first vertex of the pentagon after 1. Then the five vertices of the pentagon will be $1, \zeta, \zeta^2, \zeta^3, \zeta^4$. Note that $\zeta^4 = \zeta^{-1}$ is the complex conjugate of $\zeta$. Therefore,

$$\zeta + \zeta^4 = 2\cos 2\pi/5.$$

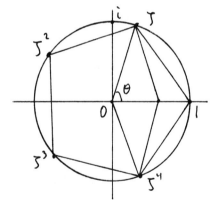

On the other hand, $\zeta$ is a root of $x^5 - 1$. This polynomial factors as

$$x^5 - 1 = (x - 1)(x^4 + x^3 + x^2 + x + 1).$$

Since $\zeta \neq 1$, it is a root of the second factor, which is called the *cyclotomic polynomial* $\Phi_5$ (cf. Section 32). Let $\alpha = \zeta + \zeta^4$. Then (remembering that $\zeta^5 = 1$)

$$\alpha^2 = \zeta^2 + 2 + \zeta^3.$$

Hence

$$\alpha^2 + \alpha = \zeta^4 + \zeta^3 + \zeta^2 + \zeta + 2 = 1,$$

using the relation $\Phi_5(\zeta) = 0$. So $\alpha$ is a root of the polynomial

$$x^2 + x - 1 = 0.$$

Using the quadratic formula,

$$x = \frac{1}{2}(-1 \pm \sqrt{5}).$$

Since our $\alpha$ is a positive real number, we conclude that $\alpha = \frac{1}{2}(\sqrt{5} - 1)$. Hence:

**Proposition 29.1**
$\cos 2\pi/5 = \frac{1}{4}(\sqrt{5} - 1)$.

**Remark 29.1.1**
This gives another algebraic proof of Euclid's construction of the regular pentagon (IV.11). Compare (13.4) and (13.5).

Now let us turn our attention to the regular 17-gon.

**Theorem 29.2** (Gauss)
*The regular 17-sided polygon is constructible with ruler and compass.*

*Proof* An abstract proof of constructibility can be given using Galois theory (29.3.1). Here, however, we will give an elementary constructive proof, close to Gauss's original method.

We follow the general idea of the method used above with the pentagon. Let

$$\zeta = \cos 2\pi/17 + i\sin 2\pi/17.$$

Let

$$\alpha = \zeta + \zeta^{-1},$$
$$\beta = \zeta + \zeta^4 + \zeta^{-1} + \zeta^{-4},$$
$$\gamma = \zeta + \zeta^2 + \zeta^4 + \zeta^8 + \zeta^{-1} + \zeta^{-2} + \zeta^{-4} + \zeta^{-8}.$$

We will show that

$$\mathbb{Q} \subseteq \mathbb{Q}(\gamma) \subseteq \mathbb{Q}(\beta) \subseteq \mathbb{Q}(\alpha),$$

and each field is quadratic over the previous one. Since $\alpha = 2\cos 2\pi/17$, this will show that the regular 17-gon is constructible.

(At this point, you may ask, where did these expressions for $\alpha, \beta, \gamma$ come from? Here are three possible answers: 1. Gauss was a genius. 2. Never mind where they came from; just follow the proof and see that it works. 3. If you look at the subgroups of order 2, 4, 8 of the Galois group of $\mathbb{Q}(\zeta)$, these are elements that will be left fixed by them—see (29.3.1).)

Let us start with $\gamma$. It is natural to consider also the element $\gamma'$ that is the sum of the other eight powers of $\zeta$, namely

$$\gamma' = \zeta^3 + \zeta^5 + \zeta^6 + \zeta^7 + \zeta^{-3} + \zeta^{-5} + \zeta^{-6} + \zeta^{-7}.$$

Since $\zeta$ is a 17th root of unity, it is a root of the cyclotomic polynomial

$$\Phi_{17} = x^{16} + x^{15} + \cdots + x + 1.$$

Thus $\gamma + \gamma' = -1$ (remember to treat exponents of $\zeta$ (mod 17), since $\zeta^{17} = 1$). We will show that $\gamma\gamma'$ is also in $\mathbb{Q}$, so that $\gamma$, which is a root of the equation

$$x^2 - (\gamma + \gamma')x + \gamma\gamma' = 0,$$

will be quadratic over $\mathbb{Q}$.

To find $\gamma\gamma'$ we must make a computation. Multiplying each term of $\gamma'$ by each term of $\gamma$, we obtain the sum of $\zeta$ raised to each of the powers shown in the accompanying table. Observe that each integer $1 \leq i \leq 16$ (mod 17) occurs exactly four times in the table. Thus

$$\gamma\gamma' = 4\left(\sum_{i=1}^{16} \zeta^i\right) = -4.$$

**Table of exponents of $\gamma\gamma'$**

| | | | | | | | |
|---|---|---|---|---|---|---|---|
| 4 | 6 | 7 | 8 | −2 | −4 | −5 | −6 |
| 5 | 7 | 8 | 9 | −1 | −3 | −4 | −5 |
| 7 | 9 | 10 | 11 | 1 | −1 | −2 | −3 |
| 11 | 13 | 14 | 15 | 5 | 3 | 2 | 1 |
| 2 | 4 | 5 | 6 | −4 | −6 | −7 | −8 |
| 1 | 3 | 4 | 5 | −5 | −7 | −8 | −9 |
| −1 | 1 | 2 | 3 | −7 | −9 | −10 | −11 |
| −5 | −3 | −2 | −1 | −11 | −13 | −14 | −15 |

So $\gamma$ is a root of the equation

$$x^2 + x - 4 = 0.$$

Using the quadratic formula, we obtain

$$x = \frac{1}{2}(-1 \pm \sqrt{17}).$$

To choose the correct sign for $\gamma$, we make an estimate. Note that in the sum for $\gamma$, the imaginary parts cancel, so $\gamma$ is real. In fact,

$$\gamma = 2(\cos 2\pi/17 + \cos 4\pi/17$$

$$+ \cos 8\pi/17 + \cos 16\pi/17).$$

Now looking at the approximate position of the powers $\zeta$, $\zeta^2$, $\zeta^4$, $\zeta^8$, it is clear that $\gamma$ is positive. Therefore,

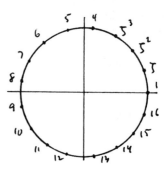

$$\gamma = \frac{1}{2}(-1 + \sqrt{17}).$$

We can also confirm this choice of sign with a calculator by evaluating the sum of cosines above and the expression with the radical. Both come to 1.56155 to five decimal places. Note, however, that the calculator check does not prove the equality: It is merely a good way to detect errors in our work.

Next we consider $\beta$. If we let

$$\beta' = \zeta^2 + \zeta^8 + \zeta^{-2} + \zeta^{-8},$$

then $\beta + \beta' = \gamma$. On the other hand, $\beta\beta'$ turns out to be the sum of all the non-zero powers of $\zeta$, so $\beta\beta' = -1$. Thus $\beta$ is a root of the equation

$$x^2 - \gamma x - 1 = 0.$$

By the quadratic formula,

$$x = \frac{1}{2}\left(\gamma \pm \sqrt{\gamma^2 + 4}\right).$$

Since $\beta$ is positive, we choose the $+$ sign, and obtain

$$\beta = \frac{1}{2}\left(\gamma + \sqrt{\gamma^2 + 4}\right).$$

Substituting for $\gamma$ and simplifying gives

$$\beta = \frac{1}{4}\left(-1 + \sqrt{17} + \sqrt{34 - 2\sqrt{17}}\right).$$

Note also that by definition,

$$\beta = 2(\cos 2\pi/17 + \cos 8\pi/17).$$

Both check with a calculator to give $\beta = 2.04948$ (to 5 decimals).

Now let us look at the field extensions created so far. Clearly, $\mathbb{Q}(\gamma) = \mathbb{Q}(\sqrt{17})$. We claim that $\mathbb{Q}(\beta) = \mathbb{Q}\left(\sqrt{34 - 2\sqrt{17}}\right)$, which contains $\mathbb{Q}(\gamma)$. It is clear that

$$\mathbb{Q}(\beta) = \mathbb{Q}\left(\sqrt{17} + \sqrt{34 - 2\sqrt{17}}\right).$$

If we let

$$x = \sqrt{17} + \sqrt{34 - 2\sqrt{17}},$$

then squaring twice to get rid of radicals gives the equation

$$x^4 - 6 \cdot 17x^2 + 8 \cdot 17x + 13 \cdot 17 = 0.$$

By Eisenstein's criterion (32.8) this is irreducible, so the degree of the field extension $\mathbb{Q}(\beta)/\mathbb{Q}$ is four. Since $\beta$ is clearly contained in the field $\mathbb{Q}\left(\sqrt{34 - 2\sqrt{17}}\right)$, which has degree at most four, we conclude that $\mathbb{Q}(\beta) = \mathbb{Q}\left(\sqrt{34 - 2\sqrt{17}}\right)$, as claimed.

Next let us conside $\alpha$. Let

$$\alpha' = \zeta^4 + \zeta^{-4}.$$

Then

$$\alpha + \alpha' = \beta$$

and

$$\alpha\alpha' = \zeta^3 + \zeta^5 + \zeta^{-3} + \zeta^{-5}.$$

Let us denote $\alpha\alpha'$ by $\beta''$, and let

$$\beta''' = \zeta^6 + \zeta^7 + \zeta^{-6} + \zeta^{-7}.$$

Then

$$\beta'' + \beta''' = \gamma'$$

and

$$\beta''\beta''' = \sum_{i=1}^{16} \zeta^i = -1.$$

Thus $\beta''$ is a root of the equation

$$x^2 - \gamma'x - 1 = 0.$$

A calculation similar to the one used for $\beta$ then gives

$$\beta'' = \frac{1}{4}\left(-1 - \sqrt{17} + \sqrt{34 + 2\sqrt{17}}\right).$$

Note that

$$\sqrt{34 - 2\sqrt{17}} \cdot \sqrt{34 + 2\sqrt{17}} = \sqrt{34^2 - 4 \cdot 17} = 8\sqrt{17}.$$

Therefore, $\beta'' \in \mathbb{Q}\left(\sqrt{34 - 2\sqrt{17}}\right) = \mathbb{Q}(\beta)$.

Now $\alpha$ satisfies the equation

$$x^2 - \beta x + \beta'' = 0,$$

so $\alpha$ is quadratic over $\mathbb{Q}(\beta)$. Using the quadratic formula, substituting, and choosing the sign correctly, we obtain

$$\alpha = 2\cos 2\pi/17$$

$$= \tfrac{1}{8}\left(-1 + \sqrt{17} + \sqrt{34 - 2\sqrt{17}} + 2\sqrt{17 + 3\sqrt{17} + \sqrt{170 - 26\sqrt{17}} - 4\sqrt{34 + 2\sqrt{17}}}\right).$$

This checks by calculator, giving $\alpha = 1.86494$. In particular, this shows that $\alpha$ is a constructible number, and hence the regular 17-gon is constructible by ruler and compass.

Although we have just completed the proof of the theorem, we have not yet shown that $\mathbb{Q}(\beta) \subseteq \mathbb{Q}(\alpha)$ as claimed earlier. To do this, write $\alpha = \zeta + \zeta^{-1}$ and then compute

$$\alpha^2 = \zeta^2 + 2 + \zeta^{-1},$$
$$\alpha^4 = \zeta^4 + 4\zeta^2 + 6 + 4\zeta^{-2} + \zeta^{-4}.$$

Therefore, $\beta$, which was defined to be $\zeta + \zeta^4 + \zeta^{-1} + \zeta^{-4}$, can be obtained as

$$\beta = \alpha^4 - 4\alpha^2 + \alpha + 2.$$

Therefore, $\beta \in \mathbb{Q}(\alpha)$, as required.

### Corollary 29.3
*The side of the regular 17-gon inscribed in the unit circle is*

$$\frac{1}{4}\sqrt{34 - \sqrt{17} - \sqrt{34 - 2\sqrt{17}} - 2\sqrt{17 + 3\sqrt{17} + \sqrt{170 - 26\sqrt{17}} - 4\sqrt{34 + 2\sqrt{17}}}}.$$

*Proof*   Use (13.3) and the expression for $\alpha$ above.

### Remark 29.3.1
If we wish to show only that the regular 17-gon is constructible, without an explicit formula for $\cos 2\pi/17$, we can proceed as follows. The question is whether $\alpha = 2\cos 2\pi/17$ is a constructible real number. Now, $\alpha$ is contained in the field $\mathbb{Q}(\zeta)$ of 17th roots of unity. By (32.7) this is a normal extension of $\mathbb{Q}$ of degree 16 and with Galois group $\mathbb{Z}_{17}^*$, which is cyclic of degree 16. Therefore, the Galois

group of the splitting field of the extension $\mathbb{Q}(\alpha)$ will be a quotient of this one, and its order will be a power of 2. By (28.7) therefore, $\alpha$ is constructible.

The proof of (29.2) given above actually illustrates the proof of (28.7) and at the same time gives a nice example of the fundamental theorem of Galois theory (32.4) by showing the correspondence between the subgroups of the Galois group and the subfields of the field. For inside the group $\mathbb{Z}_{17}^*$ we have a chain of subgroups

$$\{1, -1\} \subseteq \{1, -1, 4, -4\} \subseteq \{1, 2, 4, 8, -1, -2, -4, -8\}.$$

The fixed fields of these subgroups are precisely $\mathbb{Q}(\alpha)$, $\mathbb{Q}(\beta)$, and $\mathbb{Q}(\gamma)$. And of course, it was these subgroups of $\mathbb{Z}_{17}^*$ that provided the motivation for choosing $\alpha$, $\beta$, $\gamma$ the way we did.

### Remark 29.3.2
The explicit expressions for $\alpha$, $\beta$, $\gamma$ found in the proof of (29.2) can without too much difficulty be turned into an actual construction of the 17-gon. Here is a particularly simple one:

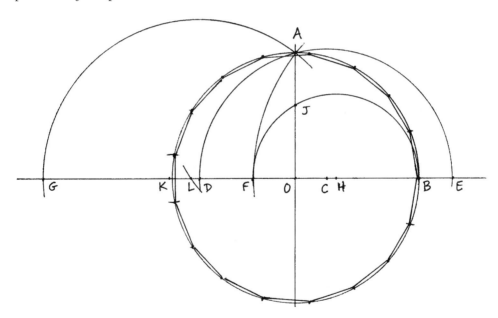

Let the circle with center $O$ be given. Let $OA$ and $OB$ be two orthogonal radii. Take $OC = \frac{1}{4} OB$. Let the circle $CA$ meeet $OB$ in $D$ and $E$. Let the circle $EA$ give $F$, and let the circle $DA$ give $G$. Take $H$ the midpoint of $BF$, and let the circle $HB$ meet $OA$ in $J$. Take $K$ to be the midpoint of $OG$. Let the circle with center $J$ and radius $OK$ meet $OB$ in $L$. Then $KL$ is the side of the inscribed 34-gon. See (Exercise 29.4).

Using the ideas of this section we can now obtain a complete determination of the possible constructible regular $n$-gons.

### Theorem 29.4
*The regular polygon of $n$ sides is constructible by ruler and compass if and only if $n$ is a number of the form*

$$n = 2^r p_1 \ldots p_s, \quad r, s \geq 0,$$

*where the $p_i$ are distinct odd primes, each of which is a prime of the form*

$$p = 2^{2^k} + 1.$$

*Proof* For any integer $n$, let

$$\zeta = \cos 2\pi/n + i \sin 2\pi/n$$

and let

$$\alpha = \zeta + \zeta^{-1}.$$

Then

$$\alpha = 2 \cos 2\pi/n,$$

and $\zeta$ is a root of the equation

$$x^2 - \alpha x + 1 = 0$$

with coefficients in $\mathbb{Q}(\alpha)$. Therefore, $\deg \mathbb{Q}(\zeta)/\mathbb{Q}(\alpha) = 2$. Since $\mathbb{Q}(\zeta)$ is the field of $n$th roots of unity, it is a normal field extension of $\mathbb{Q}$ with Galois group $\mathbb{Z}_n^*$, which is an abelian group of order $\varphi(n)$, the Euler $\varphi$-function (32.7). Since the Galois group is abelian, $\mathbb{Q}(\alpha)$ is a normal extension of $\mathbb{Q}$, and the orders of the Galois groups of $\mathbb{Q}(\alpha)$ and $\mathbb{Q}(\zeta)$ differ by 2. We conclude from (28.7) that the regular $n$-gon is constructible if and only if the order of the Galois group of $\mathbb{Q}(\zeta)$, which is $\varphi(n)$, is a power of 2.

Now let us write

$$n = 2^k p_1^{e_1} \ldots p_s^{e_s}$$

with distinct odd primes $p_i$. Then

$$\varphi(n) = 2^{k-1} \prod p_i^{e_i - 1}(p_i - 1).$$

In order for this to be a power of 2, we must have $e_i = 1$ for each odd prime occurring, and $p_i - 1$ must be a power of 2, so

$$p_i = 2^{t_i} + 1.$$

One sees easily that in order for $p_i$ to be prime, it is necessary that $t_i$ be a power of 2. Thus $p_i$ is of the form

$$p = 2^{2^k} + 1$$

as required. This argument is reversible, and so the theorem is proved.

**Remark 29.4.1**
Prime numbers of the form $p = 2^{2^k} + 1$ were studied by Fermat, who hoped he had discovered a formula for generating prime numbers. If we denote by $F_k$ the kth *Fermat number* $2^{2^k} + 1$, then $F_0 = 3$, $F_1 = 5$, $F_2 = 17$, $F_3 = 257$, and $F_4 = 65537$ are all primes. It is an open problem whether there are any further prime Fermat numbers. As of this date (April 1999), $F_k$ is known to be composite for $5 \leq k \leq 23$, so $F_{24}$ is the first unknown case.

# Exercises

29.1  To illustrate that the square root of a complex number with constructible coordinates also has constructible coordinates, write $\sqrt{5 + 2\sqrt{-5}}$ in the form $a + bi$ where $a$ and $b$ are constructible real numbers in standard form (cf. Exercise 13.2).

29.2  Let $\zeta = \cos 2\pi/7 + i \sin 2\pi/7$, and let $\alpha = \zeta + \zeta^{-1}$.

   (a)  Find the minimal polynomial for $\alpha$ over $\mathbb{Q}$ (it is a cubic).

   (b)  Show that $\mathbb{Q}(\zeta)$ contains a unique subfield $E$ of degree 2 over $\mathbb{Q}$. Find an integer $d$ for which $E = \mathbb{Q}(\sqrt{d})$.

29.3  An investigation of the Fermat number $F_{24}$. Using a *hand calculator* only, find

   (a)  How many digits does $F_{24}$ have in its decimal expansion?

   (b)  What are the first six digits of $F_{24}$?

   (c)  What are the last six digits of $F_{24}$?

   In each case, explain your method, and include a discussion of your calculator's round-off error and how reliable you believe your answer to be.

29.4  (a)  Carry out the construction of the 17-gon described in Remark 29.3.2 (par = 20 steps to get $KL$; 25 more to mark the vertices and draw the edges of the 17-gon).

   (b)  Prove that the construction works by showing (in the notation of the proof of Theorem 29.2) that $OD = \frac{1}{2}\gamma$, $OG = \beta$, $OF = \beta''$, $KL = \alpha'$. Then show that $\alpha' = \zeta^4 + \zeta^{-4}$ is equal to the side of the inscribed regular 34-gon.

29.5  If a regular polygon of $n$ sides is constructible with ruler and compass, show that it is also constructible with Hilbert's tools.

# 30  Constructions with Compass and Marked Ruler

Up to now we have studied the classical ruler and compass constructions of Euclid's *Elements*. We have seen that there are some problems that cannot be solved by ruler and compass, such as the duplication of the cube (28.3), the tri-section of the angle (28.4), or the construction of a regular 7-sided polygon (29.4). Although Euclid uses only ruler and compass, other classical authors, both before and after Euclid, used a variety of other methods for more difficult problems. Some used an auxiliary curve given in the plane, such as a parabola (Exercise 28.2), the quadratrix (Section 25), or the conchoid of Nicomedes (later in this section). As the theory of conic sections became more developed, espe-cially in the work of Apollonius, problems were solved using intersections of conics in the plane. New tools were invented, such as the marked ruler that could slide to cut off a given distance between two curves. There was even a classification of problems according to the methods needed for their solution (though the geometers of that time were not in a position to prove that any given problem belonged to a certain class). A *planar* problem was one that could be solved with ruler and compass. A *solid* problem needed the use of conic sec-tions, and a *linear* problem (not our meaning of the word!) was one that required curves (which they called lines) of higher degree.

In this section we will study one of these methods, namely the use of the marked ruler, and we will see that with its help we can trisect angles and extract cube roots. As an application we give a construction of the regular heptagon (7-sided polygon). We will also show that analytically, the use of the marked ruler between lines corresponds to finding a root of a certain quartic equation. In the next section we will discuss the associated field theory and show that the geo-metrical use of the compass and marked ruler is equivalent to the algebraic solution of cubic and quartic equations, and this in turn corresponds to the "solid" problems in the ancient classification.

First let us make clear what the marked ruler can do. You can make two marks on the ruler corresponding to a given distance, and then you can slide the ruler along so that the marks lie on two given lines, while at the same time the ruler passes through a given point.

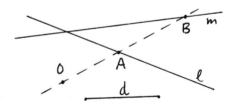

In other words, given two lines $l, m$, given a distance $d$, and given a point $O$, you can draw a line $OAB$ with $A \in l$, $B \in m$, and such that $AB = d$. This will count as one marked ruler step. Meanwhile, the ruler can still be used as an ordinary ruler in the old sense.

## Proposition 30.1
*Using compass and marked ruler, it is possible to trisect any angle.*

*Proof* Let $AOB$ be the given angle. Drop a perpendicular $AC$ from $A$ to $OB$. Draw a line $l$ through $A$ parallel to $OB$. Now use the marked ruler to draw a line

circumscribere ac circulum quindecagono inscribere ac etiam circumscribere cx.12.
13.7.14.buius plene intellectis facile perficies. ☙Et nota ꝗ quamcunꝗ figuram
equilateram circulo scimus inscribere ouplo plurium laterum circulo scimus inscri
bere ꝛ circuscribere.ꝛ ipsi circulum.oiuisis eni arcubus quibus latera eius ꝗ scitur
inscribi subtenditur.per equalia ꝛ a punctis medijs ad cxtremitates laterū ipsius
figure ouctis lineis fict intra circulum figura ouplo plurium laterum que erit equi
latera per.28.tertij.ergo ꝛ equiangula.boc enim oemonstratū est sup:a.15.buius
ꝗ omnis figura equilatera circulo inscripta est etiam equiangula.Et quia bāc cir/
culo scimus inscribere sciemus cetera tria per.12.13.7.14.buius. ☙Quia igitur sci
mus inscribere triangulum equilaterum: sciemus per boc ꝛ exagonum ꝛ per exa/
gonū ouodecagonū ac per ouodecagonuꝛ figuram.24.laterum. ꝫ sic in infinitum
ouplando.Et licet per triangulum possit vt oiximus inscribi exagonum.posuit ta/
men buius propriā oemonstrationē ex qua sequitur potissima per vtile. Et simili/
ter quia scimus ꝛ inscribere quadratum sciemus per boc inscribere omnem figurā
cuius laterum numerus est pariter par.per pentbagonum quoꝗ sciemus occago
num.ꝛ figurā.20.laterum.sicꝗ continue ouplando.idem quoꝗ intellige oe quin/
oecagono.per ipsum enim scientur figure.30.7.60.ꝛ omnium continue ouplatoꝛ
laterum.☙Ceterarum autem figurarum oe quibus ista non oocet. vel que p bas
nō babent oifficilis est scientia.ꝛ parū vtilis.vt sunt eptagona nonagena vndeca/
gona. Qō si scirem⁹ triangulū oui equaliū lateriū oesignare.cuius vterꝗ angulo
rum ad basim triplus esset ad reliquū sciremus eptagonū vt sup:a pentbagonum
circulo inscribere.ꝗ si vterꝗ quadruplus esset ad reliquū sciremus nonagonū.et
si quintuplus.vndecagonū. Idenꝗ in ceteris figuris imparium laterum.posito
vtroꝗ anguloꝛ ad basim multiplici ad reliquū.per eum numerum qui est medie/
tas.maximi paris sub impari numero laterum ipsius figure contenti.

**O**Atū angulū in tria equa oiuidere.Sit angulus oat⁹.c.volo ipsū oiuidere in
treꝛ eꝗles angulos qō sic facio.pono p̄mo.c.centrū circuli oescribendo circu
lū ꝗliterꝗꝗ cōtingat. ꝛ ꝓtrabo latera cōtinentia oatū angulū vsꝗ quo secet cir/
cūferentiā in punctis.a.ꝛ.b.tunc a puncto.c.qō est centrū circuli ouco lineā. c.d.
perpendiculariē ad lineā.c.b.ꝛ in linea.c.d.assigno punctū.c.a quo ouco lineam
ad equalitate.c.b.vsꝗ quo secet circūferentiā circuli in puncto.f. ꝛ ꝓduco .c.vsꝗ
a.oeinde protrabo lineā.g.b.equidistante.f.a.que scꝛ.g.b.transeat per centrū.ꝛ
ouco lineam.f.g.equidistante linee.c.c.ꝛ protrabo lineam.c.b.incontinuū ꝛ oire/
ctum vsꝗ ad.l.que secat lineā.f.g.ortbogonaliter in puncto.o.ꝛ per equalia. oico
ergo ꝗ arcus.l.g.est equalis arcui.b.b.propter boc.ꝗ angulus.l.g. c. est equalis
angulo.b.c.b.cū sint contra se positi.Cum igitur arcus.f.g.sit ouplus arcui.l.g.
erit etiā ouplus arcui.b.b.sed arcus.f.g.est equalis arcui.a. b.cū sint inter ouas
lineas equidistantes que sunt.f.a.ꝛ.g.b.ergo arcus.b.a.est ouplus arcui.b.b.er/
go ꝛ angulus.a.c.b.est ouplus angulo.b.c.b.oiuidam ergo angulum.a. c.b. per
equalia per lineam.c.m.ꝛ patet propositum.

**I**Ntra oatū circulū nonāgulū equilateꝛ atꝗ eꝗāgulū oesignare.qō sic fieri po/
test iuxta ooctrinā scōe bui⁹.inscribā circulo assignato triangulū ꝛolateꝛ atꝗ
eꝗangulū ꝗ sit a.b.c.ꝛ vnūqueꝗ anguloꝛ ei⁹oiuidā ꝑ tria eꝗlia ꝛ protrabā lineas
oiuidētes angulos vsꝗ ad circūferentiā ꝛ tu nc ꝗ: noue anguli locati in circulo sūt
equales oe necessitate arcus suppositi ipsis angulis sunt equales. protrabā enim

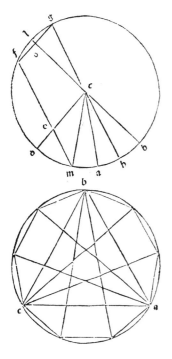

Plate X. The end of Book IV of Campanus' translation of Euclid's *Elements*, published by
Ratdolt in (1482). The text describes another method of trisecting an angle with marked
ruler. Reprinted courtesy of the Bancroft Library of the University of California at Berkeley.

**261**

$ODE$ such that $D \in AC$, $E \in l$, and $DE = 2AO$. This line will be the trisector of the original angle.

To see this, let $F$ be the midpoint of $DE$, and let $G$ be the midpoint of $AE$. Then $FG$ is perpendicular to $AE$, so by (SAS) the triangles $EFG$ and $AFG$ are congruent. Now the new angle $EOB = \alpha$ is equal to $\angle AEO$ by parallel lines and to $\angle EAF$ by congruent triangles. So $\angle AFO = 2\alpha$, since it is an exterior angle to the triangle $AEF$.

But $DE$ was constructed equal to $2AO$, so $AO = EF = AF$. Hence the triangle $AOF$ is isosceles, and so $\angle AOD = 2\alpha$. Thus the original angle $AOB$ is equal to $3\alpha$, and $\alpha$ is one-third of it, as required.

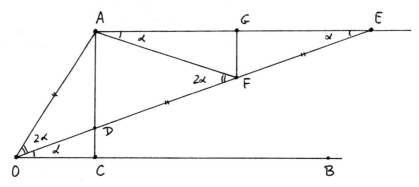

**Proposition 30.2**
*Given segments of lengths* 1 *and* $a$, *it is possible with compass and marked ruler to construct a segment of length* $\sqrt[3]{a}$.

*Proof*  Let $AB$ be the given segment of length $a$. Using the segment of length 1, choose $b = 2^{3k-1}$ for suitable $k$ such that $b > a$. Make an isosceles triangle $ABC$ with $CA = CB = b$, and extend $CA$ to $D$ with $AD = b$. Draw the line $DB$. Now use the marked ruler to draw $CEF$ with $E \in DB$ and $F \in AB$ and $EF = b$. Let $BF = y$. Then $\sqrt[3]{a} = y/2^{2k}$.

To see this, we first apply Menelaus's theorem (20.10) to the triangle $ACF$ and the transversal line $DBE$. Letting $CE = x$, starting with vertex $A$, and going clockwise, it says that

$$\frac{b}{2b} \cdot \frac{x}{b} \cdot \frac{y}{a} = 1.$$

This gives us

$$xy = 2ab.$$

Then we apply (III.36) to the circle with center $C$ and radius $b$, and the point $F$ outside, and the two chords $FBA$ and $FGH$. Note that $CG = b$, so by subtraction, $FG = x$. Thus we obtain

$$y(y + a) = x(x + 2b).$$

Eliminating $x$ from our two equations we obtain

$$y^3 = 4ab^2.$$

Substituting $b = 2^{3k-1}$ we obtain

$$y^3 = 2^{6k} \cdot a.$$

Therefore, $\sqrt[3]{a} = y/2^{2k}$ can be obtained from $y$ by bisecting $2k$ times.

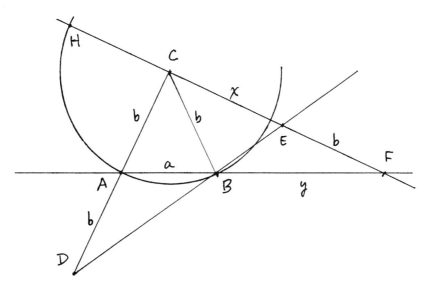

## Proposition 30.3

*Let lines l, m, a point O, and a segment of length d be given in the Cartesian plane over a field F. Suppose OAB is a line with $A \in l$, $B \in m$, and $AB = d$. Then the coordinates of A and B lie in a field $F(\alpha)$, where $\alpha$ is a root of a quartic polynomial with coefficients in F.*

*Proof* By a linear change of coordinates, we may assume that $O = (0,0)$ is the origin, and that $l$ is the line $y = b$, for some $b$. We consider the locus of all points $P$ such that $OP$ cuts $l$ in a point $Q$ and $PQ = d$. This locus is called the *conchoid of Nicomedes*. To find the line $OAB$ of the proposition is equivalent to finding the intersection of the second line $m$ with the conchoid. This we will now do analytically.

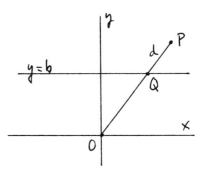

First we need to find the equation of the conchoid. Take an arbitrary line $y = ax$ through the origin. This line meets $l$ at the point $Q = (b/a, b)$. Let $P$ have variable coordinates $P = (x, y)$. Then the condition $PQ = d$ gives

$$(x - b/a)^2 + (y - b)^2 = d^2.$$

But $P$ also lies on the line $y = ax$. We use this equation to eliminate the variable $a$ from the equation: Substituting $a = y/x$ and simplifying, we obtain the equation of the conchoid,

$$(x^2 + y^2)(y - b)^2 = d^2 y^2.$$

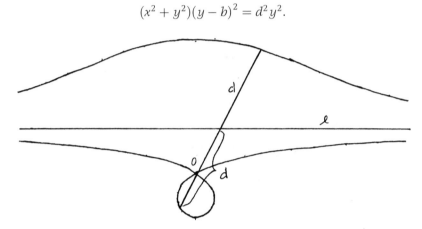

To find the intersection $B$ of the conchoid with the line $m$, we substitute the linear equation of $m$ in the equation of the conchoid. This gives a quartic equation in $x$. If $\alpha$ is a root of that equation, the coordinates of $B$ are then $x = \alpha$, and $y$ is linear expression in $\alpha$. From there we get the line $OB$ and the coordinates of $A$.

### Remark 30.3.1
It is clear from this proposition that the use of the marked ruler is equivalent to being given a single conchoid, drawn somewhere in the plane, and being allowed to intersect it with lines. Indeed, by rigid motions and similarities, any application of the marked ruler can be reduced to finding an intersection point of the conchoid with a certain line.

As an application of the marked ruler, we will give the elegant construction, due to Viète, of the regular heptagon. This he accomplished two hundred years before Gauss, working within the tradition of Euclid, without the benefit of modern algebra, but using only some simple algebraic manipulation of equations.

### Problem 30.4
Given a circle and its center, to construct with compass and marked ruler a regular heptagon inscribed in the circle.

1. *OA* get *B*.
2. Circle *AO*, get *C*, *D*.
3. Circle *DO*.
4. Circle, center *O*, radius *CD*, get *E*.
5. *CE*, get *F* (then $OF = \frac{1}{3}OA$).
6. Circle *FC*.
*7. Line *CGH*, so that *GH* = *FC*.
8. Circle *H*, radius *OA*, get *I*, *K*.
9. Circle *B*, radius *IK*, get *L*, *M*.
10. Circle *B*, radius *IM*, get *N*, *P*.
11–17. Draw sides of heptagon *BILNPMK*.

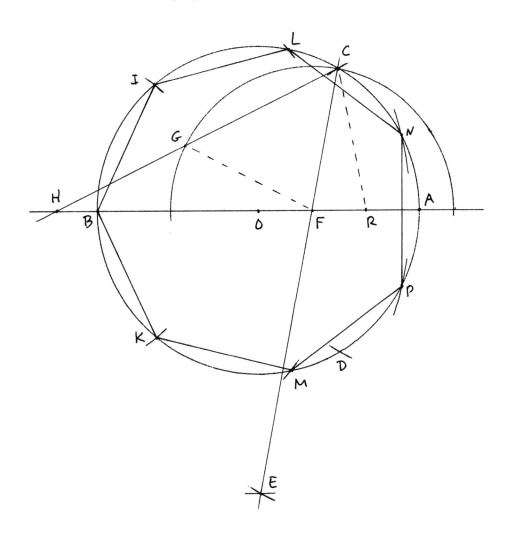

To see why this works, we will follow Viète's original proof, for which we need two lemmas.

### Lemma 30.5

*Suppose we are given two isosceles triangles based on the same line, with equal sides, and such that the vertex of one is in line with the side of the other, as shown. Then the angle at E is three times the angle at A, and if we denote the bases by x, b, and the side by r, then*

$$x^3 - 3xr^2 = br^2.$$

*Proof* Let the angle at $A$ be $\alpha$. Then using (I.5) and (I.32) several times, we obtain that the angles marked 1, 2, 3, are respectively $\alpha$, $2\alpha$, and $3\alpha$.

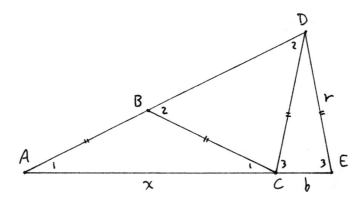

For the relation between $x$, $b$, $r$, we draw the circle with center $C$ and radius $r$, and drop perpendiculars $BF$ and $DG$ to the line $ACE$. Then $AF = \frac{1}{2}x$ and $FG = \frac{1}{2}(b + x)$, so by (VI.2) = (20.2), we get

$$\frac{y}{r} = \frac{\frac{1}{2}(b+x)}{\frac{1}{2}x},$$

and so $xy = r(b + x)$, where $y = BD$.

On the other hand, using (III.36), from the point $A$, we obtain

$$r(y + r) = AH \cdot AK = (x - r)(x + r).$$

Eliminating $y$ from these two equations, we obtain

$$x^3 - 3xr^2 = br^2$$

as required.

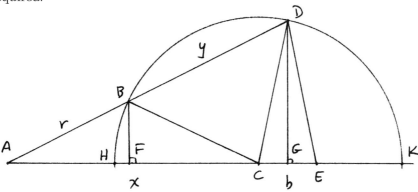

### Remark 30.5.1

If we write the trigonometric functions $\cos \alpha = x/2r$ and $\cos 3\alpha = b/2r$, then this equation translates to $\cos 3\alpha = 4\cos^3 \alpha - 3\cos \alpha$. So we see that Viète knew the equivalent of this triple angle formula, and could prove it geometrically.

### Lemma 30.6

*Suppose we are given a circle with center O, diameter AB, and a point H on the diameter extended such that*

$$HB \cdot HA^2 = HO \cdot OA^2.$$

*Let I be a point on the circle for which $HI = OA$. Then BI is a side of the inscribed regular heptagon.*

*Proof*  Let $Q$ be the other intersection of the line $HI$ with the circle. First we will show that $OQ$ is parallel to $AI$. By $(VI.2) = (20.2)$, it will be sufficient to show that

$$\frac{HQ}{HI} = \frac{HO}{HA}.$$

Using $(III.36)$ we have $HQ \cdot HI = HB \cdot HA$. Therefore,

$$\frac{HQ}{HI} = \frac{HB \cdot HA}{HI^2}.$$

But $HI = OA$, so using the hypothesis of the lemma, we obtain

$$\frac{HQ}{HI} = \frac{HB \cdot HA}{OA^2} = \frac{HO}{OA}.$$

Thus $OQ$ and $AI$ are parallel, as claimed.

Now let the angle at $A$ be $\alpha$. Then using (I.5) and (I.32) several times, we find that the angles marked 1, 2, 3 in the diagram are respectively $\alpha$, $2\alpha$, $3\alpha$. In particular, $OQI$ is an isosceles triangle whose base angles are three times the vertex angle. It follows that the angle $\alpha$ is $\frac{1}{7}(2RA)$, and so the angle $BOI = 2\alpha$ is $\frac{1}{7}(4RA)$. Thus $BI$ is a side of the inscribed regular heptagon.

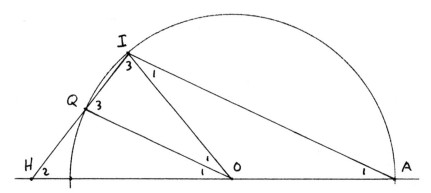

*Proof of Construction of* (30.4)    Using these two lemmas, we can prove that the construction gives a regular heptagon. If we add lines $FG$ and $CR$, where $R$ is the midpoint of $FA$ (dotted in the diagram), then we recognize triangles $HGF$ and $FCR$ satisfying the conditions of (30.5). Let us fix $OA = 1$. Then $OF = FR = \frac{1}{3}$. If $S$ is the midpoint of $FR$, then $CS = \frac{1}{2}\sqrt{3}$, because it is the altitude of the equilateral triangle $OAC$. On the other hand, $FS = \frac{1}{6}$, so by (I.47), we find that $FC = \frac{1}{3}\sqrt{7}$. So the cubic equation of (30.5) becomes

$$x^3 - \frac{7}{3}x = \frac{7}{27}.$$

To complete the proof that $BI$ is a side of the regular heptagon, by (30.6), we need to verify that $HB \cdot HA^2 = HO \cdot OA^2$. Remembering that $HF = x$, we must show that

$$\left(x - \frac{4}{3}\right) \cdot \left(x + \frac{2}{3}\right)^2 = \left(x - \frac{1}{3}\right) \cdot 1^2.$$

A simple calculation reduces this to the cubic equation for $x$ obtained from (30.5). So the condition of (30.6) is satisfied, and we conclude that $BI$ is a side of the regular heptagon. The remaining steps of the construction merely identify the other vertices of the heptagon.

**Remark 30.6.1**

In this construction, the marked ruler is used to insert the segment $GH = FC$ between a line and a circle. This is a stronger use of the marked ruler than what we described at the beginning of this section. However, since the function of this step is to trisect the angle at $F$ (Exercise 30.1), the construction could have been accomplished using (30.1), where the marked ruler is used to insert a segment between lines only, at the cost of a few extra steps.

For a more algebraic derivation of a construction of the heptagon, see Exercise 30.3.

# Exercises

30.1 Given an angle $\alpha$, with vertex $O$, draw a circle of any radius, cutting the angle in $A$ and $B$. With the marked ruler draw a line $BCD$ such that $C$ lies on the circle, $D$ lies on the line $OA$ extended, and $CD = OA$. Show that the angle at $D$ is $\frac{1}{3}\alpha$.

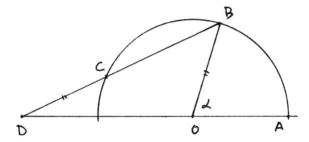

30.2 Construct a regular enneagon (9-sided polygon) using compass and marked ruler (between lines only). (Par $= 21$ steps, including one marked ruler step.)

30.3 Derive a construction of the regular heptagon, independent of the method of Problem 30.4, as follows.

(a) Find a cubic equation with root $\cos 2\pi/7$—cf. Exercise 29.2.

(b) Make a change of variables so that this equation is brought into the form $y^3 - 3y - b = 0$, and so can be solved by trisecting a certain angle—cf. Theorem 28.4.

(c) Now construct the required angle, then use the marked ruler to trisect that angle (Proposition 30.1), and thus make a construction for the regular heptagon using compass and marked ruler (par $= 28$ steps).

30.4 Show that a regular 13-gon can be constructed with compass and marked ruler as follows.

(a) Find a cubic and a quadratic equation whose successive solution will give $\cos 2\pi/13$.

(b) Show that the cubic equation in (a) can be solved by trisecting a certain angle $\theta$, and find $\cos \theta$.

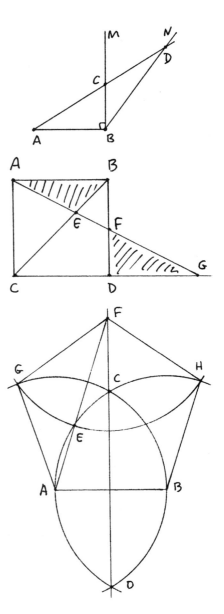

30.5 Let $AB$ be a given segment. Make $ABM$ a right angle, and $ABN = 120°$. With the marked ruler, draw $ACD$ cutting off a segment $CD = AB$ between the lines $BM$ and $BN$. Show that $AC = \sqrt[3]{2} \cdot AB$.

30.6 Here is another kind of neusis (sliding ruler) used by Archimedes in his study of the heptagon. Let $ABCD$ be a unit square with its diagonal $BC$. Rotate the ruler around the point $A$ until the areas of the two triangles $ABE$ and $DFG$ are equal. If $CG = a$ show that $-a$ satisfies the equation of Exercise 29.2a. Conclude that $a = 2\cos 2\pi/14$.

30.7 Verify the following marked-ruler construction of a regular pentagon on a given segment $AB$. (Of course, here the marked ruler is not necessary, but it gives a particularly elegant construction.)

1. Circle $AB$.
2. Circle $BA$, get $C$, $D$.
3. Line $CD$.
*4. Line $AEF$ so that $EF = AB$.
5. Circle $FE$, get $G$, $H$.
6–9. Draw pentagon $ABHFG$.

# 31   Cubic and Quartic Equations

Our purpose in this section is to investigate the roots of cubic and quartic equations, and to show that the use of the compass and marked ruler is equivalent to finding successive real roots of such equations.

In the quadratic case, there is a happy coincidence between solving quadratic equations, adjoining square roots of field elements, and considering degree-2 field extensions (28.6). For cubic and quartic equations, the situation is more complicated. A real root of a cubic equation may not be expressible in terms of cube roots (Exercise 28.5). A root of a quartic equation may give a degree-4 field extension that has no intermediate degree-2 subfield, and so $\alpha$ may not be expressible by square and fourth roots (32.5.1).

If we ignore questions of reality, it can be shown that arbitrary cubic and quartic equations can be solved by successive square roots and cube roots *of complex numbers*. But since this is a course in geometry, we are interested in what happens in the real Cartesian plane, and so we will be concerned with real roots of these equations. We will show that real roots of cubic and quartic equations can be expressed by three types of field extensions of a subfield $F \subseteq \mathbb{R}$:

(1)  $F(\sqrt{a})$ where $a \in F$, $a > 0$.
(2)  $F(\sqrt[3]{a})$ where $a \in F$.
(3)  $F\left(\cos \frac{1}{3}\theta\right)$ where $\cos \theta \in F$.

Roughly speaking, we will say these equations can be solved by taking square roots, cube roots, and trisecting angles. In this way we will see that solving cubic and quartic equations is equivalent to the use of the compass and marked ruler.

It is a remarkable tribute to the wisdom of the Greek geometers that two of the problems they highlighted, the duplication of the cube and the trisection of the angle, provide precisely the tools needed for the solution of cubic and quartic equations.

For the reader who is already curious about quintic polynomials, see Exercise 31.5.

We begin our work with some equations we do know how to solve. Throughout these discussions we fix a subfield $F$ of the real numbers. We consider polynomials with coefficients in $F$, and will be concerned with the field extensions necessary to find roots. Of course, the equation $x^3 - a = 0$ can be solved by $x = \sqrt[3]{a}$. Another type of equation we can solve by trisecting an angle.

**Proposition 31.1**
*Suppose we are given an equation*

$$x^3 - 3x - b = 0$$

*with $b \in F$ and $|b| < 2$. Let $\theta$ be an angle with $\cos \theta = \frac{1}{2}b$. Then $\alpha = 2\cos\frac{1}{3}\theta$ is a root of the equation.*

*Proof*  We have already encountered this equation twice before ((28.4) and Exercise 30.3). It is a consequence of the trigonometric identity

$$\cos 3\psi = 4\cos^3 \psi - 3\cos\psi,$$

where we put $\theta = 3\psi$, $x = 2\cos\psi$, and $b = 2\cos 3\psi$. The restriction $|b| < 2$ is necessary to find a $\theta$ with $\cos\theta = \frac{1}{2}b$.

To study a general cubic equation we will follow the method of Cardano. First note that a general cubic

$$x^3 + ax^2 + bx + c = 0$$

can be simplified by the substitution $x = y - \frac{1}{3}a$ so as to eliminate the $x^2$ term. So it will be sufficient to consider the equation

$$x^3 + px + q = 0.$$

We look for a solution of the form $x = u + v$, so we need

$$u^3 + 3u^2v + 3uv^2 + v^3 + p(u+v) + q = 0.$$

This can be accomplished by setting $p = -3uv$ and $q = -u^3 - v^3$. Then

$$u^3 + v^3 = -q,$$

$$u^3v^3 = -\left(\frac{p}{3}\right)^3,$$

so $u^3$ and $v^3$ are roots of the quadratic equation

$$y^2 + qy - \left(\frac{p}{3}\right)^3 = 0.$$

We solve this by the quadratic formula to obtain

$$y = -\frac{q}{2} \pm \sqrt{\left(\frac{q}{2}\right)^2 + \left(\frac{p}{3}\right)^3}.$$

Since $u$ and $v$ are the cube roots of these two values of $y$, we get

$$x = \sqrt[3]{\sqrt{\left(\frac{q}{2}\right)^2 + \left(\frac{p}{3}\right)^3} - \frac{q}{2}} - \sqrt[3]{\sqrt{\left(\frac{q}{2}\right)^2 + \left(\frac{p}{3}\right)^3} + \frac{q}{2}}.$$

This is the so-called formula of Cardano. In order for the roots to be real, we need $\left(\frac{q}{2}\right)^2 + \left(\frac{p}{3}\right)^3 \geq 0$. Thus we have proved the following result:

**Proposition 31.2**
*If $(q/2)^2 + (p/3)^3 \geq 0$, then a real root of the equation $x^3 + px + q = 0$ can be found by taking real square roots and cube roots.*

**Remark 31.2.1**
Although it appears that we need two cube roots, since $p = -3uv$, we have $v = -p/(3u)$, and so $x = u - p/(3u)$ can be expressed using one square root and one cube root.

**Proposition 31.3**
*If $(q/2)^2 + (p/3)^3 < 0$, then a real root of the equation $x^3 + px + q = 0$ can be found by taking a square root and trisecting an angle.*

*Proof*  The hypothesis implies that $p < 0$, so we adjoin $\sqrt{-3p}$ to our field and make a change of variables $x = \frac{1}{3}\sqrt{-3p}z$. This gives the equation

$$z^3 - 3z + b = 0,$$

where

$$b = q\sqrt{-\frac{27}{p^3}}.$$

Now, our hypothesis $(q/2)^2 + (p/3)^3 < 0$ implies $|b| < 2$, so we can use (31.1) to solve the equation by trisecting an angle.

**Remark 31.3.1**
The conditions $(q/2)^2 + (p/3)^3 \geq 0$ (resp. $< 0$) of (31.2) and (31.3) are equivalent to saying that the *discriminant* $\Delta$ is $\leq 0$ (resp. $> 0$)—cf. (Exercises 31.14, 31.15). The case of (31.3) is called the *casus irreducibilis* of the cubic equation.

**Proposition 31.4**
*If $\alpha$ is a real root of a quartic polynomial with coefficients in F, then $\alpha$ can be found by first adjoining a real root of a cubic polynomial with coefficients in F (called the* cubic resolvent *of the quartic equation), followed by successive real square roots.*

*Proof*  Here we follow Descartes's method. By a linear change of variables, we may assume that there is no $x^3$ term, so the quartic polynomial has the form

$$x^4 + px^2 + qx + r = 0.$$

If $\alpha$ is one real root, there must be another real root $\beta$ (since complex roots come in pairs), so $\alpha$ and $\beta$ are roots of a quadratic polynomial $x^2 + ax + b$, with $a, b$ in some real extension field of $F$. Then the quartic polynomial factors into

$$(x^2 + ax + b)(x^2 - ax + c)$$

for some $c \in \mathbb{R}$. From this we obtain

$$p = b + c - a^2,$$
$$q = a(c - b),$$
$$r = bc.$$

Eliminating $b$ and $c$ from these equations, we obtain

$$b = \frac{1}{2}\left(p + a^2 - \frac{q}{a}\right) \quad \text{and} \quad a^6 + 2pa^4 + (p^2 - 4r)a^2 - q^2 = 0.$$

In other words, $a^2$ is a root of the *cubic resolvent* polynomial

$$y^3 + 2py^2 + (p^2 - 4r)y - q^2 = 0,$$

whose coefficients are in $F$.

Having found $a^2$, which is a positive real root of this cubic equation, we can then find $a, b$, and $\alpha$ by successive (real) square roots.

**Theorem 31.5**
*Let $F$ be a subfield of $\mathbb{R}$ and let $\alpha \in \mathbb{R}$. The following conditions are equivalent:*

(i) *There exists a tower of subfields*

$$F = F_0 \subseteq F_1 \subseteq \cdots \subseteq F_k \subseteq \mathbb{R}$$

*with $\alpha \in F_k$, and for each $i$, $F_i$ is obtained from $F_{i-1}$ by adjoining an element $\beta_i = \beta$, where either*

(1) *$\beta = \sqrt{a}$, with $\alpha \in F_{i-1}$, $a > 0$, or*
(2) *$\beta = \sqrt[3]{a}$, with $a \in F_{i-1}$, or*
(3) *$\beta = \cos\frac{1}{3}\theta$, with $\cos\theta \in F_{i-1}$.*

(ii) *There exists a tower of subfields*

$$F = F_0 \subseteq F_1 \subseteq \cdots \subseteq F_n \subseteq \mathbb{R}$$

*with $\alpha \in F_n$, and each $F_i$ is obtained from $F_{i-1}$ by adjoining a root of a quadratic, a cubic, or a quartic polynomial.*

(iii) *The quantity $\alpha$ can be constructed by compass and marked ruler (using the marked ruler between lines only) from data with coordinates in $F$.*

*Proof* (i) $\Rightarrow$ (iii). The three types of extensions are constructible with compass and marked ruler, the first by ordinary ruler and compass construction, the second by (30.2), and the third by (30.1). (Note that $\cos\theta \in F$ does not necessarily imply that the angle $\theta$ can be realized by lines in the Cartesian plane over $F$. We may have to make a quadratic extension to obtain $\sin\theta$ first, before we have an actual angle to trisect.)

(iii) $\Rightarrow$ (ii). Regular ruler and compass constructions correspond to quadratic equations, and each use of the marked ruler between lines can be accomplished by solving a quartic equation (30.3).

(ii) $\Rightarrow$ (i). A quartic polynomial can be reduced to cubic and quadratic equations by (31.4), and cubic and quadratic polynomials can be solved by the three types of extensions listed, by (31.2) and (31.3).

**Corollary 31.6**

*If the quantity $\alpha \in \mathbb{R}$ is constructible by compass and marked ruler (used between lines only) from data in the field $F$, then $\deg F(\alpha)/F$ is $2^r 3^s$ for some $r, s \geq 0$.*

*Proof*  Since $F(\alpha) \subseteq F_k$ of (i) in the theorem, and each of the indicated extensions is of degree 2 or 3, it follows that $\deg F_k/F$, and hence also $\deg F(\alpha)/F$, is of the form $2^r 3^s$ for some $r, s \geq 0$.

In fact, a stronger result is true.

**Proposition 31.7**

*The quantity $\alpha \in \mathbb{R}$ is constructible by compass and marked ruler (used between lines only) from data in the field $F$ if and only if the Galois group of the minimal polynomial of $\alpha$ over $F$ has order $2^a 3^b$ for some $a, b \geq 0$.*

*Proof*  If $\alpha$ is so constructible, they by (31.6) the field $F(\alpha)$ has degree $2^r 3^s$ over $F$, for some $r, s \geq 0$. The same will be true for all the conjugates of $\alpha$, and these generate the splitting field, so the degree of the splitting field, which is also the order of the Galois group, will be $2^a 3^b$ for some $a, b \geq 0$.

Conversely, suppose the order of the Galois group $G$ is $2^a 3^b$. We apply the theorem of Burnside (see Hall (1959), Theorem 9.3.2) that any group of order $p^a q^b$ is solvable to conclude that $G$ is *solvable*. This means that there is a chain of subgroups, each one a normal subgroup of the next, such that all the quotients are cyclic groups of prime order. To continue the proof, we need a lemma.

**Lemma 31.8**

*Let $G$ be a solvable finite group and let $T$ be a subgroup of order 2. Then there exists a chain of subgroups*

$$T = G_1 \subseteq G_2 \subseteq \cdots \subseteq G_n = G$$

*such that for each $i$, the index of $G_i$ in $G_{i+1}$ is a prime number.*

*Proof*  We proceed by induction on the order of $G$, the case $G = T$ being trivial.

Since $G$ is solvable, it has a normal subgroup $H$ of some prime index $p$. If $T \subseteq H$, we can apply the induction hypothesis to $H$ (since any subgroup of a solvable group is solvable), and then we are done.

So suppose $T \nsubseteq H$. Let $T = \{e, \tau\}$, where $\tau$ is an element of order 2. Then $G/H$ has an element of order 2, so $p = 2$. Since $H$ is also solvable, it has a normal subgroup $K$ of prime index $q$. Since $H$ is normal in $G$, conjugation by $\tau$ will send $K$ to a normal subgroup $K^\tau = \tau K \tau^{-1}$ of $H$.

*Case 1*  If $K^\tau = K$, then $TK$ is a subgroup of $G$, and $K$ will have index 2 in $TK$. Hence $TK$ has index $q$ in $G$, and we can apply the induction hypothesis to $TK$, concluding the proof as before.

*Case 2*   If $K^\tau \neq K$, let $L = K \cap K^\tau$. Then $L$ is a normal subgroup of $H$ and the quotient $H/L$ is isomorphic to the direct product $H/K \times H/K^\tau$, which is abelian of order $q^2$. Let $\sigma \in K$ generate $K/L$. Then $\sigma^\tau = \tau \sigma \tau^{-1}$ will generate $K^\tau/L$. Let $M$ be the subgroup of $H$ generated by $L$ and $\rho = \sigma \sigma^\tau$. Now,

$$\rho^\tau = \sigma^\tau \sigma \equiv \sigma \sigma^\tau = \rho \pmod{L},$$

so $M^\tau = M$. Since $H/L$ is abelian, $M$ will be a normal subgroup of $H$ of index $q$, and we are reduced to the situation of Case 1.

*Proof of 31.7, continued*   We apply the lemma to the Galois group $G$, taking $T$ to be the subgroup generated by complex conjugation $\tau$. Then the fixed field $E$ of $T$ is the intersection of the splitting field with the real numbers. Now $\alpha \in E$, and by the fundamental theorem of Galois theory (32.4), the chain of subgroups $G_i$ will give a chain of field extensions

$$F = E_n \subseteq E_{n-1} \subseteq \cdots \subseteq E_1 = E$$

where each field has degree 2 or 3 over the previous one. So $\alpha$ can be obtained by finding real roots of a succession of quadratic and cubic polynomials, and so by (31.5) is constructible with compass and marked ruler.

### Remark 31.8.1
The condition of (31.6) is not sufficient for $\alpha$ to be constructible by compass and marked ruler, because for example, $\alpha$ could be a root of a 6th-degree equation with Galois group $S_6$, whose order is divisible by 5.

### Corollary 31.9
*A regular polygon of $n$ sides can be constructed with compass and marked ruler (between lines only) if and only if $n$ is of the form*

$$n = 2^k 3^l p_1 \ldots p_s, \quad k, l \geq 0,$$

*where the $p_i$ are distinct primes, each of the form*

$$p_i = 2^{a_i} 3^{b_i} + 1.$$

*Proof*   As in the proof of (29.4), each prime different from 2 or 3 must be of the form

$$p = 2^a 3^b + 1.$$

Conversely, if $n$ is of this form, the Galois group will be abelian of order $2^r 3^s$ for some $r, s$, and the result follows from (31.7) (whose proof is much easier in the abelian case).

**Remark 31.9.1**
As Gleason (1988) has pointed out, since the Galois group is abelian, all the polygons in (31.9) can be constructed with ruler, compass, and angle trisector. Indeed, the marked ruler steps can be reduced to two kinds, extraction of a cube root and angle trisection, and a real cube root has complex conjugates, which would contribute a nonabelian factor $S_3$ to the Galois group.

# Exercises

31.1 Show that it is not possible to construct a regular 11-gon with compass and marked ruler.

31.2 Show that it is possible to construct a regular 19-gon with compass and marked ruler.

31.3 Show that it is not possible to extract 5th roots with compass and marked ruler.

31.4 Show that it is not possible to quintisect (divide in 5) a general angle with compass and marked ruler.

31.5 Suppose, in addition to the compass and marked ruler, we were given tools to extract 5th roots and quintisect angles. Show that even with these new tools, it is still not possible to solve a general fifth-degree equation. *Hint*: Show that the Galois group of any extension obtained with these new tools is still solvable, and then take a quintic equation with Galois group $S_5$, such as Example 32.4.4.

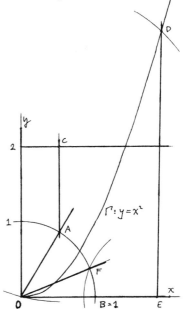

31.6 Verify that the following construction using a parabola will trisect an angle. Suppose we are given in the $xy$-plane the parabola $\Gamma$ defined by $y = x^2$.

   Suppose we are also given an angle $AOB$, with $A$ lying on the unit circle and $B = (1, 0)$. Draw a vertical line through $A$ to meet the line $y = 2$ at $C$. Draw a circle with center $C$, passing through the origin $O$, and let it meet the parabola $\Gamma$ at $D$. Drop a perpendicular $DE$ from $D$ to the $x$-axis. Draw a circle with center $E$ and radius 1 to meet the unit circle at $F$. Then $\angle FOB$ will be one-third of the angle $\angle AOB$.

31.7 Show that the constructions one can accomplish given the fixed parabola $y = x^2$ in the $xy$-plane, and being allowed to intersect it with lines and circles, are precisely equivalent of those one can achieve with compass and marked ruler.

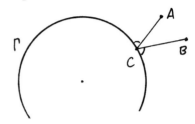

31.8 The problem of Alhazen. Given a circle $\Gamma$, and given two points $A, B$, find a point $C \in \Gamma$ such that the lines $AC$ and $BC$ make equal angles with the circle. Show that this problem leads to a quartic equation, and so can be solved with marked ruler and compass.

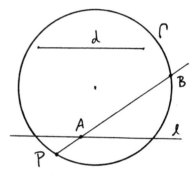

31.9 In the proofs of Propositions 5–9 of his book on spirals, Archimedes makes use of the following construction: Given a circle $\Gamma$, given a chord $l$, given a point $P$ on the circle, and given a segment $d$, to draw a line through $P$ such that the segment $AB$ cut off by the line and the circle is equal to the given segment $d$. This is not a legitimate use of the marked ruler in the limited sense we have considered. Show, however, that:

(a) This construction can be made with compass and marked ruler in our sense.

(b) If $l$ is perpendicular to the diameter of the circle passing through $P$, then the construction is possible even with ruler and compass only.

31.10 Show that those cubic equations that can be solved with square roots and an angle trisection are the equations with three real roots, while those equations needing square roots plus a cube root are the equations with one real and two complex roots.

31.11 If $f(x)$ is an irreducible cubic equation over the field $F \subseteq \mathbb{R}$ with one real and two complex roots, then its Galois group is $S_3$.

31.12 We have seen (Theorem 28.4) that $\alpha = 2\cos 20°$ is a root of the equation $x^3 - 3x - 1 = 0$. Show that the other two roots of this equation are contained in the field $\mathbb{Q}(\alpha)$, and so its Galois group is $\mathbb{Z}_3$.

31.13 You have probably seen (Exercise 29.2) that $\alpha = 2\cos 2\pi/7$ is a root of the equation $x^3 + x^2 - 2x - 1 = 0$. Show that the other two roots of this equation are also in the field $\mathbb{Q}(\alpha)$, so that it is a normal extension with Galois group $\mathbb{Z}_3$. Note that in this case, however, it is only after adjoining $\sqrt{7}$ that we can express $\alpha$ using a trisection of an angle.

31.14 *The discriminant.* Let $f(x)$ be an irreducible cubic polynomial with coefficients in the field $F \subseteq \mathbb{R}$, and let its roots be $\alpha_1, \alpha_2, \alpha_3$ in its splitting field. We define the *discriminant* of $f(x)$ to be

$$\Delta = (\alpha_1 - \alpha_2)^2 (\alpha_1 - \alpha_3)^2 (\alpha_2 - \alpha_3)^2.$$

(a) Show that $\Delta \in F$.

(b) Show that $\Delta > 0$ if and only if $f(x)$ has 3 real roots, while $\Delta < 0$ if and only if $f(x)$ has one real and two complex roots.

(c) Show that the Galois group of $f(x)$ is $\mathbb{Z}_3$ if and only if $\sqrt{\Delta} \in F$; otherwise, it is $S_3$.

31.15 If $f(x) = x^3 + px + q$, show that its discriminant is $\Delta = -4p^3 - 27q^2$, as follows: Let $\alpha$ be one root. Then show that the remaining two roots of $f(x)$ are roots of the quadratic equation

$$x^2 + \alpha x + (p + \alpha^2) = 0.$$

Solve this using the quadratic formula. Then put the three roots into the definition of $\Delta$ (Exercise 31.14) and simplify.

31.16 Consider the equation $x^3 - 3x - \frac{1}{2} = 0$ over $\mathbb{Q}$. If we take $\theta$ such that $\cos\theta = \frac{1}{4}$, then $\alpha = 2\cos\frac{1}{3}\theta$ is a root, by Proposition 31.1. Show that in this case, the other two roots of the equation are not contained in the field $\mathbb{Q}(\alpha)$, so the Galois group will be $S_3$. Show that the splitting field of $\mathbb{Q}(\alpha)$ is $\mathbb{Q}(\alpha, \sqrt{5})$.

31.17 In this exercise we investigate when two different polynomials can give rise to the same field extension. For simplicity let us consider an irreducible polynomial $f(x) = x^3 - 3x - b$ over $\mathbb{Q}$. Let $\alpha$ be a root of this polynomial, and denote by $\Delta(\alpha)$ its discriminant.

(a) If $\beta$ is any other element of the field $\mathbb{Q}(\alpha)$, $\beta \notin \mathbb{Q}$, show that $\Delta(\beta)$, the discriminant of its minimal polynomial, satisfies

$$\Delta(\beta) = \Delta(\alpha) \cdot a^6 (f(c))^2$$

for some $a, c \in \mathbb{Q}$. In particular, $\Delta(\beta)/\Delta(\alpha)$ is a square in $\mathbb{Q}$.

(b) If $f(x) = x^3 - 3x - 1$ with root $\alpha$, and $g(x) = x^3 - 3x - \frac{1}{2}$ with root $\beta$, show that $\Delta(\beta)/\Delta(\alpha)$ is not a square, and hence $\mathbb{Q}(\alpha) \neq \mathbb{Q}(\beta)$.

(c) Now let $f(x) = x^3 - 3x - 1$ have root $\alpha$ and $g(x) = x^3 - 3x - 13/7$ have root $\beta$. In this case show that $\Delta(\beta)/\Delta(\alpha)$ is a square, but nevertheless, $\mathbb{Q}(\alpha) \neq \mathbb{Q}(\beta)$. *Hint*: Use the criterion of (a) to reduce to a certain cubic Diophantine equation over $\mathbb{Z}$, and show that it has no solutions.

31.18 Show that those construction problems that can be solved by intersecting arbitrary conics in the plane—the so-called solid problems of the ancients—are exactly those described in (Theorem 31.5).

31.19 Given a unit square, find a point $E$ on $AB$ extended such that the line $CE$ cuts off a segment $EF$ equal to a given segment $b$.

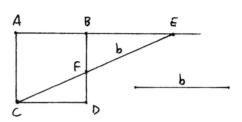

(a) Show that this marked ruler problem gives rise to a quartic equation that can be solved with square roots only. Thus $E$ can be constructed with ruler and compass.

(b) Find a ruler and compass construction (par = 7).

# 32   Appendix: Finite Field Extensions

Our purpose in this section is to review the basic facts about finite field extensions that we use in studying geometric constructions. For the most part we will not give proofs, which can be found in any standard algebra textbook. If you are learning this material for the first time, it may make sense simply to accept these results and let their use in geometry provide some motivation before you study their abstract algebraic proofs.

### Field Extensions

When one field is contained in another, $F \subseteq E$, we can think of $F$ as a subfield of $E$, or we can think of $E$ as an extension field of $F$, depending on the point of view. For example, when we see $\mathbb{Q} \subseteq \mathbb{Q}(\sqrt{2})$, the field of rational numbers $\mathbb{Q}$ is more familiar, so we will think of $\mathbb{Q}(\sqrt{2})$ as a field extension. Or when a field $F$ is given, and we are looking for an extension field $E$ with some particular property, we think of $E$ as an extension field.

If $F \subseteq E$ are two fields, and if $\alpha_1, \ldots, \alpha_n \in E$, then we denote by $F(\alpha_1, \ldots, \alpha_n)$ the smallest subfield of $E$ that contains $F$ and the elements $\alpha_1, \ldots, \alpha_n$. We call it the subfield of $E$ *generated* by the elements $\alpha_1, \ldots, \alpha_n$ over $F$.

When one field is contained in another, $F \subseteq E$, we can regard the larger field as a vector space over the smaller field. The dimension of this vector space is called the *degree* of the field extension, written $\deg E/F$.

### Example 32.0.1

Every element of $\mathbb{Q}(\sqrt{2})$ can be written uniquely in the form $a + b\sqrt{2}$ with $a, b \in \mathbb{Q}$. Thus $\mathbb{Q}(\sqrt{2})$ is a vector space of dimension 2 over $\mathbb{Q}$, so $\deg \mathbb{Q}(\sqrt{2})/\mathbb{Q} = 2$.

A field extension may have infinite degree. For example, if $\bar{\mathbb{Q}}$ is the set of all *algebraic numbers* (those complex numbers that are roots of polynomial equations with coefficients in $\mathbb{Q}$), then $\deg \bar{\mathbb{Q}}/\mathbb{Q}$ is (countably) infinite. If $\mathbb{R}$ is the real numbers, $\deg \mathbb{R}/\mathbb{Q}$ is (uncountably) infinite.

**Proposition 32.1**
*If $F \subseteq E \subseteq G$ are three fields, each contained in the next, then*

$$\deg G/F = (\deg G/E) \cdot (\deg E/F).$$

*Proof*  If $\alpha_1, \ldots, \alpha_n$ is a vector space basis for $E$ over $F$, and if $\beta_1, \ldots, \beta_m$ is a vector space basis for $G$ over $E$, then $\{\alpha_i \beta_j | i = 1, \ldots, n; j = 1, \ldots, m\}$ is a basis for $G$ over $F$.

**Example 32.1.1**
Consider $\mathbb{Q} \subseteq \mathbb{Q}(\sqrt{2}) \subseteq \mathbb{Q}(\sqrt{2}, \sqrt{3})$. Adjoining a square root of an element that was not already a square in a field makes a field extension of degree 2. We know that $\sqrt{2} \notin \mathbb{Q}$. We can also show that $\sqrt{3} \notin \mathbb{Q}(\sqrt{2})$ as follows. If $\sqrt{3} = a + b\sqrt{2}$, then

$$a^2 + 2ab\sqrt{2} + 2b^2 = 3.$$

This equation takes place in $\mathbb{Q}(\sqrt{2})$, where each element is *uniquely* written as $c + d\sqrt{2}$. Hence

$$a^2 + 2b^2 = 3,$$

$$2ab = 0.$$

Thus $a = 0$ or $b = 0$, so either $a^2 = 3$ or $2b^2 = 3$, both of which are impossible in $\mathbb{Q}$. Hence $\sqrt{3} \notin \mathbb{Q}(\sqrt{2})$, so $\mathbb{Q}(\sqrt{2}, \sqrt{3})$ is an extension of degree 4 of $\mathbb{Q}$. For a basis, we can take the elements $1, \sqrt{2}, \sqrt{3}, \sqrt{6}$. In other words, every element of this last field can be written uniquely as $a + b\sqrt{2} + c\sqrt{3} + d\sqrt{6}$, with $a, b, c, d \in \mathbb{Q}$.

**Roots of Polynomials**

Let $F$ be a field, and let

$$f(x) = a_0 x^n + a_1 x^{n-1} + \cdots + a_n$$

be a polynomial with coefficients $a_i \in F$. A *root* of $f$ means an element $\alpha$ in some extension field $E$ of $F$ for which $f(\alpha) = 0$.

**Proposition 32.2**
*For any field $F$ and any nonconstant polynomial $f(x)$ there always exists an extension field $E$ that contains a root $\alpha$ of $f(x)$. If $f(x)$ is irreducible of degree $n$, then the degree of the field extension $F(\alpha)/F$ is $n$. If $f(x)$ is irreducible, and if $\alpha_1 \in E_1$ and*

$\alpha_2 \in E_2$ *are roots of* $f(x)$ *in two extension fields of* $F$ *(which may be the same), then there is an isomorphism* $\varphi : F(\alpha_1) \to F(\alpha_2)$ *of the subfields of* $E_1$, $E_2$ *generated by* $\alpha_1$, $\alpha_2$, *respectively, that sends* $\alpha_1$ *to* $\alpha_2$ *and leaves every element of* $F$ *fixed.*

This proposition is usually proved using an abstract field extension $E$ constructed by taking the polynomial ring $F[x]$ and dividing out by the ideal $I$ generated by $f(x)$. Since $f(x)$ is irreducible, $I$ will be a maximal ideal, and so the quotient ring $E = F[x]/I$ is a field. The image $\alpha \in E$ of $x \in F[x]$ then becomes a root of $f(x)$.

In the case that will concern us most, polynomials with rational coefficients, there is another, more concrete, way of finding roots. Since $\mathbb{Q} \subseteq \mathbb{C}$, the field of complex numbers, and since every polynomial with complex coefficients has a root in $\mathbb{C}$ (this is the so-called fundamental theorem of algebra, whose proof, however, requires complex analysis), it follows that every polynomial $f(x) \in \mathbb{Q}[x]$ has a root in $\mathbb{C}$ (and sometimes even a root in $\mathbb{R}$). So if we write $\sqrt{2}$, for example, it can be assumed that we are referring to the real number $\sqrt{2} = 1.414\ldots$, and not to some abstract element in some abstract field extension of $\mathbb{Q}$ in which the polynomial $x^2 - 2$ has a root.

### Example 32.2.1
The polynomial $x^2 - 2$ is irreducible over $\mathbb{Q}$, since $\sqrt{2}$ is irrational, as was shown by Euclid (X.117). It has a root $\sqrt{2} \in \mathbb{R}$, so we can denote by $\mathbb{Q}(\sqrt{2})$ the subfield of $\mathbb{R}$ consisting of all real numbers of the form $a + b\sqrt{2}$, with $a, b \in \mathbb{Q}$. This is a field extension of degree 2 of $\mathbb{Q}$, with $1, \sqrt{2}$ as a basis. Now, the polynomial $x^2 - 2$ also has another root, $-\sqrt{2}$. The field extension generated by this root, $\mathbb{Q}(-\sqrt{2})$, is equal to $\mathbb{Q}(\sqrt{2})$. According to (32.2) there is an isomorphism (in this case an *automorphism*) $\varphi : \mathbb{Q}(\sqrt{2}) \to \mathbb{Q}(\sqrt{2})$ that leaves all rational numbers fixed and sends $\sqrt{2}$ to $-\sqrt{2}$. Indeed, you can verify that the map $\varphi$ defined by $\varphi(a + b\sqrt{2}) = a - b\sqrt{2}$ is a field automorphism of $\mathbb{Q}(\sqrt{2})$, namely, it satisfies $\varphi(\alpha + \beta) = \varphi(\alpha) + \varphi(\beta)$ and $\varphi(\alpha\beta) = \varphi(\alpha) \cdot \varphi(\beta)$ for all $\alpha, \beta \in \mathbb{Q}(\sqrt{2})$.

### Example 32.2.2
For a slightly more complicated example, consider the polynomial $x^3 - 2$ over $\mathbb{Q}$. It has a root $\sqrt[3]{2} = 1.2599\ldots$ in $\mathbb{R}$, so we can consider the field $\mathbb{Q}(\sqrt[3]{2})$. This is an extension field of degree 3 of $\mathbb{Q}$, with basis $1$, $\sqrt[3]{2}$, and $\sqrt[3]{4}$. But the equation $x^3 - 2$ also has a complex root $\omega\sqrt[3]{2}$, where $\omega = \frac{1}{2}(-1 + \sqrt{-3})$ is a cube root of unity. This gives a different extension field $\mathbb{Q}(\omega\sqrt[3]{2})$ of $\mathbb{Q}$ in which the polynomial $x^3 - 2$ has a root. This is also an extension of degree 3 of $\mathbb{Q}$, with basis $1$, $\omega\sqrt[3]{2}$, $\omega^2\sqrt[3]{4}$. According to the proposition, these two field extensions are isomorphic *as field extensions* of $\mathbb{Q}$, even though one is contained in $\mathbb{R}$ and the other is not. The isomorphism $\varphi : \mathbb{Q}(\sqrt[3]{2}) \to \mathbb{Q}(\omega\sqrt[3]{2})$ sends $1 \to 1$, $\sqrt[3]{2} \to \omega\sqrt[3]{2}$, and $\sqrt[3]{4} \to \omega^2\sqrt[3]{4}$.

**Example 32.2.3**

In some other cases we cannot write roots explicitly, but we can show that they exist. Try $x^5 - 5x - 1$ for example. Substituting $x = y + 1$ gives

$$y^5 + 5y^4 + 10y^3 + 10y^2 - 5,$$

which is irreducible by Eisenstein's criterion (32.8), so $x^5 - 5x - 1$ is irreducible. A little elementary calculus shows that the graph has a relative maximum at $(-1, 3)$ and a relative minimum at $(1, -5)$ and no other relative extremum. Therefore, it has exactly 3 real roots and 2 complex roots, though we cannot write them explicitly. Nevertheless, the roots exist. Say $\alpha_1, \alpha_2, \alpha_3$ are the real roots and $\alpha_4$, $\alpha_5$ are the complex roots. In this case the field extensions $\mathbb{Q}(\alpha_1)$, $\mathbb{Q}(\alpha_2)$, $\mathbb{Q}(\alpha_3)$, $\mathbb{Q}(\alpha_4)$, $\mathbb{Q}(\alpha_5)$ are actually five *distinct* subfields of $\mathbb{C}$, three of which are contained in $\mathbb{R}$, and which are all isomorphic as field extensions of $\mathbb{Q}$ (Exercise 32.7).

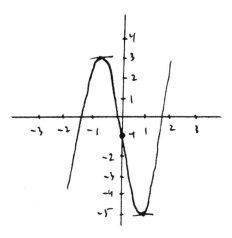

If instead of starting with the field $F$ and the polynomial $f(x)$, we start with a field extension $F \subseteq E$ and an element $\alpha \in E$, and if $E$ is a *finite* field extension, meaning $\deg E/F$ is finite, then $\alpha$ will be *algebraic* over $F$, that is, there exists a polynomial with coefficients in $F$ of which $\alpha$ is a root. Indeed, if $\deg E/F = n$, the $n + 1$ elements $1, \alpha, \alpha^2, \ldots, \alpha^n$ of $E$ must be linearly dependent over $F$. In other words, there are elements $a_0, \ldots, a_n \in F$, not all 0, such that

$$a_0 \alpha^n + a_1 \alpha^{n-1} + \cdots + a_n = 0.$$

Thus $\alpha$ is a root of the polynomial

$$f(x) = a_0 x^n + a_1 x^{n-1} + \cdots + a_n.$$

Once we know that there is some polynomial of which $\alpha$ is a root, we can find one of minimal degree, and this is called the *minimal polynomial* of $\alpha$ (usually normalized to have leading coefficient 1). If $g(x)$ is the minimal polynomial of $\alpha$, then $g(x)$ is necessarily irreducible, so by the previous proposition, $\deg F(\alpha)/F = \deg g(x)$.

**Example 32.2.4**

Take $\alpha = \frac{1}{2}\sqrt{10 - 2\sqrt{5}} \in \mathbb{R}$, for example, which is the side of the regular pentagon. Let us find its minimal polynomial. Set

$$x = \frac{1}{2}\sqrt{10 - 2\sqrt{5}}.$$

Squaring,

$$4x^2 = 10 - 2\sqrt{5}.$$

Regrouping,

$$2x^2 - 5 = -\sqrt{5}.$$

Squaring again,

$$4x^4 - 20x^2 + 25 = 5.$$

So

$$x^4 - 5x^2 + 5 = 0.$$

This procedure produces for us a polynomial of which $\alpha$ is a root. In fact, running the same calculation backward shows that the four roots of this polynomial are $\pm\frac{1}{2}\sqrt{10 \pm 2\sqrt{5}}$. To see whether this polynomial is the minimal polynomial of $\alpha$, we must decide whether it is *irreducible*. We first note that none of the four roots is rational, so that the polynomial has no linear factor. If it had quadratic factors, say

$$x^4 - 5x^2 + 5 = (x^2 + ax + b)(x^2 + cx + d),$$

then from the absence of cubic and linear terms we obtain $c = -a$ and $d = b$. Then

$$(x^2 + ax + b)(x^2 - ax + b) = x^4 + (2b - a^2)x^2 + b^2,$$

so we must have

$$2b - a^2 = -5,$$
$$b^2 = 5,$$

which is not possible for $a, b \in \mathbf{Q}$.

Therefore, $x^4 - 5x^2 + 5$ is irreducible, and it is the minimal polynomial of $\alpha$. We could also have used Eisenstein's criterion (32.8) with $p = 5$.

## Splitting Fields

Given a field $F$ and a polynomial $f(x) \in F[x]$, one can always find a root $\alpha$ in some extension field $E$ as we have seen above. Then in the polynomial ring $E[x]$ one can factor out a linear factor, so $f(x) = (x - \alpha)g(x)$, where $g(x)$ is a new polynomial, of degree one less than $f(x)$ and having coefficients in the new field $E$.

Repeating this process, one can find a root of $g(x)$ in some extension field of $E$, and split off another linear factor. In this way, one eventually finds an extension field $E'$ of $F$ in which the polynomial $f(x)$ splits into linear factors $f(x) = c\prod(x - \alpha_i)$, where $\alpha_1, \ldots, \alpha_n$ are $n = \deg f(x)$ roots of $f(x)$ in $E'$, and $c \in F$

is a constant. Note that the $\alpha_i$ need not all be distinct in general, although if $f$ is irreducible and the characteristic of the field $F$ is zero, they will be.

A *splitting field* for the polynomial $f(x)$ over the field $F$ is a field extension $E$ of $F$ such that over $E$ one can factor $f(x) = c \prod(x - \alpha_i)$ as above, with $c \in F$, $\alpha_i \in E$, and furthermore, $E = F(\alpha_1, \ldots, \alpha_n)$; in other words, the field $E$ is minimal with this property.

**Proposition 32.3**

*Let a field $F$ and a polynomial $f(x) \in F[x]$ be given. Then there exists a splitting field for $f(x)$ over $F$. Any two splitting fields are isomorphic as field extensions of $F$. If $f(x)$ is irreducible, if $E$ is a splitting field for $f(x)$ over $F$, and if $\alpha_1, \alpha_2 \in E$ are any two roots of $f(x)$ in $E$, then there exists an automorphism $\varphi : E \to E$ leaving elements of $F$ fixed and sending $\alpha_1$ to $\alpha_2$.*

**Example 32.3.1**

If $f(x) = x^2 - 2$, the two roots are $\pm\sqrt{2}$, and $\mathbb{Q}(\sqrt{2})$ is a splitting field.

**Example 32.3.2**

If $f(x) = x^3 - 2$, the roots are $\sqrt[3]{2}$, $\omega\sqrt[3]{2}$, $\omega^2\sqrt[3]{2}$, so the splitting field is $\mathbb{Q}(\sqrt[3]{2}, \omega\sqrt[3]{2}, \omega^2\sqrt[3]{2})$. It is easy to see that this field is the same as the field $\mathbb{Q}(\sqrt[3]{2}, \sqrt{-3})$, which has degree 6 over $\mathbb{Q}$. In general, as we see from the method of constructing a splitting field, the splitting field of a polynomial of degree $n$ will have degree less than or equal to $n!$. In this case we have equality: $6 = 3!$.

**Example 32.3.3**

Let $f(x) = x^4 - 5x^2 + 5$. We saw earlier that the roots of this polynomial are $\pm\frac{1}{2}\sqrt{10 \pm 2\sqrt{5}}$, so the splitting field will be $\mathbb{Q}\left(\sqrt{10 + 2\sqrt{5}}, \sqrt{10 - 2\sqrt{5}}\right)$. Since

$$\sqrt{10 + 2\sqrt{5}} \cdot \sqrt{10 - 2\sqrt{5}} = \sqrt{80} = 4\sqrt{5},$$

the splitting field is actually equal to $\mathbb{Q}\left(\sqrt{10 + 2\sqrt{5}}\right)$, because this field already contains $\sqrt{5}$. We saw earlier that this polynomial is irreducible, so the splitting field has degree 4. This is an example where the degree of the splitting field is considerably less than $4! = 24$.

**Example 32.3.4**

For $x^5 - 5x - 1$, which we looked at earlier, the splitting field has degree $120 = 5!$ over $\mathbb{Q}$ (32.4.4).

## Normal Extensions and Galois Groups

For this part, we will restrict our attention to fields of *characteristic* 0 for simplicity, since in the applications to geometry we will be dealing mainly with extension fields of $\mathbb{Q}$.

A finite field extension $E/F$ is called *normal* if it is equal to the splitting field of some polynomial $f(x) \in F[x]$. In this case we denote by $G$ the group of all automorphisms of $E$ that leave elements of $F$ fixed. $G$ is called the *Galois group* of the field extension $E/F$. Or if $E$ is the splitting field of a polynomial $f(x) \in F[x]$, $G$ is also called the Galois group of the polynomial $f(x)$. A normal extension $E/F$ is also sometimes called a Galois extension.

**Theorem 32.4** (Fundamental theorem of Galois theory)
*Let $E/F$ be a normal field extension with Galois group $G$. Then*

(a) *The order of $G$ is equal to the degree of the extension $E/F$.*
(b) *The only elements of $E$ fixed under all elements of $G$ are the elements of $F$.*
(c) *There is a 1-to-1 inclusion-reversing correspondence between subgroups $H \subseteq G$ and intermediary fields $F \subseteq K \subseteq E$ given as follows: To a subgroup $H \subseteq G$ we associate the field $E^H$ of elements of $E$ left fixed by all elements of $H$. Conversely, to an intermediate field $K$ we associate the subgroup $H \subseteq G$ of those elements of $G$ that leave all elements of $K$ fixed.*
(d) *Under the correspondence just described, the subgroup $H \subseteq G$ will be a normal subgroup of $G$ if and only if the associated field $K$ is a normal extension of $F$, and in that case the quotient group $G/H$ is isomorphic to the Galois group of $K/F$.*

Note that if $f(x)$ is an irreducible polynomial of degree $n$ over a field $F$ of characteristic 0, then its roots $\alpha_1, \ldots, \alpha_n$ are all distinct. Let $E = F(\alpha_1, \ldots, \alpha_n)$ be a splitting field, and $G$ its Galois group. For any root, say $\alpha_1$, the equation $f(\alpha_1) = 0$ must be preserved by any element of $G$. The coefficients of $f$ are elements of the base field $F$, so they are fixed by elements of $G$. Thus the image of $\alpha_1$ by an element of $G$ must be another root of $f(x)$, namely one of $\alpha_1, \ldots, \alpha_n$. In this way we see that an element of $G$ *permutes* the set $\alpha_1, \ldots, \alpha_n$ of the $n$ roots of $f(x)$. Since $E$ is generated by these roots, the action of an element of $G$ is completely determined by its action on the $\alpha_i$. In this way we see that $G$ can be regarded as a *subgroup* of the symmetric group $S_n$ of all permutations of the set $\{\alpha_1, \ldots, \alpha_n\}$.

**Example 32.4.1**
Consider the polynomial $x^2 - 2$ over $\mathbb{Q}$. Its splitting field is $\mathbb{Q}(\sqrt{2})$. This is a normal extension of $\mathbb{Q}$, and the Galois group consists of the identity and the automorphism $\varphi$ that takes $\sqrt{2}$ to $-\sqrt{2}$. In this case $G$ is equal to the symmetric group $S_2$, which is isomorphic to $\mathbb{Z}_2 = \mathbb{Z}/2\mathbb{Z}$.

**Example 32.4.2**
Consider $x^3 - 2$ over $\mathbb{Q}$. First look at the field extension $\mathbb{Q}(\sqrt[3]{2})$. Any automorphism of this field would have to take $\sqrt[3]{2}$ to another root of $x^3 - 2$. But the other two roots are complex numbers, not contained in the field $\mathbb{Q}(\sqrt[3]{2})$. Thus the only automorphism of this field is the identity. It follows from (32.4) that $\mathbb{Q}(\sqrt[3]{2})$ is *not* a normal extension of $\mathbb{Q}$. Of course, we know that it is not the

splitting field of the polynomial $x^3 - 2$, but this shows, assuming the result (32.4a), that $\mathbb{Q}(\sqrt[3]{2})$ cannot be the splitting field of any polynomial over $\mathbb{Q}$.

The splitting field of $x^3 - 2$ is $\mathbb{Q}(\sqrt[3]{2}, \sqrt{-3})$, which has degree 6 over $\mathbb{Q}$. Hence the Galois group will be equal to $S_3$. There are three intermediate fields of degree 3 over $\mathbb{Q}$, namely $\mathbb{Q}(\sqrt[3]{2})$, $\mathbb{Q}(\omega\sqrt[3]{2})$, and $\mathbb{Q}(\omega^2\sqrt[3]{2})$. These correspond to the three subgroups $\{e, (12)\}$, $\{e, (13)\}$, $\{e, (23)\}$ of order 2 of $S_3$. The field $\mathbb{Q}(\sqrt{-3})$, which is a normal extension of $\mathbb{Q}$, corresponds to the normal subgroup $A_3 = \{e, (123), (132)\}$ of $S_3$.

### Example 32.4.3

Consider again the polynomial $x^4 - 5x^2 + 5$ over $\mathbb{Q}$. We saw that it is irreducible with roots

$$\alpha_1 = \tfrac{1}{2}\sqrt{10 + 2\sqrt{5}},$$

$$\alpha_2 = \tfrac{1}{2}\sqrt{10 - 2\sqrt{5}},$$

$$\alpha_3 = -\tfrac{1}{2}\sqrt{10 + 2\sqrt{5}},$$

$$\alpha_4 = -\tfrac{1}{2}\sqrt{10 - 2\sqrt{5}}.$$

We saw that the splitting field is $\mathbb{Q}\left(\sqrt{10 + 2\sqrt{5}}\right)$, which has degree 4 over $\mathbb{Q}$. Thus the Galois group $G$ will be a subgroup of order 4 of the symmetric group $S_4$. Which subgroup of order 4 is it?

To investigate this question, we must describe explicitly the automorphisms of the field $E = \mathbb{Q}\left(\sqrt{10 + 2\sqrt{5}}\right)$. According to (32.3), there exists an element $\sigma \in G$ such that $\sigma(\alpha_1) = \alpha_2$. Then $\sigma(\alpha_1^2) = \alpha_2^2$, and since elements of $\mathbb{Q}$ are fixed, it follows that $\sigma(\sqrt{5}) = -\sqrt{5}$. Now we make use of the identity $\alpha_1\alpha_2 = \sqrt{5}$. Applying $\sigma$ we obtain $\sigma(\alpha_1) \cdot \sigma(\alpha_2) = \sigma(\sqrt{5}) = -\sqrt{5}$. But $\sigma(\alpha_1) = \alpha_2$ by choice of $\sigma$, so we obtain $\alpha_2 \cdot \sigma(\alpha_2) = -\sqrt{5}$. From the equation $\alpha_1\alpha_2 = \sqrt{5}$ it follows that $\sigma(\alpha_2) = -\alpha_1 = \alpha_3$. Finally, $\sigma(\alpha_3) = \sigma(-\alpha_1) = -\alpha_2 = \alpha_4$. Thus $\sigma = (1234)$. So $G$ is the subgroup of $S_4$ generated by $(1234)$, which is a *cyclic* group of order 4.

Perhaps from this example it is already clear that the determination of the splitting field of a polynomial and its Galois group is not a straightforward matter. It depends in each case on using some special information about that particular example, which may not be an easy task.

### Example 32.4.4

Coming back to the example $x^5 - 5x - 1$ we studied earlier, let us see whether we can determine the degree of its splitting field and its Galois group. The polynomial is irreducible, so if we adjoin one root $\alpha$, then $\mathbb{Q}(\alpha)$ will be an extension of degree 5 of $\mathbb{Q}$. Let $E$ be the splitting field. Then $\mathbb{Q} \subseteq \mathbb{Q}(\alpha) \subseteq E$. Therefore,

$\deg E/\mathbb{Q}$ is a multiple of 5. So the Galois group $G$ will be a subgroup of $S_5$, whose order is a multiple of 5. Hence (from abstract group theory), $G$ contains an element of order 5. In $S_5$, the only elements of order 5 are the 5-cycles. Thus $G$ contains a 5-cycle. On the other hand, we saw earlier that $f(x)$ has 3 real roots, and hence two conjugate complex roots. Complex conjugation in $\mathbb{C}$ then induces an automorphism of $E$ that fixes the three real roots and permutes the two complex roots. This is an element of $G$ whose image in $S_5$ is therefore a transposition (a 2-cycle). Now, one can show (again using abstract group theory) that a subgroup of $S_5$ that contains a 5-cycle and a transposition must be the full symmetric group. Hence $G = S_5$, and the degree of $E/\mathbb{Q}$ is $5! = 120$.

Note the indirect nature of the reasoning, and how we were able to prove that $G = S_5$ without having any explicit representation of the roots of $f(x)$.

### Reduction mod $p$

A useful technique for obtaining information about the Galois group of a polynomial with integer coefficients is the following.

### Proposition 32.5
*Let $f(x)$ be an irreducible monic polynomial with integer coefficients. Fix a prime $p$, and assume that the polynomial $\bar{f}(x)$ with coefficients reduced (mod $p$) has distinct roots in a splitting field $\bar{E}$ for $\bar{f}(x)$ over the prime field $\mathbb{F}_p$. Then there is a 1-to-1 correspondence between the roots of $f(x)$ in its splitting field $E$ over $\mathbb{Q}$ and the roots of $\bar{f}(x)$ in $\bar{E}$ such that the Galois group of $\bar{f}$ over $\mathbb{F}_p$, considered as a group of permutations of the roots, corresponds to a subgroup of the Galois group of $f$ over $\mathbb{Q}$.*

### Example 32.5.1
Consider the polynomial $f(x) = x^4 - x^3 - 5x^2 + 1$. It has no roots in $\mathbb{Q}$, because a root would have to be an integer dividing 1, and neither $+1$ nor $-1$ is a root. Reducing (mod 2), we obtain $\bar{f}(x) = x^4 + x^3 + x^2 + 1 = (x+1)(x^3 + x + 1)$. Note that $x^3 + x + 1$ is irreducible because it has no roots modulo 2. It follows that $f(x)$ cannot be a product of two quadratic polynomials, because then $\bar{f}(x)$ would be also. Thus $f(x)$ is irreducible, and so 4 divides the order of the Galois group.

Now let us apply the proposition. The polynomial $\bar{f}(x)$ has distinct roots (since the derivative of $x^3 + x + 1$ is nonzero), and its Galois group is cyclic of order 3. We conclude from the proposition that the Galois group of $f(x)$ contains a 3-cycle. Hence its order is divisible by 3, and we see that the Galois group of $f(x)$ must have order 12 or 24.

### Roots of Unity

We will examine in greater detail the Galois groups of the polynomials $x^n - 1$ over $\mathbb{Q}$, since these are closely related to the question of constructing regular polygons of $n$ sides.

# DE GLI ELEMENTI
## DI EVCLIDE
### LIBRO TERZO

CON LI SCHOLII ANTICHI,
ET COMMENTARII
Di Federico Commandino da Vrbino.

#### SCHOLIO.

*L'intentione di Euclide in questo luogo, è di trattare delle cose che auuengono a cerchi, hauuto rispetto alle linee rette & a gli angoli,*

#### DIFFINITIONI.
#### I.

CERCHI vguali sono quelli che hanno i diametri, ò vero i semidiametri loro vguali.

#### II.

La linea retta si dice toccare il cerchio, la quale toccandolo & prolungata non lo sega.

#### III.

I cerchi si dicano toccarsi fra loro, quali toccando si non si segano.

Le

Plate XI. The opening page of Book III in the Italian translation of Euclid by Commandino (1575). The initial capitals in this book are taken from Commandino's two editions of (1572) and (1575).

**289**

If $\alpha = 2\pi/n$ is an angle equal to one $n$th of a complete rotation, then the complex number

$$\zeta = e^{2\pi i/n} = \cos \alpha + i \sin \alpha$$

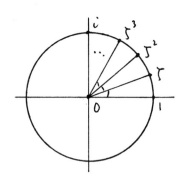

is an $n$th root of unity. In fact, $1, \zeta, \zeta^2, \ldots, \zeta^{n-1}$ are the $n$ roots of the equation $x^n - 1$ in the field $\mathbb{C}$ of complex numbers. Among these we will designate as *primitive* $n$th roots of unity those that are not a $d$th root of unity for any $d|n$.

In other words, the primitive $n$th roots of unity are those powers $\zeta^r$ of $\zeta$ for which $(r, n) = 1$. The product

$$\Phi_n(x) = \prod_{(r,n)=1} (x - \zeta^r)$$

is called the $n$th *cyclotomic polynomial*. From this description, it is not obvious where its coefficients lie, but from the expression

$$\Phi_n(x) = (x^n - 1) \bigg/ \left( \prod_{\substack{d|n \\ d \neq n}} \Phi_d(x) \right)$$

it follows inductively that $\Phi_n$ has coefficients in $\mathbb{Q}$, and in fact (using Gauss's lemma), coefficients in $\mathbb{Z}$.

The degree of the polynomial $\Phi_n(x)$ is given by the Euler $\varphi$-function: $\varphi(n) = \#\{1 \leq r < n | (r, n) = 1\}$.

### Proposition 32.6
*For any $n$, the cyclotomic polynomial $\Phi_n(x)$ is irreducible over $\mathbb{Q}$.*

### Example 32.6.1
The first few cyclotomic polynomials are

$$\Phi_1 = x - 1,$$
$$\Phi_2 = x + 1,$$
$$\Phi_3 = x^2 + x + 1,$$
$$\Phi_4 = x^2 + 1,$$
$$\Phi_5 = x^4 + x^3 + x^2 + x + 1,$$
$$\Phi_6 = x^2 - x + 1,$$
$$\Phi_7 = x^6 + x^5 + x^4 + x^3 + x^2 + x + 1,$$
$$\Phi_8 = x^4 + 1.$$

(However, do not conclude from this list that the coefficients are always $\pm 1$: There may be other integer coefficients.)

Let $\zeta = e^{2\pi i/n}$ again. Since $\Phi_n$ is irreducible, it will be the minimal polynomial of $\zeta$, and we find that degree of $\mathbb{Q}(\zeta)/\mathbb{Q}$ is $\varphi(n)$, the Euler $\varphi$-function. Furthermore, since the other roots of $\Phi_n$ are all powers of $\zeta$, the splitting field of $\Phi_n$ is this same field $\mathbb{Q}(\zeta)$. From (32.3) we find that for each $1 \le r < n$ with $(r, n) = 1$, there is an element $\varphi_r$ of the Galois group $G$ for which $\varphi_r(\zeta) = \zeta^r$. If we compose two of these, we have $\varphi_s\varphi_r(\zeta) = \varphi_s(\zeta^r) = \zeta^{rs}$. Reducing $rs$ (mod $n$) to $t$ with $1 \le t < n$, it follows that $\varphi_s\varphi_r = \varphi_t$. In this way we see that the Galois group $G$ is isomorphic to the group of integers $1 \le r < n$ relatively prime to $n$, under multiplication (mod $n$). This group is usually denoted by $\mathbb{Z}_n^*$. The field $\mathbb{Q}(\zeta)$ is called the $n$th *cyclotomic extension* of $\mathbb{Q}$, or the field of $n$th roots of unity. Summarizing, we have the following proposition.

**Proposition 32.7**
*The cyclotomic field of nth roots of unity $\mathbb{Q}(\zeta)$ is generated by $\zeta = e^{2\pi i/n}$. It has degree $\varphi(n)$ over $\mathbb{Q}$, and its Galois group is isomorphic to $\mathbb{Z}_n^*$. In particular, if $n$ is a prime number $p$, then $\mathbb{Q}(\zeta)$ has degree $p - 1$, and its Galois group is $\mathbb{Z}_p^*$, which is a cyclic group of order $p - 1$.*

**Example 32.7.1**
$n = 3$. $\Phi_3 = x^2 + x + 1$. Its roots are $\omega = \frac{1}{2}(-1 + \sqrt{-3})$ and $\omega^2 = \frac{1}{2}(-1 - \sqrt{-3})$. The cyclotomic field $\mathbb{Q}(\omega)$ is $\mathbb{Q}(\sqrt{-3})$, with Galois group isomorphic to $\mathbb{Z}_2$.

**Example 32.7.2**
$n = 4$. $\Phi_4 = x^2 + 1$. Roots are $\pm i$. The field is $\mathbb{Q}(i)$ with Galois group $\mathbb{Z}_2$.

**Example 32.7.3**
$n = 5$. $\Phi_5 = x^4 + x^3 + x^2 + x + 1$. The cyclotomic field is $\mathbb{Q}(\zeta)$, where

$$\zeta = e^{2\pi i/5} = \cos 2\pi/5 + i \sin 2\pi/5$$

$$= \tfrac{1}{4}(\sqrt{5} - 1) + i\tfrac{1}{4}\sqrt{10 + 2\sqrt{5}}.$$

The Galois group is isomorphic to $\mathbb{Z}_4$.

**Example 32.7.4**
$n = 8$. $\Phi_8 = x^4 + 1$. Its roots are $\zeta$, $\zeta^3$, $\zeta^5$, $\zeta^7$, where

$$\zeta = e^{2\pi i/8} = \frac{\sqrt{2}}{2} + i\frac{\sqrt{2}}{2}.$$

The Galois group is $\mathbb{Z}_8^*$, which is $\{1, 3, 5, 7\}$ under multiplication (mod 8). This group is isomorphic to the Klein four-group $\mathbb{Z}_2 \times \mathbb{Z}_2$.

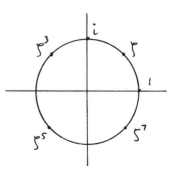

For reference we include the following useful criterion for a polynomial to be irreducible.

**Proposition 32.8** (Eisenstein)
*Let*

$$f(x) = x^n + a_1 x^{n-1} + \cdots + a_n$$

*be a monic polynomial with integer coefficients $a_i \in \mathbb{Z}$. Suppose for some prime number $p$, that $p$ divides all the $a_i$, and $p^2$ does not divide $a_n$. Then $f(x)$ is irreducible over $\mathbb{Q}$.*

# Exercises

32.1 For each of the following real numbers, find the minimal polynomial over $\mathbb{Q}$ (be sure to prove that it is irreducible), find the degree of the splitting field of the polynomial, and determine the Galois group (up to isomorphism). *Hint*: Expect to find groups of orders 2, 4, and 8.

(a) $\sqrt{2 + \sqrt{2}}$.

(b) $\sqrt{3 + \sqrt{2}}$.

(c) $\sqrt{3 + 2\sqrt{2}}$.

In the following exercises, we will investigate the *Galois group of a construction problem*, defined as follows. For a construction problem in the real Euclidean plane $\mathbb{R}^2$, starting with data defined over $\mathbb{Q}$, the construction will create various points. Let $F$ be the field $F = \mathbb{Q}(\beta_1, \ldots, \beta_m)$ obtained by adjoining the coordinates of all the points constructed. Let $E$ be the smallest normal field extension containing $F$, which can be obtained as follows. Let $f_i(x)$ be the minimal polynomial of $\beta_i$, $i = 1, \ldots, m$, let $g(x) = \prod_{i=1}^m f_i(x)$, and take $E$ to be the splitting field of $g(x)$. The Galois group $G$ of $E/\mathbb{Q}$ we will then call the Galois group of the construction problem.

32.2 Given the unit segment $A, B$, where $A = (0,0)$, $B = (1,0)$, construct an equilateral triangle with side $AB$. Determine the associated field $F$, the normal field $E$, and the Galois group $G$ of the construction problem. *Answer*: $F = E = \mathbb{Q}(\sqrt{3})$. $G \cong \mathbb{Z}_2$.

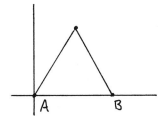

32.3 Construct a square with one corner at $(0,0)$, with sides along the $x$- and $y$-axes, and with area equal to the area of the triangle in Exercise 32.2. Find $F, E, G$ as above.

32.4 Construct a regular pentagon inscribed in the unit circle having one vertex at $(1,0)$. Find $F, E, G$ as above.

32.5 Construct an equilateral triangle with one side on the $x$-axis, and with area equal to twice the area of the triangle in Exercise 32.2 above. Find $F, E, G$ as above.

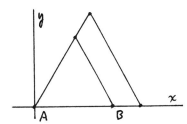

32.6 (a) Compute $\Phi_9$, the ninth cyclotomic polynomial, and find the Galois group of the cyclotomic field $\mathbb{Q}(\zeta)$ where $\zeta = e^{2\pi i/9}$.

(b) Let $\alpha = \frac{1}{2} \cos 40°$. Find the minimal polynomial of $\alpha$. Show that $\mathbb{Q}(\alpha)$ is a normal field extension of $\mathbb{Q}$ of degree 3 and that $\mathbb{Q}(\alpha) \subseteq \mathbb{Q}(\zeta)$.

(c) Show that $\sqrt{-3} \in \mathbb{Q}(\zeta)$.

(d) Finally, show that $\mathbb{Q}(\zeta) = \mathbb{Q}(\alpha, \sqrt{-3})$.

32.7 Using facts about the Galois group from Example 32.4.4, show that the five fields $\mathbb{Q}(\alpha_i)$ mentioned in Example 32.2.3 are all distinct.

32.8 Let $\mathbb{Q}(\zeta)$ be the cyclotomic field of 5th roots of unity.

(a) Show that $\sqrt{5} \in \mathbb{Q}(\zeta)$.

(b) Show that $\mathbb{Q}(\zeta) = \mathbb{Q}(\delta)$, where $\delta = \frac{i}{2}\sqrt{10 + 2\sqrt{5}}$.

(c) What is the minimal polynomial of $\delta$?

32.9 For $p = 5, 7, 11, 13, 17$, show explicitly that $\mathbb{Z}_p^*$ is cyclic by finding a suitable element and proving that it generates the group.

32.10 Consider the polynomial $x^4 - 2x^2 - 7$ over $\mathbb{Q}$.

(a) Show that its Galois group is the dihedral group $D_4$, defined by generators $a, b$ and relations $a^4 = e$, $b^2 = e$, $ba = a^{-1}b$.

(b) Find the lattice of all subgroups of $D_4$.

(c) Find all the subfields of the splitting field, and explain their correspondence with the subgroups of $D_4$.

32.11 Consider the polynomial $f(x) = x^4 + x - 3$ over $\mathbb{Q}$.

(a) Find the cubic resolvent and show that it is irreducible.

(b) Show that $f(x)$ is irreducible.

(c) By curve sketching, show that $f(x)$ has two real roots and two complex roots.

(d) Conclude that the Galois group of $f(x)$ is $S_4$.

(e) If $\alpha$ is a real root, then $\mathbb{Q}(\alpha)$ is an extension of degree 4 of $\mathbb{Q}$, but $\alpha$ is not constructible by ruler and compass.

32.12  If $f(x)$ is an irreducible quartic polynomial over $\mathbb{Q}$ whose cubic resolvent is irreducible with discriminant $\Delta$, show that the Galois group is $A_4$ if and only if $\Delta$ is a square in $\mathbb{Q}$; otherwise, the Galois group is $S_4$.

32.13  Prove the following theorem of Hölder: Let $f(x)$ be irreducible of degree $n$ over $\mathbb{Q}$, having all real roots. If at least one of these roots can be expressed by *real* radicals (of various degrees), then $n = 2^k$, and all the roots can be expressed by real square roots.

> Out two soules therefore, which are one,
> Though I must goe, endure not yet
> A breach, but an expansion,
> Like gold to ayery thinnesse beate.
>
> If they be two, they are two so
> As stiffe twin compasses are two,
> Thy soule the fixt foot, makes no show
> To move, but doth, if the'other doe.
>
> And though it in the center sit,
> Yet when the other far doth rome,
> It leanes, and hearkens after it,
> And growes erect, as that comes home.
>
> Such wilt thou be to mee, who must
> Like th'other foot, obliquely runne;
> Thy firmnes drawes my circle just,
> And makes me end, where I begunne.
>
> – from *A Valediction: Forbidding Mourning*
> by John Donne (1572–1631)

# 7 CHAPTER

# Non-Euclidean Geometry

ertainly one of the greatest mathematical discoveries of the nineteenth century was that of non-Euclidean geometry: seen but not revealed by Gauss, and developed in all its glory by Bolyai and Lobachevsky. The purpose of this chapter is to give an account of this theory, but we do not always follow the historical development. Rather, with hindsight we use those methods that seem to shed the most light on the subject. For example, continuity arguments have been replaced by a more axiomatic treatment.

There are actually three different approaches presented here. One begins with Saccheri's theory, dividing geometries into three classes, in Section 34, and the theorem of Saccheri–Legendre, using Archimedes' axiom, in Section 35. The second is the analytic model of a non-Euclidean geometry given in Section 39. Third is Hilbert's axiomatic approach based on the axiom of limiting parallels (L) in Section 40.

We start with a historical introduction to the problem of the parallels and the various futile attempts to prove Euclid's fifth postulate from the other axioms. Then we begin to explore this strange new world where the sum of the angles of a triangle can be less than two right angles. The defect of this angle sum provides a measure of area, which we exploit in Section 36.

To explain the Poincaré model of a non-Euclidean geometry, we need the Euclidean technique of circular inversion. This is developed in Section 37. It is a technique with many applications in Euclidean geometry. In particular, we

**295**

make a digression in Section 38 to show how it can provide a solution to the classical problem of Apollonius, to construct a circle tangent to three given circles.

In Section 40 we present a development of non-Euclidean geometry based on the axiom of existence of limiting parallel rays, proposed by Hilbert. This allows us to avoid the appeal to continuity invoked by the founders of the subject and free ourselves from dependence on the real numbers. Then we give Hilbert's brilliant construction of an abstract field from the set of common ends of limiting parallel rays. This allows us to characterize hyperbolic planes by their associated fields without using the techniques of projective geometry.

We follow the principle, established earlier in this book, of systematically avoiding the use of real numbers. There is a slight cost, in that some familiar results will look different here, but I believe this approach is justified by keeping the intrinsic geometry in the foreground. For example, instead of taking logarithms to define a distance function, we use a multiplicative distance function $\mu$. Then Bolyai's famous formula for the angle of parallelism $\alpha$ of a line segment $PQ$ takes the form $\tan \alpha/2 = \mu(PQ)^{-1}$ (39.13) and (41.9). The arbitrary constant $k$ that appears in some books, coming from the choice of a base for the logarithms in the distance function, is absent: In our approach, any two hyperbolic planes over the same field are isomorphic. Also, the hyperbolic trigonometric functions sinh, cosh, tanh do not appear in our formulae of hyperbolic trigonometry (42.2) and (42.3). As a result of this approach, the solution of any problem we consider can be found constructively, by ruler and compass, or, equivalently, by solving linear and quadratic equations in the coefficient field.

# 33   History of the Parallel Postulate

To set the background for the discovery of non-Euclidean geometry, a kind of geometry where there may be many lines through a point parallel to a given line, let us trace the history of attitudes toward the parallel postulate.

We have seen already that Euclid's fifth postulate, which we refer to as the parallel postulate, was of a much more sophisticated nature than the other postulates and axioms. Euclid seems to have recognized this himself, since he postponed using it as long as possible, and was careful to develop the standard congruence theorems for triangles without the parallel postulate.

Euclid was criticized for making this a postulate and not a theorem. Proclus (410–485), who represented the school of Plato in fifth-century Athens, has left an extensive commentary on the first book of Euclid's *Elements*. His opinion on the fifth postulate is unambiguous:

"This ought to be struck from the postulates altogether. For it is a theorem— one that invites many questions, which Ptolemy proposed to resolve in one of his books—and requires for its demonstration a number of definitions as well as theorems" (Proclus (1970), p. 150).

In his commentary on (I.29), Proclus gives Ptolemy's proof and points out its flaws, and then proceeds to give his own proof of the fifth postulate. First, he says, we must accept an axiom that was used earlier by Aristotle:

## Aristotle's Axiom

If from a single point two straight lines making an angle are produced indefinitely, the interval between them will exceed any finite magnitude. In other words, given any angle *BAC*, and given a segment *DE*, there exists a point *F* on the ray *AB* such that the perpendicular *FG* from *F* to the line *AC* will be greater than *DE*.

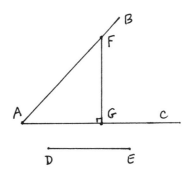

Then Proclus proposes to prove the following lemma of Proclus.

## Lemma of Proclus

*If a straight line cuts one of two parallel lines, it cuts the other also.*

His proof goes like this. If *AB* and *CD* are two parallel lines, and if *EF* cuts *AB*, with *F* on the side toward *CD*, then we apply Aristotle's axiom to the angle *BEF*. As we extend the ray *EF* indefinitely, its interval from the line *AB* will exceed the distance between the parallel lines, and so it must cut the line *CD*.

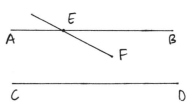

From this lemma (which is essentially the same as what we now call Playfair's axiom), Proclus easily proves the parallel postulate.

Proclus's reasoning was apparently accepted for some time, since it is reproduced without critical comment by F. Commandino in his edition of Euclid (1575).

We can observe two things about the argument of Proclus. First of all, he assumes another axiom (the axiom of Aristotle) in the course of his proof. This is not uncommon in various attempted proofs of the parallel postulate. Often, one ends up assuming (consciously or unconsciously) something else that turns out to be equivalent to the parallel postulate. In this particular case, it is not so bad: We will see that Aristotle's axiom is a consequence of Archimedes' axiom, and does not imply the parallel postulate by itself (35.6).

The more serious flaw in Proclus's argument is that he speaks of "the distance between the parallel lines" as if all the points of one line were at the same distance from the other line. Since the definition of parallel lines is lines in the same plane that do not meet, however far extended, it does not follow from the definition that they are at a constant distance from each other. In fact, this assumption of constant distance is enough to prove the parallel postulate (in the presence of Aristotle's axiom), as Proclus shows. Thus, in view of (I.34) it is equivalent to the parallel postulate.

This confusion of the definition of parallel lines as lines that do not meet with the common-sense notion of parallel lines as equidistant from each other (like railroad tracks) has persisted. For example, in the edition of Euclid's first six books by J. Peletier (1557), definition 35 says, "Parallels, or equidistant straight lines, are those which being in the same plane, and extended arbitrarily in either direction, do not meet." However, Peletier follows Euclid's proofs in Book I, and does not make use of the equidistant property.

A more striking example is the very popular edition of the *Elements of Geometry* by the Jesuit Andrea Tacquet (1612–1660), first published in 1654 and reprinted many times over the next hundred and fifty years (Tacquet (1738)). Tacquet's book is not a strict translation of Euclid, but an arrangement, to make the study of geometry easier for beginners. Though he preserves the numbering of Euclid's propositions, he takes great liberties with their proofs. For example, he says that there is no point in proving (I.16), because it is a special case of (I.32)! He apparently does not care about the fact that Euclid's proof of (I.16) is independent of the parallel postulate, while (I.32) depends on it.

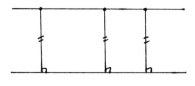

Tacquet says that since there are various species of lines (such as the hyperbola and a straight line) that approach each other indefinitely but never meet, so Euclid's definition of parallel lines does not satisfactorily reflect the nature of parallels.

He takes as his definition that two lines are parallel if the points of one are all equidistant from the other, as measured by perpendiculars from points on the first line to the second line.

There is no harm, of course, in using any definition you like of parallel lines, though this one places a great burden on the proof of existence of parallels. Tacquet misses the subtlety, however, because in the next sentence he says that you can generate parallel lines as the locus of points at a fixed distance from a given line as the perpendicular moves along. Here he is implicitly using another axiom, which was in fact stated explicitly and used earlier by Christoph Clavius (1537–1612) as a substitute for Euclid's parallel postulate:

## Clavius's Axiom

The set of points equidistant from a given line on one side of it form a straight line.

This axiom, as one can easily show, is almost equivalent to the parallel postulate that Tacquet was trying to avoid (Exercise 33.7).

The French mathematician Alexis Claude Clairaut (1713–1765) wrote an *Elémens de Géométrie* (first published in 1741) in which he tried to make geometry more accessible for students. He complained about the usual method of teaching the elements, in which "one always starts with a great number of definitions, postulates, axioms, and first principles, which appear to offer nothing but dryness to the reader." He thought that Euclid's careful reasoning was merely to satisfy a fussy audience: "That Euclid went to the trouble to prove that two circles which cut each other do not have the same center; that a triangle contained inside another triangle has the sum of its sides less than that of the triangle in which it is enclosed—one should not be surprised. For this geometer had to convince the obstinate sophists who gloried in finding fault with the most evident truths: so it was necessary that geometry, like logic, make use of proper reasoning, to close the mouths of its critics."

Clairaut's purpose is to introduce the concepts of geometry simply and naturally in the context of practical questions such as measurement of terrain. So he talks of straight lines to measure the distance between points, and how to construct perpendicular lines. Then he says, what is more easy than to use this method to construct a rectangle? One has only to take a segment $AB$, and at its endpoints raise perpendiculars $AC$ and $BD$ of equal length, and then join $CD$. From here he develops the theory of parallels. The hidden assumption is that his construction makes a rectangle. So we will call this assumption Clairaut's axiom.

## Clairaut's Axiom

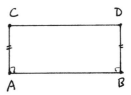

Given a segment $AB$, let $AC$ and $BD$ be equal segments perpendicular to $AB$. Then the angles at $C$ and $D$ are right angles, i.e., $ABCD$ is a rectangle.

Robert Simson, M.D. (1687–1768), professor of mathematics in the University of Glasgow, made an important edition of Euclid's elements, in Latin and in English, first published in 1756, which went through some thirty successive editions. Simson railed against the errors introduced by earlier editors, and wished to "restore the principal Books of the Elements to their original Accuracy.... This I have endeavored to do by taking away the inaccurate and false Reasonings which unskilful Editors have put into the place of some of the genuine Demonstrations of Euclid, who has ever been justly celebrated as the most accurate of Geometers, and by restoring to him those Things which Theon or others have suppressed, and which have these many ages been buried in Oblivion" (Simson (1803), Preface). Simson's restorations were not so much based on textual studies as on his faith that anything mathematically true and accurate must have been Euclid's, while anything incorrect or not rigorous must have been inserted by "some unskilful editor." About the parallel postulate, he says, "It seems not to be properly placed among the Axioms, as, indeed, it is not self-evident; but it may be demonstrated thus." Simson then introduces an axiom,

### Simson's Axiom

A straight line cannot first come nearer to another straight line, and then go further from it, before it cuts it; and, in like manner, a straight line cannot go further from another straight line, and then come nearer to it; nor can a straight line keep the same distance from another straight line, and then come nearer to it, or go further from it (Simson (1803), p. 295).

From this axiom, and implicitly making use of Archimedes' axiom, Simson proves (correctly) five propositions, of which the last is Euclid's parallel postulate.

So here we have a clear case of an author substituting another axiom that seems more natural to him, and then using it to prove the parallel postulate.

John Playfair (1748–1819), professor of natural philosophy, formerly of mathematics, in the University of Edinburgh, published a new edition of the first six books of Euclid's *Elements* that first appeared in 1795. He says that Dr. Simson has done a fine job of restoring Euclid's *Elements*, and that his purpose in presenting a new edition is to give them the form that may "render them most useful." He says, "A new axiom is also introduced in the room of the 12th [which we call the fifth postulate], for the purpose of demonstrating more easily some of the properties of parallel lines" (Playfair (1795), Preface). This is Playfair's axiom.

### Playfair's Axiom

Two straight lines that intersect one another cannot be both parallel to the same straight line.

In his notes to (I.29), Playfair has an interesting discussion of the problem of parallels. He agrees with Proclus that Euclid's postulate should be proved, and not taken as an axiom. He then reviews the three methods by which geometers "have attempted to remove this blemish from the *Elements*...

(1) by a new definition of parallel lines;
(2) by introducing a new Axiom concerning parallel lines, more obvious than Euclid's;
(3) by reasoning merely from the definition of parallels, and the properties of lines already demonstrated, without the assumption of any new Axiom."

# Exercises

Throughout these exercises, we assume the axioms of a Hilbert plane.

33.1 Show that the lemma of Proclus is equivalent to Playfair's axiom (P).

33.2 Consider the following special case of Euclid's parallel postulate, which we will call the **right triangle axiom**: Given a right angle $ABD$ and an acute angle $\alpha = CAB$ on the same side of the line $AB$, the ray $AC$ when extended will meet the ray $BD$ extended.

    Show that the right triangle axiom is equivalent to (P).

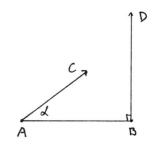

33.3 Show directly that the right triangle axiom implies the special case of Euclid's parallel postulate that says, given acute angles $\alpha = CAB$ and $\beta = ABD$ on the same side of the line $AB$, the rays $AC$ and $BD$ will meet.

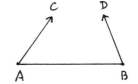

33.4 Discuss the following "proof" of the right triangle axiom due to Franceschini (1756–1840): Given $A, B, C, D$ as in Exercise 33.2, drop a perpendicular $CE$ from $C$ to the line $AB$. Since $\alpha$ is an acute angle, $E$ will lie between $A$ and $B$. Now take a point $F$ further out on the ray $AC$. Drop a perpendicular $FG$ from $F$ to $AB$. Then $G$ is between $E$ and $B$. As the point $F$ moves out the ray $AC$ without bound, so the point $G$ must move along the ray $AE$ without bound, and thus it must eventually reach $B$. Then $F$ will be the intersection of $AC$ and $BD$.

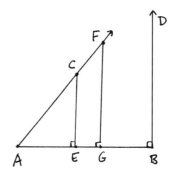

33.5 John Wallis (1616–1703) gave a proof of the parallel postulate based on the principle that to every figure there is always a similar figure of arbitrary size. To be precise, we state Wallis's axiom as follows:

**Wallis's Axiom**

Given a triangle $ABC$ and given a line segment $DE$, there exists a *similar* tri-

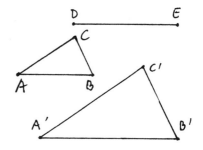

angle $A'B'C'$ (that is, a triangle with the same angles as the triangle $ABC$) having side $A'B' \geq DE$.

(a) Show that Wallis's axiom implies (P).

(b) In the non-Archimedean geometry of (18.4.3) show that there are similar triangles of different sizes, but that Wallis's axiom fails. (We will see later that in a semihyperbolic or semielliptic non-Euclidean geometry, the only similar triangles are congruent triangles (Exercise 34.4).)

33.6 In a Hilbert plane, show that opposite sides of a *rectangle* (i.e., a figure with four right angles) are equal. *Hint*: Bisect one side $AB$ at $E$, erect a perpendicular to $AB$ at $E$, and use the accompanying diagram. Your goal: to show $AB \cong CD$.

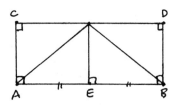

33.7 In this exercise we explore the consequences of Clavius's axiom.

(a) Let $l$ be a line, and let $m$ be a set of equidistant points, which by Clavius's axiom is a line. Thus for points $A, B, C$ in $m$, the perpendiculars $AA', BB', CC'$ to $l$ are all equal. Show that the angles at $A, B, C$ are also right angles.

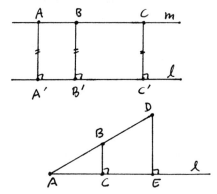

(b) Let $ABC$ be a right triangle. Extend $AB$ to $D$ so that $AB \cong BD$, and drop the perpendicular $DE$ to $AC$. Assuming Clavius's axiom, show that $DE \cong 2BC$.

(c) Show that Clavius's axiom, together with Archimedes' axiom (A), implies (P).

(d) Show that Clavius's axiom holds in the non-Archimedean plane of (18.4.3) even though (P) does not.

33.8 (a) Show that Aristotle's axiom holds in the Cartesian plane over a field $F$, even if $F$ is not Archimedean.

(b) Show that Aristotle's axiom fails in the plane of (18.4.3).

33.9 Show that Clairaut's axiom is equivalent to Clavius's axiom.

33.10 Show that Simson's axiom is equivalent to Clavius's axiom.

33.11 Farkas Bolyai, the father of János, proposed the following axiom.

**Bolyai's Axiom**
For any three noncollinear points $A, B, C$ there exists a circle containing them.

(a) Use the following construction to show that Bolyai's axiom implies Euclid's parallel postulate. Given two lines $l, m$ and a transversal $AB$, assume that the angles $\alpha, \beta$ on one side add up to less than two right angles. Let $C$ be the midpoint of $AB$. From $C$ drop perpendiculars to $l$ and $m$, and extend each an equal distance on the far side to obtain $D$ and $E$. Show that $C, D, E$ are not collinear, and then use Bolyai's axiom to prove that $l$ and $m$ must meet.

(b) Show that Bolyai's axiom holds in any Hilbert plane with (P).

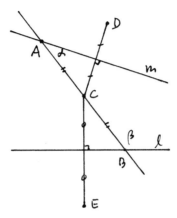

33.12 Dr. Anton Bischof in his thesis (1840) proposed to free the theory of parallels from its dependence on Euclid's parallel postulate by giving a different definition of parallel lines. Discuss his theory, which goes like this: Lines are parallel if they have the same direction.

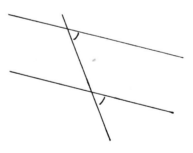

The direction of a line can be measured by the angle it makes with another line. So we define "parallelism is the equality of direction of similar lines against every other straight line." In other words, two lines are parallel if they make equal angles with every other line that meets them both.

Then it is clear that parallel lines cannot meet, because a transversal line through the point of intersection would make the same angle with both of them, so they would be equal. By the same reasoning it is clear that there can be only one parallel to a given line through a given point. If two lines make the same angle with a line that cuts them, they will be parallel. "Similarly one obtains all the other corollaries which one finds in all the textbooks."

33.13 Discuss the following "proof" that the sum of the angles of a triangle is equal to two right angles, independent of the theory of parallels, due to Thibaut (1775–1832):

Let $ABC$ be the given triangle. Take a segment $AD$ on the line $AC$, pointing away from $C$. Rotate it to the position $AE$ on the line $AB$. Then slide it along the line $AB$ into the position $BF$. Rotate to $BG$, slide to $CH$, rotate to $CI$, and slide back to $AD$. In this process, the segment $AD$ has made one complete rotation, which is 4 right angles. But the amount it has rotated is equal to the sum of the exterior angles $DAE, FBG$, and $HCI$. Replacing these by their supplementary angles, we find that the sum of the three interior angles of the triangle is equal to two right angles.

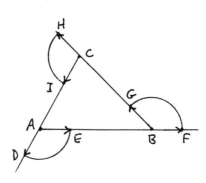

33.14 J.J. Callahan, then president of Duquesne University, in his book *Euclid or Einstein* (1931) claims to prove the parallel postulate of Euclid, and thus nullify the theories of Einstein based on non-Euclidean geometry. If you can locate a copy of his book, read his proof and find the flaw in his argument.

# 34  Neutral Geometry

Sir Henry Savile, in his public lectures on Euclid's *Elements* in Oxford in 1621, said, "In this most beautiful body of Geometry there are two moles, two blemishes, and no more, as far as I know, for whose removal and washing away, both older and more recent authors have shown much diligence." He was referring to the theory of parallels and the theory of proportion. Euclid's theory of proportion has been thoroughly vindicated, and receives its modern expression in the segment arithmetic that we have explained in Chapter 4.

The work on the theory of parallels, however, did not lead to the expected result. Instead of confirming Euclid's as the one true geometry, these researches showed that Euclid's was only one of many possible geometries. The others are what we now call non-Euclidean geometries. The story of this discovery is one of the most fascinating chapters in the history of mathematics, and has been amply told elsewhere. Here we will confine ourselves to the briefest outline.

We can distinguish four periods. The first, which we have elaborated in the previous section, might be called "dissatisfaction with Euclid." While fully accepting Euclid's *Elements* as the true geometry, critics said only that his treatment of this topic could have been better. So they tried to better Euclid, either by proving the parallel postulate, or by replacing it with some other more natural assumption.

The second period, exemplified by the work of Saccheri, Legendre, and Lambert, was based on the attitude, let us suppose the parallel postulate is false and see what conclusions we can draw. In this way they developed a collection of results that would be true if the parallel postulate were false, still expecting ultimately to find a contradiction and thus vindicate Euclid. So strong was the power of tradition that even after meticulously proving a whole series of propositions in this new geometry, each of these authors fell into error and deluded himself into thinking he had found a contradiction.

What a small step of the imagination, with what great consequences, was the transition to the third period! All it required was to think, yes it is possible to have a geometry in which the parallel postulate is false, and these are its first theorems. This step was taken independently by Carl Friedrich Gauss (1777–1855) in Germany, János Bolyai (1802–1860) in Hungary, and Nicolai Ivanovich Lobachevsky (1793–1856) in Russia. Although Gauss was the first to realize the existence of this new geometry, he published nothing of his researches, leaving Bolyai and Lobachevsky each to believe that he was the inventor of this new geometry. Bolyai exclaimed, in a letter to his father, "Out of nothing I have created a strange new universe."

The fourth period contains the confirmation of these new geometries by providing models for the axiom systems to show their consistency. This occurred only later, with the work of Beltrami, Klein, and Poincaré.

In this and the next section we will describe some work of the second period. Then in later sections we will give a model of the non-Euclidean geometry due to Poincaré, and a fuller axiomatic development of the theory, containing the results of Bolyai and Lobachevsky, in a logical framework provided by Hilbert.

A geometry satisfying Hilbert's axioms of incidence, betweenness, and congruence, in which we neither affirm nor deny the parallel axiom (P), will be called a *neutral geometry*. This is the same as a Hilbert plane, but the terminology emphasizes that we do not assume (P). Recall from Section 10 that the results of Euclid, Book I, up through (I.28), with the possible exception of (I.1) and (I.22), also hold in neutral geometry. A Hilbert plane in which (P) does not hold will be called a *non-Euclidean geometry*. We have already seen one example of a non-Euclidean geometry (18.4.3), but that one is semi-Euclidean, in the sense that the angle sum in a triangle is still equal to 2RA (two right angles) (Exercise 18.4). Now we will consider other geometries in which the angle sum of a triangle may be different from 2RA.

The results of this second period are mainly due to Girolamo Saccheri (1667–1733) and Adrien Marie Legendre (1752–1833). Saccheri's book *Euclides ab omni naevo vindicatus* was published in 1733. The title "Euclid freed of every blemish" recalls the quotation from Savile above. The first 32 propositions are a marvel of mathematical exposition. Unfortunately, after that his previously impeccable rigor lapses, and he says that he has proved the parallel postulate, because if it were false, there would be two lines having a common perpendicular at infinity, which is "repugnant to the nature of a straight line."

Saccheri's work was perhaps before its time, because it did not receive the recognition it deserved, and lay hidden in obscurity until the end of the nineteenth century. Essentially equivalent results were discovered independently half a century later by Legendre, whose book *Eléments de Géométrie* was first published in 1794. It was followed by many new editions, reprints, and translations, which had a wide influence on the teaching of geometry and revitalized interest in the question of parallels.

We start with a figure extensively studied by Saccheri, which goes back to Clavius, in his commentary on Euclid's (I.29), where he proposes the axiom that we discussed earlier (Section 33). Since it was Clavius's edition of Euclid that was recommended to Saccheri by the Jesuit mathematician Tommaso Ceva, we may assume that Saccheri was inspired by Clavius to study this figure further.

### Proposition 34.1

*In a Hilbert plane, suppose that two equal perpendiculars AC, BD stand at the ends of an interval AB, and we join CD. (This is called a* Saccheri quadrilateral.*) Then the angles at C and D are equal, and furthermore, the line joining the midpoints of AB and CD, the* midline, *is perpendicular to both.*

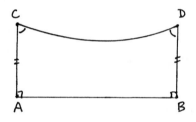

*Proof* Given *ABCD* as above, let *E* be the midpoint of *AB* and let *l* be the perpendicular to *AB* at *E*. Since *l* is the perpendicular bisector of *AB*, the points *A, C* lie on one side of *l*, while *B, D* lie on the other side. Hence *l* meets the segment *CD* in a point *F*. By (SAS) the triangles *AEF* and *BEF* are congruent. Hence the angles $\angle FAE$ and $\angle FBE$ are equal, and $AF = FB$.

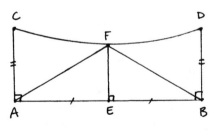

By subtraction from the right angles at *A* and *B* we find that the angles $\angle CAF$ and $\angle DBF$ are equal. So by (SAS) again, the triangles *CAF* and *DBF* are congruent. This shows that the angles at *C* and *D* are equal, and that *F* is the midpoint of *CD*.

The two pairs of congruent triangles also imply that the angles $\angle CFE$ and $\angle DFE$ are equal. So by definition, both of these angles are right angles.

**Remark 34.1.1**
From the equality of the angles at $C$ and $D$, Saccheri distinguished three cases, which he called the hypothesis of the acute angle, the hypothesis of the right angle, and the hypothesis of the obtuse angle, according to whether $C$ and $D$ were acute, right, or obtuse. He showed that if any one of these holds for one such quadrilateral, it holds for all. His proofs used continuity (in the form of the intermediate value theorem), but we will show in the following propositions that his result is also true in an arbitrary Hilbert plane.

**Proposition 34.2**
*Let ABCD be a quadrilateral with right angles at A and B, and unequal sides AC, BD. Then the angle at C is greater than the angle at D if and only if AC < BD.*

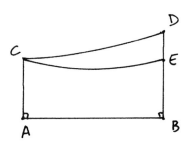

*Proof* Suppose $AC < BD$, and choose $E$ on $BD$ such that $AC = BE$. Then $ABCE$ is a Saccheri quadrilateral and $\angle ACE = \angle BEC$, by the previous proposition. Now, the angle $\angle ACD$ is bigger than $\angle ACE$, and $\angle BEC$ is bigger than the angle at $D$ by the exterior angle theorem (I.16), so we find that the angle at $C$ is bigger than the angle at $D$, as required.

On the other hand, if $AC > BD$, the same argument with roles reversed shows that the angle at $C$ is less than the angle at $D$. Hence we obtain the "if and only if" conclusion of the proposition.

**Proposition 34.3**
*Let ABCD be a Saccheri quadrilateral, let P be a point on the segment CD, and let PQ be the perpendicular to AB. Let α be the angle at C (equal to the angle at D).*

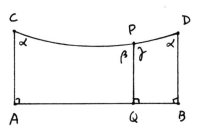

(a) *If PQ < BD, then α is acute.*
(b) *If PQ = BD, then α is right.*
(c) *If PQ > BD, then α is obtuse.*

*Proof* Let $\beta, \gamma$ be the two angles at $P$. In case (a), if $PQ < BD$, then $PQ < AC$ also, and from the previous proposition we obtain $\alpha < \beta$ and $\alpha < \gamma$. Hence $2\alpha < \beta + \gamma = 2RA$. Thus $\alpha$ is acute. The proofs of cases (b), (c) are analogous.

**Remark 34.3.1**
Once we have proved all three cases (a), (b), and (c), it follows that each one is an equivalence, not only an implication.

**Proposition 34.4**
*Again let ABCD be a Saccheri quadrilateral, but this time let P be a point on the line CD outside the interval CD. Let PQ be the perpendicular to the line AB, and let $\alpha$ be the angle at C (equal to the angle at D).*

(a) *If $PQ > BD$, then $\alpha$ is acute.*
(b) *If $PQ = BD$, then $\alpha$ is right.*
(c) *If $PQ < BD$, then $\alpha$ is obtuse.*

*Proof*  In case (a), assuming $PQ > BD$, choose $E$ in $PQ$ such that $BD = QE$. Draw $CE$ and $DE$. Then we have three Saccheri quadrilaterals. We will compare their angles. Let $\alpha, \beta, \gamma$ be the top angles of the quadrilaterals $ABCD$, $BQDE$, $AQCE$, respectively. Let $\delta = \angle EDP$. Then $\delta$ is an exterior angle of the triangle $CDE$, so by (I.16), $\delta > \angle DCE = \alpha - \gamma$. On the other hand, looking at the angles at $E$, we see that $\beta > \gamma$. Now, $2RA = \alpha + \beta + \delta > \alpha + \gamma + \alpha - \gamma = 2\alpha$, so $\alpha$ is acute.

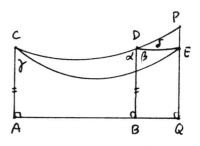

For case (b), when $PQ = BD$, then $AQCP$ is a Saccheri quadrilateral, so by (34.3b) its angle, which is equal to the angle of $ABCD$, is right.

In case (c), when $PQ < BD$, the proof is similar. Extend $PQ$ to $E$ with $BD = QE$ and join $CE, DE$. This gives three Saccheri quadrilaterals, with upper angles $\alpha, \beta, \gamma$ as marked. Let $\delta = \angle PDE$. Then by the exterior angle theorem (I.16), $\delta > \angle DCE = \gamma - \alpha$. Looking at $E$ we see that $\gamma > \beta$. On the other hand, looking at $D$ we see that $\alpha + \beta - \delta = 2RA$. So, combining these results, we obtain

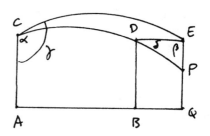

$$2RA = \alpha + \beta - \delta < \alpha + \gamma - \delta < 2\alpha.$$

Hence $\alpha$ is obtuse, as required.

**Remark 34.4.1**
As in the previous proposition, once we have proved all three cases, they each become equivalences, not just implications.

**Theorem 34.5** (Saccheri)
*In any Hilbert plane, if one Saccheri quadrilateral has acute angles, so do all Saccheri quadrilaterals. If one has right angles, so do they all. If one has obtuse angles, so do they all.*

*Proof*  We will give the proof only in the acute case, since the proofs in the two other cases are identical.

Suppose $ABCD$ is a Saccheri quadrilateral with acute angles, and let $EF$ be its midline (34.1). If $A'B'C'D'$ is another Saccheri quadrilateral with midline equal to $EF$, then it can be moved by a rigid motion to make the midlines coincide. Suppose $AB < A'B'$. We obtain a figure as shown, with $\alpha$ acute. Hence, by (34.4), $BD < B'D'$. Then by (34.3), $\alpha'$ is acute. If $AB > A'B'$, we run the same argument in the reverse order. It follows that all Saccheri quadrilaterals with midline equal to $EF$ have acute angles.

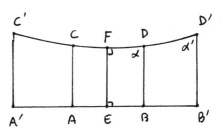

Next we show that for any other segment, there exists a Saccheri quadrilateral with acute angles and midline equal to that segment.

Lay off the given segment as $EG$ on the ray $EB$. Let the perpendicular to $AB$ at $G$ meet $CD$ in $H$. Reflect $G$ and $H$ in $EF$ to get $G_1, H_1$. Reflect $F$ and $H$ in $AB$ to get $F_2, H_2$. Now, $G_1GH_1H$ is a Saccheri quadrilateral with midline $EF$, so by the previous argument, its angle $\beta$ is acute. But then $FF_2HH_2$ is another Saccheri quadrilateral with the same acute angle $\beta$ and midline $EG$. Now by the earlier argument, every other Saccheri quadrilateral with midline equal to $EG$ has acute angles. But $EG$ was arbitrary, so the theorem is proved.

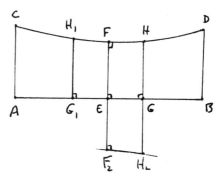

Next we will show how to interpret this result on Saccheri quadrilaterals in terms of the sum of the angles in a triangle.

## Proposition 34.6

*Given a triangle ABC, there is a Saccheri quadrilateral for which the sum of its two top angles is equal to the sum of the three angles of the triangle.*

*Proof*  Let $ABC$ be the given triangle. Let $D$ and $E$ be the midpoints of $AB$ and $AC$, and draw the line $DE$, which we call the *midline* of the triangle. Drop perpendiculars $BF, AG, CH$ to $DE$.

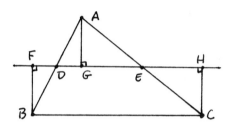

Now, $AD = DB$, and the vertical angles at $D$ are equal, so by (AAS) the triangles $ADG$ and $BDF$ are congruent. Similarly, $AE = EC$ and the vertical angles at $E$ are congruent, so the triangles $AEG$ and $CEH$ are congruent. From congruent triangles we obtain $BF = AG = CH$. The quadrilateral $FHBC$ has right angles at $F$ and $H$, so it is a Saccheri quadrilateral (upside down). The angles of the quadrilateral at $B$ and $C$ are composed of the angles of the triangle at $B$ and $C$, plus angles that are congruent to the two parts of the angle at $A$, divided by the line $AG$. Hence the angles at $B$ and $C$ of the quadrilateral equal the angle sum of the triangle. It follows that the triangle and the quadrilateral have equal defect.

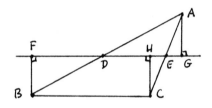

If $G$ happens to fall outside the interval $FH$, the same argument works, except that we use differences instead of sums of angles.

## Theorem 34.7

*In any Hilbert plane:*

(a) *If there exists a triangle whose angle sum is less than 2RA, then every triangle has angle sum less than 2RA.*

(b) *The following conditions are equivalent:*

    (i) *There exists a triangle with angle sum = 2RA.*
    (ii) *There exists a rectangle.*
    (iii) *Every triangle has angle sum = 2RA.*

(c) *If there exists a triangle whose angle sum is greater than 2RA, then every triangle has angle sum greater than 2RA.*

*Proof*   (a) If there exists a triangle with angle sum less than 2RA, then the associated Saccheri quadrilateral of (34.6) must have acute angles. By (34.5) it follows that every Saccheri quadrilateral has acute angles, and then by (34.6) again, every triangle must have angle sum less than 2RA.

The proof of (b) is the same, where we note that a rectangle is just the same thing as a Saccheri quadrilateral with right angles. The proof of (c) is the same as the proof of (a).

## Definition

In case (a) of the theorem, we say that the geometry is *semihyperbolic*. In case (b) we say that it is *semi-Euclidean*, and in case (c) we say that it is *semielliptic*.

## Remark 34.7.1

Note that these three cases are equivalent to what Saccheri called the hypothesis of the acute angle, the hypothesis of the right angle, and the hypothesis of the obtuse angle. Thus all Hilbert planes can be divided into these three classes. Of course, a Euclidean plane, or more generally any Hilbert plane satisfying (P), is semi-Euclidean, by (I.32). On the other hand, we have seen an example of a semi-Euclidean plane that does not satisfy (P) in Exercise 18.4.

We reserve the term *hyperbolic* for geometries satisfying Hilbert's hyperbolic axiom (cf. Section 40). Those geometries will be semihyperbolic, but there are also semihyperbolic geometries that are not hyperbolic (Exercise 39.24).

As for the semielliptic case, these were first discovered in 1900 by Dehn, who called them non-Legendrean. The term elliptic is usually applied to geometries like a projective plane in which there are no parallel lines at all. These do not satisfy Hilbert's axioms, so fall outside our realm of inquiry. However, a suitably small patch of a spherical geometry over a non-Archimedean field gives an example of a semielliptic Hilbert plane (Exercise 34.14).

## Definition

We say that a triangle is *Euclidean* if the sum of its angles is equal to 2RA. Otherwise, we call it *non-Euclidean*. To measure the divergence of a triangle from the Euclidean case, we define the *defect* of any triangle to be 2RA−(sum of angles in the triangle). Thus $\delta = 0$ for a Euclidean triangle, $\delta$ is a positive angle for a triangle in a semihyperbolic plane, and $\delta$ is the negative of an angle for a triangle in a semielliptic plane.

## Lemma 34.8

*If a triangle ABC is cut into two triangles by a single transversal BD, the defect of the big triangle is equal to the sum of the defects of the two small triangles:*

$$\delta(ABC) = \delta(ABD) + \delta(BCD).$$

*Proof*   Label the angles as shown in the diagram. Then

$$\delta(ABD) = 2RA - \alpha - \beta_1 - \delta_1,$$

$$\delta(BCD) = 2RA - \beta_2 - \delta_2 - \gamma.$$

Since $\delta_1 + \delta_2 = 2RA$, by adding we obtain

$$\delta(ABD) + \delta(BCD)$$

$$= 2RA - \alpha - \beta_1 - \beta_2 - \gamma = \delta(ABC),$$

as required.

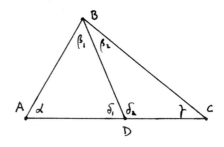

### The Theory of Parallels in Neutral Geometry

Given a line $l$ and a point $P$ not on $l$, we know from (I.31) that there exists a line through $P$ parallel to $l$. If the Hilbert plane satisfies Playfair's axiom (P), that parallel is unique. But in the non-Euclidean case, there may be more than one parallel to $l$ through $P$. Among all these parallels, there may be one that is closer to $l$ than all the others on one side. To make a formal definition, it matters which end of the line we look at, so we will phrase it in terms of rays.

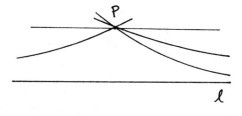

We denote a ray by the symbol $Aa$, where $A$ is its endpoint, and $a$ denotes the line carrying the ray, together with a choice of one of the two directions on

the line. Two rays are *coterminal* if they lie on the same line and "go in the same direction." This can be made precise by saying that one ray is a subset of the other. Thus if $Aa$ is a ray and $A'$ is another point on the line carrying $a$, we denote by $A'a$ the corresponding coterminal ray.

### Definition
A ray $Aa$ is *limiting parallel* to a ray $Bb$ if either they are coterminal, or if they lie on distinct lines not equal to the line $AB$, they do not meet, and every ray in the interior of the angle $BAa$ meets the ray $Bb$. In symbols we write $Aa \, ||| \, Bb$.

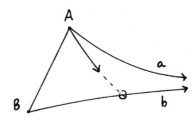

It requires some work, in the following propositions, to show that this notion is an equivalence relation. Note that we say nothing about the existence of such limiting parallels. All the following results should be understood in the sense that they hold whenever the limiting parallels exist. Later, in Section 40, we will introduce the hyperbolic axiom, which postulates the existence of limiting parallels from any point to any given ray.

### Proposition 34.9

*If $Aa \,\|\hspace{-2pt}\|\, Bb$, and if $A'a$, $B'b$ are rays coterminal to $Aa$, $Bb$ respectively, then $A'a \,\|\hspace{-2pt}\|\, B'b$.*

*Proof*  It is sufficient to replace one ray at a time by a coterminal ray. So first, suppose that $A'$ is on the ray $Aa$. We must show that every ray $n$ in the interior of the angle $BA'a$ meets the ray $Bb$. Take a point $P$ on the ray $n$, different from $A'$. Then the ray $\overrightarrow{AP}$ lies in the interior of the angle $BAa$, so by hypothesis it meets the ray $Bb$ in a point $C$. Now, the ray $n$ cuts one side of the triangle $ABC$, so by Pasch's axiom (B4) it must cut another. The side $AB$ is impossible, so $n$ meets $BC$, which is contained in the ray $Bb$, as required.

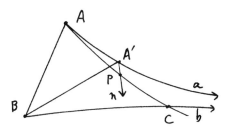

Next, suppose $A'$ is on the line $a$, but not in the ray $Aa$. Let $A'n$ be a ray in the angle $BA'a$, and take a point $P$ on the line $n$, but not in the ray $A'n$. Then the ray $\overrightarrow{PA}$, after it passes through $A$, is in the interior of the angle $BAa$, so meets $Bb$ in a point $C$. By the crossbar theorem (7.3) $A'n$ will meet $AB$, and then by Pasch's axiom it will meet $BC$.

If we replace $B$ by a point $B'$ in the ray $Bb$, or by a point $B''$ on the line $b$ outside the ray $Bb$, the proof is easier. Any ray from $A$ in the interior of the appropriate angle must meet the ray $B'b$ or $B''b$ either by the crossbar theorem or by the property $Aa \,\|\hspace{-2pt}\|\, Bb$.

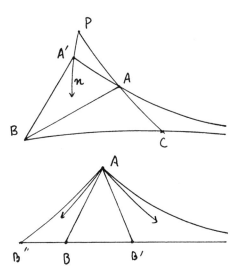

In this proof we passed over in silence a small point, namely to show that after replacing $Aa, Bb$ by coterminal rays $A'a, B'b$, we still have satisfied the condition that the rays $A'a$ and $B'b$ do not meet, and they lie on lines not equal to $A'B'$. For this it is sufficient to show that if $Aa \parallel\!\!\!\parallel Bb$, then the lines supporting those rays do not meet. We leave this as Exercise 34.6.

### Proposition 34.10
*If a ray $Aa$ is limiting parallel to another ray $Bb$, then also $Bb$ is limiting parallel to $Aa$.*

*Proof* If the rays are coterminal, this is trivial, so we may assume that $a$ and $b$ are distinct lines. Drop a perpendicular $AB'$ to the line $b$. Then by the previous proposition, $Aa$ is limiting parallel to $B'b$, and it will be sufficient to prove $B'b$ limiting parallel to $Aa$. In other words, changing notation, we may assume that the angle $ABb$ is a right angle.

We must show that any ray $Bn$ in the interior of the angle at $B$ meets the ray $Aa$. Suppose it does not. Drop the perpendicular $AC$ from $A$ to $n$. Since the angle $ABn$ is acute, by the exterior angle theorem, $C$ must lie on the ray $Bn$, not on the other side of $B$. In the triangle $ABC$, the angle at $C$ is right, while the other two angles are acute. Hence by (I.19), $AC < AB$. (Why is the angle at $A$ acute? Because it is less than the angle $BAa$, and this angle must be less than or equal to RA. Otherwise, the perpendicular to $BA$ at $A$ would lie inside the angle $BAa$ and be parallel to $Bb$, contradicting our hypothesis.)

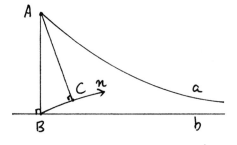

Rotate $C, n,$ and $a$ around the point $A$ until $C$ lands on a point $C'$ of $AB$, and $n', a'$ are the images of $n, a$. Then $Aa'$ will meet $Bb$, and $n'$ will be parallel to $b$, so by Pasch's axiom, it will meet $a'$. Rotating back, we find that $n$ meets $Aa$, a contradiction.

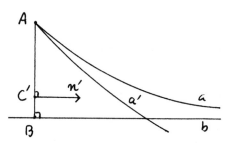

### Proposition 34.11
*Given three rays $Aa, Bb, Cc$, if $Aa \parallel\!\!\!\parallel Bb$ and $Bb \parallel\!\!\!\parallel Cc$, then $Aa \parallel\!\!\!\parallel Cc$.*

*Proof* If any two are coterminal, the result follows from the previous propositions, so we may assume that they lie on distinct lines.

**Lemma 34.12**

*Given three rays Aa, Bb, Cc lying on distinct lines, with Aa ‖‖ Bb and Bb ‖‖ Cc, after replacing one by a coterminal ray if necessary, we may assume that A, B, and C are collinear.*

*Proof*  If $A, C$ lie on opposite sides of the line $b$, then the segment $AC$ meets the line $b$ in a point $B'$. Replacing $Bb$ by the coterminal ray $B'b$, we have $A, B', C$ collinear.

If $A, C$ lie on the same side of the line $b$, we consider the angles $ABb$ and $CBb$. If these angles are equal, then $A, B, C$ are collinear. If they are not equal, one must be smaller, say $CBb$ is smaller. Then the ray $\overrightarrow{BC}$ is in the interior of the angle $ABb$, and $Bb \,\|\| Aa$ by (34.10), so the ray $\overrightarrow{BC}$ meets $Aa$ in a point $A'$. Replacing $Aa$ by $A'a$ we have $A', B, C$ collinear. If $ABb$ is smaller, the same argument works replacing $C$ by a point $C'$.

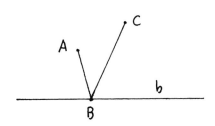

*Proof of (34.11), continued*  By the lemma, we may assume $A, B, C$ collinear. It follows immediately from the hypotheses that the rays $Aa, Bb, Cc$ are all on the same side of the line $ABC$.

*Case 1*  If $B$ is between $A$ and $C$, take any ray $An$ in the interior of the angle $CAa$. Since $Aa \,\|\| Bb$, this ray meets $Bb$ in a point $B'$. Then $B'b \,\|\| Cc$ by (34.9), so the continuation of that ray will meet $Cc$. Hence $Aa \,\|\| Cc$.

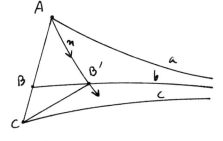

*Case 2*  $C$ is between $A$ and $B$. In this case a ray $An$ in the interior of the angle $CAa$ meets $b$ in a point $B'$. Then $Cc$ must meet $n$ by Pasch's axiom.

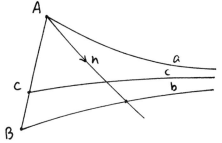

*Case 3*   *A* is between *C* and *B*. The proof is the same, taking into account $Cc \,|||\, Bb$ by (34.10).

### Remark 34.12.1
The proof of Case 2 of (34.11) actually shows a stronger result: If $Aa \,|||\, Bb$, if *C* is between *A* and *B*, and if *Cc* is any ray entirely in the interior of the angles *BAa* and *ABb*, then *Cc* is also limiting parallel to *Aa* and *Bb*.

### Corollary 34.13
*The relation "limiting parallel" for rays is an equivalence relation, which includes the equivalence relation of being coterminal. We define an* end *to be an equivalence class of limiting parallel rays.*

# Exercises

34.1  If *ABCD* is a Saccheri quadrilateral, show that $CD > AB$ if and only if the angles at *C, D* are acute.

34.2  Define a *Lambert quadrilateral* to be a quadrilateral *ABCD* with right angles at *A, B, C*. Show that the fourth angle *D* is acute, right, or obtuse according as the geometry is semihyperbolic, semi-Euclidean, or semielliptic.

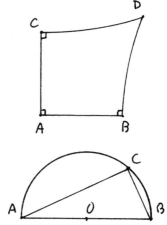

34.3  Let *AB* be the diameter of a circle, and let *ABC* be a triangle inscribed in the semicircle. Show that the angle at *C* is acute, right, or obtuse, according as the geometry is semihyperbolic, semi-Euclidean, or semielliptic.

34.4  In a semihyperbolic or a semielliptic plane, prove the (AAA) congruence theorem for triangles: If two triangles *ABC* and *A'B'C'* have $\angle A = \angle A'$, $\angle B = \angle B'$, $\angle C = \angle C'$, then the two triangles are congruent. (*Hint*: Use Lemma 34.8.)

34.5  In a semihyperbolic or a semielliptic plane, show that for any line *l* and any point *A* not on *l*, there are infinitely many lines through *A* parallel to *l*. (*Hint*: Use Saccheri quadrilaterals.)

34.6  In *Aa* and *Bb* are limiting parallel rays lying on distinct lines, show directly from the definition that the lines carrying these rays do not meet.

34.7 In a Hilbert plane satisfying Dedekind's axiom (D), show that for any point $A$ and any ray $Bb$, there exists a ray $Aa$ from $A$, limiting parallel to $Bb$.

34.8 In the Hilbert plane of (18.4.3) show that there do not exist any pairs of limiting parallel rays lying on distinct lines.

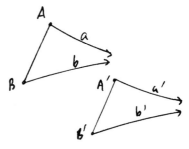

34.9 (ASAL) Given four rays $Aa$, $Bb$, $A'a'$, $B'b'$, assume that $\angle BAa = \angle B'A'a'$, $AB = A'B'$, and $\angle ABb = \angle A'B'b'$. Show that $Aa \parallel\!\parallel Bb$ if and only if $A'a' \parallel\!\parallel B'b'$.

34.10 (ASL) Given $Aa \parallel\!\parallel Bb$ and $A'a' \parallel\!\parallel B'b'$, assume $\angle BAa = \angle B'A'a'$ and $AB = A'B'$. Then $\angle ABb = \angle A'B'b'$. We call the figure consisting of the segment $AB$ and the two limiting parallel rays $Aa$ and $Bb$ a *limit triangle*.

34.11 Given a limit triangle $aABb$, construct its *midline* as follows. Let the angle bisectors at $A, B$ meet at a point $C$. Drop perpendiculars $CD, CE$ from $C$ to $a, b$. Join $DE$, and let $c$ be the perpendicular from $C$ to $DE$.

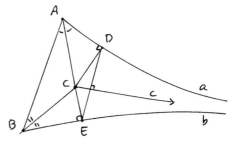

(a) Show that $Cc$ is limiting parallel to $Aa$ and $Bb$.

(b) Show that reflection in the line $c$ interchanges the lines $a$ and $b$. Thus $c$ plays a role for the rays $Aa$ and $Bb$ similar to the role of the angle bisector of an angle, which interchanges the two sides of an angle by reflection. So we can think of $C$ as the intersection of the three (generalized) angle bisectors of the limit triangle.

34.12 Show that the analogue of Pasch's axiom (B4) holds for a limit triangle $aABb$: If $l$ is a line that does not contain $A$ or $B$, and does not contain a ray limiting parallel to $Aa$ or $Bb$, and if $l$ meets one side $AB, Aa$, or $Bb$, then it must meet a second side, but not all three.

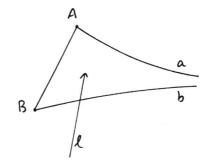

34.13 *Spherical geometry.* Let $F$ be a Euclidean ordered field. In the Cartesian 3-space over $F$ consider the sphere of radius $r$ given by the equation $x^2 + y^2 + z^2 = r^2$. We can make a geometry, called *spherical geometry*, as follows. Our *s-points* are the points of $F^3$ on the surface of the sphere. Our *s-lines* are *great circles* on the sphere, that is to say, the intersections of the sphere with planes of $F^3$ passing through the origin $O = (0,0,0)$. On any piece of an *s*-line that is less than half of a great circle, we can define betweenness by projecting the points from $O$ into any plane. We say that two segments of *s*-lines are congruent if the chords joining their endpoints, as line segments of $F^3$ inside the sphere, are congruent. We say that angles are congruent if the projected angles on the tangent planes to the sphere at their vertices are congruent.

Which of Hilbert's axioms hold in this geometry? You will see right away that (I1) fails and betweenness does not make very good sense, so it is not a Hilbert plane. Show, however, that the congruence axioms (C1)–(C6) and (ERM) do hold.

34.14 (a) Now suppose that we take $F$ to be a non-Archimedean field, such as the one in (18.4). Let $t$ be an infinite element in $F$, take the sphere of radius $t$, and take our geometry $\Pi_0$ to consist of only those points on the surface of the sphere that are at finitely bounded distance from a fixed point $A$ on the sphere. Show that this geometry satisfies all of Hilbert's axioms, so it is a Hilbert plane. Show also that the sum of the angles of any triangle in this geometry is *greater* than two right angles. This is an example of a semi-elliptic Hilbert plane.

(b) Again take $F$ to be a non-Archimedean field, and let $\Pi_1$ be the set of points on a sphere of radius 1 whose distance from a fixed point $A$ is infinitesimal. Show that $\Pi_1$ is another semielliptic Hilbert plane, and show that $\Pi_1$ is not isomorphic to the plane $\Pi_0$ of part (a). *Hint*: cf. Exercise 18.6.

34.15 In any Hilbert plane, show that the three angle bisectors of a triangle meet in a point.

34.16 In any Hilbert plane, if two of the perpendicular bisectors of the sides of a triangle meet, then all three perpendicular bisectors meet in the same point.

34.17 We say that two lines in a Hilbert plane are *strictly parallel* if every transversal line makes equal alternate interior angles. Show that the following conditions are equivalent:

 (i) The plane is semi-Euclidean.

 (ii) For every point $P$ and every line $l$, there exists a unique line $m$ through $P$ strictly parallel to $l$.

(iii) There exists at least one pair of distinct strictly parallel lines.

34.18 Show that strictly parallel lines (Exercise 34.17) behave in many of the same ways as parallel lines in Euclidean geometry:

(a) If $l$ is strictly parallel to $m$, and $m$ strictly parallel to $n$, then $l$ is strictly parallel to $n$ (analogue of (I.30)).

(b) If both pairs of opposite sides of a quadrilateral are strictly parallel, then opposite sides and opposite angles are equal (analogue of (I.34)).

34.19 In a semi-Euclidean plane, show that if two of the altitudes of a triangle meet, then all three altitudes meet in the same point.

34.20 In a semi-Euclidean plane, show that the medians of a triangle all meet in a point.

34.21 In any Hilbert plane, show that the line joining the midpoints of two sides of a triangle is orthogonal to the perpendicular bisector of the third side.

# 35    Archimedean Neutral Geometry

If we add Archimedes' axiom to the axioms of neutral geometry, we have the remarkable fact that the angle sum of a triangle is always less than or equal to two right angles. In other words, the semielliptic case is impossible. Saccheri's proof of this result uses a continuity argument, so we prefer the method of Legendre, using a repeated application of the construction Euclid used in (I.16) for the proof of the exterior angle theorem. In either case, the proof makes essential use of Archimedes' axiom. To begin with, we show that the analogue of Archimedes' axiom holds for angles.

**Lemma 35.1**

*In a Hilbert plane with* (A), *let* $\alpha, \beta$ *be given angles. Then there exists an integer* $n > 0$ *such that* $n\alpha > \beta$, *or else* $n\alpha$ *becomes undefined by exceeding* 2RA.

*Proof*    First we make a reduction. Given the angle $\beta$ at $O$, measure off equal segments $OA$ and $OB$ on the two arms, and draw $AB$. The line $OC$ joining $O$ to the midpoint of $AB$ will bisect the angle $\beta$ and will make a right angle at $C$.

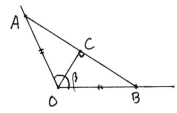

Since it is just as good to prove the lemma for $\frac{1}{2}\beta$, we reduce to studying the case of an angle contained in a right triangle.

So now let $OAB$ be a right triangle with the angle $\beta$ at $O$ and a right angle at $A$. Suppose, by way of contradiction, that $n\alpha \leq \beta$ for all $n$. Lay off the angle $\alpha$ inside the triangle, and let that angle cut off a segment $AA_1$ on the line $AB$. Again lay off the angle $\alpha$ at $O$ to cut a segment $A_1A_2$ on the line $AB$. Continuing in this manner, we obtain a sequence of points $A_1, A_2, A_3, \ldots$ on $AB$, with each successive segment $A_iA_{i+1}$ subtending an angle $\alpha$ to $O$.

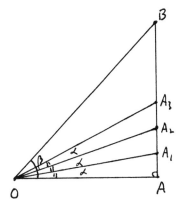

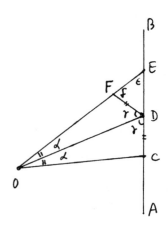

I claim that $AA_1 < A_1A_2 < A_2A_3 < \cdots$. Consider any three consecutive $A_i$ and call them $C, D, E$. Let $DF$ be drawn such that $\angle CDO = \angle FDO$. Then the triangles $\triangle CDO$ and $\triangle FDO$ are congruent by (ASA), and hence $CD = DF$. Now, the angle $\delta = \angle EFD$ is an exterior angle to the triangle $\triangle FDO$, and so $\delta > \gamma = \angle FDO = \angle CDO$, by (I.16). On the other hand, $\gamma = \angle CDO$ is an exterior angle to the triangle $\triangle DEO$, and so $\gamma > \varepsilon = \angle DEO$, by (I.16). We conclude that $\delta > \varepsilon$. Then by (I.19), the larger angle subtends the larger side, so $DE > DF = CD$, as required.

By the way, in order for the drawing to be accurate, i.e., for $F$ to lie in the segment $OE$, we were tacitly assuming that $\gamma$ is acute. This is true for the first triangle $OAA_1$, and follows inductively for the rest from (I.16).

Now, the axiom of Archimedes (A) implies that there is some integer $n$ such that $n \cdot AA_1 > AB$. Since the successive segments $A_1A_2, A_2A_3, \ldots$ are each bigger than the one before, it follows for a stronger reason that $AA_n > AB$. But this contradicts the supposition that all the points $A_i$ were in the segment $AB$, and thus proves the result.

**Theorem 35.2** (Saccheri–Legendre)
*In a Hilbert plane with Archimedes' axiom* (A), *the sum of the angles of a triangle is less than or equal to two right angles.*

*Proof*   Suppose to the contrary that there is a triangle $\triangle ABC$ whose angle sum is greater than two right angles, say two right angles plus $\varepsilon$, where $\varepsilon$ is some non-zero angle. Then we will get a contradiction by replacing the triangle $\triangle ABC$ by another triangle $\triangle DEF$, which has the same angle sum as $\triangle ABC$, and furthermore has one angle $\alpha$ very small, less than $\varepsilon$. Then the remaining two angles will be more than two right angles, which contradicts (I.17).

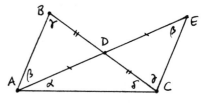

As a first step we show, given any triangle $\triangle ABC$, and having chosen one of its angles, say $\angle A$, that there is another triangle $\triangle AEC$ having the same angle sum as $\triangle ABC$, and one of whose angles is less than or equal to $\frac{1}{2}$(angle A).

We use the construction of (I.16). Let $D$ bisect $BC$, draw $AD$ and extend that line to a point $E$ such that $AD \cong DE$, and draw $EC$. Now the vertical angles at $D$ are equal (I.15) and $BD \cong DC$ and $AD \cong DE$ by construction, so by (SAS) the triangles $\triangle ABD$ and $\triangle ECD$ are congruent. In particular, angle $\beta = \angle BAD$ is equal to $\angle DEC$. Also, $\gamma = \angle ABD = \angle DCE$.

Now $\triangle ABC$ and $\triangle AEC$ both have the same angle sum, which is equal to $\alpha + \beta + \gamma + \delta$, where $\alpha = \angle DAC$ and $\gamma = \angle DCA$. On the other hand, the angle at $A$ is $\alpha + \beta$, so of the two angles $\alpha, \beta$ in the new triangle, one of them satisfies $\alpha$ (or $\beta$) $\leq \frac{1}{2}$(angle $A$).

Now let us go back to our original triangle $\triangle ABC$ having angle sum equal to two right angles plus $\varepsilon$. By applying the above process each time to the smallest angle of the preceding triangle, we can obtain a sequence of triangles

$$T_0 = \triangle ABC, T_1, T_2, T_3, \ldots$$

each of which has angle sum the same as $\triangle ABC$, and where $T_n$ has one angle less than or equal to $1/2^n(\angle A)$. Now using Archimedes' principle for angles (35.1) we see that for some $n$, $T_n$ will have one angle less than $\varepsilon$, which gives the desired contradiction to (I.17).

### Remark 35.2.1
In the above construction, two angles of the triangle $T_{n+1}$ are less than the chosen angle in triangle $T_n$. Thus $T_{n+1}$ has two angles less than or equal to $(1/2^n)(\angle A)$.

### Corollary 35.3
*In any triangle, the exterior angle is greater than or equal to the sum of the opposite interior angles.*

*Proof* Indeed, the exterior angle plus the third interior angle is equal to two right angles. Since the sum of all the angles is less than or equal to 2 right angles, the sum of the two opposite angles is less than or equal to the exterior angle.

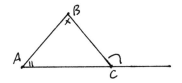

From the preceding theorem of Saccheri and Legendre, we see that the semi-elliptic case is impossible in an Archimedean Hilbert plane. Now we will show more, namely, if an Archimedean Hilbert plane is semi-Euclidean, then already (P) holds.

### Proposition 35.4
*In a Hilbert plane with* (A), *if every triangle is Euclidean, then* (P) *holds.*

*Proof*  We will prove the contrapositive, namely, if not (P), i.e., if there is a line $l$ and a point $P$ not on $l$ through which there are two or more lines parallel to $l$, then there exists a non-Euclidean triangle.

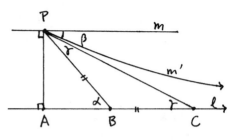

Given such a line $l$ and point $P$, drop a perpendicular from $P$ to $l$ at $A$, and let $m$ be the line through $P$ perpendicular to that perpendicular. Then $m$ is parallel to $l$ (I.27). Now let $m'$ be another line through $P$ parallel to $l$. Then on one side of $AP$, the ray $m'$ will be inside the right angle formed by $AP$ and $m$. Let it form an angle $\beta$ with $m$ at $P$.

Now take any point $B$ on $l$ on the same side of $AP$ as $m'$. Let the angle $\angle PBA$ be $\alpha$. Choose $C$ on $l$ such that $BC \cong PB$, and draw $PC$. Then we have an isosceles triangle $BPC$, so the angles $\gamma$ shown are equal. But $\alpha$ is an exterior angle to the triangle $BPC$, so $\alpha \geq 2\gamma$ by (35.3). Repeating this process $n$ times, we can find a point $F$ further out on the ray $l$ such that the angle $PFA$ is less than or equal to $(1/2^n)\alpha$. By (35.1) we can take $n$ large enough so that $\angle PFA < \beta$.

Consider the triangle $PFA$. It has one right angle, one angle less than $\beta$, and one angle, $\angle APF$, that is less than $RA - \beta$, because the ray $m'$, being parallel to $l$, cannot lie inside the triangle $PFA$. Thus the angle sum of the triangle $PFA$ is strictly less than $2RA$, and this proves the result.

Now we can explain how Legendre thought to prove the parallel postulate. He suggested another postulate, which we call Legendre's axiom.

### Legendre's Axiom
Given an angle $\alpha$ and given a point $P$ in the interior of the angle $\alpha$, there exists a line $l$ through $P$ that meets both sides of the angle.

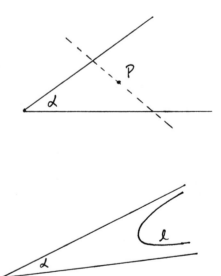

He shows that this implies the parallel postulate (see (35.5) below). While this axiom seems reasonable enough in itself, he shows further that if it fails, then for any angle $\alpha$, however small, there would have to be a line $l$ entirely contained inside the angle (Exercise 35.4).

Then he says "or, il répugne à la nature de la ligne droite"—it is repugnant to the nature of a straight line—that such a line should be entirely contained within an angle!

pouvait être nié. (1) C'est cette considération qui nous a fait revenir dans la 9ème édition, à la simple marche d'Euclide, en renvoyant aux notes pour la démonstration rigoureuse.

En examinant les choses avec plus d'attention nous sommes resté convaincu que pour démontrer complètement notre *postulatum* il fallait déduire de la définition de la ligne droite une propriété caractéristique de cette ligne qui exclût toute ressemblance avec la forme d'une hyperbole comprise entre ses deux asymptotes. Voici quel est à cet égard le résultat de nos recherches.

*Soit* BAC *un angle donné, et* M *un point donné au dedans* fig. 274. *de cet angle; divisez l'angle* BAC *en deux également par la droite* AD *et du point* M *menez* MP *perpendiculaire sur* AD : *je dis que la droite* MP *prolongée dans un sens et dans l'autre, rencontrera nécessairement les deux côtés de l'angle* BAC.

Car si elle rencontre un des côtés de cet angle, elle rencontrera l'autre, tout étant égal des deux côtés à partir du point P; si elle ne rencontrait pas un côté, elle ne rencontrerait pas l'autre par la même raison; ainsi, dans ce dernier cas elle devrait être renfermée tout entière dans l'espace compris entre les côtés de l'angle BAC; or, il répugne à la nature de la ligne droite qu'une telle ligne, indéfiniment prolongée, puisse être renfermée dans un angle.

En effet, toute ligne droite AB tracée sur un plan, et in- fig. 275. définiment prolongée dans les deux sens, divise ce plan en deux parties qui étant superposées coïncident dans toute leur étendue et sont parfaitement égales. La partie AMB du plan total, située d'un côté de AB, est égale en tout à la partie AM ~~~~ ~~ l'autre côté; car si l'on prend un point

(1) On voit dans un article du *Philosophical magazine* de mars 1822, qu'un savant géomètre a essayé de perfectionner cette démonstration et de la rendre indépendante de tout *postulatum* ; mais la construction employée pour démontrer la seconde partie consiste à mener d'un point donné différentes droites à tous les sommets d'une ligne qu'on doit considérer comme polygonale, pour raisonner dans l'hypothèse de celui qui nie la proposition : or la convexité de cette ligne, si elle avait lieu, ne permettrait pas de continuer indéfiniment la construction de l'auteur, comme il le faudrait pour l'exactitude de sa démonstration.

Plate XII. Legendre's "proof" that through any point in the interior of an angle there is a straight line meeting both sides. From his *Eléments de Géométrie* (1823).

**Proposition 35.5**

*In a Hilbert plane with* (A), *if Legendre's axiom holds for a single angle* α, *then also* (P) *holds.*

*Proof* Let the vertex of the angle be $A$, take any two points $B, C$ on the sides of the angle, and draw $BC$.

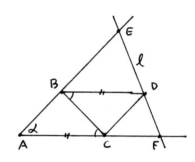

Now repeat the angle $ACB$ at $B$, lay off $BD = AC$, and join $CD$. Then by (SAS) the new triangle $BCD$ is congruent to the triangle $ABC$. Applying Legendre's axiom for the angle α to the point $D$, let $l$ be a line through $D$ meeting the sides of the angle α at $E$ and $F$. By (I.27), $AB$ is parallel to $CD$ and $BD$ is parallel to $AC$. This implies that $E$ and $F$ must lie beyond $B$ and $C$, so that the new triangle $AEF$ contains both the triangles $ABC$ and $BCD$.

Suppose that (P) does not hold. Then by (35.4) and (34.7) the triangle $ABC$ has a positive defect $\delta > 0$. By the additivity of defect (34.8), the new triangle $AEF$ will have defect greater than $2\delta$. Repeating this process and using (35.1) we would eventually have a triangle with defect greater than two right angles, which is absurd. Therefore, (P) must hold.

**Proposition 35.6**

*In a Hilbert plane with* (A), *Aristotle's axiom holds, namely, given any acute angle, the perpendicular from a point on one arm of the angle to the other arm can be made to exceed any given segment.*

*Proof* Using (A), it will be sufficient to show that if $BC$ is one perpendicular from one arm of the angle to the other, than there exists another such perpendicular with $DE \geq 2BC$. To do this, we proceed as follows. First mark off $BD = AB$, and drop the perpendicular $DE$. Then extend $BC$ and drop a perpendicular $DF$ to the extended line. The vertical angles at $B$ are equal, so by (AAS), the triangles $ABC$ and $DBF$ are congruent, and it follows that $CF = 2BC$.

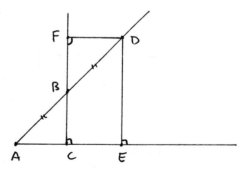

Since we have assumed (A), the angle at $D$ of the quadrilateral $FCDE$ must be acute, by (35.2), (34.7), and Exercise 34.2. Then by (34.2), $DE \geq FC = 2BC$, as required.

# Exercises

Unless specified otherwise, the following problems take place in a non-Euclidean (i.e., (P) is false) Hilbert plane satisfying Archimedes' axiom (A).

35.1 Prove: Given any angle $\varepsilon > 0$, there exists a triangle with defect $\delta < \varepsilon$.

35.2 Discuss the following "proof," due to Legendre, that the angle sum of a triangle is two right angles: We have seen that for any triangle $ABC$, there is a triangle $T_n$ with the same angle sum as $ABC$, and where $T_n$ has two angles less than or equal to $(1/2^n)(\angle A)$ (35.2.1). In the limit, the two small angles will become zero, so the triangle becomes a straight line, and the third angle will be 2RA. Thus the angle sum of the original triangle must be 2RA.

35.3 Given an angle $aAb$, show that there exists a point $B$ in the ray $Ab$ such that the perpendicular to $b$ at $B$ does not meet $a$.

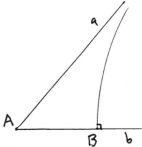

35.4 Show that for any angle $\alpha$, however small, there exists a line $l$ entirely contained in the inside of the angle. *Hint*: Apply Exercise 35.3 to the angle bisector of $\alpha$.

35.5 Given an angle $\alpha$ with vertex $A$, and given a ray $l$ inside $\alpha$, show that there exists a point $P$ on the ray $l$ such that for any two points $B, C$ on the two arms of $\alpha$, the line $BC$ meets the ray $l$ inside the interval $AP$.

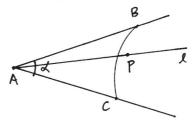

35.6 Given an angle $\varepsilon > 0$, show that there exists a triangle with angles $\alpha, \beta, \gamma$, all three smaller than $\varepsilon$. *Hint*: Use Exercise 35.4.

35.7 Show that Lemma 35.1 is false in the Cartesian plane over a non-Archimedean field.

35.8 If $Aa \,\|\|\, Bb$, show that the perpendicular distance from a point $P \in a$ to the line $b$ is strictly decreasing as $P$ moves away from $A$, e.g., in the diagram $PQ > P'Q'$.

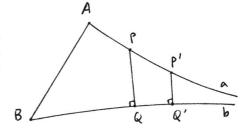

35.9  Prove the converse of Exercise 35.8, namely, if $Aa$ and $Bb$ are two rays that do not meet on the same side of the line $AB$, and if the perpendicular distance from a point $P \in a$ to $b$ is strictly decreasing as $P$ moves along $a$ away from $A$, then $Aa$ is limiting parallel to $Bb$. *Hint*: Use Proposition 35.6.

# 36  Non-Euclidean Area

The discussion of area arose quite early in the development of non-Euclidean geometry. Gauss had already thought about area as early as 1799 (see Exercise 36.1), and Bolyai's fundamental treatise of 1832 has a number of results on area. The defect of a triangle, as we will see, provides a natural measure of area, which is *absolute* in that it does not depend on any arbitrary choice of unit interval, as is the case for the Euclidean area function.

To make our treatment rigorous, we will follow Hilbert's method, which we used already for Euclidean area in Sections 22, 23. So recall that in any Hilbert plane, two figures are called *equidecomposable* if they can be written as the unions of congruent triangles. Two figures have *equal content* if one can add to them equidecomposable figures so that the whole becomes equidecomposable. We showed in Section 22 that both of these notions are equivalence relations.

In Section 23 we defined the notion of a measure of area function. In most texts this is taken to have values in the real numbers, but true to our principle of not imposing the real numbers on geometry, we prefer to have it take values in a group that arises naturally. In the Euclidean case (23.2) we used the additive group of the field of segment arithmetic. For the non-Euclidean case, we will use an ordered abelian group (Section 23) whose elements are constructed out of finite sums of angles.

To be precise, we proceed as follows. Recall first that in our development of a Hilbert plane, an *angle* is simply two rays, emanating from a point, that do not lie on the same line. Thus there is no zero angle, and every angle is less than two right angles. We have defined addition of angles only when the sum is less than 2RA (cf. Section 9). We define a set

$$A = \{0\} \cup \{\text{angles less than RA}\},$$

and we take

$$G = \mathbb{Z} \times A$$

to be the direct product set. Define addition on $G$ by

$$(n_1, \alpha_1) + (n_2, \alpha_2) = \begin{cases} (n_1 + n_2, \alpha_1 + \alpha_2) & \text{if } \alpha_1 + \alpha_2 < \text{RA}, \\ (n_1 + n_2 + 1, \alpha_1 + \alpha_2 - \text{RA}) & \text{if } \alpha_1 + \alpha_2 \geq \text{RA}. \end{cases}$$

In this definition, if either $\alpha_1$ or $\alpha_2$ is the symbol 0, we interpret it by setting $0 + \alpha_2 = \alpha_2$, and $0 + 0 = 0$, etc. Define the lexicographic order on the set $G$ by setting

$$(n_1, \alpha_1) < (n_2, \alpha_2)$$

if either $n_1 < n_2$, or $n_1 = n_2$ and $\alpha_1 < \alpha_2$.

### Proposition 36.1

*In any Hilbert plane, the set $G$ defined above, with the operation $+$ and the relation $<$, is an ordered abelian group.*

*Proof*  This is pretty much obvious, using the known properties of angles from Section 9. The element $(0,0)$ acts as zero element for the group. The inverse of an element $(n, \alpha)$ is $(-n, 0)$ if $\alpha = 0$; otherwise, it is $(-n-1, \mathrm{RA} - \alpha)$. The trichotomy for the order relation follows from the corresponding fact for angles (9.5).

### Remark 36.1.1

There is a natural homomorphism of the group $G$ to the group $R$ of rotations around a fixed point $O$ of the plane, defined as follows. Fix a ray $OA$ emanating from $O$, and let the ray $OC$ make a right angle with $OA$. For any angle $\alpha < \mathrm{RA}$, choose a ray $OB$ inside the right angle to represent $\alpha$. Now send $(1, 0) \in G$ to the rotation that sends $OA$ to $OC$; send $(0, \alpha) \in G$ to the rotation that sends $OA$ to $OB$, and extend by linearity 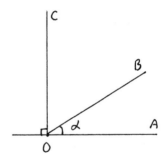 (cf. Exercise 17.4 for facts about rotations). This homomorphism is surjective, with kernel the subgroup $\mathbb{Z}$ generated by $(4, 0)$. Thus the elements of $G$ correspond to "rotations with winding number." We will call $G$ the *unwound circle group* of the given Hilbert plane (cf. Exercise 17.6).

### Example 36.1.2

In the real Cartesian plane, the group $G$ is isomorphic to $(\mathbb{R}, +)$, by sending $(1, 0)$ to $\pi/2$ and $(0, \alpha)$ to the radian measure of $\alpha$.

Now that we have a suitable group for it to take values in, we can show the existence of a measure of area function in non-Euclidean geometry. Recall (Section 34) that the *defect* of a triangle is 2RA minus the sum of the three angles of the triangle. For any angle $\delta$, we identify it with $(0, \delta) \in G$ if $\delta < \mathrm{RA}$, or $(1, \delta - \mathrm{RA}) \in G$ if $\delta \geq \mathrm{RA}$.

**Theorem 36.2**

*In a semihyperbolic Hilbert plane there is a measure of area function α with values in the unwound circle group G of the plane. It is uniquely determined by the additional condition that for any triangle T, its value is equal to the defect δ of the triangle.*

*Proof*   For any figure $P$, we write $P$ as a union of triangles $T_i$, and define $\alpha(P) = \Sigma\delta(T_i)$, where $\delta$ denotes the defect of a triangle and the sum is taken in the group $G$. Because of the semihyperbolic hypothesis this will be a positive element of the group $G$. It gives the same value for congruent triangles, and it is additive for nonoverlapping figures, so the only problem is to show that it is well-defined. Any two triangulations of a figure can be refined by a third, so as in (23.5), we need only show that $\alpha(P)$ is additive for an arbitrary subdivision of a triangle into smaller triangles. The case of a triangle cut in two by a single transversal is given in (34.8). This corresponds to Step 1 in the proof of (23.4). The remaining steps of the proof of (23.4) are valid in our case also, and thus $\alpha(P)$ is well-defined.

**Remark 36.2.1**

It now follows from (23.1) that equidecomposable figures and figures of equal content have the same area. The property (Z) of Section 22, de Zolt's axiom, also holds.

**Remark 36.2.2**

We will see later (Exercise 42.10) that in a hyperbolic plane, for any angle $\delta$, there exists a triangle with area $= \delta$, so that the image of the measure of area function $\alpha$ is just the set of positive elements of $G$.

To illustrate the theory of area, we will give the neutral geometry analogue of Euclid's (I.37), that "triangles on the same base and in the same parallels are equal." First we compare a triangle to a Saccheri quadrilateral.

**Proposition 36.3**

*In a Hilbert plane, any triangle has equal content to a suitable Saccheri quadrilateral. If furthermore we assume (A), the two are equidecomposable.*

*Proof*   We use the method of (34.6). The construction given there shows that the triangle $ABC$ is equidecomposable with the Saccheri quadrilateral $FHBC$ if $G$ lies between $F$ and $H$. If $G$ lies outside the interval $FH$, the proof given shows that they have equal content.

Now let us assume (A), and let $ABC$ be any given triangle, with associated quadrilateral $FHBC$. Join $CD$ and extend to $A'$ so that $CD = DA'$. Join $A'B$. Then by (SAS) the triangles $ACD$ and $BA'D$ are congruent. Leaving $BCD$ fixed, this gives a dissection of $ABC$ into $A'BC$. Furthermore, they have the same midline, and the same Saccheri quadrilateral, and $D'D = DE = \frac{1}{2}FH$ (Exercise 36.6a).

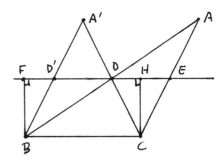

Repeating this process a finite number of times, and using (A), we transform the original triangle $ABC$ by a finite succession of dissections into a new triangle $A^*BC$ such that the foot $G^*$ of the perpendicular from $A^*$ to $FH$ lies between $F$ and $H$. Then the method of (34.6) gives a dissection to the Saccheri quadrilateral.

### Corollary 36.4
*Triangles on the same base and with the same midline (cf. proof of (34.6)) have equal content. Furthermore, assuming (A), they are equidecomposable.*

*Proof* Let $ABC$ and $A'BC$ be two triangles with the same base $BC$ and the same midline $l$. Drop perpendiculars $BF$, $CH$ to $l$. Then by the proposition, both triangles have content equal to the Saccheri quadrilateral $FHBC$, so they have equal content to each other (22.3). In the Archimedean case, they are equidecomposable.

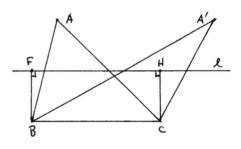

### Remark 36.4.1
Note how the Euclidean hypothesis "lying in the same parallels" of (I.37) has been replaced by "having the same midline." Of course, in the Euclidean case this is equivalent, because $A$ and $A'$ will lie on the same line parallel to $BC$. But in non-Euclidean geometry, the locus of points $A$ for which the triangle $ABC$ has midline $l$ is not a straight line: It is the set of points having the same distance from $l$ as $B$ and $C$, but on the opposite side of the line $l$. Given a line $l$ and a segment $d$, we call the set of points at distance $d$ from $l$ an *equidistant curve* (or *hypercycle*). It has two branches, one on either side of $l$. So we could rephrase this result as "Triangles on the same base $BC$, and having all three vertices $A, B, C$ on the same equidistant curve, $B, C$ on one branch, $A$ on the other, have equal content."

Using ideas from the above proofs, we can now establish the non-Euclidean analogue of the theorem of Bolyai–Gerwien (24.7). We start with the case of triangles.

**Proposition 36.5**
*In a semihyperbolic Hilbert plane satisfying* (E), *any two triangles of equal area have equal content. If furthermore we assume* (A), *they are equidecomposable.*

*Proof*  We use the area function of (36.2), so that equal area means equal defect. First let us consider the case where one side of the first triangle is equal to one side of the second triangle. We label the equal sides $BC$, so we can call the triangles $ABC$ and $A'BC$. By (36.3) we can find Saccheri quadrilaterals $FHBC$ and $F'H'BC$ that have equal content with the given triangles. Furthermore, the sum of the two angles of these quadrilaterals at $B, C$ are equal to the angle sums of the triangles (34.6), hence equal to each other. Since the top angles of a Saccheri quadrilateral are equal to each other (34.1), it follows that the angles at $B$ and $C$ of the two quadrilaterals are the same, so the sides of the quadrilaterals lie on the same lines.

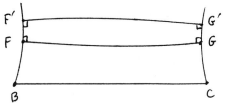

If the line $FG$ is not equal to the line $F'G'$, then $FGF'H'$ is a rectangle, which is impossible in the semihyperbolic case (34.7). So $FG = F'G'$, the two Saccheri quadrilaterals are equal, and the two triangles have equal content.

Now consider the case of two arbitrary triangles $ABC$ and $A'B'C'$. Suppose $AB < A'B'$. Then $\frac{1}{2}A'B' > BD$, so using the axiom (E) we can find a point $D^*$ on the midline $DE$ such that $BD^* = \frac{1}{2}A'B'$. Extend $BD^*$ to $A^*$ such that $BD^* = D^*A^*$. Join $A^*C$.

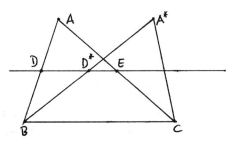

Then $ABC$ and $A^*BC$ are triangles with the same base $BC$ and the same midline. Therefore, they have equal content (36.4), and also they have the same defect. On the other hand, $A^*BC$ and $A'B'C'$ have one side equal $A^*B = A'B'$, and they have the same defect, so by the previous case, they have equal content.

If we assume (A), the same proof works with equal content replaced by equidecomposable.

## Theorem 36.6

*In a semihyperbolic Hilbert plane with* (E), *any two rectilineal figures with the same area have equal content. If furthermore we assume* (A), *they are equidecomposable.*

*Proof* Suppose rectilineal figures $P$ and $P'$ have the same area (36.2). Each one can be written as a finite union of triangles. Subdividing some triangles if necessary to increase the total number, we may assume that for some $n$, both $P$ and $P'$ are subdivided into exactly $2^n$ triangles. Let $P = \bigcup_{i=1}^{2^n} T_i$. Taking the triangles two at a time, and using the lemma below, we can find other triangles $T'_j, j = 1, \ldots, 2^{n-1}$, such that $P$ has equal content with $2\bigcup_{i=1}^{2^{n-1}} T'_j$. Repeating this process $n$ times, eventually we find a single triangle $T_0$ such that $P$ has equal content with $2^n T_0$.

Do the same with $P'$. Then there is a triangle $T'_0$ such that $P'$ has equal content with $2^n T'_0$.

Since $P$ and $P'$ have the same area by hypothesis, it follows that $T_0$ and $T'_0$ have the same area. Then by (36.5), $T_0$ and $T'_0$ have equal content. Hence $P$ and $P'$ have equal content.

If we assume (A), the same proof works for equidecomposable.

## Lemma 36.7

*In any Hilbert plane with* (E), *if $T_1$ and $T_2$ are two triangles, then there exists a triangle $T'$ such that $T_1 \cup T_2$ has equal content with $2T'$ (or, in case we assume* (A), *$T_1 \cup T_2$ is equidecomposable with $2T'$).*

*Proof* By the method of proof of (36.5), we can replace one of the triangles by another, so that now the two triangles have a side in common (and for this step we use (E)). Next we will show that we can replace each triangle by an isosceles triangle with the same base.

Given a triangle $ABC$, consider the associated Saccheri quadrilateral $FHBC$, and let $KL$ be the line joining the midpoints of the top and bottom (34.1). Choose $D'$ on $KF$ such that $KD' = \frac{1}{2}DE$. Then extend $BD'$ to $A'$ such that $BD' = D'A'$, and join $A'C$. As we have seen above, $ABC$ and $A'BC$ have equal content (or assuming (A), are equidecomposable). But furthermore, by construction, $A'BC$ is isosceles (Exercise 36.6b).

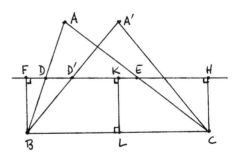

Applying this construction to both $T_1$ and $T_2$, we can assume that they are both isosceles on the same base. We put them together along their common base (say $T_1 = ABC, T_2 = A'BC$). Then the line $AA'$ divides $T_1 \cup T_2$ into two congruent triangles, so we take $T' = ABA'$. Then $T_1 \cup T_2$ is equidecomposable with $2T'$.

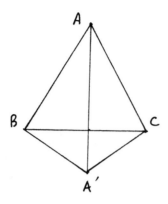

**Remark 36.7.1**

It follows from (36.6) that if two figures in a semihyperbolic Hilbert plane with (A) and (E) have equal content, they are equidecomposable — cf. (22.1.4). Indeed, they will have the same area (23.1), using the area function of (36.2), and then the theorem applies.

# Exercises

36.1 Gauss, in a letter to W. (= Farkas) Bolyai in 1799, wrote, "If one could prove that there is a triangle whose area is bigger than any given figure, then I could prove the whole geometry." Let us call this **Gauss's axiom**, that for any rectilineal figure, there exists a triangle whose content is bigger than that of the figure. In a Hilbert plane with (A), show that Gauss's axiom implies (P), as follows:

(a) Show that there exist convex rectilineal figures with arbitrarily large area.

(b) Use (35.2) and (35.4) to get a contradiction if (P) does not hold.

36.2 Let $ABC$ be an isosceles triangle, let $D, E$ be the midpoints of the sides, and let $AG$ be the altitude that bisects $DE$ and $BC$. In a semihyperbolic plane with (A):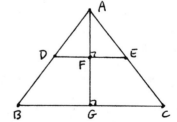

(a) Show that $AG < 2AF$.

(b) Show that $BC > 2DE$.

(c) In the Euclidean case, the area of $ABC$ would be four times the area of $ADE$. In this case, show that there is no such multiple estimate for comparing the two areas, in the following sense: For any given constant $k > 1$, there exists an isosceles triangle $ABC$ as above such that (area of $ABC$) $< k \cdot$ (area of $ADE$). (In order for multiplication by $k$ to make sense in the unwound circle group $G$, take $k$ to be a *dyadic rational* number, i.e., of the form $a/2^b$ for $a, b$ integers.)

36.3 Let $P$ be a polygon inscribed in a circle of radius $r$ in a semihyperbolic plane. Show that there is a polygon $P'$ inscribed in a circle of radius $2r$ having twice as many sides as $P$ and whose area is at least twice the area of $P$. This shows that the area of a circle (if properly defined!) can become arbitrarily large.

36.4 In a semihyperbolic plane, if a Saccheri quadrilateral $ABCD$ is cut by a diagonal $AD$, show that the upper triangle $ACD$ has area greater than the lower triangle $ABD$. Hint: First show that $\beta' < \beta$.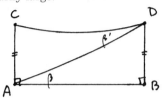

36.5 In a semihyperbolic plane, let a triangle $ABC$ be cut by its median $AD$. Assuming that the triangle is not isosceles, show that the half (in this drawing $ABD$) containing the acute angle $\alpha$ at $D$ has greater area than the other half.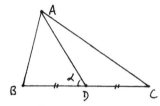

36.6 (a) In the situation of Proposition 34.6, show that $DE = \frac{1}{2}FH$.

(b) In the proof of Theorem 36.6, verify that the triangle claimed to be isosceles is indeed so.

36.7 We have seen that in the real Cartesian plane, the unwound circle group $G$ is isomorphic to $(\mathbb{R}, +)$ (36.1.2). This group is also isomorphic to the multiplicative group $(\mathbb{R}_{>0}, \cdot)$, via the exponential map.

In the case of an abstract ordered field $F$, we have three ordered abelian groups naturally associated with $F$: the additive group $(F, +)$, the multiplicative group $(F_{>0}, \cdot)$, and the unwound circle group $G$ of the Cartesian plane over $F$. We can ask which of these may be isomorphic, as ordered abelian groups.

(a) Let $F$ be the constructible field (16.4.1). Show that $(F, +)$ is not isomorphic to either $(F_{>0}, \cdot)$ or $G$ as an ordered abelian group. Hint: Think of dividing by 3 in each group.

(b) Again with $F$ the constructible field, show that if $(F_{>0}, \cdot)$ is isomorphic to $G$, then there can be only finitely many Fermat primes (29.4.1). (I have no reason to believe that these two groups should be isomorphic, so do not get your hopes up that this might be a way to prove the finiteness of the set of Fermat primes!)

36.8 Show that a Hilbert plane satisfies (A) if and only if its unwound circle group $G$ satisfies (A'): For any two elements $a, b > 0$ in $G$, there is an integer $n$ such that $na > b$.

36.9 Modify the results of this section to derive a theory of area and content for a semielliptic Hilbert plane, by taking the area of a triangle to be its *excess* $\delta =$ (sum of angles of triangle) $- 2RA$.

# 37   Circular Inversion

In this section we will study circular inversion, which is a kind of transformation of the plane that leaves the points of a given circle fixed, and sends points inside the circle to points outside and vice versa. While this study belongs in the context of Euclidean geometry, it is a technique not used by Euclid. Perhaps this is because the idea of a *transformation* of the plane, moving points to other points, was foreign to the way of thinking of the Greeks. Euclid does use the "method of superposition" to compare triangles in his proof of (SAS), but there is no evidence that he thought of a rigid motion moving the whole plane onto itself. Given that perspective it seems even less likely that the Greeks would have seen any value in a transformation of the plane that does not even preserve distances, in fact that does not even preserve proportion in figures. It seems that the notion of transformation in geometry is a relatively recent notion, which has come to serve very important roles, as we have seen in the usefulness of the existence of rigid motion (ERM).

The theory of circular inversion can be developed purely geometrically using the results of Euclid's Book III, but it will be more efficient to use the theory of proportion (similar triangles) in our proofs. So in this section we work for convenience in the Cartesian plane over a Euclidean ordered field $F$. Thus we have Hilbert's axioms, including (P) and (E), and we can use the theory of similar triangles (Section 20). The Euclidean hypothesis on $F$ can be slightly relaxed (Exercises 37.16, 37.17).

**Definition**

Let $\Gamma$ be a fixed circle in the plane (over the field $F$ as above...), with center $O$ and radius $r$. For any point $A \neq O$, draw the ray $OA$, and let $A'$ be the unique point on the ray $OA$ such that $OA \cdot OA' = r^2$. (The dot in this equation means products of lengths in the field $F$.) Then we say that $A'$ is obtained from $A$ by *circular inversion* with respect to the circle $\Gamma$.

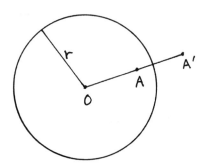

Briefly, we will say that $A'$ is the *inverse* of $A$ (with respect to the circle $\Gamma$). Since the condition $OA \cdot OA' = r^2$ is symmetric in $A$ and $A'$, this notion is reciprocal: $A'$ is the inverse of $A$ if and only if $A$ is the inverse of $A'$. We think of circular inversion in $\Gamma$ as a transformation $\rho = \rho_\Gamma$ defined for all points $A \neq O$, which thus transforms the plane $\Pi - \{O\}$ into itself by sending each point to its inverse. From the definition it is clear that any point *on* $\Gamma$ is sent to itself. Points inside $\Gamma$ are sent to points outside $\Gamma$, and vice versa. As a point $A$ approaches the center of the circle $O$, its inverse gets farther and farther away, so in the limit, the point $O$ would have to go to infinity. Since we do not have infinity in our geometry, we simply say that $\rho$ is *undefined* at $O$. (Or if you like, you can imagine completing our plane by a single point called infinity, and then $O$ and $\infty$ are interchanged—cf. Exercise 37.1 on stereographic projection for another interpretation of this idea.)

### Proposition 37.1
*Let $A$ be a point inside the circle $\Gamma$. Draw the ray $OA$. Let $PQ$ be the chord of the circle perpendicular to $OA$ at $A$. Then the tangents to $\Gamma$ at $P$ and $Q$ will meet the ray $OA$ at the point $A'$ that is the inverse of $A$ with respect to $\Gamma$.*

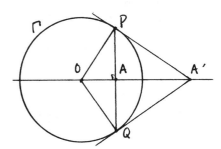

*Proof*    First note that the two tangents will both meet $OA$ in the same point $A'$, by symmetry. Now, the right triangles $\triangle OAP$ and $\triangle OPA'$ have the angle at $O$ in common, so they are similar. Hence corresponding sides are proportional. In particular,

$$\frac{OA}{OP} = \frac{OP}{OA'}.$$

Cross multiplying, we obtain

$$OA \cdot OA' = OP^2 = r^2.$$

Hence $A$ and $A'$ are circular inverses in $\Gamma$.

### Remark 37.1.1
This proposition gives us a method of constructing circular inverses by ruler and compass: If $A$ is given inside $\Gamma$, draw $OA$, construct the perpendicular to $OA$ at $A$, let it meet $\Gamma$ at $P$, draw the radius $OP$, draw the perpendicular to $OP$ at $P$, which will be the tangent line, and let this line meet $OA$ at $A'$ (9 steps). Conversely, if the point $A'$ is given outside the circle, draw the two tangent lines from $A'$ to $\Gamma$, join their points of tangency $P, Q$, and let the line $PQ$ meet $OA'$ at $A$ (6 steps).

Next we will investigate what circular inversion does to lines and circles in the plane.

### Proposition 37.2
*A line through $O$ is transformed into itself by circular inversion. A line not passing through $O$ will be transformed into a circle passing through $O$, and conversely.*

*Proof* A line through $O$ is transformed into itself by definition of circular inversion.

Now let $l$ be a line not through $O$. Let $OA$ be the perpendicular from $O$ to $l$. Let $A'$ be the inverse of $A$, and let $\gamma$ be the circle with diameter $OA'$. I claim that the inverses of points on $l$ all lie on $\gamma$ and vice versa. So let $B$ be any point on $l$. Draw $OB$ and let it meet $\gamma$ at $B'$. Then $OB'A'$ is a right triangle (III.31). It has the angle at $O$ in common with the right triangle $OAB$, so we have similar triangles. Therefore, the sides are proportional:

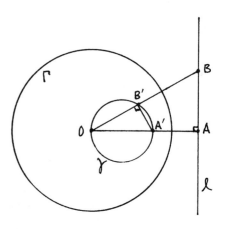

$$\frac{OB'}{OA'} = \frac{OA}{OB}.$$

Cross multiplying, we obtain

$$OB \cdot OB' = OA \cdot OA'.$$

But $A'$ was chosen to be the inverse of $A$, so $OA \cdot OA' = r^2$. Hence also $OB \cdot OB' = r^2$, so $B$ and $B'$ are inverse to each other. This shows that the circular inversion transforms the points of the line $l$ to the points of the circle $\gamma$ (except $O$) and vice versa.

### Definition
When two circles meet (or when a circle meets a line) by the *angle* between them we mean the angle between their tangent lines at that point (resp. the angle between the tangent line and the other line).

Note that when two circles meet in two points, the angle between them is the same at both points, because the two circles are symmetrical about the line joining their two centers.

### Proposition 37.3
*If a circle $\gamma$ is perpendicular to $\Gamma$ (at its intersection points), then $\gamma$ is transformed into itself by circular inversion in $\Gamma$. Conversely, if a circle $\gamma$ contains a single pair $A, A'$ of inverse points, then $\gamma$ is perpendicular to $\Gamma$ and is sent into itself.*

*Proof* First suppose that $\gamma$ is perpendicular to $\Gamma$, and let $\gamma$ meet $\Gamma$ at $P$ and $Q$. Then the radius $OP$ is tangent to $\gamma$, because radius and tangent of any circle are perpendicular (III.18). Let $A$ be another point of $\gamma$ and let $OA$ meet $\gamma$ again at $A'$. Applying (III.36) to $\gamma$ we obtain $OP^2 = OA \cdot OA'$. (Actually, Euclid meant that the square on $OP$ has content equal to the rectangle formed by $OA$ and $OA'$, but since we are working over the field $F$, we interpret this statement as lengths and products (20.9).) Since $OP = r$, this shows that $A$ and $A'$ are inverses. This holds for any $A$ on $\gamma$, so $\gamma$ is sent into itself.

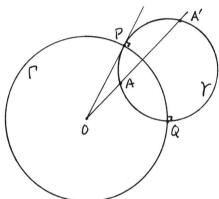

Now (using the same picture) suppose, conversely, that $\gamma$ is any circle passing through some pair of inverse points $A, A'$. Let $\gamma$ meet $\Gamma$ at $P$, and draw $OP$. Since $OP$ is a radius and $A, A'$ are inverse points, we have $OA \cdot OA' = OP^2$. But now by (III.37), the converse of (III.36), it follows that $OP$ is tangent to $\gamma$, which means that $\gamma$ and $\Gamma$ are perpendicular at $P$ (and hence also at their other point of intersection $Q$).

**Proposition 37.4**

*If $\gamma$ is a circle not passing through the center $O$ of $\Gamma$, then the transform of $\gamma$ by circular inversion is another circle $\gamma'$.*

*Proof* This result is not so easy to prove directly (you can try if you like), so we will resort to a trick. Suppose we are given a circle $\gamma$ not passing through $O$, and assume that $O$ is outside $\gamma$, for the moment.

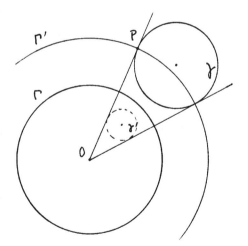

Draw $OP$ tangent to $\gamma$, and let $\Gamma'$ be a new circle with center $O$, passing through $P$. Then by construction $\gamma$ is sent into itself by $\rho_{\Gamma'}$. Thus $\rho_\Gamma(\gamma) = \rho_\Gamma \cdot \rho_{\Gamma'}(\gamma)$, and we are led to consider the new transformation of the plane $\theta = \rho_\Gamma \cdot \rho_{\Gamma'}$. Let $r =$ radius of $\Gamma$, and $r' =$ radius of $\Gamma'$. For any point $A$, let $A' = \rho_{\Gamma'}(A), A'' = \rho_\Gamma(A')$. Then $\theta(A) = A''$. By definition of inversion, $OA \cdot OA' = r'^2$ and $OA' \cdot OA'' = r^2$. Dividing, we find that

$$\frac{OA''}{OA} = \frac{r^2}{r'^2},$$

so

$$OA'' = k \cdot OA, \qquad \text{where} \quad k = \frac{r^2}{r'^2}.$$

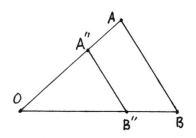

This is a transformation that leaves $O$ fixed, and stretches points toward (or away from) $O$ in a fixed ratio $k$. It is called a *dilation* with center $O$ and ratio $k$. In rectangular coordinates with center at the origin it would be expressed by

$$\begin{cases} x' = kx, \\ y' = ky. \end{cases}$$

It follows, either from thinking of the distance formula in terms of coordinates, or by using the (SAS) criterion for similar triangles (20.4), that all distances are changed by the same ratio $k$.

In particular, a dilation sends any circle (and its center) into another circle and its center. It follows that $\rho_\Gamma(\gamma) = \rho_\Gamma \cdot \rho_{\Gamma'}(\gamma) = \theta(\gamma)$ is a circle. (Warning: Even though $\rho_\Gamma(\gamma)$ is a circle, in general $\rho_\Gamma$ does not send the center of $\gamma$ to the center of $\gamma'$.)

In this proof we were assuming $O$ outside $\gamma$. If $O$ is inside $\gamma$, we leave you to construct an analogous proof in Exercise 37.4.

Now that we have seen that circular inversion preserves lines and circles (every line or circle is transformed into another line or circle), sometimes turning a line into a circle, or a circle into a line, the next step is to show that circular inversion is *conformal*, i.e., preserves angles.

**Proposition 37.5**

*Circular inversion is* conformal: *Whenever two curves meet (here "curve" means line or circle), their transforms under circular inversion meet again at the same angles.*

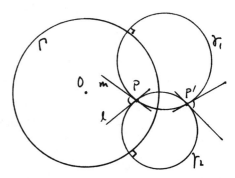

*Proof*  First suppose that $P \notin \Gamma$, and that two curves (not shown) meet at $P$ with tangent lines $l, m$. Let $P'$ be the inverse of $P$. Then we can find a circle $\gamma_1$ through $P, P'$ and with tangent $m$ at $P$; and we can find a circle $\gamma_2$ through $P, P'$ with tangent line $l$ at $P$. Now, by (37.3) $\gamma_1$ and $\gamma_2$ are transformed into themselves. Therefore, the original curves are transformed into curves at $P'$ tangent to $\gamma_1$ and $\gamma_2$, so they make the same angle as at $P$, because when two circles $\gamma_1, \gamma_2$ intersect they have the same angle at both intersections. For this proof we need to observe that a line and a circle, or two circles, are tangent if and only if they have just one point in common. Hence the property of tangency is preserved by inversion.

If $P \in \Gamma$, we leave the special case to you (Exercise 37.5).

For our last general result about the properties of circular inversion, we look at what happens to distances. Of course, distances are not preserved, because a very small distance near $O$ will be transformed into a very large distance far away. Even ratios of distances are not preserved, as you can see by simple examples. However, a remarkable fact is that if we take four points, a certain ratio of ratios of distances, called their cross-ratio, is preserved.

**Definition**
Let $A, B, P, Q$ be four distinct points in the Cartesian plane. Their *cross-ratio* (an element of the field $F$) is defined to be the ratio of ratios

$$(AB, PQ) = \frac{AP}{AQ} \div \frac{BP}{BQ},$$

which can also be written

$$\frac{AP}{AQ} \cdot \frac{BQ}{BP}.$$

**Proposition 37.6**
*Let $A, B, P, Q$ be four distinct points in the plane, different from $O$. Then circular inversion in $\Gamma$ preserves the cross-ratio: If their inverses are $A', B', P', Q'$, then*

$$(AB, PQ) = (A'B', P'Q').$$

*Proof*  Given two points $A, P$ and their inverses $A', P'$, we know by definition that

$$OA \cdot OA' = r^2 = OP \cdot OP'.$$

Thus

$$\frac{OA}{OP} = \frac{OP'}{OA'}.$$

*Case 1*  Suppose $O, A, P$ are not collinear. Since the triangles $\triangle OAP$ and $\triangle OP'A'$ have the angle at $O$ in common, they are similar (20.4), and we conclude that

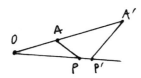

$$\frac{AP}{A'P'} = \frac{OA}{OP'}. \qquad (1)$$

*Case 2*  If $O, A, P$ are collinear, then $AP = OP - OA$ and $A'P' = OA' - OP'$. Using the fact that in a field $F$,

$$\frac{a}{b} = \frac{c}{d} \Rightarrow \frac{a}{b} = \frac{c-a}{d-b},$$

we conclude the same result (1).

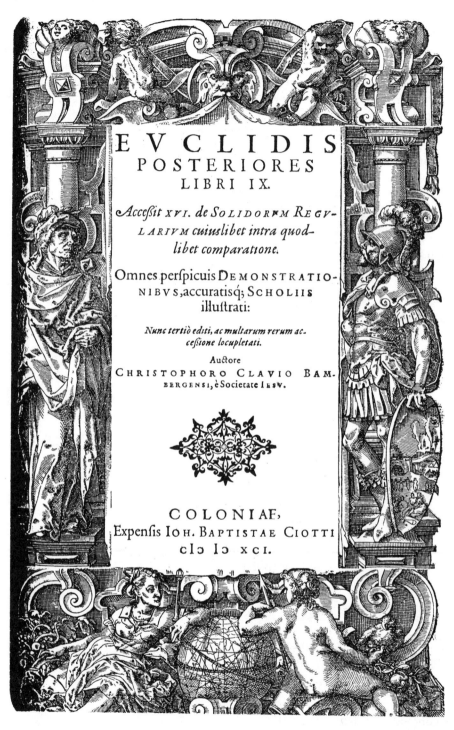

EVCLIDIS
POSTERIORES
LIBRI IX.

*Accessit XVI. de SOLIDORVM REGV-
LARIVM cuiuslibet intra quodlibet comparatione.*

Omnes perfpicuis DEMONSTRATIO-
NIBVS, accuratisq; SCHOLIIS
illuftrati:

*Nunc tertiò editi, ac multarum rerum ac-
cessione locupletati.*

Auctore
CHRISTOPHORO CLAVIO BAM-
BERGENSI, è Societate IESV.

COLONIAE,
Expenfis IOH. BAPTISTAE CIOTTI
cIɔ Iɔ XCI.

Plate XIII. Title page of volume II of the important edition of the *Elements* by the Jesuit mathematician Christopher Clavius (1591).

Now if $Q$ is another point, we find similarly that

$$\frac{AQ}{A'Q'} = \frac{OA}{OQ'}.$$  (2)

Dividing, we get

$$\frac{AP}{A'P'} \div \frac{AQ}{A'Q'} = \frac{OQ'}{OP'}.$$  (3)

Now let $B$ be another point. Working with $P$ and $Q$ as before, we obtain similarly

$$\frac{BP}{B'P'} \div \frac{BQ}{B'Q'} = \frac{OQ'}{OP'}.$$  (4)

So the expressions (3) and (4) are equal. Moving the primed letters to one side and the unprimed letters to the other side shows that the cross-ratios $(AB, PQ)$ and $(A'B', P'Q')$ are equal.

## Remark 37.6.1

At this point I can just hear someone asking, "What is the geometrical significance of the cross-ratio?" Although I first encountered cross-ratios as a senior in high school, and have dealt with them many times since then, I must say frankly that I cannot visualize a cross-ratio geometrically. If you like, it is magic. Here is this algebraic quantity whose significance it is impossible to understand, and yet it turns out to do something very useful. It works. You might say it was a triumph of algebra to invent this quantity that turns out to be so valuable and could not be imagined geometrically. Or if you are a geometer at heart, you may say that it is an invention of the devil and hate it all your life.

Let me say a few words in defense of the poor cross-ratio.

In the present context of transformations of the Euclidean plane, there are rigid motions, which preserve distance. Then there are dilations, which do not preserve distance, but do preserve ratios of distances. Then there is circular inversion, which does not preserve distances or even ratios of distances. Since it does preserve the cross-ratio, that particular ratio of ratios, it is the best we can do. It is something to hang on to, a pillar of support, when the distances and their ratios are changing all around us. In Section 39 we will use the cross-ratio to define the notion of distance in the Poincaré model of non-Euclidean geometry: It plays an essential role there.

In projective geometry the cross-ratio is also important. A *projectivity* from one line to another is defined as a composition of a finite number of projections from one line to another in the plane. A projectivity preserves neither distances nor ratios of distances, but it does preserve the cross-ratio (Exercise 37.14). In fact, a fundamental theorem of projective geometry is that a transformation of one line to another in the projective plane is a projectivity *if and only if* it preserves the cross-ratio of every set of four distinct points on the line.

If you have studied complex variables, the projectivities of the projective line over the complex numbers correspond to the fractional linear transformations of the Riemann sphere $\mathbb{C} \cup \{\infty\}$, given by

$$z' = \frac{az+b}{cz+d}, \quad ad - bc \neq 0.$$

If you are given four points $A, B, C, D$, there is always a fractional linear transformation sending $A, B, C$ to $0, 1, \infty$. In that case the image of $D$, say $\lambda$, is the cross-ratio of the original four points, in a suitable order.

Finally, if you have four points on a line, and you take signed distances ($+$ or $-$ depending on a chosen preferred direction), then the cross-ratio is equal to $-1$ if and only if the four points form a set of *four harmonic points* (Exercise 37.15). This notion of harmonic points is also important in projective geometry.

The notion of cross-ratio already occurs in the work of Pappus (300 A.D.). It came into prominence again in the early nineteenth century with the projective geometry of Poncelet and Monge.

# Exercises

Unless otherwise noted, the following exercises take place in the Cartesian plane over a Euclidean ordered field $F$.

37.1 *Stereographic projection.* In three-dimensional space, imagine our plane $\Pi$ and a circle $\Gamma$ of radius $r$ and center $O$. Now take a sphere of radius $\frac{1}{2}r$, and set it on the plane $\Pi$ so that its south pole is at $O$. Then *stereographic projection* associates to each point $B$ of the sphere, $B \neq N =$ north pole, that point of $\Pi$ obtained by drawing the line $NB$ and intersecting with $\Pi$. (In the limit, $N$ would go to infinity, so you can think of the sphere as a completion of the plane by adding the point $N$.) Under this projection, the equator of the sphere is mapped to the circle $\Gamma$.

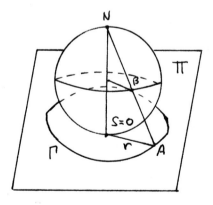

Show that circular inversion in the circle $\Gamma$ corresponds to the operation of *reflection in the equator* of the sphere, which interchanges the northern and southern hemispheres. In other words, if $B$ is a point on the sphere, and $B'$ its reflection in the equator (same longitude, but latitude has changed from north to south or from south to north), then the projected points $A, A'$ are inverses under inversion in $\Gamma$.

37.2 Prove that the following construction with compass alone gives the inverse of $A$ in $\Gamma$ (provided that $OA > \frac{1}{2}r$): Draw a circle with center $A$ through $O$ to meet $\Gamma$ at $P$ and $Q$. Then draw circles with centers $P$ and $Q$ through $O$ to meet at $A'$ (3 steps). (The diagram shows $A$ inside $\Gamma$, but the construction works equally well if $A$ is outside $\Gamma$.)

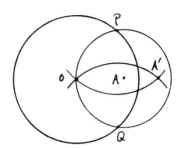

37.3 Let $l$ be a line that meets the circle $\Gamma$ in two points $A, B$. Let $\gamma$ be the (unique) circle through $O, A, B$. Prove that $\gamma$ is the transform of $l$ under inversion in $\Gamma$.

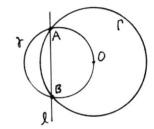

37.4 Prove the other case of (37.4), namely, if $\gamma$ is a circle containing $O$, then $\rho_{\Gamma}(\gamma)$ is a circle. For any points $A, B$ on $\gamma$, let $A', B'$ be their inverses in $\Gamma$, and let $A'', B''$ be the points where the lines $OA, OB$ meet $\gamma$ again. By (III.35) (cf. (20.8)),

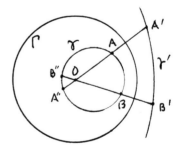

$$OA \cdot OA'' = OB \cdot OB'' = c$$

is a constant independent of the points $A, B$, depending only on $O, \gamma$.

Let us use signed lengths from $O$, so that $c < 0$. Since $OA \cdot OA' = r^2$, then show that $OA' = k \cdot OA''$ for a certain constant $k < 0$. Thus $\gamma'$ is obtained from the circle $\gamma$ by dilation with a negative constant $k$. Conclude that $\rho_{\Gamma}(\gamma) = \gamma'$ is a circle.

37.5 If two lines or circles meet at a point $P \in \Gamma$, show that their two transforms by circular inversion in $\Gamma$ meet at the same angle at $P$.

37.6 If we identify the real Euclidean plane $\mathbb{R}^2$ with the complex numbers $\mathbb{C}$, show that the transformation $z' = 1/\bar{z}$ (where $z = a + bi$, $\bar{z} = a - bi$) is just inversion in the unit circle $|z| = 1$.

37.7 If $A$ is a point inside the circle $\Gamma$, improve the ruler and compass construction of the inverse of $A$ given in (37.1.1) by constructing the circle through $O, P, Q$ instead of constructing the tangent line at $P$ (par $= 7$ steps).

37.8 Given the circle $\Gamma$ and its center $O$, and given a line $l$ not containing $O$, give a ruler and compass construction of the circle $\rho_\Gamma(l)$ (par $= 7$ steps).

37.9 Given $\Gamma$, given its center $O$, and given a circle $\gamma$ passing through $O$ (but not given the center of $\gamma$), construct the line $\rho_\Gamma(\gamma)$ (par $= 15$ steps).

37.10 Given $\Gamma$ and its center $O$, and given a circle $\gamma$ not through $O$, construct the circle $\rho_\Gamma(\gamma)$ (par $= 15$ steps).

37.11 Verify the following ruler-only construction of the inverse of a point $A$ (7 steps): Draw $OA$, get $R, S$. Draw any line $l$ through $A$ meeting $\Gamma$ in $P, Q$. Draw $RP$ and $SQ$ to meet at $T$. Draw $RQ$ and $PS$ to meet at $U$. Draw $TU$ to meet $OA$ at $A'$. Show also that $TU$ is perpendicular to $OA$.

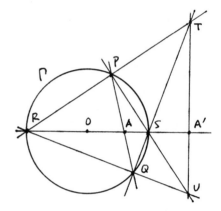

37.12 Verify the following 5-step construction for the inverse of a point $A$ with respect to the circle $\Gamma$. Take a circle of any radius with center $A$, to meet $\Gamma$ at $P$ and $Q$. Let $AP$ and $AQ$ meet $\Gamma$ in further points $R, S$. Join $PS$ and $RQ$. Their intersection is $A'$. (This works equally well if $A$ is inside $\Gamma$.)

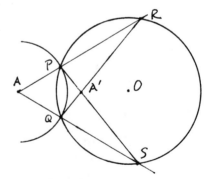

37.13 (a) Given four points $A, B, P, Q$, if you permute $A$ and $B$, or if you permute $P$ and $Q$, the cross-ratio is replaced by its inverse: $(BA, PQ) = (AB, QP) = (AB, PQ)^{-1}$.

(b) More generally, if $A, B, P, Q$ are four points on a line, and if $(AB, PQ) = \lambda$, then the 24 possible permutations of the points give rise to 6 possible values of the cross-ratio, namely

$$\lambda, \quad \frac{1}{\lambda}, \quad 1 - \lambda, \quad \frac{1}{1-\lambda}, \quad \frac{\lambda - 1}{\lambda}, \quad \frac{\lambda}{\lambda - 1}.$$

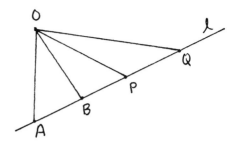

37.14 (a) Given four points on a line $l$, and given a point $O$ not on $l$, let the angles at $O$ subtended by $AP, AQ, BP, BQ$ be $\alpha_P, \alpha_Q, \beta_P, \beta_Q$. Use the law of sines to show that

$$(AB, PQ) = \frac{\sin \alpha_P}{\sin \alpha_Q} \div \frac{\sin \beta_P}{\sin \beta_Q}.$$

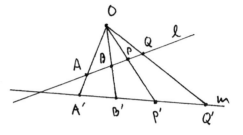

(b) If the four points $A, B, P, Q$ on $l$ are projected from $O$ to four points $A', B', P', Q'$ on another line $m$, then the cross-ratio is preserved:

$$(AB, PQ) = (A'B', P'Q').$$

Conclude that cross-ratio is preserved by any projectivity, that is, a finite succession of projections from one line to another.

37.15 We say that four points $A, B, P, Q$ on a line form a set of *four harmonic points* if their cross-ratio $(AB, PQ)$ is equal to $-1$.

(a) Given $A, B, P$, show that the fourth harmonic point $Q$ is uniquely determined.

(b) Verify the following ruler-only construction of the fourth harmonic point: Given $A, B, P$ on a line $l$, take a point $X$ not in the line. Draw $XA$, $XB$, $XP$. Take any point $Y$ on $AX$. Draw $BY$, get $W$. Draw $AW$, get $Z$. Draw $YZ$, get $Q$. *Hint*: Project the four points $A, B, P, Q$ from $X$ to the line $YQ$, and then from $W$ back to the original line $l$, and use Exercise 37.14.

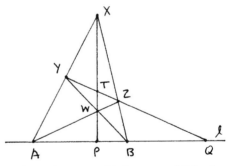

(c) If $A, B, P, Q$ are four harmonic points, show that $Q$ is the inverse of $P$ in the circle with diameter $AB$.

37.16 If $F$ is a Pythagorean ordered field, we can still define inversion in a circle $\Gamma$ by the same method as at the beginning of this section. If $\gamma$ is a circle with $\rho_\Gamma(\gamma) = \gamma$, show that $\gamma$ still meets $\Gamma$ in two points, even though we do not have the axiom (E).

37.17 Let $F$ be a Pythagorean ordered field, and let $d$ be a positive element of $F$ that has no square root in $F$. We consider the *virtual circle* $\Gamma$ defined by the equation $x^2 + y^2 = d$. Since $F$ is Pythagorean and $\sqrt{d} \notin F$, this equation has no solutions. So $\Gamma$ has no points. Nevertheless, it is useful to refer to $\Gamma$ as a virtual circle, because we can still define circular inversion $\rho$ in $\Gamma$ by the formula $OA \cdot OA' = d$.

(a) Show that the results (Propositions 37.2, 37.4, 37.5, 37.6) still hold for $\rho$.

(b) Show that a part of Proposition 37.3 holds: A circle $\gamma$ is sent to itself by $\rho$ if and only if it contains a pair of inverse points. In this case we say by abuse of language that $\gamma$ is orthogonal to $\Gamma$.

(c) Given $O$ and given one pair $A, A'$ of inverse points by $\rho$, give a construction with Hilbert's tools (Section 10) for the inverse $B'$ of a point $B$. (Par = 4 if $O, A, B$ are not collinear.)

37.18 Give a new proof of Exercise 1.15 by doing a circular inversion with center $P$ and suitable radius, and solving the transformed problem.

Poor John has lost his ruler. Can you help him do his construction problems (following) using his compass alone?

37.19 Given two points $A, B$ construct a third collinear point $C$ with $AB = BC$ (par = 4 steps).

37.20 Given two points $A, B$, construct the midpoint $C$ of the segment $AB$ (par = 7 steps). *Hint*: Use Exercise 37.2.

37.21 Given two points $A, B$, construct $C, D$ such that $ABCD$ will be a square (par = 8 steps).

37.22 Given points $A, B, C, O$, with $O, A, B$ not collinear, construct the intersection points of the line $AB$ with the circle $OC$ (assuming that they meet) (par = 4 steps).

37.23 Given noncollinear points $A, B, C$, construct the foot of the perpendicular from $C$ to the line $AB$ (par = 9 steps).

37.24 Given noncollinear points $O, A, B$, show that it is possible to construct the intersection of the circle $OA$ with the line $OB$ using compass alone. *Hint*: Perform a circular inversion that leaves the circle $OA$ fixed and transforms the line $OB$ into a circle. (Par = 13 to get one intersection point.)

37.25 Given points $A, B, C, D$ show that it is possible to construct the intersection point of the lines $AB$ and $CD$ using compass alone. *Hint*: Use a circular inversion to transform the two lines into circles. (Par = 13 steps if the points are in favorable position; otherwise 18 steps.)

37.26 Using the experience gained in the previous exercises, prove the following theorem of Mascheroni: Any point that can be constructed from given data by ruler and compass construction can also be constructed using compass alone.

# 38   Digression: Circles Determined by Three Conditions

In order to specify a circle in the plane, you give its center, which is a point, and its radius, which is a line segment, or distance. A point moves in a 2-dimensional plane, while the length of a line segment is a 1-dimensional quantity, so we can say roughly that the set of all circles forms a three-dimensional family, or that a circle depends on three parameters. If we are working in the Cartesian plane over a Euclidean field $F$, then a general circle has an equation

$$(x-a)^2 + (y-b)^2 = r^2,$$

where $a, b, r$ are elements of the field $F$. In this case the circle is determined by the three quantities $a, b, r \in F$, so again it makes sense to say that a circle depends on three parameters.

From this informal discussion it is reasonable that we can impose three conditions and then expect to find a unique circle satisfying those three conditions. Of course, there will be situations where this does not work. For example, you cannot find a circle passing through three collinear points (unless you allow the line containing the points to count as a limiting case of a circle). Another example is, you cannot find a circle tangent to two given parallel lines $l, m$ and passing through a point $P$ not between the two lines.

In general, however, we can expect to find a circle satisfying three conditions. Among conditions we can impose are:

(P)  to require that the circle pass through a given point $A$;
(L)  to require that the circle be tangent to a given line $l$;
(C)  to require that the circle be tangent to a given circle $\gamma$.

Furthermore, it is natural to expect that the required circle be constructible by ruler and compass. In this way, taking all possible combinations of three conditions of types (P), (L), and (C), we obtain the following ten construction problems.

PPP. To construct a circle through three given points $A, B, C$. This is the *circumscribed* circle to the triangle $ABC$.

LLL. To construct a circle tangent to three given lines $l, m, n$. This is the *inscribed* circle or one of the exscribed circles of the triangle formed by $l, m, n$.

The remaining eight problems we designate as PPL, PLL, PPC, PLC, PCC, LLC, LCC, and CCC, according to the conditions imposed. The last one, to find a circle tangent to three given circles, is classically known as the problem of Apollonius. Apollonius of Perga (c. 262–c. 200 B.C.) is best known for his book on conics, but he also wrote a book *On Tangencies*, which is now lost, in which he discussed this problem of the three circles. We know of Apollonius's book from the commentary of Pappus (1876), Book VII, Sections 11, 12. Based on this, Viète made a restitution of the lost book of Apollonius, and later Camerer (1795) edited both Pappus's commentary and Viète's restitution with added work of his own. One of Pappus's constructions was given in (5.11).

In this section we give another method for solving these problems, using circular inversion.

Advice to the reader: The best way to study this material is to stop reading now, and attempt to solve each of these problems yourself. Then come back for hints if you need some help. If you read on further, you will find my solutions, but I expect you will find other, perhaps even better, solutions on your own.

As hints let me list the general techniques we will use. First of all, we use the basic miniconstructions of three or four steps each that we have often used before. With these techniques you should be able to solve problems PPP, PPL, PLL, and LLL.

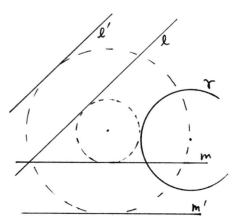

For the problems involving circles, there is sometimes an easy method to reduce one problem to another. For example, in LLC, given two lines $l, m$ and a circle $\gamma$ with center $O$, we are trying to find the small dotted circle tangent to $l, m$, and $\gamma$. Let $l'$ and $m'$ be lines parallel to $l$ and $m$, and at a distance from $l$ and $m$ equal to the radius of $\gamma$. Then the circle through $O$ (the center of $\gamma$) and tangent to $l'$ and $m'$ will have the same center as the circle we are looking for (verify!). Thus we can reduce LLC to PLL.

In the most difficult problems we can make use of circular inversion. For example, in CCC, the problem of Apollonius, by adding or subtracting the same quantity to the radius of each circle, we reduce to a problem where two circles meet. Let $O$ be an intersection point of two circles, and let $\Gamma$ be a circle with center $O$. Perform a circular inversion in the circle $\Gamma$. Then the two circles through $O$ are transformed into straight lines, while the third circle becomes another circle. By LLC we can find a circle tangent to the two lines and circle obtained by circular inversion: Applying a circular inversion $\Gamma$ to this newly constructed circle, we obtain a solution to the original problem.

Of course, these constructions may get rather long, and in carrying them out on a piece of paper with pencil or pen, the accumulated error may be greater than what one could achieve by guessing. But that is not the point. The point is to have a correct theoretical procedure that gives a mathematically exact answer when carried out in ideal conditions.

**Construction 38.1** (PPP)
To construct a circle through three given points $A, B, C$. The center of this circle must lie on the perpendicular bisectors of $AB$ and $BC$. So construct the perpendicular bisectors (3 steps each) to get the center $O$, and then draw the circle (7 steps altogether). However, we can save one step by making one of our circles do double duty:

1. Circle center $B$, radius $BC$.
2. Circle center $A$, same radius. Get $E, D$.
3. Draw line $ED$.
4. Circle center $C$, same radius. Get $F, G$.
5. Line $FG$. Get $O$.
6. Circle center $O$, radius $OA$, passes through $B, C$.

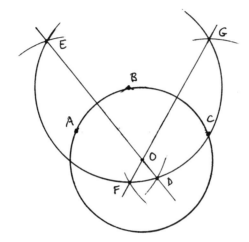

**Construction 38.2** (LLL)

To find a circle tangent to three given lines $l, m, n$. Let $ABC$ be the triangle formed by the three lines. The center of the inscribed circle must lie on the angle bisectors of the three angles. So we bisect the angles at $A$ and $B$ to get the center of the circle $O$. Then we drop a perpendicular from $O$ to one side to find the radius, then draw the circle. Adding miniconstructions comes to 13 steps. But with care, we reduce the construction to 10 steps:

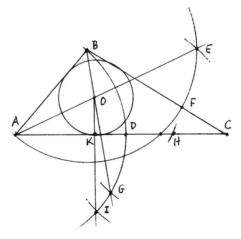

Bisect angle $A$, using circles centered at $A, B, D$, all of radius $AB$ (4 steps).

Bisect angle at $B$, using $F, G$ (2 more steps).

Perpendicular from $O$ to $AC$, using $H, I, K$ (3 more steps).

Circle center $O$, radius $OK$ (1 step).

**Construction 38.3** (PPL)

To construct a circle through two given points $A, B$, tangent to a given line $l$.

First draw the line $AB$ to meet $l$ at $P$ (1 step). If we imagine the circle already drawn being tangent to $l$ at $Q$, then $PQ$ is a tangent to the circle and $PAB$ cuts the circle, so by (III.36) $PA \cdot PB = PQ^2$. So our method will be to use this fact to find $Q$, and then draw the circle. According to (III.36) again, if we take any circle through $AB$, the tangent from $P$ to that circle will have the same length as $PQ$. So let us bisect $AB$, using $CD$ (3 steps) and get its midpoint $E$. Draw the circle with center $E$ passing through $A$ and $B$ (1 step). To find the tangent from $P$ to this circle, bisect $PE$, using $F, G$ (3 steps) and get $H$. Draw the circle with center $H$ through $P, E$ (1 step) and let it meet the circle through $AB$ at $K$. Then $PK$ will be tangent to the circle. Find $Q$ and $l$ with $PQ = PK$ (1 step). Now to get the circle through $A, B, Q$, bisect $AQ$ (2 more steps), get $O$, and draw the circle with center $O$ through $A, B, Q$ (1 step). Total: 13 steps.

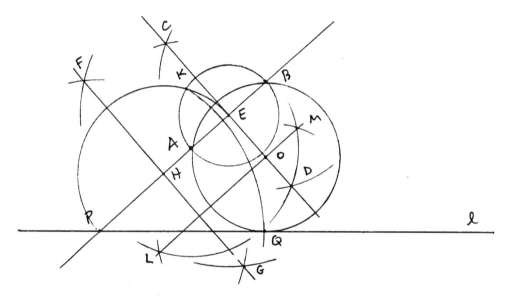

**Construction 38.4** (PLL)

To construct a circle through a given point $A$ and tangent to two given lines $l, m$.

One method is to reduce to PPL as follows: Imagine a reflection in the angle bisector of the angle at $O$. This will give a new point $A'$, symmetrically located, and any circle through $A$ tangent to $l$ and $m$ must also pass through $A'$. Thus we need only construct the circle through $A, A'$ tangent to $l$ by PPL. We can construct $A'$ in 2 steps: Circle center $O$ through $A$, get $B, C$. Circle through $C$, radius $AB$. Get $A'$. Then do PPL. Total: 15 steps.

Another method is to use the fact that circles tangent to two lines $l, m$ are all related by dilations from center $O$ where $l$ and $m$ meet. So this time our strategy is to construct any other circle tangent to $l$ and $m$, and then scale it down or up so as to pass through $A$. First bisect the angle at $O$, using $B, C, D$ and circles all of the same radius (4 steps). Get $E, F$ where the circle with center $B$ through $O$ meets $l, m$. Draw $EF$ (1 step), which will be perpendicular to $l$, since it is contained in a semicircle (III.31). Get $G$, and draw a circle with center $G$ and radius $GF$ (1 step). This is our comparison circle $\gamma'$. Draw $OA$ to meet $\gamma'$ at $A'$ (1 step). The desired circle $\gamma$ will have center $K$ on $OD$, and with $KA$ parallel to $GA'$. So construct $AH$ parallel to $GA'$ (3 steps), and get $K$. Draw the circle $\gamma$ with center $K$ and radius $KA$ (1 step). Total: 11 steps.

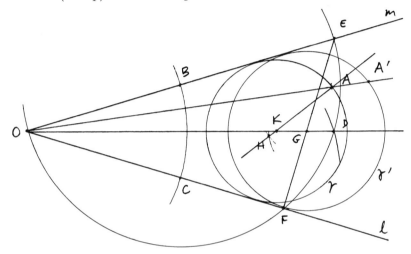

**Construction 38.5** (PPC)

As an application of the use of circular inversion, we will construct a circle passing through two given points $A, B$, and tangent to a given circle $\gamma$. Since circular inversion sends circles to circles (or lines) and preserves incidence and tangencies, any circular inversion will transform one of our ten problems into another of our ten problems, and solving one is equivalent to solving the other, since the constructed circle can also be transformed by the same circular inversion. The point is to make a judicious choice of which circular inversion to perform so that our job becomes as simple as possible.

In this case, let $A, B$ be the given points, and let $\gamma$ be the given circle. Draw the perpendicular bisector of $AB$ (using $C, D$, 3 steps) and let it meet $\gamma$ at $O$. (If it does not meet $\gamma$, then the construction will be longer.) Take the circle $\Gamma$ with center $O$, passing through $A, B$ (1 step) as our circle to invert in. Let $\Gamma$ meet $\gamma$ at $P, R$. Draw the line $l$ through $P, R$ (1 step). Since $\gamma$ passes through $O, l$ will be the inverse of the circle $\gamma$, while $A, B$ are fixed, since they lie on $\Gamma$. So we have the new problem to find the circle $\delta$ through $A, B$, tangent to $l$. Suppose this is solved by PPL (13 steps, not shown), and let $Q$ be the point of tangency of $\delta$ with $l$. We find the inverse $Q'$ of $Q$ by drawing the line $OQ$ and intersecting with $\gamma$ (1 step). So the desired circle is the circle through $A, B$, and $Q'$. This is PPP. We need only construct the perpendicular bisector of $AQ'$ (using $E, F$, 3 steps) to find $G$, and then draw the circle with center $G$ through $A, B, Q'$ (1 step). Total: 23 steps. Actually, in applying PPL, we already have the perpendicular bisector of $AB$, so that saves 3 steps. Also, we do not need the actual circle $\gamma$, only its point of tangency $Q$ with $l$. That saves 3 more steps, so the count becomes 17 steps.

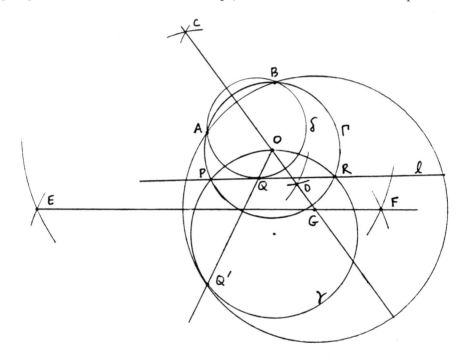

To complete this section, we will do a special case of LLC, and then show how to use that to solve CCC, the problem of Apollonius. The remaining problems we will leave as exercises for the reader.

**Construction 38.6**

Given two parallel lines $l, m$, and given a circle $\gamma$ with center $O$, to find a circle tangent to $l, m$, and $\gamma$.

The idea is to construct the line parallel to $l$ and $m$ that is halfway between $l$ and $m$. The center of the desired circle will be a point on this midline whose distance from the center $O$ of $\gamma$ is equal to (its radius) $\pm$ (radius $\gamma$) depending on whether we want a circle containing $\gamma$ or not.

Pick any point $A$ on $l$, and draw circle with large enough radius to intersect $m$ at $B, C$ (1 step). Bisect $BC$ (3 steps) using $A, D$, get $E$. Circle with center $E$, radius $AB$ to get $F, G$ (1 step). Draw $FG$ (1 step) and get $H$. This is the midline.

Make $HI$ = radius of $\gamma$ (1 step). Center $O$, radius $EI$, find $K$ on the midline (1 step). Center $K$, radius $EH$, draw the required circle $\sigma_1$ (1 step). Total: 9 steps. To get another solution $\sigma_2$, center $O$, radius $AI$ to get $L$. Center $L$, radius $EH$ is $\sigma_2$. Of course, there are two more solutions with centers $K', L'$ on the other side of $O$.

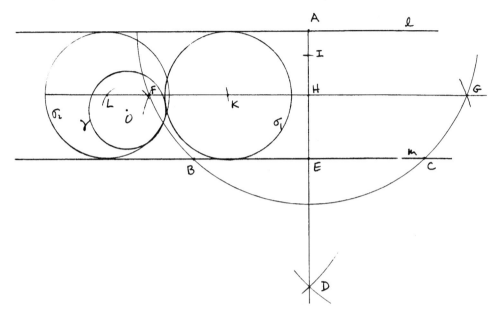

**Construction 38.7** (The problem of Apollonius)
Given three circles $\gamma_1, \gamma_2$, and $\gamma_3$, with their centers $O_1, O_2, O_3$, to find a circle tangent to all three given circles.

Note first that if we were not given the centers of the circles, we could find them with little constructions of 5 steps each, so to make life a little easier for ourselves, we suppose $O_1, O_2, O_3$ also given.

The idea is to expand the three circles by adding a fixed length to each of their radii. This will not change the center of the circle in the middle tangent to all three. (If we wanted a circle containing one of the $\gamma_i$, then we should subtract from its radius.) Doing this expansion carefully, we can arrange that two of the new circles are tangent to each other.

This first part of the construction goes as follows: Draw $O_1O_2$ (1 step), get $A, B$. Bisect $AB$, using $C, D$ (3 steps), get $E$. Draw $O_3E$, get $F'$ (1 step). Make $FF' = AE$ (1 step). Now draw the new circles $\gamma_1'$: center $O_1$, through $E$; $\gamma_2'$: center $O_2$, through $E$; and $\gamma_3'$: center $O_3$, through $F$ (3 steps). Then, by this operation, we have reduced to a special case of CCC where two of the circles are tangent (9 steps).

For the next stage of the construction, we will perform a circular inversion in a circle $\Gamma$ with center $E$. This will transform $\gamma_1'$ and $\gamma_2'$ into two parallel straight lines, and $\gamma_3'$ will become another circle. It seems worthwhile to choose

$\Gamma$ orthogonal to $\gamma'_3$, because then $\gamma'_3$ will be sent into itself, saving us the trouble of finding its image under the circular inversion.

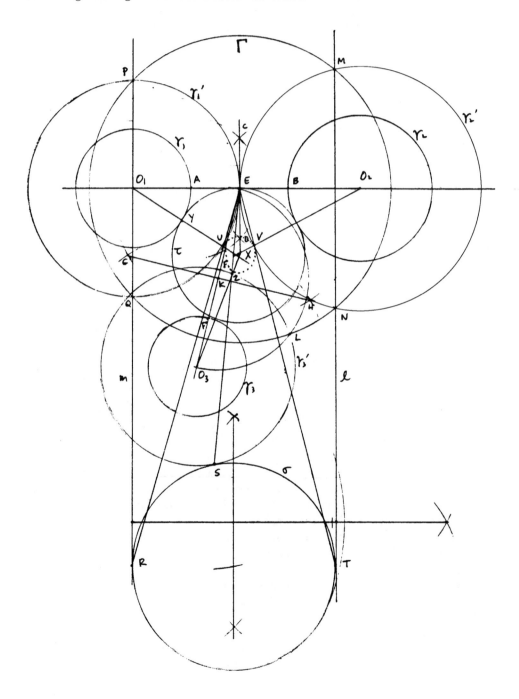

So we proceed as follows: Bisect $O_3E$ using $G, H$ (3 steps), and get $K$. Draw the circle with center $K$, radius $KE$ (1 step), and let it intersect $\gamma_3'$ at $L$. Draw the circle $\Gamma$ with center $E$ through $L$ (1 step). Then $\Gamma$ will be orthogonal to $\gamma_3'$, and so will be left fixed by circular inversion in $\Gamma$. Let $\Gamma$ meet $\gamma_2'$ in $M, N$, and draw the line $l = MN$ (1 step). Let $\Gamma$ meet $\gamma_1'$ in $P, Q$, and draw the line $m = PQ$ (1 step). Thus we have transformed $\gamma_1', \gamma_2', \gamma_3'$ into $l, m$, and $\gamma_3'$, and we have the new problem of finding a circle tangent to these three. Furthermore, since $\gamma_1'$ and $\gamma_2'$ were tangent to $O$, their transforms $l, m$ do not meet. In other words, $l$ and $m$ are parallel, so we have a case of (38.6) treated above. This portion of our construction was 7 steps.

Now perform (38.6) to find a circle $\sigma$ tangent to $l, m, \gamma_3'$, and let the points of tangency be $R, S, T$ (9 steps). Actually, since we already have a line $O, O_2$ perpendicular to $l$ and $m$, we can get the midline in 3 steps instead of 6, thus saving 3 steps. So this part of the construction counts 6 steps.

The last stage of the construction is to transport back $\sigma$ by the circular inversion in $\Gamma$ to get a circle tangent to $\gamma_1', \gamma_2', \gamma_3'$. Then for the same center we can draw the desired circle $\tau$ tangent to $\gamma_1, \gamma_2, \gamma_3$.

It is actually sufficient to pull back two of the points of tangency. Draw $ER$ and let it intersect $\gamma_1'$ at $U$ (1 step). Then $U$ is the inverse of $R$ in $\Gamma$. Draw $ET$ and let it meet $\gamma_2'$ at $V$ (1 step). Now $U, V$ are two of the points of tangency of a circle (dotted) tangent to $\gamma_1', \gamma_2', \gamma_3'$. To find its center, draw $O_1U$ and $O_2V$ and let them meet at $X$ (2 steps). Now the circle $\tau$ with center $X$ and radius $XY$ is the desired circle (1 step). This last part of the construction is 5 steps. (In the drawing I also found the inverse $Z$ of $S$ and drew $O_3Z$ to check for accuracy, but this is not really part of the theoretical construction.) Total: 27 steps.

## Exercises

Carry out the following ruler and compass constructions.

38.1  PLC. Treat as a special case of LCC.

38.2  LLC. Follow hint given earlier in text.

38.3  PCC. Treat as a special case of CCC.

38.4  LCC. Use a technique similar to the one we used for CCC to reduce to (38.6).

38.5  PPC. Do the general case, where the perpendicular bisector of $AB$ does not meet $\gamma$.

38.6  Describe how you would construct all eight solutions to the problem of Apollonius.

## 39  The Poincaré Model

In this section we will show the *existence* of a non-Euclidean geometry, and hence the *consistency* of the axioms of non-Euclidean geometry, by exhibiting a model for a non-Euclidean geometry. Ironically, our model of a non-Euclidean geometry will be constructed within the logical framework of Euclidean geometry. So what we must do is to give an *interpretation* of the undefined notions of geometry in the model: point, line, betweenness, and congruence for line segments and angles, and then we must prove that the axioms all hold in this interpretation.

Our starting point will be the Cartesian plane $\Pi$ over a Euclidean ordered field $F$. In this plane we fix a circle $\Gamma$ with center $O$. (For a weakening of the Euclidean hypothesis on $F$, see Exercises 39.25 ff.)

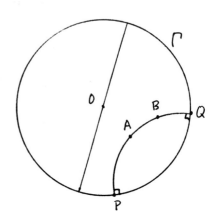

The *points* of our model (which we will call P-*points*) will be the set of points of $\Pi$ *inside* $\Gamma$, not counting the points on $\Gamma$. A P-*line* will be the set of all P-points lying on a circle $\gamma$ that is orthogonal to $\Gamma$, or that lie on a line through $O$. (To keep our language straight, the words point, line, circle will refer to the Euclidean notions in $\Pi$, and we will prefix a P to any word to mean the corresponding concept in the model we are building.)

Having thus defined the P-points and P-lines of our model, we can verify the incidence axioms (I1), (I2), (I3).

## Proposition 39.1
*The* P-*model satisfies* (I1), (I2), *and* (I3).

*Proof* For (I1), suppose we are given two P-points, $A, B$. If the line $AB$ passes through $O$, then it is a P-line containing them and is the only such. If $A, B$, and $O$ are not collinear, let $A'$ be the inverse of $A$ under inversion in the circle $\Gamma$ (cf. Section 37). Then there is a unique circle $\gamma$ passing through $A, A'$, and $B$. By (37.3), $\gamma$ is orthogonal to $\Gamma$, so that portion of $\gamma$ that is inside $\Gamma$ becomes a P-line containing $A$ and $B$. It is unique, because again by (37.3), any circle $\gamma$ orthogonal to $\Gamma$ that contains $A$ also contains $A'$.

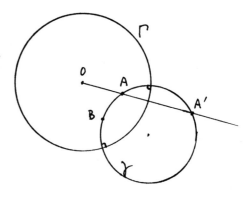

The other two axioms (I2), existence of at least two points on a line and (I3) existence of three noncollinear points, are obvious.

We see immediately that this geometry will be non-Euclidean because the parallel axiom (P) does not hold.

### Proposition 39.2

*The parallel axiom* (P) *does not hold in the P-model: There is a P-line $\gamma$ and a P-point A such that there is more than one P-line through A that is P-parallel to $\gamma$. (Of course, P-parallel means that two P-lines do not intersect.)*

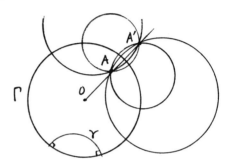

*Proof*   Take a P-line $\gamma$ in one part of our P-plane and take a point $A$ far away. Let $A'$ be the inverse of $A$. Then (by 37.3)), any circle through $A$ and $A'$ will be orthogonal to $\Gamma$, so it gives a P-line passing through $A$. There are many of these that do not meet $\gamma$, and these are all P-lines through $A$ that are P-parallel to $\gamma$.

### Definition

If $A, B, C$ are P-points on a P-line $\gamma$, we define the P-*betweenness* relation $A * B * C$ as follows. Let $O'$ be the center of $\gamma$ (which is always outside $\Gamma$), draw the line $PQ$, and project the points $A, B, C$ to points $A', B', C' \in PQ$ from the point $O'$. Then we will say $A * B * C$ (P-betweenness) if and only if $A' * B' * C'$ on the line $PQ$ (usual betweenness). If $A, B, C$ lie on a P-line that is an ordinary line through $O$, we take the usual notion of betweenness.

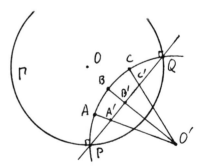

### Proposition 39.3

*The notion of P-betweenness for P-points satisfies axioms* (B1)–(B4).

*Proof*   Axioms (B1), (B2), and (B3) follow immediately from the corresponding statements for ordinary betweenness on the line $PQ$. For (B4), taking into account the circle–circle intersection property (E) in $\Pi$ and noting that two circles orthogonal to $\Gamma$ can meet at most once inside $\Gamma$, we see that to say that P-points $A, B$ are on the same P-side of a P-line $\gamma$ is equivalent to saying that $A, B$ as ordinary points are either both inside $\gamma$ or both outside $\gamma$. Thus we can define the inside of a P-triangle, and (B4) is clear.

### Definition

We define *congruence* in our P-model as follows. Two P-angles are *P-congruent* if the Euclidean angles they define are congruent in the usual sense. For line segments, we proceed as follows. Given two P-points, let the P-line joining them be the circle $\gamma$ orthogonal to $\Gamma$. Let $\gamma$ meet $\Gamma$ in two points $P, Q$, and label them so that $P$ is the one closer to $A$. For another pair of points $A', B'$ lying on a P-line $\gamma'$, label $P', Q'$ similarly. Then we say that the P-segment $AB$ is *P-congruent* to the P-segment $A'B'$ if the cross-ratio $(AB, PQ)$ is equal to the cross-ratio $(A'B', P'Q')$ (cf. Section 37 for cross-ratios).

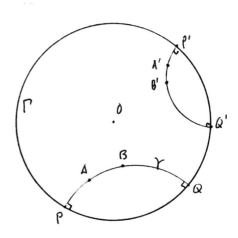

Now the real work begins, to verify the congruence axioms. We start with the easy ones.

### Proposition 39.4

*P-congruence satisfies axioms* (C2)–(C5).

*Proof* (C2) is obvious from the definition, since congruence of segments is defined by equivalence of associated quantities in the field.

(C3) requires a calculation. From the definition of cross-ratio it follows that $(AB, PQ) \cdot (BC, PQ) = (AC, PQ)$ (verify!). So when two segments are added together, the associated cross-ratios multiply. From this (C3) follows immediately.

To prove (C4), laying off angles, first suppose that we are given a point $A$ inside $\Gamma$ and a line $m$ through $A$. Let $A'$ be the inverse of $A$. Then there exists a unique circle $\gamma$ passing through $A$ and $A'$ and tangent to the line $m$. By (37.3) $\gamma$ is orthogonal to $\Gamma$. This shows that there exists a P-line at $A$ with any given tangent direction. Now, if an angle $\alpha$ is given and a P-line $\delta$ given at $A$, by (C4)

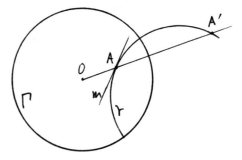

in $\Pi$ there is a unique line $m$ forming the angle $\alpha$ with $\delta$ at $A$ (and on a given side of $\delta$). Then the P-line with tangent $m$ gives the required P-angle at $A$, and is unique.

(C5) follows from the same statemnt for Euclidean angles, because congruence of angles is the same.

Before proceeding to a discussion of the remaining axioms (C1), (C6), (E), (A), and (D), we will establish the existence of rigid motion (ERM) in this model. Recall from Section 17 that a *rigid motion* is a transformation of the geometry that preserves the undefined notions of point, line, betweenness, and congruence. In our case, a P-*rigid motion* will be a transformation of the set of points inside $\Gamma$ that is 1-to-1 and onto, sends P-lines to P-lines, and preserves P-betweenness and P-congruence of angles and segments.

**Proposition 39.5** (Existence of rigid motions (ERM) for the Poincaré model)
*There are enough* P-*rigid motions of the Poincaré model so that*:

(1) *For any two* P-*points* $A, A'$, *there is a* P-*rigid motion sending* $A$ *to* $A'$.
(2) *Given* P-*points* $A, B, B'$, *there is a* P-*rigid motion leaving* $A$ *fixed and sending the ray* $\overrightarrow{AB}$ *to the ray* $\overrightarrow{AB}'$.
(3) *For any* P-*line* $\gamma$ *there is a* P-*rigid motion leaving all the points of* $\gamma$ *fixed and interchanging the two sides of* $\gamma$.

*Proof*  We start with the last property. Given a P-line $\gamma$, let $\rho_\gamma$ be the circular inversion in $\gamma$. Since $\Gamma$ is orthogonal to $\gamma$, $\rho_\gamma$ sends $\Gamma$ to itself (37.3). Also, the inside of $\Gamma$ is sent to the inside of $\Gamma$, so that the P-plane is mapped to itself, in a way that is clearly 1-to-1 and onto. Since circular inversion sends circles into circles (37.4) and is conformal (37.5), a circle orthogonal to $\Gamma$ will be sent to another circle orthogonal to $\Gamma$, in other words, $\rho_\gamma$ sends P-lines into P-lines. (Note that this works also for the limiting case of a line through $O$, which is also orthogonal to $\Gamma$.)

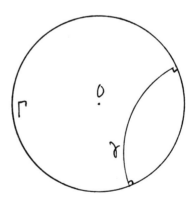

Circular inversion clearly preserves betweenness (Exercise 39.1). It preserves P-congruence of angles because this is the same as usual congruence of angles, and inversion is conformal (37.5). Also, $\rho_\gamma$ preserves P-congruence of P-segments, because this is defined by the cross-ratio, which is invariant under circular inversion (37.6). Finally, note that $\rho_\gamma$ interchanges that part of the P-plane that is inside $\gamma$ with that part that is outside $\gamma$, so $\rho_\gamma$ is a P-rigid motion as required for the third statement of (ERM). Since it leaves the points of $\gamma$ fixed and interchanges the sides of $\gamma$, it is the P-*reflection* in $\gamma$ (Section 17).

Next we will show that for any
$A \neq O$, there is a circle $\gamma$ orthogonal to $\Gamma$
(a P-line) such that the P-reflection in $\gamma$
interchanges $O$ and $A$. Let $A'$ be the inverse of $A$; let $\gamma$ be the circle with center
$A'$ that is orthogonal to $\Gamma$. Then the
construction (37.1) for the circle $\gamma$, using the same diagram (!), shows that inversion in $\gamma$ sends $A$ to $O$. Thus the P-
reflection in $\gamma$ interchanges $A$ and $O$.

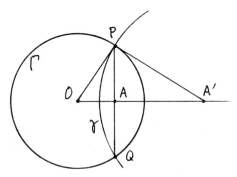

Now, since a composition of P-rigid motions is again a P-rigid motion, given
two points $A, A'$, we can first send $A$ to $O$ as above, then send $O$ to $A'$. The composition of these two reflections will be a P-rigid motion sending $A$ to $A'$, which
proves (1).

Now suppose that we are given three points $A, B, B'$. Let $\rho$ be a P-rigid motion
taking $A$ to $O$, and let $\rho(B) = C$, $\rho(B') = C'$. If we can solve problem (2) for
$O, C, C'$, in other words, if there is a P-rigid motion $\theta$ leaving $O$ fixed and sending
the ray $\overrightarrow{OC}$ to the ray $\overrightarrow{OC'}$, then $\rho^{-1}\theta\rho$ will solve the problem (2) for $A, B, B'$. So
we reduce to solving the problem for $O, C, C'$.

Let $l$ be the angle bisector of angle
$COC'$. Then $l$ is a line through $O$, which
is also a P-line. The ordinary reflection
in $l$ is clearly a P-rigid motion that
leaves $O$ fixed and sends the ray $\overrightarrow{OC}$ to
the ray $\overrightarrow{OC'}$.

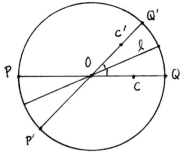

This completes the proof of (ERM) for the Poincaré model.

**Proposition 39.6**
*Axioms* (C1) *and* (C6) *hold in the Poincaré model.*

*Proof* Suppose it is required to find a point $B'$ on a P-ray emanating from a
point $A'$ such that $A'B'$ is P-congruent to a given P-segment $AB$. By (ERM) =
(39.5), there is a P-rigid motion $\varphi$ taking $A$ to $A'$. There is also a P-rigid motion
$\psi$ taking the ray $\varphi(\overrightarrow{AB})$ to the given ray from $A'$. Then $B' = \psi\varphi(A')$ is a point
on the given ray, and $AB \cong A'B'$ because rigid motions preserve congruence.
Thus (C1) holds in the Poincaré model. The uniqueness of (C1) follows from the
uniqueness of (C1) in the Euclidean plane, using (39.7) below.

To show that (C6) holds, since we have already established (C1)–(C5), we simply apply (17.1), which shows that under those circumstances, (ERM) implies (C6).

In order to discuss (E) in the Poincaré model, we first need to identify what is a P-circle. By definition, of course, it is the set of all P-points $B'$ such that the P-segment $A'B'$, for a certain fixed point $A'$, is P-congruent to a given P-segment $AB$. Since the definition of P-congruence of segments is not very intuitive, it is not easy to see immediately what kind of curves these are. First we need a lemma.

**Lemma 39.7**
*If $C, C'$ are two points inside $\Gamma$, not equal to the center of $\Gamma, O$, then the P-segment $OC$ (which is equal to the Euclidean segment $OC$, since the P-line joining $O$ and $C$ is just the usual line $OC$) is P-congruent to the P-segment $OC'$ if and only if $OC$ is congruent to $OC'$ in the ambient Euclidean plane $\Pi$.*

*Proof* Let $P$ and $Q$ be the endpoints of the diameter of $\Gamma$ passing through $O$ and $C$. Then the P-congruence of $OC$ is determined by the cross-ratio

$$(OC, PQ) = \frac{OP}{OQ} \div \frac{CP}{CQ}.$$

Let $r =$ radius of $\Gamma$ and let $x =$ Euclidean distance from $O$ to $C$. Then the cross-ratio is

$$\frac{r}{r} \div \frac{r+x}{r-x} = \frac{r-x}{r+x}.$$

If $C'$ is another point, and if the distance from $O$ to $C'$ is $y$, then we obtain similarly

$$(OC', P'Q') = \frac{r-y}{r+y}.$$

Thus, to say that $OC$ is P-congruent to $OC'$ is to say that

$$\frac{r-x}{r+x} = \frac{r-y}{r+y}.$$

Cross multiplying, we obtain

$$r^2 - rx + ry - xy = r^2 + rx - ry - xy,$$

so

$$2rx = 2ry.$$

Since our field has characteristic 0, this is equivalent to $x = y$, i.e., $OC$ is congruent to $OC'$ in the usual sense.

**Proposition 39.8**
*Every P-circle is an ordinary circle that is entirely contained in the inside of $\Gamma$, and conversely, every circle entirely inside $\Gamma$ is a P-circle.* (Warning: *The P-center of a P-circle is usually not equal to its ordinary center.*)

*Proof*  Given a P-circle $\zeta$ with P-center $A'$, consider a rigid motion $\theta$ that takes $A'$ to $O$. This will transform $\zeta$ into a P-circle with P-center $O$. Since P-congruence and ordinary congruence are the same for segments beginning at $O$ by the lemma, this image $\theta(\zeta)$ is an ordinary circle with center $O$. Then $\theta^{-1}$ will carry this ordinary circle back to the given P-circle $\zeta$. Now observe that in the proof of (ERM), all the rigid motions we needed were made out of compositions of P-reflections (which are circular inversions in suitable circles) or reflections in a line through $O$. Since all of these transformations send circles into circles (37.4), it follows that $\zeta$ is a circle. Since the transformed circle was a circle around $O$ entirely contained inside $\Gamma$, the image is also entirely contained inside $\Gamma$.

Conversely, given an ordinary circle $\zeta$ completely contained inside $\Gamma$, with (ordinary) center $O'$, draw $OO'$. Let it meet $\zeta$ at $A, B$. P-bisect the segment $AB$ at $A'$, and choose a P-reflection $\rho_\gamma$ that sends $A'$ to $O$. Then $\rho_\gamma(\zeta)$ will be a circle, the images of $A$ and $B$ will be equidistant from $O$, and this circle will be symmetric about the line $l = OO'$, which is sent into itself by $\rho_\gamma$. Hence $\rho_\gamma(\zeta)$ is a circle with center $O$, which is also a P-circle. Applying $\rho_\gamma^{-1}$, it follows that the original circle $\zeta$ is a P-circle with P-center $C$.

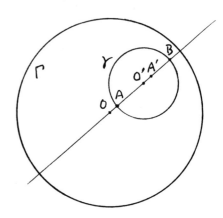

### Proposition 39.9
*The circle–circle intersection property* (E) *holds in the Poincaré model over a Euclidean ordered field F.*

*Proof*  Since P-lines and P-circles are all either usual circles or lines through $O$, and since betweenness is the same in the P-model as in the ambient Euclidean space, (E) in the P-model follows directly from (E) in the Cartesian plane $\Pi$, and this in turn follows from the Euclidean hypothesis on $F$ (16.2). Since P-circles are usual circles entirely contained inside $\Gamma$, there is no problem about any of the intersections falling outside $\Gamma$.

For the next proposition it will be convenient to introduce the notion of a distance function. In ordinary Euclidean geometry the distance function assigns to each interval a positive number, and adding segments corresponds to adding numbers. More generally, we make the following definition.

### Definition
A *distance function* on a Hilbert plane is a function $d$ that to each segment assigns an element of an ordered abelian group $G$ such that

(1)  $d(AB) > 0$ for any segment $AB$.
(2)  $d(AB) = d(A'B')$ if and only if $AB \cong A'B'$.
(3)  if $A * B * C$, then $d(AC) = d(AB) + d(BC)$.

If the group happens to be written multiplicatively, we will call it a *multiplicative* distance function. The usual distance function on the Cartesian plane over a field $F$ (Section 16) is an additive distance function with values in the additive group of the field $(F, +)$

### Lemma 39.10

*In the Poincaré model over a field $F$, the function $\mu(AB) = (AB, PQ)^{-1}$ is a multiplicative distance function with values in the multiplicative group of the field $(F_{>0}, \cdot)$.*

*Proof*   Because of our convention that $P$ is the endpoint closer to $A$, the cross-ratio $(AB, PQ)$ is in the interval $(0, 1)$ in $F$. Therefore, $\mu(AB) > 1$. We have already used it to define congruence, and we have seen that it is multiplicative (proof of 39.4). Hence $\mu$ is a multiplicative distance function.

### Proposition 39.11

*Archimedes' axiom (A) will hold in the P-model if we assume Archimedes' axiom (A')
for the field $F$. Similarly, Dedekind's axiom (D) will hold if we assume (D') in the field. (Cf. (15.4) for (A') and (D').)*

*Proof*   Using the multiplicative distance function $\mu$ of (39.10), Archimedes' axiom in the P-plane is equivalent to the following statement in $F$: Given $c, d \in F$, $c, d > 1$, $\exists n > 0$ such that $c^n > d$.

We will show that this property is a consequence of Archimedes' axiom (A') for $F$. Write $c = 1 + x$, so $x \in F, x > 0$. Then

$$c^n = (1 + x)^n = 1 + nx + \text{positive terms} \geq 1 + nx.$$

Now (A') says that for some $n, nx > d$. Hence also $c^n > d$, as required.

For Dedekind's axiom, (D') in $F$ implies (D) in II (15.4), and this clearly implies (D) in the P-plane because of the way we defined betweenness by projecting onto a line segment. (For a converse to (39.11), see Exercise 39.7.)

### Proposition 39.12

*For any point $A$ and any ray $Bb$ in the Poincaré model, there exists a limiting parallel ray (cf. Section 34) $Aa$ to $Bb$.*

*Proof*   Let the P-ray $Bb$ meet the defining circle $\Gamma$ of the Poincaré model in a point $Q$. Let $A'$ be the circular inverse of $A$ in $\Gamma$, and let $\gamma$ be the circle through

$A, Q, A'$. Then $\gamma$ defines a *P*-line, and we take $Aa$ to be the *P*-ray of that *P*-line having $Q$ at its end. Then it is clear that $Aa$ and $Bb$ are limiting parallel rays in the Poincaré model.

Using a little Euclidean geometry in the ambient Cartesian plane, we can derive a marvelous relationship between the length of a segment and the angle it makes with a limiting parallel.

**Proposition 39.13** (Bolyai's formula)
*Suppose we are given in the Poincaré model a point P, a line l, the perpendicular PQ to l, and a limiting parallel line m, making an angle α with PQ.*
   *Then*

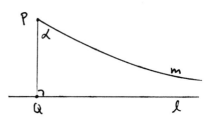

$$\tan \frac{\alpha}{2} = \mu(PQ)^{-1},$$

*where the tangent is understood to be of the corresponding Euclidean angle, and μ is the multiplicative distance function. The equality takes place in the field F.*

*Proof*   We may assume that the Poincaré model is made with a circle $\Gamma$ of radius 1 (cf. Exercise 39.23). We can move $P, Q, l, m$ so that $Q$ becomes the center of $\Gamma$, the line $l$ becomes a radius $QA$, and $P$ lies on an orthogonal radius $QB$. The limiting parallel through $P$ to $l$ will be part of a circle $\Delta$, orthogonal to $\Gamma$ at $A$. Its center therefore is at a point $C = (1, c)$ on the line $x = 1$. Let $P$ be the point $(0, y)$. Then $CP = CA$, so

$$c^2 = (c - y)^2 + 1.$$

Therefore,

$$c = \frac{1 + y^2}{2y}. \tag{1}$$

Draw a diameter $EF$ of $\Delta$ parallel to the *x*-axis. Then the angle $\alpha$ between our limiting parallel and $PQ$, called the *angle of parallelism* of the segment $PQ$, is equal to the angle $PCF$. If we draw $EP$, then the angle $PEF = \alpha/2$ (III.20). Now

$$\tan \frac{\alpha}{2} = DP/DE = \frac{c - y}{c + 1}.$$

Substituting from (1) we obtain

$$\tan \frac{\alpha}{2} = \frac{1-y}{1+y}. \tag{2}$$

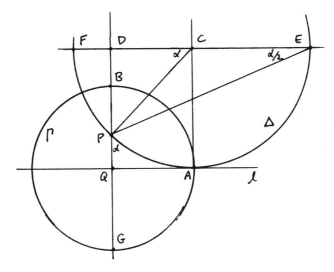

On the other hand, the multiplicative distance function is

$$\mu(PQ) = (PQ, BG)^{-1}$$
$$= \left(\frac{PB}{PG} \div \frac{QB}{QG}\right)^{-1}$$
$$= \left(\frac{1-y}{1+y} \div \frac{1}{1}\right)^{-1}$$
$$= \frac{1+y}{1-y}. \tag{3}$$

From (2) and (3) we conclude that

$$\tan \frac{\alpha}{2} = \mu(PQ)^{-1},$$

as required.

### Remark 39.13.1

From this it follows that given any angle $\alpha$ less than a right angle, there exists a segment $PQ$ with angle of parallelism equal to $\alpha$. Indeed, $\tan(\alpha/2)$ will be an element of the field $F$, and then we can find a $y \in F$ satisfying (2) above. In particular if we take $\alpha = \frac{1}{2}$ RA (one-half right angle), there will be a corresponding segment $PQ$ uniquely determined up to congruence. In this sense there is an absolute standard of length in the Poincaré model, whereas in Euclidean geometry the choice of unit length is arbitrary.

# Exercises

All exercises take place in the Poincaré model over a Euclidean ordered field $F$, unless otherwise noted. Proofs should be based on the Euclidean geometry of the Cartesian plane over $F$. In particular, do not use any of the results of Section 34 or Section 35 that depend on Archimedes' axiom.

39.1 Verify that circular inversion preserves betweenness in the Poincaré model (cf. proof of Proposition 39.5).

39.2 Show that the angle sum of any triangle in the Poincaré model is less than 2RA so this geometry is semihyperbolic (Section 34).

39.3 For any angle $\alpha$, show the existence of a line entirely contained inside the angle $\alpha$ (cf. Exercise 35.4).

39.4 Show that for any angle $\alpha < 60°$ there exists an equilateral triangle with all of its angles equal to $\alpha$.

39.5 If an equilateral triangle has sides equal to $AB$ and angles equal to $\alpha$, show that

$$\frac{2a}{1+a^2} = \frac{2t^2}{1-t^2},$$

where $a = \mu(AB)$ is its multiplicative length, and where $t = \tan(\alpha/2)$ (cf. Example 42.3.2).

39.6 Given any three angles $\alpha, \beta, \gamma$ with $\alpha + \beta + \gamma < 2RA$, show that there exists a triangle with angles $\alpha, \beta, \gamma$ in the Poincaré model. *Hint*: First show in the Cartesian plane that you can find an angle $\alpha$ meeting a circle at angles $\beta$ and $\gamma$. Then shrink or expand this figure so that it becomes a triangle in the Poincaré model.

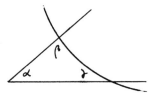

39.7 Prove the converse of Proposition 39.11, namely, if (A) or (D) holds in the Poincaré model, then (A') resp. (D') holds in $F$.

39.8 If two lines are parallel, but not limiting parallel, then they have a unique common orthogonal line.

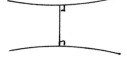

39.9 For any angle $\alpha$, there is an *enclosing line*, which is a line limiting parallel to both arms of $\alpha$.

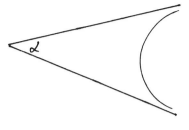

39.10 Give an alternative proof of (C1) in the Poincaré model, without using rigid motions, as follows. Given a point $A$, a P-line $\gamma$, and given a quantity $b \in F$, $0 < b < 1$, we need to find a point $B \in \gamma$ such that

$$(AB, PQ) = b.$$

Do this by showing that in Euclidean geometry, the locus of points $B$ such that $BP/BQ$ is a given ratio $k \in F$ is a *circle*. Then use (E), in the Cartesian plane, to show that this circle intersects $\gamma$ and thus find the required $B$.

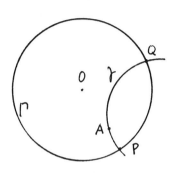

39.11 Given the circle $\Gamma$, its center $O$, and another circle $\zeta$ entirely contained inside $\Gamma$, give a ruler and compass construction (in the ambient Euclidean plane) of the P-center $\zeta$ regarded as a P-circle (cf. Proposition 39.8).

39.12 (Euclidean geometry). Find all possible ways of filling the entire Euclidean plane with triangles satisfying the following conditions:

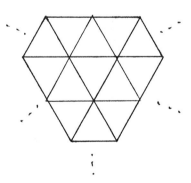

(a) The triangles are all congruent to each other. There is no overlap, and they fill the entire plane.

(b) At each vertex of the triangulation, all the angles are the same (though they may be different from the angles at a different vertex).

We consider two "ways" of filling the plane "the same" if one can be moved to the other by a dilation followed by a rigid motion.

One such triangulation is shown, where the angles at each vertex are all 60°. This is the only possibility if all angles are equal. Expect to find three more ways, allowing angles at different vertices to be different, and *prove* that you have found all possibilities.

39.13 In the Poincaré model of non-Euclidean geometry, show, in contrast to the Euclidean situation described in Exercise 39.12 above, that there are infinitely many different ways to cover the P-plane by congruent P-triangles satisfying properties (a) and (b).

In particular, prove that the plane can be covered by equilateral triangles with all angles equal to 45° and with eight meeting at each vertex. If $AB$ is a side of one of these triangles, find $\mu(AB)$.

Draw a big circle $\Gamma$ on a piece of paper, and then accurately draw enough of these P-triangles inside $\Gamma$ to show how they cover the whole P-plane. (This drawing can be accomplished entirely by ruler and compass, but don't bother listing the steps, except to show how you got the first triangle.)

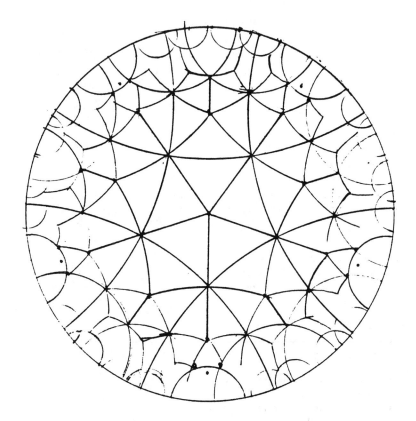

Congruent, isosceles, 72°–45°–45° triangles, filling up the Poincaré model of the non-Euclidean plane (cf. Exercise 39.13).

39.14 In the Poincaré model made inside a circle $\Gamma$ in the Cartesian plane over $F$, we have seen that any Euclidean circle $\gamma$ entirely contained inside $\Gamma$ is a P-circle (Proposition 39.8).

(a) If $\gamma$ is a Euclidean circle inside $\Gamma$ and tangent to $\Gamma$, show that there is a pencil of limiting parallel lines (a *pencil* means the set of all lines that are mutually limiting parallels at one end) such that the curve $\gamma$ is orthogonal to all the lines of the pencil. Such a curve is called a *horocycle* in the Poincaré model.

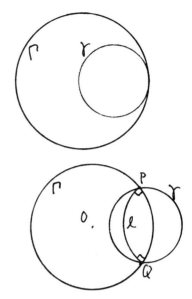

(b) If $\gamma$ is a Euclidean circle that cuts $\Gamma$ at points $P, Q$, let $l$ be the P-line having the endpoints $P, Q$. Show that the points of $\gamma$ inside $\Gamma$ form a curve of points equidistant from the P-line $l$. Such a curve is called an *equidistant curve* or *hypercycle*.

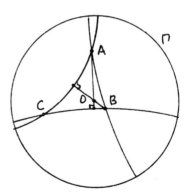

(c) Given any three distinct points $A, B, C$ in the Poincaré model, show that they are contained in a unique P-line or P-circle or horocycle or hypercycle. (Contrast to Euclidean geometry, where only the first two possibilities occur.)

39.15  Show in the Poincaré model that it is in general not possible to trisect an angle (i.e., if $\alpha$ is an angle, the angle $\frac{1}{3}\alpha$ may not exist) (cf. Section 28).

39.16  Show in the Poincaré model, in contrast to the Euclidean case (Exercise 2.14), that it is in general not possible to trisect a line segment (i.e., the 3-division points may not exist).

39.17  In the Poincaré model, show that if two altitudes of a triangle meet in a point, then the third altitude also passes through that point. Here is a method. Let the triangle be $ABC$, and suppose that the altitudes from $A$ and $B$ meet. By a rigid motion of the Poincaré plane we move that meeting point to the center $O$ of the defining circle $\Gamma$. Then those altitudes become Euclidean lines through $O$. We must show that the line $OC$ is orthogonal to the side $AB$.

The P-lines $AB, AC, BC$ are Euclidean circles orthogonal to $\Gamma$. Let $D, E, F$ be the centers of these circles. Show that the altitudes of the P-triangle $ABC$ are at the same time altitudes of the Euclidean triangle $DEF$. Then use the Euclidean theorem that the altitudes of a triangle meet (Proposition 5.6) to finish the proof.

*Note*: This is a curious method, whereby the Euclidean result is used to show (via Euclidean geometry) that the same result holds in the non-Euclidean Poincaré model. Since we now know that this result holds in both Euclidean and non-Euclidean geometry, it would be nice to have a single proof in neutral geometry that applies to both cases—cf. Exercise 40.14 and Theorem 43.15.

39.18 Show that the result of Exercise 1.15 is also valid in the Poincaré model, by moving the figure so that $P$ becomes the center of $\Gamma$ and using the Euclidean result already proved. Can you find a proof in neutral geometry that will cover both cases at once?

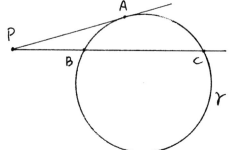

39.19 Prove a non-Euclidean analogue of (III.36) in the Poincaré model, as follows. Let $P$ be a point outside a circle $\gamma$, let $PA$ be a tangent to $\gamma$, and let $PBC$ be a secant. Let $a = \mu(PA)$, $b = \mu(PB)$, and $c = \mu(PC)$. Then

$$\left(\frac{a-1}{a+1}\right)^2 = \left(\frac{b-1}{b+1}\right)\left(\frac{c-1}{c+1}\right).$$

*Hint*: Move $P$ to the center $O$ of the Poincaré model, use the Euclidean (III.36)—cf. Proposition 20.9—and compute $\mu$ as in the proof of Proposition 39.13.

39.20 In the Poincaré model, if three circles each meet the others in two points, show that the three radical axes (Exercise 20.4) meet in a point.

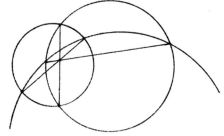

(a) One method is to suppose that two of the radical axes meet in a point $A$. Move that point to $O$, and use the Euclidean result (Exercise 20.5).

(b) Another method is to use Exercise 39.19 to define the power of a point with respect to a circle, and imitate the proofs of Exercises 20.4, 20.5.

39.21 There is another model of a non-Euclidean geometry, due to Felix Klein, constructed as follows. In the Cartesian plane over a field $F$, fix a circle $\Delta$. Then the K-points are the points inside $\Delta$, and the K-lines are chords of Euclidean lines contained inside $\Delta$. In this model the incidence axioms (I1)–(I3) and the betweenness axioms (B1)–(B4) are immediate, taking betweenness to be the same as in the Cartesian plane.

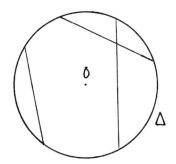

However, the model is not conformal (i.e., angles are not the same as Euclidean angles), so the definition and properties of congruence for line segments and for angles are more complicated. Rather than doing this directly, we will show in this exercise how to obtain the Klein model from the Poincaré model.

Let $\Delta$ be a circle of radius 1 centered at the origin, and in the Cartesian 3-space, place a sphere of radius 1 on the plane, with its south pole at the origin (cf. Exercise 37.1). Let $\Gamma$ be the circle of radius 2 centered at the origin. For each K-point inside $\Delta$, project it straight up to obtain a point of the southern hemisphere of the sphere, and then use the stereographic projection (Exercise 37.1) from the north pole to obtain a P-point inside $\Gamma$.

Show that this transformation gives a 1-to-1 correspondence between the points of the K-plane inside $\Delta$ with the points of the P-plane inside $\Gamma$, which sends K-lines to P-lines and vice versa. Then we can transport the notions of congruence for P-segments and P-angles to the K-plane, so that the K-plane becomes a model of a non-Euclidean Hilbert plane, isomorphic to the Poincaré model.

39.22 If $ABC$ is a triangle having a circumscribed circle, prove that the medians of $ABC$ meet in a point, as follows. Use the Klein model (Exercise 39.21) and place the center of the circumscribed circle at the center $O$ of the circle $\Delta$. Then the perpendicular bisectors of the sides of $ABC$ become diameters of the circle $\Delta$. Conclude that the $K$-midpoints of the sides of the triangle are equal to the Euclidean midpoints, and then use the Euclidean theorem about medians in the ambient plane.

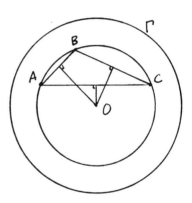

39.23 In the Cartesian plane over the field $F$, let $\Gamma$ be a circle of radius $r$ centered at the origin, and let $\Gamma'$ be a concentric circle of radius $r'$. Consider the map $\varphi$ from the set of points inside $\Gamma$ to the set of points inside $\Gamma'$ given by

$$\begin{cases} x' = kx, \\ y' = ky, \end{cases}$$

where $k = r'/r$. Show that $\varphi$ gives an isomorphism of the Poincaré model made with $\Gamma$ to the Poincaré model made with $\Gamma'$, which preserves the multiplicative distance function of Lemma 39.10. Conclude that if $\Gamma$ and $\Gamma'$ are *any* two circles in the Cartesian plane over $F$, the associated Poincaré models are isomorphic Hilbert planes.

39.24 Let $F$ be a non-Archimedean Euclidean field such as the one described in Proposition 18.4. Let $\Pi$ be the Poincaré model over $F$ and let $\Pi_0$ be the subset of points that are at finitely bounded multiplicative distance $\mu$ from some fixed point $O$. Show that $\Pi_0$ is a non-Euclidean Hilbert plane with properties (a) and (b) below.

(a) The angle sum of any triangle is less than 2RA, so it is semihyperbolic.

(b) Limiting parallel rays on distinct lines do not exist.

(c) Let $\Pi_1$ be the subset of those points of $\Pi$ whose distance from $O$ is infinitesimal. Show that $\Pi_1$ is another Hilbert plane satisfying (a) and (b) above.

(d) Show that $\Pi_0$ and $\Pi_1$ are not isomorphic Hilbert planes.

Compare Exercises 18.3–18.6.

39.25 In this and the following exercises we investigate the Poincaré model over a field that need not be Euclidean. Let $F$ be a Pythagorean ordered field, let $d \in F$, and let $\Gamma$ be the circle $x^2 + y^2 = d$, which may be a virtual circle if $\sqrt{d} \notin F$ (Exercise 37.17). We define the *Poincaré model* in $\Gamma$ as in the text. The *interior* of $\Gamma$ is the set of points $(x, y)$ with $x^2 + y^2 < d$. These are the P-points. The P-lines are segments of circles $\gamma$ orthogonal to $\Gamma$ (which means stable under circular inversion in $\Gamma$ (Exercises 37.16, 37.17)) as before.

(a) Show that the incidence axioms (I1)–(I3) holds, as in Proposition 39.1.

(b) If $\gamma$ is a P-line, the intersection points $P, Q$ of $\gamma$ with $\Gamma$ may not exist, but the line $PQ$ is still well-defined: It is the perpendicular to $OO'$ at the inverse of $O'$, where $O'$ is the center of $\gamma$. So we can define betweenness as before. Show that betweenness satisfies axioms (B1)–(B3) as in the text.

39.26 With hypotheses as in Exercise 39.25, now suppose that $F$ satisfies the additional condition $(*d)$: For any $a \in F$, if $a^2 - d > 0$, then $\sqrt{a^2 - d} \in F$.

(a) Show that the circle–circle intersection property (E) holds for circles $\gamma, \delta$ orthogonal to $\Gamma$. *Hint*: Write the equations of $\gamma, \delta$, and show that the square root needed to find their intersection exists because of condition $(*d)$.

(b) Conclude that axiom (B4) also holds in this model.

39.27 Continuing with the situation of the two previous exercises, if $\gamma$ is a P-line, the points of intersection $P, Q$ with $\Gamma$ do not exist, but at least they have coordinates in the field $F(\sqrt{d})$. Hence we can compute the cross-ratio $(AB, PQ)$ in that field, and define congruence of angles and segments as in the text.

(a) Using condition $(*d)$, show that for any point $A'$ outside $\Gamma$, there exists a circle $\gamma$ with center $A'$ and orthogonal to $\Gamma$.

(b) Verify that Propositions 39.4, 39.5, 39.6 hold in this model, so it is a Hilbert plane. We call it the Poincaré model in the (virtual) circle $x^2 + y^2 = d$. You will need part (a) of the proof of Proposition 39.5.

39.28 In the model of Exercise 39.27, if $\sqrt{d} \notin F$, show that there are no limiting parallel rays on distinct lines, but that any two parallel lines have a common orthogonal.

39.29 For an example of a field $F$ satisfying the conditions of Exercises 39.25–39.28, let $K$ be a Pythagorean ordered field, for example the field of constructible real numbers; let $F = K((z))$ be the field of Laurent series over $K$ (Exercise 18.9); and let $d = z$. Verify that $d > 0$, $\sqrt{d} \notin F$, and that $F$ satisfies condition $(*d)$.

39.30 For an Archimedean example of a field as in Exercise 39.29, let $F$ be the field of all those real numbers that can be expressed using rational numbers and a finite number of operations $+, -, \cdot, \div, a \mapsto \sqrt{1 + a^2}$, and $a \mapsto \sqrt{a^2 - \sqrt{2}}$, provided that $a^2 - \sqrt{2} > 0$.

(a) $F$ is a Pythagorean ordered field, $d = \sqrt{2}$ is in $F$, and $F$ satisfies condition $(*d)$ of Exercise 39.26 for $d = \sqrt{2}$.

(b) Let $\varphi : \mathbb{Q}(\sqrt{2}) \to \mathbb{R}$ be the homomorphism that makes $\varphi(\sqrt{2}) = -\sqrt{2}$. Show inductively that $\varphi$ extends to a homomorphism $\varphi$ of $F$ to $\mathbb{R}$.

(c) Since $\varphi(\sqrt{2}) < 0$, conclude that $\sqrt{2}$ cannot be a square in $F$.

39.31 Show that in the Poincaré model in the virtual circle $x^2 + y^2 = \sqrt{2}$ over the field $F$ of Exercise 39.30, not every segment can be the side of an equilateral triangle, as follows.

(a) If $x \in F$ with $0 < x$ and $x^2 < \sqrt{2}$, let $AB$ be the segment from $(0, 0)$ to $(x, 0)$ in the Poincaré model, and show that

$$\mu(AB) = \frac{\sqrt[4]{2} + x}{\sqrt[4]{2} - x}.$$

(b) If there is an equilateral triangle with side $AB$, let the angle at a vertex be $\alpha$, and let $t = \tan(\alpha/2)$. Use Exercise 39.5 to show that

$$t = \sqrt{\frac{\sqrt{2} - x^2}{3\sqrt{2} + x^2}} = \frac{1}{3\sqrt{2} + x^2} \sqrt{6 - 2x^2\sqrt{2} - x^4}.$$

(c) Now take a suitable $x$, such as $x = \sqrt{3} - 1$, and use an argument similar to the previous exercise to show that the corresponding $t$ is not in $F$. Hence the equilateral triangle with side $AB$ does not exist. *Hint*: For these two exercises, it may be useful to review the techniques used in Exercises 16.10–16.14.

# 40   Hyperbolic Geometry

In the earlier sections of this chapter we have seen something of the development of neutral geometry and the study of the angle sum of a triangle using Archimedes' axiom. We have also seen the Poincaré model of a non-Euclidean geometry over a field. For the full development of the geometry of Bolyai and Lobachevsky, we need the limiting parallels. The existence of these limiting parallels, which we have seen in the Poincaré model (39.12), does not follow in the axiomatic treatment from what we have done so far (Exercises 39.24, 39.28). Therefore, following Hilbert, we will take the existence of the limiting parallels as an axiom. This axiom is quite strong. It will allow us to develop non-Euclidean geometry independently of Archimedes' axiom. It also allows the construction of an ordered field out of the geometry (Section 41), and a proof that the abstract geometry is isomorphic to the Poincaré model over this field (Section 43). Using coordinates from this field we can develop non-Euclidean analytic geometry and trigonometry (Section 42).

So at this point we start the axiomatic development of hyperbolic geometry, which is essentially the "classic" non-Euclidean geometry of Bolyai and Lobachevsky, freed from hypotheses of continuity. In particular, we will not use the circle–circle intersection axiom (E) nor Archimedes' axiom (A). Instead, we use Hilbert's axioms of incidence, betweenness, and congruence plus the following *hyperbolic axiom* (L):

**L.** For each line *l* and each point *A* not on *l*, there are two rays *Aa* and *Aa'* from *A*, not lying on the same line, and not meeting *l*, such that any ray *An* in the interior of the angle *aAa'* meets *l*.

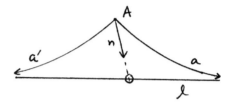

Note that (L) immediately implies that the geometry is non-Euclidean, because the two rays *Aa* and *Aa'* lie on distinct lines through *A* that will both be parallel to *l*.

**Definition**

A Hilbert plane satisfying (L) will be called a *hyperbolic* plane, or a *hyperbolic geometry*.

We will see shortly (40.3) that the angle sum of a triangle in a hyperbolic *plane* is less than 2RA, so this terminology is consistent with the term semi-hyperbolic introduced earlier (Section 34).

Recalling the definition of limiting parallel rays from Section 34, we see that if we pick any point *B* on *l* and let *Bb*, *Bb'* be the two rays from *B* lying on *l*, then *Aa* will be limiting parallel to *Bb* and *Aa'* limiting parallel to *Bb'*. Thus (L) implies that for any point *A* and any ray *Bb*, there exists a limiting parallel *Aa* to *Bb*. We define an *end* to be an equivalence class of limiting parallel rays (34.13).

**Definition**

For any segment *AB*, let *b* be a line perpendicular to *AB* at *B*; choose one ray *Bb* on the line *b*, and let *Aa* be the limiting parallel ray to *Bb*, which exists by (L). Then we call $\alpha = \angle BAa$ the *angle of parallelism* of the segment *AB*, and we denote it by $\alpha(AB)$. (Lobachevsky uses the notation $\Pi(AB)$.)

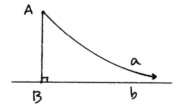

Note that the angle of parallelism in well-defined: If we reflect $Aa$ in the line $AB$, then clearly it will be limiting parallel to the other ray on $b$, so that the angle $\alpha$ is independent of which ray we chose on the line $b$. Note also that the angle of parallelism $\alpha$ is necessarily *acute*, because the two limiting parallels from $A$ to $b$ do not lie on the same line, by (L).

### Proposition 40.1

*The angle of parallelism varies inversely with the segment*:

(a)  $AB < A'B' \Leftrightarrow \alpha(AB) > \alpha(A'B')$.
(b)  $AB \cong A'B' \Leftrightarrow \alpha(AB) = \alpha(A'B')$.

*Proof*  First suppose that $AB \cong A'B'$. Then by the (ASL) congruence theorem for limit triangles (Exercise 34.10) it follows that $\alpha(AB) = \alpha(A'B')$.

Next, suppose $AB < A'B'$. Mark off $C$ on the ray $\overrightarrow{AB}$ such that $AC = A'B'$, draw the perpendicular $c$ to $AC$ at $C$, and let $Aa'$ be the limiting parallel from $A$ to $Cc$. Then $\alpha' = \angle CAa'$ is $\alpha(AC) =$ $\alpha(A'B')$.

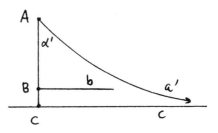

Let $Bb$ be the ray perpendicular to $AB$ at $B$ on the same side of $AC$ as $a'$ and $c$. I claim that $Bb$ meets $a'$. If not, then the ray $Bb$ would be in the interior of the angles $CAa'$ and $ACc$, meeting neither the ray $a'$ nor $c$, and so it would be also limiting parallel to $Aa'$ and $Cc$ by (34.12.1). But this contradicts the fact that the angle of parallelism is always acute, since $Bb \parallel Cc$ and the angles at $B$ and $C$ are both right angles.

So $Bb$ meets $Aa'$, and this implies that the limiting parallel from $A$ to $Bb$ makes an angle $\alpha$ greater than $\alpha'$, i.e., $\alpha(AB) > \alpha(A'B')$.

Reversing the roles of $AB$ and $A'B'$ we find that if $AB > A'B'$, then $\alpha(AB) < \alpha(A'B')$. Combining all three results now gives the desired reverse implications.

### Remark 40.1.1

We will see later (40.7) that for every acute angle $\alpha$, there exists a segment $AB$ with $\alpha(AB) = \alpha$.

Our next goal is to establish some results about limiting parallel rays, limit triangles, and parallel lines that are not limiting parallel. We have already seen two congruence results (ASL) = (Exercise 34.10) and (ASAL) = (Exercise 34.9). We will prove some others, analogous to those for ordinary triangles in Euclid's *Elements*, Book I.

**Proposition 40.2** (Exterior angle theorem)
*If AB is a segment, with limiting parallel rays emanating from A and B, then the exterior angle β at B is greater than the interior angle α at A.*

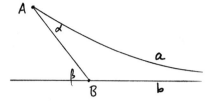

*Proof*   Because the ray through $A$ making an angle $\beta$ with $AB$ is parallel to $l$ (I.27) we know at least that $\alpha \leq \beta$.

So suppose $\alpha = \beta$. Let $a'$ and $b'$ be the opposite rays to $a$ and $b$. The supplementary angles at $A$ and $B$ will also be equal. Since $AB$ is equal to itself, we can apply (ASAL) = (Exercise 34.9) to $AB, a, b$, and $BA, b', a'$. We conclude that $a'$ is also limiting parallel to $b'$.

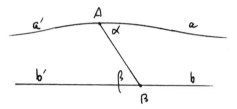

But this contradicts the axiom (L), which says the two limiting parallels from $A$ to $b$ do not lie on the same line. Therefore, $\alpha < \beta$, as required.

**Corollary 40.3**
*In a hyperbolic plane, the sum of the angles of any triangle is less than two right angles.*

*Proof*   According to (34.6), for any triangle there is a Saccheri quadrilateral whose top two angles are equal to the angle sum of the triangle. So we have only to prove that the top two equal angles of any Saccheri quadrilateral are acute.

Let the Saccheri quadrilateral be $ABCD$, with base $AB = l$. Draw limiting parallels from $C$ and $D$ to $l$, with end $\omega$ by axiom (L). Then by (40.1) the angles of parallelism $\alpha$ are equal.

Looking at the limit triangle $CD\omega$, by the exterior angle theorem (40.2), $\beta > \gamma$. On the other hand, by (34.1), the top angles $\alpha + \gamma$ and $\delta$ of the Saccheri quadrilateral are equal. We conclude that $\alpha + \beta > \alpha + \gamma = \delta$, and so $\delta$ must be acute.

**Remark 40.3.1**
Note how different this proof is from the proof of the Saccheri–Legendre theorem (35.2), which reaches the same conclusion under different hypotheses.

There we made use of Archimedes' axiom and a countable limiting process. Here we do not need (A), but we use instead the powerful axiom (L) on the existence of limiting parallels. This result says that a hyperbolic plane is semi-hyperbolic, thus justifying the terminology introduced earlier (Section 34).

**Proposition 40.4** (AAL)

*Given two limit triangles ABlm and A'B'l'm', suppose that the angles at A and B are equal respectively to the angles at A' and B'. Then also the sides AB and A'B' are equal.*

*Proof*  If not, let us suppose that $AB > A'B'$. Choose a point $C$ on $AB$ such that $CB = A'B'$, and draw a ray $n$ at $C$, on the same side of $AB$ as $l$ and $m$, making an angle equal to the angle at $A'$, which is also equal to the angle at $A$. Now comparing $C, B, n, m$ to the limit triangle $A'B'l'm'$, it follows from (ASAL) = (Exercise 34.9) that $n$ is limiting parallel to $m$.

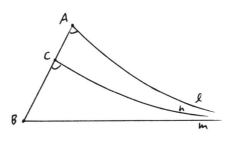

Then by transitivity (34.11) it follows also that $l$ is limiting parallel to $n$. But this contradicts the exterior angle theorem (40.2) because the angle at $C$, which is exterior to the limit triangle $ACln$, is equal to the angle at $A$.

We conclude that $AB = A'B'$, as required.

**Remark 40.4.1**

For some results about triangles with two or three "limit angles," see Exercises 40.2, 40.8.

**Theorem 40.5**

*In a hyperbolic plane, if l and m are two parallel lines that are not limiting parallels, then there is a unique line in the plane that is perpendicular to both of them.*

*Proof*  Let $l$ and $m$ be two parallel lines that are not limiting parallels. Let $AB$ and $CD$ be two perpendiculars from points $A, C$ on $l$ to $m$. If $AB = CD$, then $DBCA$ is a Saccheri quadrilateral, and hence the line joining the midpoints of $AC$ and $BD$ will be perpendicular to both $l$ and $m$, by (34.1).

If $AB \neq CD$, we may assume $CD > AB$, and we proceed as follows. Take $E$ on $CD$ such that $AB = ED$. Let $n$ be a ray through $E$ making the same angle with $ED$ as $l$ makes with $AB$. I claim that $n$ will meet $l$ in a point $F$. Indeed, let $p$ be a limiting parallel from $B$ to $l$. Since by hypothesis $l$ and $m$ are not limiting parallels, this ray does not lie on the line $m$. Let $q$ be the ray through $D$ making the same angle with $m$ as $p$ does at $B$. Then $q$ is parallel to $p$ by (I.28), but not a limiting parallel, by the exterior angle theorem (40.2). On the other hand, applying (ASAL) to *ABlp* and *EDnq*, we find that *q* is limiting parallel to *n*. Therefore, *n* is not limiting parallel to *p*, and hence *n* must meet *l* at some point *F*. (In the figure we put *F* on the far side of *A* from *C*, but the proof works equally well if *F* is between *A* and *C*.) Let *FG* be perpendicular to *m*.

Now take *H* on *l* such that *AH* = *EF*, and take *K* on *m* such that *BK* = *DG*. Then comparing the quadrilaterals *EFDG* and *AHBK*, two applications of (SAS) show that *FG* = *HK* and *HK* is perpendicular to *m*. Thus *GKFH* is a Saccheri quadrilateral, and the line joining the midpoints of *FH* and *GK* will be perpendicular to both *l* and *m*.

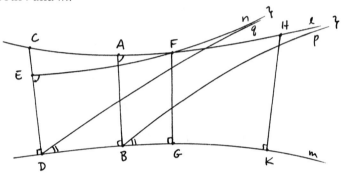

It remains to show the uniqueness of the line perpendicular to both *l* and *m*. Suppose to the contrary that *AB* and *CD* were two common perpendiculars to *l* and *m*. Then *ABCD* would be a rectangle, which is impossible—cf. (40.3) and (34.7).

## Proposition 40.6

*Given an angle in the hyperbolic plane, there is a unique line (called the* enclosing line *of the angle) that is limiting parallel to both arms of the angle.*

*Proof*  Let *O* be the vertex of the angle, and choose points *A, B* on the two arms of the angle, at equal distance from *O*. It will be convenient at this point to introduce a new notation. We denote by α the end of the ray *OA*, that is, the equivalence class of all rays limiting parallel to *OA*. Then we may draw the line *B*α, meaning, let *B*α be the ray through *B* limiting parallel to *OA*. We may also speak of the limit triangle *AB*α, consisting of the segment *AB* plus the two limiting parallel rays *A*α and *B*α.

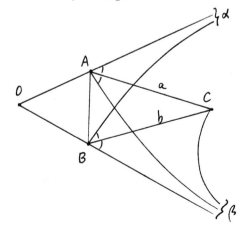

To continue our proof, let α be the end of $OA$, and let β be the end of $OB$. Draw $B\alpha$ and $A\beta$. Let $a$ be the ray bisecting the angle $\alpha A\beta$, and let $b$ be the ray bisecting the angle $\alpha B\beta$. Note by symmetry (!) that the bisected angles at $A$ and $B$ are equal. We distinguish three cases.

*Case 1*   The lines $a$ and $b$ meet at a point $C$. By symmetry (!) $AC = BC$. Draw the line $C\beta$. Then by (ASL) = (Exercise 34.10) applied to the limit triangle $AC\beta$ and $BC\beta$, the angles at $C$ of these two triangles are equal. But this is clearly not so, so this case cannot occur. (See diagram on previous page.)

*Case 2*   The rays $a$ and $b$ are limiting parallel with an end $\gamma$. In this case the ray $B\gamma$ is in the interior of the angle $AB\beta$, so it meets $A\beta$ in a point $C$. By (AAL) = (40.4) applied to the limit triangles $AC\gamma$ and $BC\beta$, the sides $AC$ and $BC$ are equal. Therefore, by (I.5) the angles $BAC$ and $ABC$ are equal. But this is not so, because the angle $BAC$ is also equal to the angle $AB\alpha$, which is properly contained in the angle $ABC$. So this case cannot occur either.

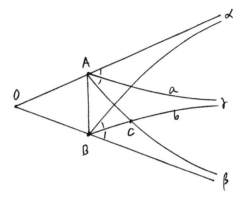

*Case 3*   The only remaining possibility is that $a$ and $b$ are parallel but not limiting parallels. Then by (40.5) there is a common perpendicular line $l$, meeting $a$ at $C$ and $b$ at $D$. I claim that $l$ is the required enclosing line, i.e., $l$ has the ends α and β.

By symmetry it is enough to show that $l$ has end β. If not, draw the lines $C\beta$ and $D\beta$, which will be distinct from $l$. We compare the limit triangles $AC\beta$ and $BD\beta$. The angles at $A$ and $B$ are equal, by construction. The sides $AC$ and $BD$ are equal by symmetry (!), so by (ASL) the angles at $C$ and $D$ are equal. It follows that $C\beta$ and $D\beta$ make equal angles with $l$ at $C$ and $D$, which contradicts the exterior angle theorem (40.2). We conclude that $l$ has ends α and β, as required.

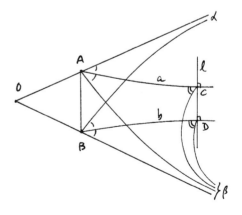

The uniqueness of the enclosing line is clear, because by (L) we cannot have two distinct lines that are limiting parallel at both ends.

### Corollary 40.7

*For any acute angle α, there exists a line that is limiting parallel to one arm of the angle and orthogonal to the other arm of the angle. In particular, there is a segment whose angle of parallelism is equal to α.*

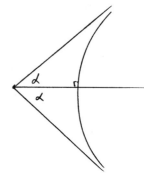

*Proof* Given the acute angle α, we double it, and consider the enclosing line (40.6) of the angle 2α. This will be orthogonal to the angle bisector of 2α, which is one arm of the original angle α. Thus α becomes the angle of parallelism of the segment cut off on that arm of the angle.

### Remark 40.7.1

Combining with (40.1), we see that there is a one-to-one correspondence between the set of congruence equivalence classes of line segments and the set of congruence equivalence classes of acute angles, given by associating a segment *AB* to its angle of parallelism α. In particular, there is a uniquely determined standard or absolute segment size corresponding to one-half of a right angle.

Be careful, however, because this correspondence does not send sums of segments into sums of angles. There is a more complex relationship that we will see later (Exercise 42.7).

### Proposition 40.8

*In a hyperbolic plane, Aristotle's axiom holds, namely, given an angle α and a segment AB, there exists a point C on one arm of the angle such that the perpendicular CD from C to the other arm of the angle is greater than AB.*

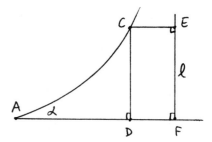

*Proof* Given the angle α at *A*, let *l* be a line limiting parallel to one arm of α and meeting the other arm at right angles at a point *F* (40.7). Take *E* on *l* such that *EF = AB*. Draw a perpendicular to *l* at *E*, and let it meet the other arm of the angle α at *C* (cf. Exercise 34.12). Drop a perpendicular *CD* from *C* to *AF*.

# Volgen des Ersten Buch

### Euclidis Proposition/Das ist/Fürgä=
### ben/oder Schlußreden.

## I. Die erst Proposition/ Lehr

**Auff ein gerade linj so einer bekhantten lenge ist / ainen gleichseittigen Triangel stellen.**

### Warnung an den Leser.

Reündelicher lieber leser/dieweil die Demonstrationes(das ist grundtlich vnnd onwider sprechliche beweysungen deß yenigen/so Euclides in seinen Proposition als warhafftig fürgibt)nit von jme dem Euclide selbs/sondern von andern hoch gelerten khüstreichen menneren/als Theone/Hypsicle/Campano/rc.hinzügesetzt worden: Zü dem bemelte Demonstration/ ettwa schwärlich von vngelertten mögen vernommen vnnd begriffen werden/vnnd dann ein einfeltiger Teütscher liebhaber diser khünsten woll begnügt vnnd zefriden ist/So er die sach versteht/ob er schon vrsach vnnd den grund desselben nitt allmal erkhent:Hab ich solche Demonstration zü zeitten außgelassen/vnnd (welchs ich dem leser nützer vnd angenemmer vermain) an statt derselben/den gebrauch vnnd nutz solcher Proposition/wa ich das füglich sein vermaint angezaigt/vnnd mit Exemplen vnnd der ziffer zimlich erklertt.

Vnd seittenmal die Schlußreden oder Proposition Euclidis zwaierlai seind Dann etliche schlecht ein aigenschafft anzaigen/welche zübeweisen andere ain grund ist (Als da seind in disem ersten büch die 4.5.6 propositiones/rc.) Etlich aber lehren ettwas auß solchem grund machen/alß dise Erste ist/vnd darnach die 9.10.11/rc. Hab ich allmal mit fleiß anzaigt/wie sollichs zümachen sei so Euclides leeret/vnnd zü vnderschaid allweg zü der zaal der Proposition geschriben/Lehrt.Auch alß dann gewonlich den grund sollicher handlung ainfeltiger weiß mit angezaigt.Auch hab ich zü zeitten in den figuren/büchstaben oder ziffern gebraucht/zü zeitten nicht/nach dem ich hab mögen erachten füglich vnd verstendtlich sein woll der Leser für gütl haben.Waß forthin in disem büch von linien geredt/versthe alles von rechten gestracken linien.

*Theoremata.*

*Problemata*

### Figur vnnd Erklärung der ersten Proposition.

Wiewol dise proposition leichtlich mag verstanden werden/auß beigesetzter figur/will ich sich je doch(dieweil sy die erst)weitleffig erklerē.Die für geben linj darauff ich den triangell soll machen/ist bezaichnet mit denn büchstaben a b/sollicher linj lenge begreiff ich mitt einem zirckel/vnnd setz den ainen füß in den puncten a/vnd reiß mit dem andern den zirckel b c d/darnach setz den ainen füß in den puncten b/vnnd reiß den zirckel

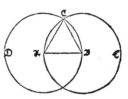

a c c/dise zwen zirckell werden on zweiffel gleich sein/dann sy baid mit onuer ruckhten zirckel/in ainer weittin beschriben/Vnnd bey den puncten c/gehn sy durcheinander/vnd machen ain creütz.demnach zeuch von den puncten c/gegen dem a ain rechte linj/vnnd dergleichen aine gegen dem b/ so hastu den tria ngel

Consider the quadrilateral *DFCE*. Because of (40.3), the angle at *C* must be acute. Therefore, *CD* > *EF* = *AB* (34.2), as required.

### Remark 40.8.1
In fact, a stronger result is true, namely, given α and *AB* as above, one can find *C* such that *CD* = *AB*. The proof uses hyperbolic trigonometry (Exercise 42.8).

Now, as an illustration of the techniques of this section, we will give the hyperbolic version of a familiar Euclidean theorem on the angle bisectors of a triangle. The fact that the (internal) angle bisectors of a triangle meet in a point is true in neutral geometry, hence both in Euclidean and hyperbolic geometry, as we have seen before (Exercise 11.6). The following result has to do with the external angle bisectors of a triangle.

### Proposition 40.9
*In a hyperbolic plane, let ABC be a triangle, and consider the (internal) angle bisector at A and the external angle bisectors at B and C.*

(a) *If two of these angle bisectors meet in a point, so does the third.*
(b) *If two of these angle bisectors have a common perpendicular line l, then the third is also perpendicular to l.*
(c) *If two of these angle bisectors are limiting parallels, so is the third, at the same end.*

*Proof* (a) If two of them meet in a point *Y*, then *Y* is equidistant from all three sides of the triangle; hence it lies on the third angle bisector. The proof in this case is the same as the Euclidean case (IV.4).

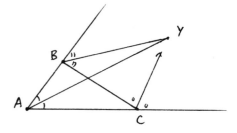

(b) Suppose that the angle bisectors at *A* and *B* have a common perpendicular line *l*.

We first claim that *l* cannot meet any side of the triangle. If it meets one side, then by reflecting in the two angle bisectors, it will meet the other two sides, and it will meet all three at the same angle. Two out of three of these intersections (in the diagram *V*, *W*) will have the angles in corresponding positions, so that by (I.28) the lines *BC* and *AC* will be parallel. This contradicts their meeting at the point *C*. Thus *l* cannot meet any side of the triangle.

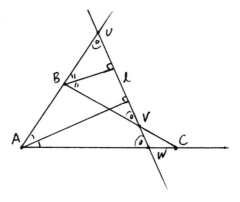

Secondly, we note that $l$ cannot be a limiting parallel to any side of the triangle. If it were, then by reflecting in the angle bisectors, it would be limiting parallel to the other two sides, and so would have three ends, which is absurd.

So $l$ neither meets nor is limiting parallel to any side of the triangle; hence by (40.5) it has a common perpendicular with each side of the triangle. Using the lemma below, the first and second of these common perpendiculars are equal. Similarly, the first and third are equal, because the angle bisectors at $A$ and $B$ are orthogonal to $l$. Therefore, the second and third are equal, and using the lemma in the other direction, we see that the angle bisector at $C$ is also perpendicular to $l$.

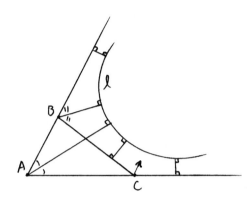

(c) This third case of the proposition follows by elimination. Suppose two angle bisectors are limiting parallel. If the third is not, then it either meets one of the others or has a common perpendicular, which puts us in case (a) or (b), contradicting the first two being limiting parallel.

**Lemma 40.10**
*Consider a five-sided figure ABCDE with right angles at A, B, C, D. Then AC = BD if and only if the angle bisector at E meets the opposite side at a point F at right angles.*

*Proof* First suppose that the angle bisector at $E$ meets $AB$ at a point $F$, making a right angle there. Then reflection in the line $EF$ sends the line $AB$ into itself and interchanges the lines $CE$ and $DE$. So the segments $AC$ and $BD$ are interchanged, because they are the unique common perpendiculars (40.5) between the lines $AB$ and $CE$ and $AB$ and $DE$. Hence $AC = DB$.

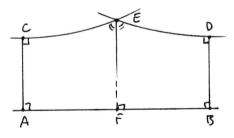

Conversely, suppose $AC = DB$. Draw the line $CD$. Then $ABCD$ is a Saccheri quadrilateral, and the angles at $C$ and $D$ are equal (34.1). It follows that the base angles of the triangle $CDE$ are equal. Hence it is an isosceles triangle, and the angle bisector at $E$ will meet $CD$ at its midpoint at right angles. Now it follows from (34.1) that this line continued will meet $AB$ at its midpoint $F$, at right angles.

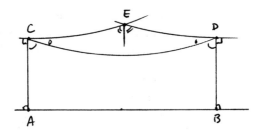

### Remark 40.10.1

To make a more unified statement of (40.9) we will define an *ideal point* $P^*$ to be an equivalence class of lines, all of which have a common perpendicular line $p$. We will say that $P^*$ lies on a line $l$, if $l \perp p$. We define a *generalized point* to be either a usual point, or an end, or an ideal point. Using this language, we can say that the three angle bisectors of (40.9) meet in a common generalized point.

## Exercises

The following exercises all take place in a hyperbolic plane, that is, a Hilbert plane satisfying (L).

40.1 If two lines $l$, $m$ have a transversal $n$ that makes equal alternate interior angles, then $l$, $m$ are parallel but not limiting parallel. Furthermore, in that case there is a unique point $P$ such that every transversal that makes equal alternate interior angles to $l$ and $m$ passes through $P$.

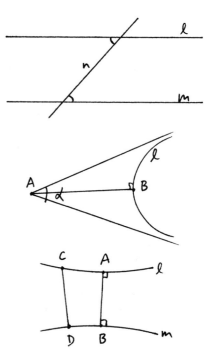

40.2 (ALL) Suppose we are given equal angles at two points $A$ and $A'$, and let $l$ and $l'$ be their enclosing lines. Show that the perpendicular $AB$ from $A$ to $l$ is equal to the perpendicular $A'B'$ from $A'$ to $l'$.

40.3 If $l$ and $m$ are two parallel, but not limiting parallel, lines, show that their common perpendicular $AB$ is the shortest distance between the two lines. Namely, show for any other points $C \in l$ and $D \in m$ that $CD > AB$.

40.4 Show that ends of lines behave somewhat like points, as follows.

(a) Given a point $P$ and an end $\alpha$, there exists a unique line $l$ passing through $P$ and having end $\alpha$.

(b) Given two distinct ends $\alpha, \beta$, there exists a unique line $l$ having ends $\alpha$ and $\beta$.

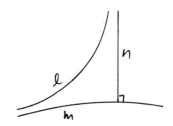

40.5 Given two lines $l$ and $m$, limiting parallel at one end, show that there exists a line $n$, limiting parallel to (the other end of) $l$, and orthogonal to $m$.

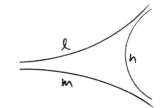

40.6 Given two lines $l$, $m$, limiting parallel at one end, show that there exists a third line $n$, limiting parallel to the other ends of both $l$ and $m$.

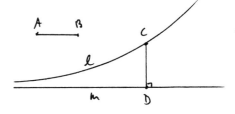

40.7 Given two lines $l$, $m$, limiting parallel at one end, and given a segment $AB$, no matter how large or how small, there exists a point $C$ on $l$ such that the perpendicular $CD$ to $m$ is equal to $AB$. *Hint*: Take $m'$ perpendicular to $AB$ through $B$ and let $l'$ be the limiting parallel to $m'$ through $A$. Apply Exercise 40.5 to both the pair $l$, $m$ and the pair $l', m'$, and compare.

40.8 (LLL). Let $l, m, n$ be three lines, each limiting parallel to the other two at opposite ends.

(a) Show that the three midlines (Exercise 34.11) to the three pairs of limiting parallel rays are orthogonal to the opposite sides of the trilimit triangle $l, m, n$, and all meet in a single point $A$, which is equidistant from $l, m, n$.

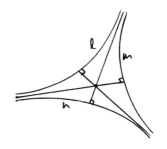

(b) If $l', m', n'$ is another such trilimit triangle, with corresponding point $A'$, show that the distance from $A$ to the three sides $l, m, n$ is equal to the distance from $A'$ to the three sides $l', m', n'$ of the second triangle.

(c) Given any point $P$ on one side of the trilimit triangle, show that the perpendiculars $PQ$, $PR$ from $P$ to the other two sides make a right angle at $P$.

40.9 Given two angles $\alpha, \beta$, with $\alpha + \beta < 2$RA, show that there exists a limit triangle with angles $\alpha, \beta$.

40.10 A *limit quadrilateral* is a figure consisting of four lines $l, m, n, p$, with each limit parallel at opposite ends to the next, in cyclic order.

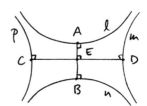

(a) If *lmnp* is a limit quadrilateral, show that opposite sides are parallel but not limit parallel.

(b) Show that the common orthogonals $AB$ of $l$ and $n$ and $CD$ of $m$ and $p$ meet at right angles at a point $E$.

(c) Show that there exists a limit quadrilateral with $AB$ equal to any prescribed segment.

(d) Two such limit quadrilaterals can be moved one to the other by a rigid motion of the plane if and only if the segment $AB$ of the first is equal to one of the segments $A'B'$ or $C'D'$ of the second.

40.11 Show that ideal points (40.10.1) behave somewhat like regular points, as follows.

(a) Given a (regular) point $P$ and an ideal point $Q^*$, there is a unique line containing them both.

(b) Given an end $\alpha$ and an ideal point $Q^*$, and assuming that $\alpha$ is not an end of the defining line $q$ of $Q^*$, then there is a unique line containing $Q^*$ with end $\alpha$.

(c) Any two distinct lines have a unique generalized point in common.

40.12 You may have noticed while doing Exercise 40.11 that two ideal points do not necessarily lie on a line. So we define a *generalized line* to be either.

(1) a regular line, together with its two ends and ideal points, or

(2) a *limit line*, which consists of an end $\alpha$, together with all ideal points $P^*$ whose defining line $p$ contains $\alpha$, or

(3) an *ideal line*, which consists of all ideal points $P^*$ whose defining line $p$ contains a fixed (regular) point $L$.

Show that the set of all generalized points of the hyperbolic plane, together with the subsets of generalized lines, forms a *projective plane* (Exercise 6.3). In particular, any two generalized points lie on a unique generalized line, and any two generalized lines meet in a unique generalized point.

40.13 Let $ABC$ be any triangle. Show that the external angle bisectors at $A$, $B$, $C$ form a "generalized triangle," i.e., a set of three lines meeting in generalized points $X$, $Y$, $Z$. Show that the internal angle bisectors of $ABC$, which meet at a point $W$, are the altitudes of the new triangle $XYZ$.

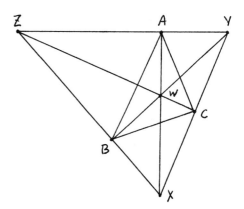

40.14 Reverse the argument of Exercise 40.13 to prove that in any triangle $ABC$, the three altitudes will meet in a generalized point.
*Hint*: Let $BD$ and $CE$ be two altitudes. Reflect the line $DE$ in $AB$ and in $AC$ to get two new lines, which meet at a generalized point $F$. Show that $B$ is equidistant from the three sides of the (generalized) triangle $DEF$, and from this conclude that $F$ is a real point (not an end or an ideal point). Now apply Exercise 40.13 to the triangle $DEF$. Conclude that $BD$, $AF$, $CE$ meet in a generalized point $G$, and that $F$ lies on $BC$, and $AF$ is orthogonal to $BC$, so in fact, $AF$ is the third altitude of the original triangle.

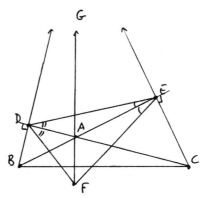

Note that if we assume that two altitudes of the triangle meet in a (regular) point, then the entire proof can be carried out in a Hilbert plane with no further hypothesis, i.e., in neutral geometry.

40.15 Extend the theorem on (internal) angle bisectors of a triangle as follows. Consider a generalized triangle, consisting of three nonconcurrent lines (meaning they have no generalized point in common). Let the vertices be generalized points $A$, $B$, $C$.

(a) Define the analogue of an angle bisector for two lines meeting at an end or an ideal point.

(b) Show that the three (internal) angle bisectors of the generalized triangle $ABC$ always meet at a (regular) point $W$.

(c) Show that $W$ is the center of an *inscribed circle* that is tangent to the three sides of the triangle.

40.16 Prove the results of Exercises 35.8, 35.9 in a hyperbolic plane, without using Archimedes' axiom.

# 41   Hilbert's Arithmetic of Ends

We come now to one of the most beautiful parts of the theory of non-Euclidean geometry, which is another illustration of the usefulness of abstract algebra. This is Hilbert's tour de force, the creation of an abstract field out of the geometry of a hyperbolic plane. In the same way that the field of segment arithmetic (Section 19) helped us to understand Euclidean geometry, this field will help us in our study of non-Euclidean geometry. Using it we can prove results such as Bolyai's parallel construction, or the theorem on the three altitudes of a triangle. We also set up a hyperbolic analytic geometry and hyperbolic trigonometry, whereby any geometric problem can (in principle) be translated into an algebraic problem in the field. This is analogous to ordinary analytic geometry, but the particulars are all different, having "suffered a sea change into something rich and strange." Finally we will be able to show that the hyperbolic plane is uniquely characterized by its associated field, and is in fact isomorphic to the Poincaré model over that field.

   We start with a hyperbolic plane, as in the previous section, which is a Hilbert plane satisfying the axiom of limiting parallels (L).

### Proposition 41.1
*Let A, B, C be three noncollinear points in a hyperbolic plane and consider the three perpendicular bisectors l, m, n of the sides of the triangle ABC.*

(a) *If two of the lines l, m, n meet at a point P, then the third line also passes through P, and in this case A, B, C all lie on a circle with center P.*

(b) *If two of the lines l, m, n have a common perpendicular p, then the third is also perpendicular to p, and the three points A, B, C are equidistant from the line p.*

(c) *If two of the lines l, m, n are limiting parallel, the third is also, and all three have a common end.*

*Proof*  (a) This is essentially what Euclid proves in (IV.5), the only difference being that we must assume that two of the lines meet.

   (b) Suppose that two of the lines, say *l* and *m*, have a common orthogonal line *p*. Dropping perpendiculars *AG* and *BH* to *p*, one application of (SAS) and one application of (AAS) show that *AG = BH* (see second diagram). In other words, *A* and *B* are equidistant from *p*.

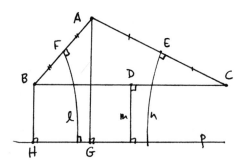

The same reasoning applied to $m$ shows that $B$ and $C$ are equidistant from $p$. Therefore, $A$ and $C$ are also equidistant from $p$, and then it follows from (34.1) that $n$ is orthogonal to $p$, as required.

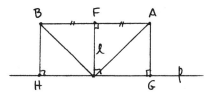

(c) Now suppose that $l$ and $m$ are limiting parallels. If the third line $n$ meets $l$ at a point, we apply (a) above to find that $l$ and $m$ also meet, which contradicts our hypothesis. If $n$ and $l$ have a common perpendicular, then by (b) so do $l$ and $m$, which again contradicts our hypothesis. It follows from (40.5) that the only remaining possibility is that $n$ and $l$ are limiting parallels. (We leave to the reader to figure out why all three lines are limiting parallel at the same end.)

**Proposition 41.2** (Theorem of three reflections)
*Given three lines $a$, $b$, $c$ in the hyperbolic plane, with a common end $\omega$, there exists a fourth line $d$ with end $\omega$ such that reflection in $d$ is equal to the product of the reflections in $a$, $b$, $c$:*

$$\sigma_c \sigma_b \sigma_a = \sigma_d.$$

*Here $\sigma_l$ for any line $l$ denotes reflection in the line $l$.*

*Proof*  Take any point $A$ on the line $a$. Let $B$ be its reflection in the line $b$. Let $C$ be the reflection of $B$ in $c$. Draw $AC$, and let $d$ be the perpendicular bisector of $AC$. Then $d$ will be the required fourth line.

To show this, it is equivalent to show that the product

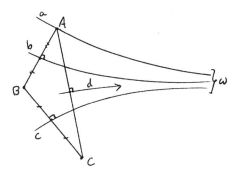

$$\varphi = \sigma_d \sigma_c \sigma_b \sigma_a$$

is equal to the identity. Note first that $\varphi(A) = A$ by construction of $d$. Note also that $b$ and $c$ are two perpendicular bisectors of sides of the triangle $ABC$ and they are limiting parallels with end $\omega$. It follows from (41.1) that $d$ also has end $\omega$. Therefore, $\varphi$ preserves the end $\omega$, and because of this $\varphi$ fixes every point of the line $a$. Therefore, $\varphi$ is either the identity or reflection in the line $a$ (cf. (17.4) and Exercise 17.3).

Suppose that $\varphi = \sigma_a$. Then $\sigma_d \sigma_c \sigma_b$ is equal to the identity. Then for any point $P \in b$ we would have $\sigma_d \sigma_c(P) = P$, so $\sigma_c(P) = \sigma_d(P)$, and this implies $c = d$, which is absurd. We conclude that $\varphi = $ identity, as required.

**Remark 41.2.1**
This result shows for a pencil of lines with a common end what we have already seen for a pencil of lines through a point, or a pencil of lines with a common orthogonal line (Exercise 17.14).

Now we are ready to define the arithmetic of ends. Fix a hyperbolic plane $\Pi$. Recall that an *end* is an equivalence class of limiting parallel rays (34.13). Fix one line and label its ends 0 and $\infty$. We let $F$ be the set of all ends in the plane $\Pi$ different from $\infty$, and then we set $F' = F \cup \{\infty\}$, so that $F'$ is the set of all ends of the plane. We will make the set $F$ into an ordered field by defining arithmetic operations $+, \cdot$, and an ordering on it.

Note that given any end $\alpha$ and any point $P$, there exists a unique line $m$ passing through $P$ and with end $\alpha$. Just take $l$ a line having end $\alpha$, and use the axiom (L) to find a line $m$ through $P$ limiting parallel to $l$. Similarly, given two ends $\alpha, \beta$, there exists a unique line $l$ having ends $\alpha$ and $\beta$. Just take any point $P$, consider the angle formed by the rays $P\alpha$ and $P\beta$, and let $l$ be the enclosing line of the angle $\alpha P\beta$ (40.6). We will denote the line with ends $\alpha, \beta$ by $(\alpha, \beta)$.

**Definition**
Given two ends $\alpha, \beta$ not equal to $\infty$, we define their sum $\alpha + \beta$ as follows. Take any point $C$ on the line $(0, \infty)$. Let $A$ be its reflection in the line $(\alpha, \infty)$. Let $B$ be its reflection in the line $(\beta, \infty)$. Then $\alpha + \beta$ is the end of the perpendicular bisector of $AB$ other than $\infty$.

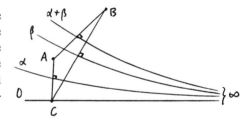

**Proposition 41.3**
*The addition of ends is well-defined, and makes the set $(F, +)$ into an abelian group with additive identity 0.*

*Proof*   First note that by (41.1) the perpendicular bisector of $AB$ has one end $\infty$, so the definition makes sense. Second, note by (41.2) that the sum $\alpha + \beta$ is characterized by the property

$$\sigma_{\alpha+\beta} = \sigma_\beta \sigma_0 \sigma_\alpha,$$

where for any end $\alpha$, $\sigma_\alpha$ denotes reflection in the line $(\alpha, \infty)$. Thus $\alpha + \beta$ is independent of the choice of $C$, and so is well-defined.

From the definition, it is clear that $\alpha + \beta = \beta + \alpha$. It is also clear that $0 + \alpha = \alpha$ for any $\alpha$. If we denote by $-\alpha$ the reflection of $\alpha$ in the line $(0, \infty)$, then

$\alpha + (-\alpha) = 0$, so we have additive inverses. For the associative law, just note that

$$\sigma_{(\alpha+\beta)+\gamma} = \sigma_\gamma \sigma_0 \sigma_{\alpha+\beta} = \sigma_\gamma \sigma_0 \sigma_\beta \sigma_0 \sigma_\alpha,$$

which is independent of the order of the operations.

### Definition

In order to define multiplication of ends, we first fix a line perpendicular to the line $(0, \infty)$, and label one of its ends 1. Note that the other end is then $-1$. Given ends $\alpha, \beta$, draw the lines $(\alpha, -\alpha)$ and $(\beta, -\beta)$, which will meet the line $(0, \infty)$ at right angles at points $A, B$. Let $O$ be the point where the line $(1, -1)$ meets $(0, \infty)$. Find $C$ on the line $(0, \infty)$ such that $OC = OA + OB$, treating these as *signed* distances: If $A$ is on the ray $O\infty$, then $OA$ is positive; if $A$ is on the ray $O0$, then $OA$ is negative.

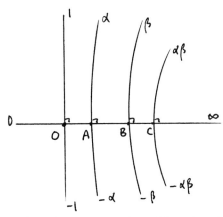

Take the line through $C$, perpendicular to $(0, \infty)$, and its ends will be $\alpha\beta$ and $-\alpha\beta$. Adjust the sign as follows: Ends on the same side of $(0, \infty)$ as 1 will be called *positive*, and ends on the same side of $(0, \infty)$ as $-1$ will be *negative*. We choose the sign so that pos $\times$ pos = pos, pos $\times$ neg = neg, and neg $\times$ neg = pos, as usual.

### Theorem 41.4

*In a hyperbolic plane $\Pi$, fix two perpendicular lines and label their ends $0, \infty, 1, -1$ as above. Let $F$ be the set of all ends of $\Pi$ different from $\infty$. Then $F$, with the two operations $+$, $\cdot$ and the notion of positive elements defined above, is a Euclidean ordered field.*

*Proof*   We have already seen (41.3) that $(F, +)$ is an abelian group with identity 0.

From the definition of multiplication we see immediately that $(F \backslash \{0\}, \cdot)$ is an abelian group with identity 1. Indeed, multiplication of ends corresponds to addition of signed segments on the line $(0, \infty)$, which is an abelian group. Reflection in the line $(1, -1)$ sends $\alpha$ to $\alpha^{-1}$.

Multiplication by zero was not defined, so we define $0 \cdot \alpha = 0$ for all $\alpha \in F$.

For the distributive law, we proceed as follows. Given an end $\gamma$, which we may assume to be positive, let the line $(\gamma, -\gamma)$ meet $(0, \infty)$ at $C$. We define a rigid motion $\tau$ of the plane, called *translation along the line $(0, \infty)$ by $OC$*, as follows. For any point $P$, let $PQ$ be the orthogonal to $(0, \infty)$. Choose $Q'$ on $(0, \infty)$ such that $OC = QQ'$ as signed intervals. Take $P'$ on the line through $Q'$, orthogonal to $(0, \infty)$, on the same side as $P$, such that $PQ = P'Q'$.

One can verify easily that $\tau$ is a rigid motion of the plane (Exercise 41.1). It leaves the ends 0 and $\infty$ fixed, and sends any other end $\alpha$ to $\gamma\alpha$. Applying this rigid motion to the diagram used to define $\alpha + \beta$, the diagram is sent to another diagram with the same properties, and from this it is clear that $\gamma(\alpha + \beta) = \gamma\alpha + \gamma\beta$.

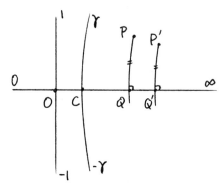

We have defined the positive ends to be those on the same side of the line $(0, \infty)$ as 1. From our definitions it is clear that sums and products of positive ends are positive. It is also clear by definition that for any $\alpha \in F$, either $\alpha$ is positive, or $\alpha = 0$, or $-\alpha$ is positive. Hence $F$ is an ordered field.

To show that $F$ is Euclidean, let $\alpha$ be a positive end, let the line $(\alpha, -\alpha)$ meet $(0, \infty)$ at $A$, and let $B$ be the midpoint of $OA$. Then the line perpendicular to $(0, \infty)$ at $B$ will have an end $\beta$ with the property $\beta^2 = \alpha$. Thus $\alpha$ has a square root in $F$.

To relate this newly constructed field of ends with the geometry of the plane, our first step will be to describe some useful rigid motions and their effect on the ends. Also, to simplify notation, we will now write elements of the field $F$ with Latin letters, even though as ends, they were originally written with Greek letters.

**Proposition 41.5**

*Let $\Pi$ be a hyperbolic plane and $F$ its field of ends, as above. We describe various rigid motions and their effect on ends.*

(a) *The reflection in $(0, \infty)$ sends $x \in F$ to $-x$. So we write $x' = -x$.*
(b) *The reflection in $(1, -1)$ gives $x' = 1/x$.*
(c) *Translation along $(0, \infty)$ is represented by $x' = ax$ for any $a \in F$, $a > 0$.*
(d) *For any $a \in F$, there is a rigid motion $\sigma_{\frac{1}{2}a}\sigma_0$, the composition of reflection in the line $(0, \infty)$ and reflection in the line $\left(\frac{1}{2}a, \infty\right)$, which we call "rotation around $\infty$," which gives $x' = x + a$.*

(e) *The rotation around the point O, which sends 0 to a for any a ∈ F, gives*

$$x' = \frac{x+a}{-ax+1}$$

*on ends. The rotation around O sending 0 to ∞ gives* $x' = -\dfrac{1}{x}$.

*Proof*   (a) We saw this in the definition of addition.
(b) We saw this in the definition of multiplication.
(c) We saw this in the proof of the distributive law in (41.4).
(d) From the definition of addition, we see that $\sigma_a$ sends 0 to $2a$, and $\sigma_b$ sends 0 to $2b$. Then $\sigma_{a+b}$ sends $2a$ to $2b$. Changing variables, let $c = a + b$ and $d = 2a$. Then $\sigma_c$ sends $d$ to $2c - d$. Using these relations, we see that $\sigma_{\frac{1}{2}a}\sigma_0$ sends any $x$ to $x + a$, as required.
(e) We give an indirect proof that rotation around $O$ has the given effect on ends. If $a = 0$ we have the identity. If $a = \infty$, we have the composition of reflection in $(0, \infty)$ and reflection in $(1, -1)$, which is a rotation of 2RA. So we may assume $a \neq 0, \infty$.
Note that

$$\frac{x+a}{-ax+1} = \frac{1}{-\left(\dfrac{a}{a+a^{-1}}\right)x + \dfrac{1}{a+a^{-1}}} - a^{-1},$$

which is a composition of operations of the types given in (a), (b), (c), (d) above. So there is a rigid motion having the effect

$$x' = \frac{x+a}{-ax+1}$$

on ends. Next we will show that any rigid motion having that effect on ends must be the rotation around $O$ taking 0 to $a$. Indeed, substituting 0 and ∞ in this expression, we find that 0 goes to $a$ and ∞ goes to $-a^{-1}$, so the line $(0, \infty)$ goes to $(a, -a^{-1})$, which passes through $O$.
Next, we compute and find that the line $(1, -1)$ goes to the line $((1+a)/(-a+1), (-1+a)/(a+1))$, which also passes through $O$. Indeed, a line $(c, d)$ passes through $O$ if and only if $d = -c^{-1}$, as we see by applying reflections in $(0, \infty)$ and $(1, -1)$. Hence the point $O$, which is the intersection of the two lines $(0, \infty)$ and $(1, -1)$, goes to the intersection of the images of those lines, which is $O$.
Finally, if we try to solve the equation $x = (x+a)/(-ax+1)$, we obtain $x^2 = -1$, which has no solution because $F$ is an ordered field. Thus no end is fixed under this transformation. Now, a rigid motion that fixes a point $O$ and has no fixed ends must be a rotation around $O$ (Exercise 17.4), so we are done.

**Proposition 41.6**

Let $\Pi$ *be a hyperbolic plane with field of ends F, as above.*

   (a) *A line is represented by an unordered pair* $(u_1, u_2)$ *of distinct elements of* $F' = F \cup \{\infty\}$.

   (b) *A point is given by the equation of all lines containing it, which is of the form*

$$u_1 u_2 - b(u_1 + u_2) + a^2 = 0$$

*with* $a, b \in F$, $a > 0$, *and* $|b| < a$.

*Proof* (a) Indeed, a line is uniquely determined by its two ends, which are distinct elements of $F'$, and any two ends lie on a line.

   (b) First consider the point $O$. A line $(u_1, u_2)$ contains $O$ if and only if $u_1 u_2 = -1$. Indeed, such a line is stable under the rotation around $O$ by 2RA, which sends $x' = -1/x$ by (41.5).

   Now we can move $O$ to any other point by first making a translation along $(0, \infty)$ and then a rotation around $\infty$. A translation along $(0, \infty)$ gives $x' = cx$, so we get a new equation $u_1 u_2 = -c^2$. Then a rotation around $\infty$ gives $x' = x - b$, so we get a new equation

$$(u_1 - b)(u_2 - b) + c^2 = 0,$$

or

$$u_1 u_2 - b(u_1 + u_2) + b^2 + c^2 = 0.$$

Here $b$ is any element of $F$, and we can set $b^2 + c^2 = a^2$ with $a$ positive, $a > |b|$.

**Remark 41.6.1**

We can think of the point with equation

$$u_1 u_2 - b(u_1 + u_2) + a^2 = 0$$

as having coordinates $(a, b)$. It is the intersection of the lines $(a, -a)$ and $(b, \infty)$. However, we have found that calculations seem to work out better if we continue to think of a line as given by coordinates $(u_1, u_2)$, and a point as given by an equation. This is the opposite of the analytic geometry we are used to, where a point has coordinates and a line has an equation.

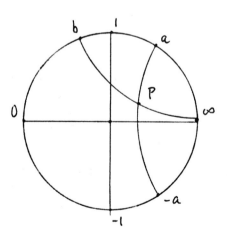

**Proposition 41.7**

*For any segment AB in the hyperbolic plane, lay off a congruent segment OC on the ray* $O\infty$, *and let the line perpendicular to* $(0, \infty)$ *at C be* $(a, -a)$ *with* $a > 0$. *We de-*

fine $\mu(AB) = a$. Then $\mu$ is a multiplicative distance function *on the plane* (*Section 39*) *with values in the multiplicative group of positive elements of the field F, namely*:

(a) $\mu(AB) > 1$.
(b) $AB \cong A'B'$ if and only if $\mu(AB) = \mu(A'B')$.
(c) $AB < A'B'$ if and only if $\mu(AB) < \mu(A'B')$.
(d) $\mu(AB + CD) = \mu(AB) \cdot \mu(CD)$.

*Proof*  These properties are all immediate from the definition of multiplication in $F$.

**Proposition 41.8**
*For any angle $\theta$ in the hyperbolic plane, lay out an equal angle centered at O, and reaching from 0 to a on the positive side of $(0, \infty)$. Then we define $\tan \theta/2 = a$. This tangent function has the following properties.*

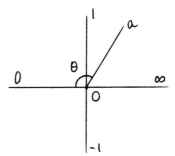

(a) $\tan \theta/2 \in F$, and $\tan \theta/2 > 0$.
(b) $\tan \theta/2 = \tan \psi/2$ if and only if $\theta = \psi$.
(c) *If $\theta < \psi$ then $\tan \theta/2 < \tan \psi/2$.*
(d) *If $\theta = $ RA, $\tan \theta/2 = 1$.*
(e) *If $\theta$ and $\psi$ are two angles for which $\theta + \psi$ is defined, then*

$$\tan \frac{(\theta + \psi)}{2} = \frac{\tan \dfrac{\theta}{2} + \tan \dfrac{\psi}{2}}{1 - \tan \dfrac{\theta}{2} \tan \dfrac{\psi}{2}}.$$

*Proof*  (a), (b), (c), (d) are immediate by construction. For (e), let $\tan \theta/2 = a$, $\tan \psi/2 = b$, and $\tan(\theta + \psi)/2 = c$. Then the rotation around $O$ from 0 to $c$ is the composition of the rotations from 0 to $a$ and 0 to $b$. Using the formula of (41.5), we have

$$\frac{x + c}{-cx + 1} = \frac{\left(\dfrac{x + a}{-ax + 1}\right) + b}{-b\left(\dfrac{x + a}{-ax + 1}\right) + 1}.$$

A brief calculation gives

$$c = \frac{a + b}{1 - ab},$$

which is the desired result.

**Remark 41.8.1**
We cannot define the tangent of an angle as in Euclidean geometry by using right triangles, because in non-Euclidean geometry, right triangles of different sizes are not similar to each other. Therefore, we use this ad hoc definition. However, the terminology $\tan\theta/2$ for this function is justified by the properties it enjoys. In the presence of Archimedes' axiom, one can define the radian measure of an angle by a limiting process using dyadic rational multiples of a right angle. Then, viewing the field $F$ as a subfield of $\mathbb{R}$ (15.5) it is easy to prove that this tangent function is the same as the usual one as a function from $\mathbb{R}$ to $\mathbb{R}$ (Exercise 41.16).

**Proposition 41.9** (Bolyai's formula)
*If $\alpha$ is the angle of parallelism of a segment AB, then using the distance function and the tangent function defined above, we have*

$$\tan\frac{\alpha}{2} = \mu(AB)^{-1}.$$

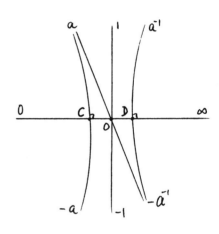

*Proof*  Lay out the angle $\alpha$ as $0Oa$. Then the line $(a, -a)$ will meet the ray $O0$ at right angles at a point $C$, and $\alpha$ will be the angle of parallelism of the segment $OC$ equal to $AB$. To find $\mu(OC)$, reflect in the line $(1, -1)$ to get $OD$ and the line $(a^{-1}, -a^{-1})$. Then $\mu(AB) = a^{-1}$ and $\tan\alpha/2 = a$, which gives the desired result.

**Remark 41.9.1**
We recover the same result as (39.13) under different hypotheses and different definitions of tan and $\mu$.

Now we have enough basic results to be able to apply this "hyperbolic analytic geometry" to problems in the hyperbolic plane. By way of illustration of these techniques, we will give Bolyai's parallel construction, and prove the theorem about the altitudes of a triangle.

**Proposition 41.10** (Bolyai's parallel construction)
*Suppose we are given a line l and a point P not on l in the hyperbolic plane. Let PQ be the perpendicular to l. Let m be a line through P, perpendicular to PQ. Choose any point R on l, and let RS be the perpendicular to m. Then the circle of radius QR*

*around P will meet the segment RS at a point T, and the ray n = PT will be the limiting parallel ray to l through P.*

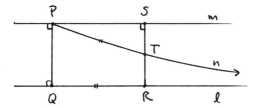

*Proof*  To prove this, we use the field of ends introduced above. By a rigid motion of the plane, we may assume that $P$ is the center of our coordinate system, and $PQ$ is the line $(0, \infty)$ and $PS$ is the line $(1, -1)$. We will take $n$ to be the limiting parallel to $l$, and let it meet $RS$ at $T$. Then we must prove that $QR = PT$. To show this, we will apply rigid motions to move each of these segments to the line $(0, \infty)$, with one end at $P$. Then we will verify that the other ends of the segments land in the same place. For this we need a lemma.

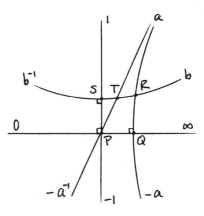

**Lemma 41.11**
*Two lines $(u_1, u_2)$ and $(v_1, v_2)$ in the hyperbolic plane meet the line $(0, \infty)$ at the same point if and only if $u_1 u_2 = v_1 v_2$, and $u_1 u_2$ is negative.*

*Proof*  Indeed, we have seen in the proof of (41.6) that a point on the line $(0, \infty)$ has an equation of the form $u_1 u_2 = -c^2$ for some $c \in F$. So the two lines $(u_1, u_2)$ and $(v_1, v_2)$ pass through the same point of $(0, \infty)$ if and only if $u_1 u_2 = -c^2 = v_1 v_2$. Since every positive element of $F$ is a square (41.4), such a $c$ will exist if and only if $u_1 u_2 = v_1 v_2$ and $u_1 u_2$ is negative.

*Proof of (41.10), continued*  Let us start with the segment $PT$. We let the ends of the line $l = QR$ be $a$ and $-a$. The line $RS$ we let have ends $b$ and $b^{-1}$. Then $n$ is the line $(a, -a^{-1})$. We use the rotation around $P$ that sends $a$ to 0. Its effect on ends, by (41.5), is

$$x' = \frac{x - a}{ax + 1}.$$

Hence it sends the line $(b, b^{-1})$ to the line

$$(u_1, u_2) = \left(\frac{b-a}{ab+1}, \frac{b^{-1}-a}{ab^{-1}+1}\right).$$

The image of $T$ will be the point where this line meets the line $(0, \infty)$. In order to apply the lemma, we compute

$$u_1 u_2 = \left(\frac{b-a}{ab+1}\right)\left(\frac{b^{-1}-a}{ab^{-1}+1}\right) = \frac{(b-a)(1-ab)}{(1+ab)(a+b)}.$$

Now consider the segment $QR$. First we make a translation along $(0, \infty)$ to send $Q$ to $P$. Its effect on ends will be $x' = a^{-1}x$, so it will send the line $(b, b^{-1})$ to the line $(a^{-1}b, a^{-1}b^{-1})$, and the line $(a, a^{-1})$ will become $(1, -1)$. Next we do a rotation around $P$ sending 1 to 0, whose effect on ends is

$$x' = \frac{x-1}{x+1}.$$

So this line just mentioned will go to

$$(v_1, v_2) = \left(\frac{a^{-1}b-1}{a^{-1}b+1}, \frac{a^{-1}b^{-1}-1}{a^{-1}b^{-1}+1}\right).$$

The image of $R$ under the two rigid motions will be the intersection of $(v_1, v_2)$ with $(0, \infty)$. So we compute

$$v_1 v_2 = \left(\frac{a^{-1}b-1}{a^{-1}b+1}\right)\left(\frac{a^{-1}b^{-1}-1}{a^{-1}b^{-1}+1}\right) = \frac{(b-a)(1-ab)}{(b+a)(1+ab)}.$$

Observing that $u_1 u_2 = v_1 v_2$ and $u_1 u_2 < 0$, since $b > a > 1$ in our situation, and using the lemma, we conclude that $QR \cong PT$, as required.

### Remarks 41.11.1

This remarkable result gives a ruler and compass construction for the limiting parallel line. In other words, by making constructions such as dropping a perpendicular from a point to a line, or intersecting a line with a circle — constructions that are possible in any Hilbert plane with axiom (E) — one obtains the limiting parallel ray from a point $P$ to a line $l$. However, the curious feature of our proof is that we prove that this construction works only by first assuming (via axiom (L)) that the object we wish to construct already exists. And in fact, without assuming the existence of the limiting parallel line ahead of time, this result may fail. For example, in the non-Archimedean geometries of (18.4.3) or Exercise 39.24, the construction gives a line $n$ that is not limiting parallel to $l$. These examples suggest that this construction may work in a Hilbert plane satisfying (A) and (E). Indeed, it follows from the classification of Hilbert planes due to Pejas that any non-Euclidean Hilbert plane satisfying (A) and (E) is hyperbolic (43.8), and so Bolyai's construction works. Greenberg (1993, p. 222) says that if you could find a direct geometric proof of this result, you would probably receive an instant Ph.D.

**Proposition 41.12**

*If $(u_1, u_2)$ and $(v_1, v_2)$ are any two lines that meet in the plane, and if $\theta$ is the angle between the rays $u_1$ and $v_1$, and $u_1 < v_1$, then*

$$\tan \theta/2 = \sqrt{-(v_1, v_2; u_1, u_2)},$$

*where the expression under the radical is the* cross-ratio *defined as*

$$(v_1, v_2; u_1, u_2) = \frac{v_1 - u_1}{v_1 - u_2} \cdot \frac{v_2 - u_2}{v_2 - u_1}.$$

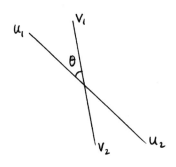

*Proof*  First consider the case $(u_1, u_2) = (0, \infty)$ and $(v_1, v_2) = (a, -a^{-1})$. In this case we know that $\tan \theta/2 = a$ by definition of the tangent function. We compute the cross-ratio

$$(a, -a^{-1}; 0, \infty) = \frac{a - 0}{a - \infty} \cdot \frac{-a^{-1} - \infty}{-a^{-1} - 0} = -a^2.$$

So the formula holds in this case.

In the general case, we consider a rigid motion that takes $(u_1, u_2)$ to $(0, \infty)$ and $(v_1, v_2)$ to $(a, -a^{-1})$ for suitable $a \in F$. Since a rigid motion preserves angles, we need to show only that a rigid motion preserves cross-ratios of ends. In fact, we can accomplish the rigid motion we need using a composition of rigid motions of types (a)–(e) of (41.5), and type (e) is itself a composition of the four earlier types, so we need only verify that the cross-ratio of four ends is stable under the transformations of types (a) $x' = -x$; (b) $x' = x^{-1}$; (c) $x' = ax$ for $a \in F$, $a > 0$; and (d) $x' = x + a$. These verifications are immediate from the definition of cross-ratio, so we are done.

**Proposition 41.13** (Altitudes)

*If two of the altitudes of a triangle meet in a point, so does the third. If two of the altitudes are limiting parallels, then all three have a common end. If two of the altitudes have a common perpendicular, then all three have the same common perpendicular.*

*Proof*  By a rigid motion of the plane, we may assume that our triangle $ABC$ is so placed that $AB$ is the line $(1, -1)$, and $C$ lies on the line $(0, \infty)$, so that $OC$ is one of the three altitudes. Our method is to let $(u_1, u_2)$ be one of the other altitudes, find the intersection of this line with $(0, \infty)$, and then show that the third altitude passes through this same point.

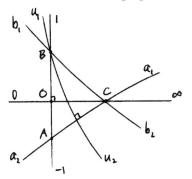

Let $AC$ be $(a_1, a_2)$ and let $BC$ be $(b_1, b_2)$. Since these two lines meet at $C$ on the line $(0, \infty)$, we must have $a_1 a_2 = b_1 b_2 < 0$ by (41.11).

First we will find the equation of the point $B$. By (41.6) it is of the form

$$u_1 u_2 - d(u_1 + u_2) + e^2 = 0.$$

Since $B$ lies on the lines $(1, -1)$ and $(b_1, b_2)$ we obtain

$$-1 - d \cdot 0 + e^2 = 0,$$

$$b_1 b_2 - d(b_1 + b_2) + e^2 = 0.$$

Solving for $d$ and $e^2$, we find that $e^2 = 1$, and $d = (1 + b_1 b_2)/(b_1 + b_2)$, so the equation of $B$ is

$$u_1 u_2 - \frac{1 + b_1 b_2}{b_1 + b_2}(u_1 + u_2) + 1 = 0. \tag{1}$$

Now, the altitude $(u_1, u_2)$ through $B$ satisfies this equation. It is also orthogonal to $(a_1, a_2)$, so by (41.12) the cross-ratio

$$(u_1, u_2; a_1, a_2) = -1.$$

This gives

$$2u_1 u_2 - (a_1 + a_2)(u_1 + u_2) + 2a_1 a_2 = 0. \tag{2}$$

We solve the equation (2) for $u_1 + u_2$ and substitute in the equation (1). We obtain

$$u_1 u_2 - \frac{2(1 + b_1 b_2)(u_1 u_2 + a_1 a_2)}{(a_1 + a_2)(b_1 + b_2)} + 1 = 0. \tag{3}$$

Since $a_1 a_2 = b_1 b_2$ as noted above, this equation is stable under interchanging $a$'s and $b$'s. So if $(v_1, v_2)$ is the third altitude of the triangle, then $v_1 v_2$ satisfies the same equation (3), and so $u_1 u_2 = v_1 v_2$.

Now, if the altitude $(u_1, u_2)$ meets the line $(0, \infty)$, then by (41.11), $u_1 u_2 < 0$, and $(v_1, v_2)$ meets $(0, \infty)$ in the same point. If $(u_1, u_2)$ is limiting parallel to $(0, \infty)$, then one of $u_1$ or $u_2$ is equal to 0 or $\infty$, in which case $u_1 u_2 = 0$ or $\infty$. The same is true of $(v_1, v_2)$, so one of $v_1, v_2$ is equal to 0 or $\infty$, respectively, so all three have a common end.

We leave as an exercise (Exercise 41.2) to show that if $u_1 u_2 = v_1 v_2 > 0$, then all three altitudes have a common perpendicular.

## Exercises

These exercises all take place in a hyperbolic plane.

41.1 Verify that translation along a line, defined in the proof of Theorem 41.4, is a rigid motion.

41.2 In the proof of Proposition 41.13, show that if two of the altitudes have a common perpendicular, then all three altitudes have a common perpendicular.

41.3 Consider the pencil (set) of all lines $(u_1, u_2)$ having a fixed common orthogonal line. Show that the lines of such a pencil satisfy an equation of the form

$$u_1 u_2 - b(u_1 + u_2) + a = 0$$

with $a < |b|$, or an equation of the form

$$u_1 + u_2 = c$$

for some $c \in F$.

41.4 Find the two ends (as elements of $F$) of the angle bisector of the angle $1O\infty$.

41.5 Find the multiplicative distance $\mu$ between the points $O = (1, 0)$ and $A = (1, a)$, using the point coordinates of (41.6.1).

41.6 In the trilimit triangle of Exercise 40.8, find the multiplicative distance $\mu$ from the center point $A$ to one of the sides.

41.7 In the limit quadrilateral of Exercise 40.10, express $\mu(CD)$ in terms of $\mu(AB)$.

41.8 Prove that an angle of $\frac{1}{3}$RA exists in the hyperbolic plane.

41.9 Prove that an angle of $\frac{1}{5}$RA exists in the hyperbolic plane.

41.10 In this exercise we describe the group $G$ of all rigid motions of the hyperbolic plane, as follows.

(a) Show that every rigid motion of the plane can be expressed as a composition of motions of the four types (a), (b), (c), (d) of Proposition 41.5. In other words, elements of those four types generate the group $G$.

(b) Show that an element of $G$ is uniquely determined by its effect on the set $F'$ of ends.

(c) If $\varphi \in G$ is any rigid motion, then $\varphi$ acts on $F'$ as a *fractional linear transformation* of the form

$$x' = \frac{ax + b}{cx + d}$$

for suitable $a, b, c, d \in F'$, with $ad - bc \neq 0$.

(d) Every fractional linear transformation of $F'$ arises as the action of some element $\varphi \in G$ on the set of ends. Thus $G$ is isomorphic to the group of fractional linear transformations of $F'$.

Because of this result, we can say that the group of rigid motions of the hyperbolic plane is composed of four components: a copy of the additive group of the field corresponding to (41.5d); a copy of the multiplicative group of positive elements of the field (41.5c); a copy of the circle group of the field (41.5e), cf. Exercise 41.11 below; and possibly a reflection.

This is in contrast to the Euclidean situation, where the group of rigid motions is made up of two copies of the additive group of the field, corresponding to the translations, and one copy of the circle group for rotations, plus a possible reflection.

41.11 To each rotation $\rho$ around the point $O$, let us associate that element $a \in F'$ to which $\rho$ sends 0. Show that this correspondence gives an isomorphism of the group of rotations around $O$ with the circle group of the field $F$, as defined in Exercise 17.6. In particular, we find the curious result that the group of rotations around a point in the hyperbolic plane is isomorphic to the group of rotations around a point in the Cartesian plane over the field $F$.

41.12 Let $l, m$ be two lines with a common end $\alpha$. Generalizing (41.5d), we call the composition $\sigma_l \sigma_m$ of the reflections in $l$ and $m$, a "rotation around the end $\alpha$." Fix an end $\alpha$ and a point $A$. Then the set of all points $\rho(A)$, where $\rho$ ranges over all the rotations around the end $\alpha$, will be called a *horocycle*. We think of it as analogous to a circle, but one whose center is an end instead of a point.

(a) Show that a horocycle could also be defined as the set of all points $\sigma(A)$, where $\sigma$ ranges over all the reflections in lines with end $\alpha$.

(b) If $m$ is the line through $A$, orthogonal to the line $l = A\alpha$, then $m$ meets the horocycle only at $A$. We call $m$ the *tangent line* to the horocycle at $A$. Show that any other line through $A$ different from $l$ or $m$ will meet the horocycle in one other point $B \neq A$.

(c) In the case of the Poincaré model, show that the horocycles defined here are the same as those defined in Exercise 39.14.

41.13 Let $(u_1, u_2)$ and $(v_1, v_2)$ be two parallel, not limiting parallel, lines. Show that the multiplicative distance between the two lines along their common perpendicular is

$$\mu = \frac{a-1}{1-a},$$

where $a$ is the square root of the cross-ratio

$$a = \sqrt{(u_1, u_2; v_1, v_2)}.$$

41.14 If $F$ is a subfield of the field of real numbers $\mathbb{R}$, we can define a function $\lambda(AB) = \ln(\mu(AB))$ where $\mu$ is the multiplicative distance function of Proposition 41.7 and ln is the natural logarithm.

(a) Show that $\lambda$ is an *additive distance function* (Section 39) with values in the group $(\mathbb{R}, +)$, namely,

(1) for any interval $AB$, we have $\lambda(AB) > 0$;
(2) $AB \cong A'B'$ if and only if $\lambda(AB) = \lambda(A'B')$;
(3) $\lambda(AB + CD) = \lambda(AB) + \lambda(CD)$.

Note, however, that the values of the function $\lambda$ may not belong to $F$.

(b) Show that Bolyai's formula (Proposition 41.9) then becomes

$$\tan \alpha/2 = e^{-\lambda(AB)},$$

where $\alpha$ is the angle of parallelism of the segment $AB$.

41.15 Suppose again that $F$ is a subfield of $\mathbb{R}$. Show that if $\lambda$ is any additive distance function with values in $(\mathbb{R}, +)$, in the sense of Exercise 41.14, then there is a real constant $k > 0$ such that for all $AB$,

$$\lambda(AB) = k \ln(\mu(AB)).$$

41.16  Assume that $F$ is a subfield of $\mathbb{R}$. Replacing angles by their radian measure (41.8.1), show that any function having the properties (a)–(e) of Proposition 41.8 is the same as the usual tangent function.

# 42   Hyperbolic Trigonometry

The usual trigonometry that we learn in high school could be described as a collection of relationships between the sides and angles of a triangle, together with rules of operation (the trigonometric identities) that allow us to compute all the parts of a triangle from a few given parts.

To be more precise, if $ABC$ is a right triangle, with angles $\alpha$, $\beta$ at $A$ and $B$, and sides $a, b, c$ opposite the vertices $A, B, C$, then we have the relations

$$\sin \alpha = \frac{a}{c},$$

$$\cos \alpha = \frac{b}{c},$$

$$\tan \alpha = \frac{a}{b},$$

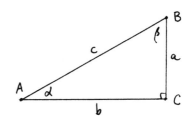

and trigonometric identities such as

$$\sin^2 \alpha + \cos^2 \alpha = 1$$

and

$$\sin(\mathrm{RA} - \alpha) = \cos \alpha,$$

and many others.

Using these relations, if we are given any two sides of a right triangle, or one side and one angle, then we can compute the remaining sides and angles of the triangle. Substituting the formulas for $\sin \alpha$ and $\cos \alpha$ in the relation $\sin^2 \alpha + \cos^2 \alpha = 1$ we recover the Pythagorean theorem $a^2 + b^2 = c^2$ (20.6).

The hyperbolic trigonometry that we develop in this section accomplishes the same thing in the hyperbolic plane. We develop a series of relationships between the sides and the angles of a right triangle, together with rules of operation on the trigonometric functions, so that if any two of the quantities $a, b, c, \alpha, \beta$ are given, we can compute the others from those two. This is a slightly stronger result than in the Euclidean case, because in hyperbolic geometry there are no similar triangles, so even giving the two angles $\alpha, \beta$ uniquely determines the triangle. The situation is also somewhat complicated by the fact that in hyperbolic geometry we have only $\alpha + \beta < $ RA, whereas in Euclidean geometry, $\beta$ is uniquely determined by $\alpha$, since $\alpha + \beta = $ RA.

Since we use the multiplicative distance function and the tangent function defined in the previous section, which take values in the field of ends $F$ instead of the real numbers, our formulae look different from the formulae in most books on non-Euclidean geometry. The advantage of this method is that everything takes place in a field naturally associated with our geometry, and works also in the non-Archimedean case. See Exercise 42.15 for a translation into the usual terminology for the case $F \subseteq \mathbb{R}$.

To derive the formulas of hyperbolic trigonometry, we put a right triangle in special position as shown, with $A$ at the center of our coordinate system and $C$ on the line $(0, \infty)$. Then the side $AB$ describes the angle $\alpha$, so that it is the line $(t, -t^{-1})$, where $t = \tan \alpha/2$ (41.8). We let $a, b, c$ represent the multiplicative lengths of the sides opposite $A, B, C$, so that $BC$ is then the line $(b^{-1}, -b^{-1})$ (41.7).

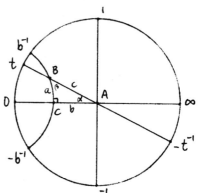

We will use suitable rigid motions to compute the lengths $a$ and $c$ in terms of $b$ and $t$. For $a$, we first perform a translation along $(0, \infty)$ to move $C$ to $A$. This acts by $x' = bx$ on ends, so the line $(t, -t^{-1})$ goes to $(bt, -bt^{-1})$. Now do a rotation by a right angle that sends 1 to $\infty$, and 0 to 1. This acts on ends by $x' = (x+1)/(-x+1)$ (41.5), so the line $(bt, -bt^{-1})$ becomes the line

$$(u_1, u_2) = \left( \frac{bt+1}{-bt+1}, \frac{-bt^{-1}+1}{bt^{-1}+1} \right).$$

This line will meet $(0, \infty)$ at the point that is the image of $B$ under these two maps. To find the multiplicative length $a$ of the segment $BC$, we need to find the line $(a, -a)$ that meets $(0, \infty)$ in the same point as $(u_1, u_2)$. Therefore, by (41.11), $u_1 u_2 = -a^2$. Substituting for $u_1$ and $u_2$ and simplifying, we obtain

$$a^2 = \frac{(1+bt)(b-t)}{(1-bt)(b+t)}. \tag{1}$$

Next, to find $c$, we perform a rotation around $A$ that sends $t$ to 0. This acts by $x' = (x-t)/(tx+1)$ on ends, so that the line $(b^{-1}, -b^{-1})$ goes to

$$(v_1, v_2) = \left( \frac{b^{-1} - t}{b^{-1}t + 1}, \frac{-b^{-1} - t}{-b^{-1}t + 1} \right).$$

If $c$ is the multiplicative length of $AB$, the line $(c^{-1}, -c^{-1})$ will meet $(0, \infty)$ at the same point as $(v_1, v_2)$. Again using (41.11), we obtain $v_1 v_2 = -c^{-2}$, from which we obtain

$$c^2 = \frac{(b + t)(b - t)}{(1 + bt)(1 - bt)}. \tag{2}$$

To compute the angle $\beta$, we use (41.12). If $s = \tan \beta/2$, then $-s^2 = (b^{-1}, -b^{-1}; t, -t^{-1})$. Writing out the cross-ratio and simplifying, we obtain

$$s^2 = \frac{(1 - bt)(b - t)}{(b + t)(1 + bt)}. \tag{3}$$

In some sense, the formulae (1), (2), (3) accomplish our purpose, because thereby the quantities $a, b, c, t$, and $s$ are all related to one another. However, to bring this information into a form that is easier to manage, we will transform these equations.

From (1) an elementary calculation gives

$$\frac{1 - a^2}{1 + a^2} = \frac{2t}{1 - t^2} \cdot \frac{1 - b^2}{2b}. \tag{4}$$

From (2) we obtain

$$\frac{1 - c^2}{1 + c^2} = \frac{1 - b^2}{1 + b^2} \cdot \frac{1 + t^2}{1 - t^2}, \tag{5}$$

and from (3),

$$\frac{1 - s^2}{1 + s^2} = \frac{2t}{1 + t^2} \cdot \frac{1 - b^2}{2b}. \tag{6}$$

These formulae, which separate the variables in question, suggest the following definition (cf. Exercise 17.6).

**Definition**
For any angle $\alpha$, if $t = \tan \alpha/2$, define

$$\sin \alpha = \frac{2t}{1 + t^2} \quad \text{and} \quad \cos \alpha = \frac{1 - t^2}{1 + t^2}.$$

**Proposition 42.1**
*With these definitions, the functions* $\sin \alpha$ *and* $\cos \alpha$ *for angles in the hyperbolic plane enjoy the usual identities of the trigonometric functions. In particular,*

(a) $\tan \alpha = (\sin \alpha)/(\cos \alpha)$, *and*
(b) $\sin^2 \alpha + \cos^2 \alpha = 1$.

*Proof*  From (41.8e) it follows that $\tan \alpha = (2t)/(1 - t^2)$, which gives (a). The formula (b) is a simple computation from the definitions.

### Proposition 42.2
*In the hyperbolic plane, let ABC be a right triangle, with angles $\alpha, \beta$ at A, B, and let $\bar{a}$, $\bar{b}$, $\bar{c}$ be the angles of parallelism of the sides opposite A, B, C. Then*

(a)  $\tan \alpha = \cos \bar{a} \tan \bar{b}$,
(b)  $\cos \bar{b} = \cos \alpha \cos \bar{c}$,
(c)  $\sin \alpha = \cos \beta \sin \bar{b}$.

*Proof*  By (41.9) we have $\tan \bar{a}/2 = a^{-1}$ (where as before, $a$ is the multiplicative length of the side opposite $A$). Therefore,

$$\sin \bar{a} = \frac{2a^{-1}}{1 + a^{-2}} = \frac{2a}{1 + a^2},$$

$$\cos \bar{a} = \frac{1 - a^{-2}}{1 + a^{-2}} = \frac{a^2 - 1}{a^2 + 1},$$

and

$$\tan \bar{a} = \frac{2a^{-1}}{1 - a^{-2}} = \frac{2a}{a^2 - 1}.$$

There are similar expressions for $\bar{b}$ and $\bar{c}$. We use these to interpret the equations (4), (5), (6), which then become (a), (b), (c).

### Proposition 42.3
*With the same hypothesis as (42.2) we have the further relations*
(d)  $\tan \bar{c} = \sin \alpha \tan \bar{a}$,
(e)  $\sin \bar{c} = \tan \alpha \tan \beta$,
(f)  $\sin \bar{c} = \sin \bar{a} \sin \bar{b}$.

*Proof*  We will prove (f) and leave (d), (e) as Exercise 42.5. To do this we eliminate $\alpha$ from (a) and (b) of (42.2). From (a),

$$\tan \alpha = \cos \bar{a} \tan \bar{b},$$

and from (b),

$$\cos \alpha = \cos \bar{b} \cos^{-1} \bar{c}.$$

Multiplying, we obtain

$$\sin \alpha = \cos \bar{a} \sin \bar{b} \cos^{-1} \bar{c}.$$

Substituting these expressions in $\cos^2 \alpha + \sin^2 \alpha = 1$, we obtain

$$\cos^2 \bar{b} \cos^{-2} \bar{c} + \cos^2 \bar{a} \sin^2 \bar{b} \cos^{-2} \bar{c} = 1.$$

Multiplying by $\cos^2 \bar{c}$, we get

$$\cos^2 \bar{b} + \cos^2 \bar{a} \sin^2 \bar{b} = \cos^2 \bar{c}.$$

Now use $\cos^2 = 1 - \sin^2$ on $\bar{a}, \bar{b}, \bar{c}$. We get

$$1 - \sin^2 \bar{b} + (1 - \sin^2 \bar{a}) \sin^2 \bar{b} = 1 - \sin^2 \bar{c}.$$

This simplifies to

$$\sin^2 \bar{c} = \sin^2 \bar{a} \sin^2 \bar{b}.$$

Since the sin function on angles is always positive, we conclude that

$$\sin \bar{c} = \sin \bar{a} \sin \bar{b},$$

as required.

**Remark 42.3.1**
The statements of (42.2) and (42.3) plus four more obtained from (a), (b), (c), (d) by reversing the roles of $A$ and $B$ give one equation for each subset of three of the five quantities $a, b, c, \alpha, \beta$. Thus given any two of these, we can calculate the others.

**Example 42.3.2**
As an application, let us find the relation between the sides and the angles of an equilateral triangle. Let $CD$ be an altitude, let $\alpha$ be the angle at each vertex, and let $c$ be the multiplicative length of the side. Then $ACD$ is a right triangle with $\beta = \alpha/2$. Using (42.3e) we obtain

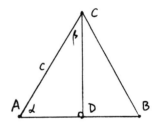

$$\sin \bar{c} = \tan \alpha \tan \alpha/2.$$

Translating back in terms of $t = \tan \alpha/2$ and $c$, this gives

$$\frac{2c}{1 + c^2} = \frac{2t^2}{1 - t^2},$$

which we obtained by a different method in the Poincaré model (Exercise 39.5).

As another application, we will derive a formula for the area of a circle.

**Proposition 42.4**
*Let $\Pi$ be a hyperbolic plane whose field of ends $F$ is a subfield of the real numbers $\mathbb{R}$. To each angle we associate its radian measure (41.8.1), so as to obtain a measure of area function (36.2) with values in $\mathbb{R}$. Let $\Gamma$ be a circle whose radius has multi-*

*plicative length r. We define the area of* Γ *to be the limit of the areas of regular in-scribed n-gons as n* → ∞. *This limit A exists (as a real number), and*

$$A = \frac{\pi(r-1)^2}{r}.$$

*Proof* First we will express the area of a right triangle in a particular form. Let $ABC$ be a right triangle, with our usual notation, and let $\delta$ be its area (as an angle). Then I claim that

$$\sin \delta = \frac{\sin \alpha \cos \alpha}{\sin \bar{b}}(1 - \sin \bar{c}).$$

Indeed, since it is a right triangle, $\delta = \pi/2 - \alpha - \beta$, and

$$\sin \delta = \cos(\alpha + \beta) = \cos \alpha \cos \beta - \sin \alpha \sin \beta.$$

We first use (c) and (d) of (42.2) and (42.3) to replace $\cos \beta$ and $\sin \beta$ by expressions in $\alpha, \bar{b}$, and $\bar{a}$. Then use (f) to replace $\sin \bar{a}$ by an expression in $\bar{b}$ and $\bar{c}$. This gives the formula above.

Now consider a regular $n$-gon inscribed in the circle, and consider the triangle formed by the radius to one of the vertices and the orthogonal to the midpoint of one side. Its angle $\alpha$ is $\pi/n$, its hypotenuse is $r$, and its side we call $b_n$. Its area is $A_n/2n$, where $A_n$ is the area of the inscribed polygon. Hence

$$\sin \frac{A_n}{2n} = \frac{\sin \pi/n \cos \pi/n}{\sin \bar{b}_n}(1 - \sin \bar{r}).$$

Multiply both sides of this equation by $2n$ and take the limit as $n \to \infty$. The limit of $A_n$ will be $A$, the area of the circle. The limit of $b_n$ is $r$. We use the convenient results from calculus that

$$\lim_{x \to 0} \frac{\sin x}{x} = 1$$

and

$$\lim_{x \to 0} \cos x = 1.$$

In the limit we obtain

$$A = 2\pi \left(\frac{1 - \sin \bar{r}}{\sin \bar{r}}\right).$$

Substituting

$$\sin \bar{r} = \frac{2r}{1 + r^2},$$

we obtain

$$A = \frac{\pi(r-1)^2}{r}$$

as required. Note that before we took the limit, all our terms were elements of the field $F$. However, in the limit we obtain quantities $\pi$ and $A$ that are real numbers, but not necessarily in $F$.

**Remark 42.4.1**
Using this result, we can study the question of "squaring the circle" in non-Euclidean geometry. Of course, there are no squares in a hyperbolic plane, so the appropriate question would be, given a circle in the hyperbolic plane, does there exist a rectilineal figure with area equal to the area of the circle?

In the Euclidean case, if we work over the field $K$ of constructible numbers, for example, the answer is never. Indeed, the formula $A = \pi r^2$ and the fact that $\pi$ is transcendental imply that if $r \in K$, then $A \notin K$; and conversely, if $A \in K$, then $r \notin K$.

In the non-Euclidean case, that is, in a hyperbolic plane whose field of ends $F$ is a subfield of $\mathbb{R}$, we find the surprising answer, sometimes yes and sometimes no. Consider a circle of radius $r \in F$. The question is whether the real number $A$ can be written as a sum of real numbers that are the radian measures of angles in our geometry. For every angle less than a right angle can be made equal to the area of some right triangle (Exercise 42.9).

If, for example, the quantity $x = (r-1)^2/r \in F$ is a dyadic rational number, then $\pi x$ will be the sum of radian measures of angles, because we can bisect a right angle any number of times. Thus if $r = 2$, or if $r$ is any power of 2, we obtain circles each of whose area is that of a rectilineal figure.

On the other hand, if $F$ is the field of constructible numbers, and if $x = \frac{1}{7}$, say, we can find an $r \in F$ with $x = (r-1)^2/r$, for example $r = \frac{1}{14}(15 + \sqrt{29})$, and a circle of that radius has area not corresponding to any rectilineal figure, because an angle of $\pi/7$ does not exist in that plane (cf. (29.4) and Exercise 41.11).

Thus the problem of "squaring the circle" is solvable for some circles, and unsolvable for others.

As another application of our trigonometric formulae, we will give Engel's *associated triangles*. Given a right triangle $ABC$, with right angle at $C$, we denote by $a, b, c$ the multiplicative lengths of the sides opposite $A, B, C$, and by $\alpha, \beta$ the angles at $A, B$. We say that the triangle has the five elements $(a, b, c, \alpha, \beta)$. We will use the following notation: For any segment $a$, the corresponding angle of parallelism is $\bar{a}$, and conversely, for any acute angle $\alpha$, the segment having that angle of parallelism is $\bar{\alpha}$ (40.7.1). Also, for an acute angle $\alpha$ we denote by

$\alpha' = RA - \alpha$ its *complementary* angle. If $a$ is a segment, then by abuse of notation we write $a'$ for $((\bar{a})')^-$.

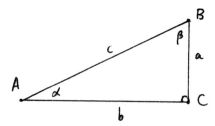

**Proposition 42.5**

*In a hyperbolic plane, given a right triangle with elements $(a, b, c, \alpha, \beta)$, there is an associated right triangle with elements $(b, \bar{\beta}', \bar{\alpha}, \bar{a}', \bar{c})$.*

*Proof* From (42.2c) we have

$$\sin \alpha = \cos \beta \sin \bar{b}.$$

Replacing $\beta$ by its complementary angle $\beta'$, we can write this as

$$\sin \alpha = \sin \beta' \sin \bar{b}.$$

Consider a right triangle with sides $b$ and $\bar{\beta}'$ around the right angle. Then by (42.3f) applied to this new triangle, the third side $x$ will satisfy

$$\sin \bar{x} = \sin \bar{b} \sin \beta'.$$

Thus $\sin \bar{x} = \sin \alpha$. Since the angles $\bar{x}$ and $\alpha$ are both acute, it follows that $\bar{x} = \alpha$, and so $x = \bar{\alpha}$. Thus our new triangle has sides $b, \bar{\beta}', \bar{\alpha}$ as required.

To find the angles of the new triangle, call them $\lambda, \mu$ for the moment. Then by (42.2a) applied to the new triangle,

$$\tan \lambda = \cos \bar{b} \tan \beta'.$$

Since the tangent of a complementary angle is the inverse of the tangent of the angle, we can rewrite this as

$$\tan \beta = \cos \bar{b} \tan \lambda'.$$

But (42.2a) for the original triangle, with the roles of $a$ and $b$ reversed, gives

$$\tan \beta = \cos \bar{b} \tan \bar{a}.$$

Hence $\tan \lambda' = \tan \bar{a}$, so $\lambda' = \bar{a}$ and $\lambda = \bar{a}'$.

To find $\mu$, use (42.2a) with $a, b$ reversed:

$$\tan \mu = \cos \beta' \tan \bar{b}.$$

Now, $\cos \beta' = \sin \beta$, so this gives

$$\tan \mu = \sin \beta \tan \bar{b}.$$

But (42.3d) with $a, b$ reversed gives

$$\tan \bar{c} = \sin \beta \tan \bar{b}.$$

So $\tan \mu = \tan \bar{c}$, and $\mu = \bar{c}$, as required.

### Remark 42.5.1

If we repeat the operation of taking the associated triangle five times, we recover the original triangle (Exercise 42.22). Thus one triangle gives rise to a cycle of five associated triangles. Our definition of the associated triangle depends on the ordering of the vertices $A, B$. If we use the reverse ordering, we will go around the cycle of five triangles in the reverse order.

There is also a geometrical construction of the associated triangle, which depends on Bolyai's parallel construction (Exercise 42.23).

## Exercises

The following exercises all take place in a hyperbolic plane.

42.1  Find the sides and angles of the equilateral triangle formed by the feet of the altitudes of the trilimit triangle of Exercise 40.8.

42.2  Find the side of an equilateral triangle with angles of $45°$.

42.3  Find the sides of an isosceles right triangle with angles $30°$, $30°$, $90°$.

42.4  Find the side of an equilateral pentagon with all right angles.

42.5  Derive the formulae of Proposition 42.3d,e from Proposition 42.2.

42.6  Prove the law of sines: In an arbitrary triangle $ABC$, with angles $\alpha, \beta, \gamma$ and opposite sides $a, b, c$,

$$\sin \alpha \tan \bar{a} = \sin \beta \tan \bar{b}$$
$$= \sin \gamma \tan \bar{c}.$$

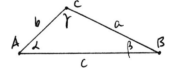

*Hint*: Use an altitude to divide the triangle into two right triangles.

42.7  Let $AB$ and $BC$ be two consecutive segments on a line, so that $AC = AB + BC$ as segments. Let $\alpha, \beta, \gamma$ be the angles of parallelism associated to $AB$, $BC$, and $AC$, respectively.

(a) Show that

$$\sin \gamma = \frac{\sin \alpha \sin \beta}{\cos \alpha \cos \beta + 1}.$$

(b) Derive analogous formulae for $\cos \gamma$ and $\tan \gamma$.

(c) Verify the following formula, due to Lobachevsky, analogous to the law of cosines in Euclidean geometry. Let $ABC$ be any triangle, with angles $\alpha, \beta, \gamma$, and opposite sides $a, b, c$. Then

$$\cos \gamma \cos \bar{a} \cos \bar{b} + \frac{\sin \bar{a} \sin \bar{b}}{\sin \bar{c}} = 1.$$

Note that if $\gamma$ is a right angle, this reduces to Proposition 42.3f.

42.8 Given an acute angle $\alpha$ at $A$, and given a segment $DE$, show that there exists a point $B$ on one arm of the angle such that the perpendicular $BC$ to the other arm is equal to $DE$. This is a strengthening of Aristotle's axiom—cf. (40.8.1).

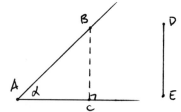

42.9 Given two angles $\alpha, \beta$ with $\alpha + \beta < $ RA, show that there exists a right triangle with angles $\alpha, \beta$.

42.10 Given three angles $\alpha, \beta, \gamma$ whose sum is less than 2RA, prove that there exists a triangle with angles equal to $\alpha, \beta, \gamma$. *Hint*: Glue together two suitable right triangles, and make them fit by solving a quadratic equation.

42.11 Let $T$ be a right triangle with angles $\alpha, \beta$ and sides $a, b, c$. Let $\delta$ be the area of $T$ (Theorem 36.2). Show that

$$\tan \delta = \frac{\cos \bar{a} \cos \bar{b}}{\sin \bar{a} + \sin \bar{b}}.$$

42.12 Given a triangle $ABC$, let $\delta$ be its area, and suppose we are given an angle $\delta'$ with $0 < \delta' < \delta$. Show that there is a point $D$ between $B$ and $C$ such that the area of $ABD$ is equal to $\delta'$. *Hint*: First draw an altitude from $A$ to $BC$, and thus reduce to the case of a right triangle. Then show that the point $D$ can be found by solving a suitable quadratic equation in $F$.

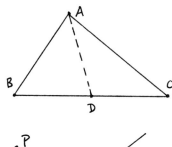

42.13 Given a point $P$, a line $l$, and an acute angle $\alpha$, show that there is a line $m$ through $P$ making an angle $\alpha$ with the line $l$.

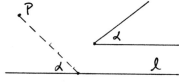

42.14 Given a point $A$ on a line $l$ and an end $\lambda$ of $l$, let $\Gamma$ be the horocycle defined by $A$ and $\lambda$ (Exercise 41.12). Let $m$ be a line that meets the ray $A\lambda$ in a point $C \neq A$. Prove that the line $m$ meets the horocycle in two points $B, B'$. *Hint*: First reduce to the case $m \perp \lambda$. Then find $B$ using trigonometry and solving a quadratic equation.

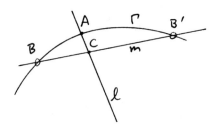

42.15 In this exercise we show how to translate the formulae of hyperbolic trigonometry into relations among the hyperbolic trigonometric functions, in the case where our field $F$ is a subfield of the real numbers.

For any $x \in \mathbb{R}$, we define the hyperbolic trigonometric functions

$$\sinh x = \frac{1}{2}(e^x - e^{-x}),$$

$$\cosh x = \frac{1}{2}(e^x + e^{-x}),$$

$$\tanh x = \sinh x / \cosh x.$$

(a) Assume $F \subseteq \mathbb{R}$, and let $\lambda$ be the additive distance function of Exercise 41.14. With our usual notation for a triangle, let $\tilde{a}, \tilde{b}, \tilde{c}$ be the additive lengths $\lambda$ of the sides opposite $A, B, C$. Show that

$$\sin \bar{a} = 1/\cosh \tilde{a},$$

$$\cos \bar{a} = \tanh \tilde{a},$$

$$\tan \bar{a} = 1/\sinh \tilde{a}.$$

(b) Use these to translate the formulae of Proposition 42.2 and Proposition 42.3. In particular, show that

$$\cos \alpha = \frac{\tanh \tilde{b}}{\tanh \tilde{c}} \quad \text{and} \quad \cosh \tilde{a} = \frac{\cos \alpha}{\sin \beta}.$$

42.16 Again assume $F \subseteq \mathbb{R}$ and let $\Gamma$ be a circle of radius $r$, and let $x = \ln r$ be the additive length of the radius (Exercise 42.15).

(a) Show that the area can be written as

$$A = 4\pi \sinh^2(x/2).$$

(b) Expand in power series to show that

$$A = \pi \left( x^2 + \frac{1}{12} x^4 + \cdots \right).$$

Thus the area of a hyperbolic circle is "bigger" than the area of a Euclidean circle with the same radius.

42.17 Use a limiting process similar to the one in Proposition 42.4 to define the multiplicative length $p$ of the circumference of a circle of radius $r$, and show that

$$\ln(p) = \frac{2\pi}{\tan \bar{r}}.$$

If $F$ is the field of constructible numbers, do you think that the circumference of a circle can ever be *rectifiable*, i.e., have length equal to the length of a segment in the plane?

42.18 Let $T$ be an equilateral triangle with all angles equal to 30°. Find, and express using square roots in standard form (Exercise 13.2),

(a) the radius of the inscribed circle;

(b) the radius of the circumscribed circle;

(c) the radius of the circle with the same area.

*Check*: Decimal answers to two of the above are 1.6093 and 8.1266.

42.19 Let $S$ be a regular quadrilateral (all 4 sides equal and all 4 angles equal) whose area is equal to the area of a circle of radius 3.

(a) find the radius of the inscribed circle

(b) find the radius of the circumscribed circle

(c) find the side of the quadrilateral.

*Check*: Decimal equivalents to two of these are 7.3276 and 27.8205.

42.20 Find an element $R \in F$ such that a circle of radius $r$ admits a circumscribed hexagon if and only if $r < R$.

42.21 Use Proposition 42.3f and Exercise 42.7c to derive the non-Euclidean (III.36) formula of Exercise 39.19. If $d = \mu(\frac{1}{2}PA)$, $e = \mu(\frac{1}{2}PB)$, and $f = \mu(\frac{1}{2}PC)$, show that this formula can be written $\cos^2 \bar{d} = \cos \bar{e} \cos \bar{f}$.

42.22 Show that the operation of taking the Engel associated triangle (Proposition 42.5) five times gives back the original triangle.

42.23 In the figure of Bolyai's parallel construction (Proposition 41.10), consider the right triangle $ABC = TPS$.

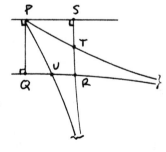

(a) Show that any right triangle $ABC$ can be embedded as the triangle $TPS$ in the figure of Bolyai's construction.

(b) Draw the limiting parallel ray from $P$ to the ray $\overrightarrow{SR}$, and let it meet $QR$ at $U$. Show that the triangle $A'B'C' = PUQ$ is the associated triangle to $ABC$ (Proposition 42.5). *Hint*: Get two elements of the new triangle from the figure; then use the formulae of Propositions 42.2 and 42.3 to get the others.

42.24 If $ABC$ is a right triangle with elements $(a, b, c, \alpha, \beta)$, and if $a, b, c \in \mathbb{Q}$, show that also $\bar{\alpha}, \bar{\beta}, \bar{\alpha}', \bar{\beta}', a', b', c'$ are in $\mathbb{Q}$ (using the notation of Proposition 42.5). *Hint*: Use the formula $\tan \alpha/2 = \sin \alpha/(1 + \cos \alpha)$.

42.25 In Euclidean geometry, if we have a right triangle whose sides $a, b, c$ are integers, then $a^2 + b^2 = c^2$, and we call the triple $(a, b, c)$ a *Pythagorean triple*. For example $(3, 4, 5)$ and $(5, 12, 13)$ are Pythagorean triples. If we have a right triangle with $a, b, c$ rational numbers, then multiplying by a common denominator, we can replace it by a similar triangle with integer sides.

In hyperbolic geometry, if we have a right triangle with sides $a, b, c \in \mathbb{Q}$, we will call this a *non-Euclidean Pythagorean triple*. Since there are no similar triangles, we cannot reduce these necessarily to integers.

(a) Show that $(a, b, c)$ is a non-Euclidean Pythagorean triple if and only if $a, b, c \in \mathbb{Q}$, $a, b, c > 1$, and

$$\frac{2a}{1 + a^2} \cdot \frac{2b}{1 + b^2} = \frac{2c}{1 + c^2}.$$

(b) Verify that $\left(\frac{5}{4}, \frac{7}{6}, \frac{21}{16}\right)$ satisfies the equation of (a).

(c) Use Engel's associated triangles to find four more solutions to (a).

(d) Show that (a) has no solutions in integers greater than 1. *Hint*: Estimate the sizes of the quantities involved.

(e) Find more solutions to (a), besides the ones in (b) and (c).

# 43  Characterization of Hilbert Planes

Thinking back to our study of Euclidean geometry, recall that we pursued two different logical paths. One was the abstract development of a geometry from axioms. The other was the analytic approach given by the Cartesian plane over an ordered field $F$. We brought these two paths together by introducing the field of segment arithmetic into the abstract geometry (Section 19) and then showing that any Hilbert plane with (P) is isomorphic to the Cartesian plane over its field of segment arithmetic (Section 21). To express this in other words, a Hilbert plane with (P) is *characterized* as the Cartesian plane over a certain Pythagorean ordered field $F$. It follows that two Hilbert planes with (P) will be isomorphic, as abstract geometries, if and only if their associated fields are isomorphic, as ordered fields.

In this section we will do the same thing for non-Euclidean geometry. For Hilbert planes satisfying (L), which we have called hyperbolic planes, we will prove a coordinatization theorem analogous to the one in the Euclidean case. For more general Hilbert planes we will discuss the theorem of Pejas, some ideas of its proof and some consequences, but we cannot enter into full details. We also describe the calculus of reflections initiated by Hjelmslev, and use it to give a proof of the three altitudes of a triangle theorem in neutral geometry.

We have developed some properties of the abstract hyperbolic planes in Section 40. On the other hand, we have introduced the Poincaré model over a field in Section 39, and have used the ambient Cartesian geometry to investigate some of its properties. Then in Section 41 and Section 42 we have introduced the field of ends into an abstract hyperbolic plane, and have used the field to create a sort of hyperbolic analytic geometry and trigonometry.

The last step in this logical progression is to characterize the hyperbolic planes by showing that a hyperbolic plane is determined, up to isomorphism, by its associated field of ends, and that any hyperbolic plane is isomorphic to the Poincaré model over its associated field.

### Proposition 43.1

*Given a hyperbolic plane* $\Pi$, *the field of ends is uniquely determined, up to isomorphism. Two hyperbolic planes* $\Pi_1$ *and* $\Pi_2$ *are isomorphic as Hilbert planes if and only if the associated fields* $F_1$ *and* $F_2$ *are isomorphic as ordered fields.*

*Proof*  Of course, the set of ends in the plane $\Pi$ is uniquely determined. But in order to define the field structure on this set (41.4), we made a choice of two orthogonal lines, and labeled their ends $0, \infty, 1, -1$. If we made a different choice, a different end might become the zero element of the field, so clearly the field structure on the set of ends is not unique. But we will show that the field is unique up to isomorphism.

So suppose $l_1, m_1$ are two orthogonal lines with ends labeled $0_1, \infty_1, 1_1, -1_1$, giving rise to a field structure $F_1$ on the set of ends (minus $\infty_1$). Suppose $l_2, m_2$ is a second choice of orthogonal lines with ends labeled $0_2, \infty_2, 1_2, -1_2$, giving a second field structure $F_2$. We can find a rigid motion $\varphi$ of the plane that takes $l_1$ to $l_2$ and $m_1$ to $m_2$, in such a way that the ends $0_1, \infty_1, 1_1, -1_1$ are sent to the corresponding ends $0_2, \infty_2, 1_2, -1_2$. This rigid motion induces a one-to-one correspondence from the set of ends of $\Pi$ to itself, which therefore gives a one-to-one map of $F_1$ to $F_2$ sending the elements $0_1, 1_1, -1_1$ of $F_1$ to the corresponding elements of $F_2$. The constructions that we used to define addition and multiplication in $F_1$ are now carried over to the corresponding constructions for $F_2$, and the ordering is preserved. Therefore, $\varphi$ induces an isomorphism of $F_1$ and $F_2$ as ordered fields, which shows that the field associated to $\Pi$ is unique up to isomorphism.

Now suppose that $\Pi_1$ and $\Pi_2$ are isomorphic hyperbolic planes. If $\varphi : \Pi_1 \to \Pi_2$ is an isomorphism, we can choose orthogonal lines $l_1$ and $m_1$ in $\Pi_1$ with which to construct the field of ends $F_1$ of $\Pi_1$, and then take $l_2 = \varphi(l_1)$, $m_2 = \varphi(m_1)$ to construct the field of ends $F_2$ to $\Pi_2$. Then it is clear that the induced map $\varphi' : F_1 \to F_2$ on ends will give an isomorphism of fields.

Finally, let $\Pi_1$ and $\Pi_2$ be hyperbolic planes, and suppose that we are given an isomorphism $\psi : F_1 \to F_2$ of the associated fields of ends. We wish to show

# ÉLÉMENS
## DE
# *GÉOMÉTRIE.*

---

## PREMIERE PARTIE.

*Des moyens qu'il étoit le plus na-*
*turel d'employer pour parvenir*
*à la mesure des Terreins.*

CE qu'il semble qu'on a dû mesurer
d'abord, ce sont les longueurs & les
distances.

### I.

POUR mesurer une longueur quelconque, l'expédient que fournit une

A

Plate XV. The beginning of Clairaut's *Elémens de Géométrie* (1775).

that $\Pi_1$ and $\Pi_2$ are isomorphic Hilbert planes. That is to say, there is a map $\varphi : \Pi_1 \to \Pi_2$ of the points that is one-to-one and onto, and $\varphi$ preserves lines, betweenness, and congruence line segments and angles.

We construct $\varphi$ as follows. First extend the map $\psi : F_1 \to F_2$ to the set of all ends by setting $\psi(\infty_1) = \infty_2$. Any line of $\Pi_1$ is given by an unordered pair $(\alpha, \beta)$ of distinct elements of $F_1' = F_1 \cup \{\infty_1\}$. So by sending $(\alpha, \beta)$ to $(\psi(\alpha), \psi(\beta))$, we obtain a one-to-one map of the set of lines of $\Pi_1$ onto the set of lines of $\Pi_2$. A point $P$ of $\Pi_1$ is determined by the set of all lines passing through it, and this set of lines satisfies an equation of the form

$$u_1 u_2 - b(u_1 + u_2) + a^2 = 0$$

with $a, b \in F_1$, $a > 0$, and $|b| < a$ (41.6). Since $\psi$ is an isomorphism of ordered fields, the set of images by $\varphi$ of the lines containing $P$ will satisfy a similar equation in $F_2$, and so they will define a unique point that we denote by $\varphi(P)$. Then by construction $\varphi$ is a one-to-one map of the set of points of $\Pi_1$ onto the set of points of $\Pi_2$, sending lines into lines.

Since betweenness of points can be expressed in terms of the ordering of the field of ends, and $\psi$ is an isomorphism of ordered fields, $\varphi$ preserves betweenness.

Congruence of line segments can be measured by the multiplicative distance function $\mu$ (41.7), and congruence of angles can be measured by the tangent function (41.8). Therefore, the map $\varphi : \Pi_1 \to \Pi_2$ also preserves congruence of line segments and angles. Thus $\varphi$ is an isomorphism of Hilbert planes, as required.

### Theorem 43.2

*Let $F$ be a Euclidean ordered field, and let $\Pi$ be the Poincaré model constructed over the field $F$ (Section 39). Let $F_1$ be the field of ends of $\Pi$. Then $F$ and $F_1$ are isomorphic ordered fields.*

*Proof*  Our strategy is to establish a one-to-one correspondence between the sets $F$ and $F_1$, and then carry out the constructions of addition and multiplication in $F_1$ in the geometry of the Poincaré model, to show that the field structures are isomorphic.

We may assume that the Poincaré model is constructed using the unit circle $\Gamma$ in the Cartesian plane over $F$ (Exercise 39.23). Let us choose the

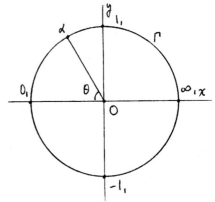

*x*-axis and *y*-axis to be the orthogonal lines used in the construction of the field of ends $F_1$. Recall that the points of the Poincaré model are those points of the Cartesian plane inside $\Gamma$. It follows that the ends of the Poincaré model II are exactly the Cartesian points of the circle $\Gamma$. We label the ends of our two axes so that

$$0_1 = (-1, 0),$$

$$\infty_1 = (1, 0),$$

$$1_1 = (0, 1),$$

$$-1_1 = (0, -1),$$

in Cartesian coordinates.

The next step is to give a mapping of sets between the elements of the fields $F$ and $F_1$. The elements of $F_1$ are all the points of the circle $\Gamma$ except $\infty = (1, 0)$. We define a map $\varphi : F_1 \to F$ as follows. First we set $\varphi(0_1) = 0$. Next, for any end $\alpha$ in the upper semicircle of $\Gamma$, we set $\varphi(\alpha) = \tan \theta/2$, where $\theta$ is the angle from $0_1$ to $\alpha$ subtended at the center of the circle. If $-\alpha$ is the reflection of this point in the *x*-axis, we set $\varphi(-\alpha) = -\varphi(\alpha)$. Here we understand $\theta$ to be the angle in the Cartesian plane, and the tangent function has its usual meaning (Section 16). As $\alpha$ ranges over the upper semicircle, $\theta$ ranges over all possible angles (between 0 and 2RA), so the function $\tan \theta/2$ ranges over all positive elements of the field $F$. Given any element $a \in F$, $a > 0$, there is an angle $\theta$ such that $\tan \theta/2 = a$, so this mapping $\varphi$ is a one-to-one correspondence between the sets $F_1$ and $F$. Clearly, $\varphi$ preserves the ordering on the two sets.

The hard work we must do is to show that $\varphi$ is compatible with the operations of addition and multiplication in the two fields. We start by computing the Cartesian coordinates of an end $\alpha \in F_1$, which is a point of the circle $\Gamma$, in terms of $\varphi(\alpha)$. We will do all our calculations for points $\alpha$ in the upper semicircle, since the results for their negatives will follow immediately.

So let $\alpha \in F_1$, $\alpha > 0$. Let $\varphi(\alpha) = a = \tan \theta/2$. It follows that the Cartesian coordinates $(x, y)$ of $\alpha$ are given by

$$\begin{cases} x = -\cos \theta = \dfrac{a^2 - 1}{1 + a^2}, \\[2mm] y = \sin \theta = \dfrac{2a}{1 + a^2}. \end{cases}$$

Here we use the usual trigonometric formulae expressing $\sin \theta$ and $\cos \theta$ in terms of $\tan \theta/2$ (cf. the calculations of Exercise 17.6).

To study addition, we will compute the reflection $\sigma_\alpha$ in a line $(\alpha, \infty)$. In the Poincaré model, reflection is given by circular inversion in the corresponding Cartesian circle (39.5). So given a point $\alpha$ on $\Gamma$, let $\Delta$ be the $P$-line $(\alpha, \infty)$, which is a circle orthogonal to $\Gamma$ at the points $\alpha$ and $\infty$. Let $A$ be the center of this circle. Then by a little elementary Euclidean geometry we see that the angle $OA\infty_1$ is equal to $\theta/2$, so the coordinates of $A$ are $(1, a^{-1})$.

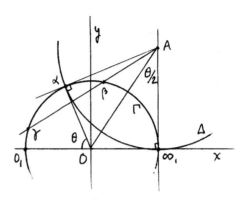

The $P$-reflection in $\Delta$ is circular inversion in $\Delta$ in the Cartesian plane. To study its effect on ends, let $\beta$ be any end, and let $\gamma$ be its image under reflection in $\Delta$. Since $\Gamma$ and $\Delta$ are orthogonal circles, $\gamma$ is just the intersection of the line $A\beta$ with $\Gamma$. Let $\varphi(\beta) = b$ and $\varphi(\gamma) = c$. Then $\beta$ and $\gamma$ have Cartesian coordinates

$$\beta = \left(\frac{b^2 - 1}{1 + b^2}, \frac{2b}{1 + b^2}\right),$$

$$\gamma = \left(\frac{c^2 - 1}{1 + c^2}, \frac{2c}{1 + c^2}\right).$$

To express the fact that $A, \beta, \gamma$ are collinear, we set the slopes of $A\beta$ and $A\gamma$ equal to each other. This gives

$$\frac{\dfrac{2b}{1 + b^2} - a^{-1}}{\dfrac{b^2 - 1}{1 + b^2} - 1} = \frac{\dfrac{2c}{1 + c^2} - a^{-1}}{\dfrac{c^2 - 1}{1 + c^2} - 1}.$$

Simplifying, we get

$$2a(b - c) = b^2 - c^2.$$

Assuming $\beta \neq \gamma$, so $b \neq c$, we can divide out $b - c$, and so

$$c = 2a - b.$$

Thus the reflection in $\Delta$, which is $\sigma_\alpha$, has an effect on ends in $F_1$, which is transformed by $\varphi$ into the transformation

$$x' = 2\varphi(\alpha) - x$$

for elements of $F$.

Recall that addition of ends is characterized by

$$\sigma_{\alpha+\beta} = \sigma_\beta \sigma_0 \sigma_\alpha$$

using reflections (41.3). Via the mapping $\varphi : F_1 \to F$, therefore, $\sigma_{\alpha+\beta}$ becomes

$$x' = 2\varphi(\alpha + \beta) - x.$$

On the other hand, $\sigma_\beta \sigma_0 \sigma_\alpha$ becomes a transformation that first sends $x$ to

$$2\varphi(\alpha) - x,$$

then to

$$2\varphi(0) - (2\varphi(\alpha) - x),$$

then to

$$2\varphi(\beta) - [2\varphi(0) - (2\varphi(\alpha) - x)] = 2\varphi(\alpha) + 2\varphi(\beta) - x.$$

We conclude that

$$\varphi(\alpha + \beta) = \varphi(\alpha) + \varphi(\beta),$$

so that $\varphi$ is a homomorphism for addition.

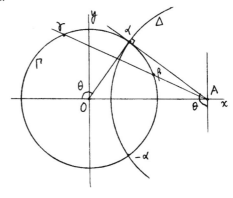

Now let us consider multiplication. To do this (changing notation) we consider the reflection $\sigma_\alpha$ in a line $(\alpha, -\alpha)$ in the Poincaré model. This is given by circular inversion in a circle $\Delta$, orthogonal to $\Gamma$, passing through the points $\alpha$ and $-\alpha$. Let $A$ be the center of this circle. Then the angle $OA\alpha$ is equal to $\theta - RA$ in our diagram. Thus $OA = -1/\cos\theta$. If $\varphi(\alpha) = a = \tan\theta/2$, then

$$A = \left(\frac{1+a^2}{a^2-1}, 0\right).$$

Now let $\beta$ be another end, and let $\gamma$ be its image under $\sigma_\alpha$, which is circular inversion in $\Delta$. Then $A, \beta, \gamma$ are collinear. Letting $\varphi(\beta) = b$ and $\varphi(\gamma) = c$ as before, we express the collinearity by equating the slopes of the lines $A\beta$ and $A\gamma$. This time we get

$$\frac{\dfrac{2b}{1+b^2}}{\dfrac{b^2-1}{1+b^2} - \dfrac{1+a^2}{a^2-1}} = \frac{\dfrac{2c}{1+c^2}}{\dfrac{c^2-1}{1+c^2} - \dfrac{1+a^2}{a^2-1}}.$$

Simplifying gives

$$bc(b - c) = a^2(b - c).$$

Assuming $b \neq c$, we get

$$c = a^2 b^{-1}.$$

Thus the reflection $\sigma_\alpha$ is transformed by $\varphi$ into the transformation

$$x' = \varphi(\alpha)^2 x^{-1}$$

in $F$.

Going back to the hyperbolic plane, it is easy to see that the composition $\sigma_\alpha \sigma_1$ is equal to a translation along the line $(0, \infty)$ (cf. proof of $(41.4)$) that sends $1$ to $\alpha^2$. On the other hand, multiplication is defined by composing translations. So multiplication is characterized by the formula

$$\sigma_{\alpha\beta}\sigma_1 = \sigma_\beta \sigma_1 \sigma_\alpha \sigma_1,$$

or simply

$$\sigma_{\alpha\beta} = \sigma_\beta \sigma_1 \sigma_\alpha.$$

Transporting this by $\varphi$, we find that for elements of $F$, the transformation

$$x' = \varphi(\alpha\beta)^2 x^{-1}$$

is equal to

$$x' = \varphi(\beta)^2 \varphi(\alpha)^2 x^{-1}.$$

We conclude that $\varphi(\alpha\beta) = \varphi(\alpha)\varphi(\beta)$ as required.

We have now shown that $\varphi : F_1 \to F$ is a one-to-one transformation compatible with addition, multiplication, and the ordering. Hence $F_1$ and $F$ are isomorphic as ordered fields.

### Corollary 43.3

*If $\Pi$ is a hyperbolic plane with associated field of ends $F$, then $\Pi$ is isomorphic to the Poincaré model over the field $F$.*

*Proof*  Indeed, the field $F$ is Euclidean $(41.4)$, and the plane $\Pi$ and the Poincaré model over $F$ both have isomorphic fields of ends by $(43.2)$, so by $(43.1)$ they are isomorphic planes.

### Corollary 43.4

*The circle–circle intersection property* (E) *holds in a hyperbolic plane.*

*Proof*  Indeed, the field of ends is Euclidean $(41.4)$, the hyperbolic plane is isomorphic to the Poincaré model over that field, and (E) holds in the Poincaré model $(39.9)$.

**Remark 43.4.1**
Thus in a Hilbert plane, the axiom (L) of existence of limiting parallels implies (E).

**Remark 43.4.2**
In fact, a stronger result holds. Namely, in any hyperbolic plane, given any two curves each of which is either a line or a circle or a horocycle or a hypercycle, they will intersect if a certain betweenness condition is satisfied (which we leave to the reader to make explicit). Indeed, these all correspond to various Euclidean circles in the Poincaré model (Exercise 39.14), and these will meet by (E) in the ambient Cartesian plane.

## The Classification of Hilbert Planes According to Pejas

So far in this book we have seen two classification theorems. One was that any Hilbert plane with (P) is isomorphic to the Cartesian plane over a Pythagorean ordered field (21.1), and in this section we have seen (43.3) that any hyperbolic plane is isomorphic to the Poincaré model over a Euclidean ordered field. These are both special cases of a more general theorem due to Pejas (1961), which gives an algebraic model of any Hilbert plane. The value of a classification theorem is that it allows one to prove theorems essentially by checking what happens in all possible planes, even when one does not have a direct proof. Examples of such reasoning that we have already used are (21.2), that (LCI) is equivalent to (E) in a Hilbert plane with (P), and (43.4), that (E) holds in any hyperbolic plane. A consequence of Pejas's general theorem is that (LCI) is equivalent to (E) in any Hilbert plane.

To explain properly the complete statement of Pejas's theorem would carry us too far beyond the realm of the present book, so I will give only some partial statements of the theorem, some applications, and some comments on the main ideas of the proof. For full details, see the paper of Pejas (1961) and the books of Bachmann (1959) and Hessenberg–Diller (1967).

To begin with, let us consider a general method of constructing subplanes of a given Hilbert plane.

## Definition
A *full subplane* of a Hilbert plane $\Pi$ is a Hilbert plane $\Pi_0$ whose points are a subset of the points of $\Pi$, whose lines are the intersections of the lines of $\Pi$ with the points of $\Pi_0$ whenever that intersection is nonempty, and whose betweenness and congruence are induced from the ambient plane. (See, for example, (18.4.3), Exercise 18.5, and Exercise 39.24.)

For any Hilbert plane we will consider the group $G$ of *segment addition* of the plane. This is an ordered abelian group whose positive elements are the congruence equivalence classes of line segments (cf. (19.1) for the addition of segments, and the additive part of (19.3) for the existence of the group $G$). For example, in the Cartesian plane over a field $F$, the group $G$ is just the additive group of the field $(F, +)$. In the Poincaré model over a Euclidean field $F$, it is the group of positive elements of the field under multiplication $(F_{>0}, \cdot)$ using the multiplicative distance function (39.10).

We say that a subgroup $M$ of an ordered abelian group $G$ is *convex* if $a, b \in M$ and $a < c < b$ in $G$ implies $c \in M$.

### Proposition 43.5

*Let $\Pi$ be any Hilbert plane with group of segment addition $G$.*

(a) *If $\Pi_0 \subseteq \Pi$ is a full subplane, then the group of segment addition of $\Pi_0$ is a nonzero convex subgroup $M$ of $G$.*

(b) *Each nonzero convex subgroup $M$ of $G$ arises from a full subplane as in* (a).

(c) *Two full subplanes $\Pi_0$ and $\Pi_1$ of $\Pi$ give the same subgroup $M \subseteq G$ if and only if there is a rigid motion of $\Pi$ taking $\Pi_0$ to $\Pi_1$.*

*Proof* (a) Obviously, $M$ is a nonzero subgroup of $G$. To show that $M$ is convex it is sufficient to show that $\Pi_0$ is convex, namely, if $A, B$ are points of $\Pi_0$ and $C$ is a point of $\Pi$ that lies between $A$ and $B$, then $C$ is in $\Pi_0$.

To do this, take points $D, E$ in $\Pi_0$ such that $D$ is not on the line $AB$, and $E$ is between $A$ and $D$. Let $l$ be the line $EC$ of $\Pi$. Since $\Pi_0$ is a full subplane of $\Pi$ and the line $l$ contains a point $E$ of $\Pi_0$, it follows that $l' = l \cap \Pi_0$ is a line of $\Pi_0$. Now, $l'$ meets one side of the triangle $ABD$. It cannot meet $BD$, since $l$ does not, so by Pasch's axiom it must meet $AB$. The intersection point is $C$, so $C \in \Pi_0$.

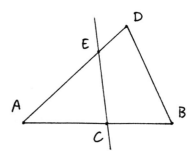

(b) Given $M$ a nonzero convex subgroup of $G$, fix a point $O \in \Pi$, let $\Pi_0$ be the set of points $A$ of $\Pi$ such that the segment class $[OA]$ is in $M$, and take for lines of $\Pi_0$ the nonempty intersection of lines in $\Pi$ with $\Pi_0$. We must verify that $\Pi_0$ satisfies all the axioms of a Hilbert plane. They all follow immediately from the corresponding axioms of $\Pi$ or from (C1), which is the only nontrivial one.

To verify (C1), let $AB$ be a segment in $\Pi_0$, and suppose we are given a point $C$ and a ray $r$ emanating from $C$. By the axiom (C1) in $\Pi$, there is a unique point $D \in \Pi$ on the ray $r$ such that $AB \cong CD$. We have only to show that $D \in \Pi_0$. From (I.20) it follows that $AB \leq OA + OB$. Since $A, B \in \Pi_0$, we have $[OA] \in M$ and $[OB] \in M$. But $M$ is a convex subgroup of $G$; therefore, $[AB] \in M$ also. This shows that $M$ is equal to the group of segment addition of $\Pi_0$.

To continue the proof of (C1), use (I.20) again to see that $OD \leq OC + CD$. Now, $C \in \Pi_0$ implies $[OC] \in M$. Also, $CD \cong AB$ implies $[CD] \in M$. Since $M$ is a convex subgroup of $G$, it follows that $[OD] \in M$, so $D \in \Pi_0$. We leave to the reader to check special cases where some points coincide with $O$.

(c) If there is a rigid motion taking $\Pi_0$ to $\Pi_1$, then they clearly give the same subgroup $M \subseteq G$. Conversely, suppose $\Pi_0$ and $\Pi_1$ give the same subgroup $M$. Fix a point $A_0 \in \Pi_0$ and a point $A_1 \in \Pi_1$. Then $\Pi_0$ is equal to the set of points $B \in \Pi$ for which $[A_0B] \in M$, and similarly for $\Pi_1$. Thus any rigid motion of $\Pi$ that takes $A_0$ to $A_1$ will take $\Pi_0$ to $\Pi_1$.

### Corollary 43.6
*If a Hilbert plane is Archimedean, it has no proper full subplanes.*

*Proof*  The group of segment addition of an Archimedean plane satisfies the Archimedean property $(A')$, namely, for any $a, b \in G$, $a, b > 0$, there exists an integer $n$ such that $na > b$. In this case the only nonzero convex subgroup is the whole group, so there can be no proper subplanes.

Now we can state (without proof) a somewhat restricted form of the theorem of Pejas (1961), using the language of full subplanes.

### Theorem 43.7 (Pejas)
(a) *Any semi-Euclidean Hilbert plane is a full subplane of the Cartesian plane over a Pythagorean ordered field F.*

(b) *Any semielliptic Hilbert plane satisfying* (E) *is a full subplane of a plane of the form given in Exercise 34.14b over a non-Archimedean Euclidean ordered field F.*

(c) *Any semihyperbolic Hilbert plane satisfying* (E) *is a full subplane of the Poincaré model in the unit circle over a Euclidean ordered field F.*

For Hilbert planes that do not satisfy the circle axiom (E), the statements are more complicated. For example, in the semihyperbolic case one must allow Poincaré models in (possibly virtual) circles of the form $x^2 + y^2 = d$ over Pythagorean ordered fields satisfying additional conditions similar to $(*d)$ of Exercise 39.26. We omit the details.

For the cases treated in (43.7) we see using (43.5) that to give a complete description, we need only specify the field giving the main model and then a nonzero convex subgroup $M$ of its group of segment addition $G$. In the semi-Euclidean case, $G$ is the additive group of the field $(F, +)$. In the semielliptic case, $G$ is the subgroup of infinitesimal elements in the circle group of the field (Exercise 17.6), since arcs on the sphere are measured by the angles they subtend at the center of the sphere. In the semihyperbolic case, $G$ is the multiplicative group of positive elements $(F_{>0}, \cdot)$ of the field.

In the original paper of Pejas, he gave two apparently unrelated constructions for the semihyperbolic planes, which Hessenberg and Diller (1967, Section 68) called the modular and the nonmodular case. In our formulation there is a single construction for both—the distinction being whether $M$ consists entirely of infinitesimal elements or not (Exercise 43.6).

Now let us consider some consequences of this theorem.

**Corollary 43.8**
*A Hilbert plane satisfying* (A) *and* (E) *is either the Cartesian plane or the Poincaré model over a Euclidean ordered field F.*

*Proof*  From the hypothesis (A), it follows that the field $F$ of the theorem is Archimedean (Exercise 43.9). This rules out the semielliptic case, so our plane is a full subplane of a Euclidean or hyperbolic plane. But by (43.6) there are no proper full subplanes. So our plane must be Euclidean or hyperbolic.

In the semi-Euclidean case this result follows from our earlier work without using Pejas's theorem (35.4). In the semihyperbolic case, the corollary gives the remarkable implication

$$(A) + (E) + (\sim P) \Rightarrow (L).$$

There is no direct proof (without classification) for this result, which is essentially the same as the problem we discussed earlier of proving Bolyai's parallel construction without assuming beforehand the existence of the limiting parallel ray (41.11.1).

See the exercises for further consequences of Pejas's theorem.

Now let us say a few words about the proof of Pejas's theorem. It proceeds in three stages. The first stage is to extend a given Hilbert plane by the addition of new "ideal" points and "ideal" lines, so that the original plane is embedded in a projective plane. We have already seen two examples of this procedure. One was in Exercise 6.7, where we add ideal points to an affine plane, or for example a Euclidean Hilbert plane, to obtain a projective plane. The other was in (40.10.1) and Exercises 40.11, 40.12, where a similar but more subtle construction is given for a hyperbolic plane.

The idea of introducing ideal points seems to go back to von Staudt (1847) for the Euclidean plane, and to Klein for the hyperbolic case. Pasch (1882) carried out this procedure using properties of three-dimensional space. The first person to succeed using plane geometry only, in the form of Hilbert's axioms, was Hjelmslev (1907). Making extensive use of reflections and the group of all rigid motions of the plane, he showed how to embed an arbitrary Hilbert plane in a projective plane, and he was able to prove that this projective plane satisfies "Pascal's theorem," which is a projective analogue of Pappus's theorem (14.4).

The second stage is to introduce coordinates into the projective plane. This problem also has its roots in the early nineteenth century. Again, von Staudt was probably the first, with his theory of "Würfe," to introduce a rational net of points, and thus by continuity to obtain real-number coordinates. But by the end of the nineteenth century there was growing interest in building foundations of geometry without continuity. The significance of Pascal's theorem was made clear by Hilbert (1899), who showed that it was precisely the condition for the field of segment arithmetic to be commutative. The construction was generalized by Schwan, and finally reached its modern form in the book of Artin (1957), where the elements of the field appear as operators on the group of translations of an affine plane.

These two stages are explained in detail in the book of Hessenberg–Diller (1967), Sections 37–44 for the work of Hjelmslev, and Sections 55–58 for the introduction of coordinates in the projective plane.

The third stage, which was accomplished by Pejas in his thesis, is to identify those subsets of the projective planes that give Hilbert planes. We will not give any details here, except to point out that Pejas's work was done in the context of the "metric planes" of Bachmann (1959).

Bachmann's observation was that the work of Hjelmselv took place almost entirely in the group of rigid motions of the plane and made little use of the order relation. So Bachmann defined the notion of a "metric plane" in which you retain the properties of incidence and orthogonality, but forget order and congruence. To any line we associate the reflection in that line, and to any point we associate the reflection in that point. Then we forget the original points and lines, and axiomatize geometry purely within the group of rigid motions (see Bachmann (1959), Section 3.2, for his axiom set).

Bachmann's metric planes include all Hilbert planes, but also include elliptic geometry, some finite geometries, and many others. Generalizing the work of Hjelmslev, Bachmann is able to embed any one of his metric planes in a projective metric plane. The Hilbert planes appear as those metric planes with an order relation satisfying Hilbert's axioms (B1)–(B4) and having free mobility— essentially what we call (ERM).

It is based on Bachmann's formulation of this whole theory that Pejas proves his theorem.

## The Calculus of Reflections

To illustrate this new approach to geometry, we will give some elementary results in the *calculus of reflections*, initiated by Hjelmslev, and elevated by Bachmann to a position of central importance in the theory of metric planes. This calculus of reflections is analogous to analytic geometry, in that it gives an algebraic method of treating geometric problems. It has advantages over the usual analytic geometry in that there is no arbitrary choice of coordinate axes and it works in an arbitrary Hilbert plane.

Fix a Hilbert plane. We denote by $G$ the group of all rigid motions of the plane, and by $S$ the subset of $G$ consisting of reflections in a line. We know that for every line $a$ there is a reflection $\sigma_a$ in that line, and these reflections generate the group of rigid motions (cf. proof of (17.4) and Exercise 17.3). For simplicity we will denote $\sigma_a$ simply by $a \in S$, and for any point $A$, we denote by $A \in G$ the point reflection ($=$ rotation through 2RA) around that point.

### Proposition 43.9

*If $a, b$ are two distinct lines, then $ab = ba$ (in the group $G$) if and only if $a \perp b$. In that case the product $A = ab$ is the point reflection in the intersection point $A$ of $a$ and $b$.*

*Proof*  If $a \perp b$, then clearly $A = ab$ is the point reflection and $ab = ba$.

Conversely, suppose $ab = ba$. Let $P$ be any point of $a$ not lying on $b$. Let $ba(P) = b(P) = P'$ be the reflection of $P$ in $b$. Then $ab(P) = a(P') = P'$. So $P'$ also lies in $a$. Since $P \notin b$, $P \neq P'$. Then $b$ is the perpendicular bisector of the segment $\overline{PP'}$, which lies on the line $a$, so $a \perp b$.

### Proposition 43.10

*If four lines $a, b, c, d$ pass through a point, then $ab = cd$ if and only if the acute (or right) angles formed by $a, b$ and $c, d$ are equal and the (acute) rotation from $a$ to $b$ has the same orientation as from $c$ to $d$. In that case also $ba = dc$, $ac = bd$, and $ca = db$.*

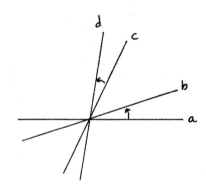

*Proof*  The motion $ab$ is a rotation through twice the acute angle between $a$ and $b$, in the opposite direction (cf. Exercise 17.4). So if $ab = cd$, they give the same rotations, and the result is clear. If we take the same rotations in the reverse direction we get $ba = dc$. If we add (or subtract) the rotation from $b$ to $c$, we get $ac = bd$ and $ca = db$.

### Proposition 43.11

*If $a, b, c, d$ are four lines perpendicular to a line $l$, and meeting $l$ in $A, B, C, D$, then $ab = cd$ if and only if the segments $\overline{AB}$ and $\overline{CD}$ are equal and have the same orientation on $l$. In that case also $ba = dc$, $ac = bd$, and $ca = db$.*

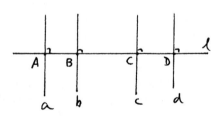

*Proof*  Note that $ab$ is a translation along the line $l$ (cf. proof of (41.4)) of twice the segment $\overline{AB}$, in the opposite direction. So the proof is analogous to the previous proof.

**Proposition 43.12**

*Let $a, b$ be lines, and let $C, D$ be distinct points. Then $aC = bD$ if and only if $a, b$ are perpendicular to the line $\overline{CD}$, and the segments $\overline{AC}$ and $\overline{BD}$ are equal, in the same orientation, where $A, B$ are the points $a \cap \overline{CD}, b \cap \overline{CD}$.*

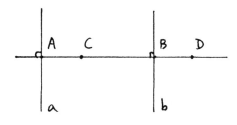

*Proof*  Note that $aC$ is a glide reflection along the line through $C$, perpendicular to $a$, in an amount equal to twice the segment $\overline{AC}$. So if $aC = bD$, they have the same axis, which must be the line $CD$. The rest is clear.

**Proposition 43.13**

*Let $a, b$ be lines and $C, D$ distinct points. Then $aC = Db$ if and only if $a, b$ are perpendicular to the line $\overline{CD}$, and the segments $\overline{AC}, \overline{BD}$ are equal, in reverse order.*

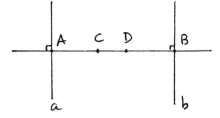

*Proof*  Similar to previous proof.

**Proposition 43.14** (Hjelmslev)

*In any Hilbert plane, given a quadrilateral ABCD with right angles at B, D, draw the diagonals e, f, and drop perpendiculars AG, CH from A and C to f as shown. Then (in the notation of the diagram):*

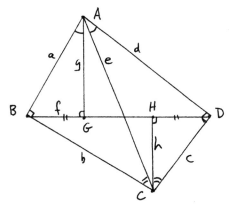

(1)  *$ag = ed$.*
(2)  *$be = hc$.*
(3)  *$Bg = hD$.*

*In particular, the angles marked at A and at C are equal, and the segments $\overline{BG}, \overline{HD}$ are equal.*

*Proof*  By the theorem of three reflections (Exercise 17.14), we can find lines $g', h'$ such that $ag' = ed$ and $be = h'c$. Using $B = ba$ and $D = cd$ (43.9) and substituting, we obtain

$$Bg' = bag' = bed = h'cd = h'D.$$

Now, by (43.13), $g'$ and $h'$ are perpendicular to the line $\overline{BD} = f$, so $g = g', h = h'$.

This gives (1) and (2) and (3) above. The equalities of angles and segments follow from (43.10) and (43.13).

**Remark 43.14.1**

We call $ABCD$ as above a *Hjelmslev quadrilateral*. This important result plays a role in neutral geometry analogous to the role of cyclic quadrilaterals in Euclidean geometry. In fact, it can also be proved easily in Euclidean geometry using cyclic quadrilaterals (Exercise 5.19).

Note that the result and its proof work equally well if $A, C$ are on the same side of $BD$.

As an application, we prove the theorem on the intersection of the altitudes of a triangle in neutral geometry.

**Theorem 43.15**

*In any Hilbert plane, if two of the altitudes of a triangle meet, then all three meet at the same point.*

*Proof*   In the triangle $ABC$, let the altitudes $BE, CD$ meet at $H$. Draw the line $DE$, and drop perpendiculars $x, y, z, w$ from $B, C, A, H$ to that line. Let $AH$ be the line $f$. We want to show $f \perp a$.

We have a Hjelmslev quadrilateral $AEHD$, which by (43.14) gives $fe = dw$ at $H$, $bz = fc$ at $A$, and $\overline{DW} = \overline{ZE}$. There is another Hjelmslev quadrilateral $BECD$, which gives $cx = ae$ at $B$, $ad = by$ at $C$, and $\overline{XD} = \overline{EY}$. Combining the last statements of each and adding $\overline{ZW}$, we obtain $\overline{ZX} = \overline{YW}$, so $zx = yw$ by (43.11). Now let us calculate:

$$afe = adw = byw = bzx = fcx = fae.$$

Canceling $e$ gives $af = fa$, so by (43.9), $f$ is perpendicular to $a$, as required.

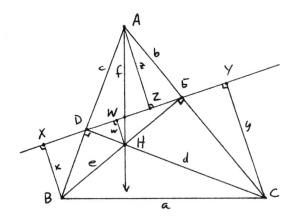

# Exercises

43.1 Show that the result of Exercise 1.15 holds in an arbitrary hyperbolic plane.

43.2 If a semihyperbolic plane $\Pi$ satisfies Dedekind's axiom (D), show (without using Theorem 43.7) that $\Pi$ is isomorphic to the Poincaré model over the real numbers $\mathbb{R}$.

43.3 (a) Use Theorem 43.7 to show that in a semi-Euclidean plane, Aristotle's axiom (Section 33) implies (P).

(b) Now prove the same result without using Theorem 43.7, by strengthening the proof of Proposition 35.4 to replace (A) by Aristotle's axiom.

43.4 Use Theorem 43.7 to show that in any semi-Euclidean Hilbert plane, the following three conditions are equivalent:

(i) (LCI) holds.

(ii) The field $F$ is Euclidean.

(iii) (E) holds.

43.5 Use Theorem 43.7 to prove that in a semi-Euclidean plane, given any segment $AB$, there exists an equilateral triangle with side $AB$. Can you prove this without using Pejas's theorem?

43.6 We say that the *Lotschnitt axiom* of Bachmann holds in a Hilbert plane if for any four lines $a, b, c, d$ with $a \perp b, b \perp c, c \perp d$, it follows that $a$ meets $d$.

(a) Show that the Lotschnitt axiom holds in any semi-Euclidean or semielliptic plane.

(b) If $\Pi_0$ is a full subplane of the Poincaré model $\Pi$ over a Euclidean ordered field $F$, corresponding to a nonzero convex subgroup $M$ of the group $G = (F_{>0}, \cdot)$, show that the following conditions are equivalent:

(i) The Lotschnitt axiom holds in $\Pi_0$.

(ii) All elements of $M$ are infinitesimal (i.e., of the form $1 + x$ for $x \in F$ infinitesimal).

(iii) The angle sum of any triangle differs from 2RA by an infinitesimal angle.

These conditions describe what Hessenberg–Diller call the "modular" case; otherwise, $\Pi_0$ is called nonmodular. Thus we see that the Lotschnitt axiom characterizes geometries in which the angle sum of a triangle differs at most infinitesimally from 2RA.

43.7 Show that the Lotschnitt axiom (Exercise 43.6) is equivalent to Legendre's axiom (cf. Section 35) for a right angle: Namely, for any point $P$ in the interior of a right angle, there exists a line meeting both sides of the angle.

43.8 Show in any Hilbert plane (without using classification) that (A) plus Lotschnitt implies (P). (Pambuccian (1994) strengthens this result to Aristotle's axiom plus Lotschnitt implies (P).)

43.9 Let $F$ be a non-Archimedean ordered field, and let $M$ be a nonzero convex subgroup of either $(F, +)$ or $(F_{>0}, \cdot)$ or the circle group of $F$ (Exercise 17.6). Show that $M$ cannot be Archimedean.

43.10 Use Theorem 43.7 to prove the theorem of Greenberg (1988) that a Hilbert plane satisfying (E) plus Aristotle's axiom must be Euclidean or hyperbolic.

43.11 (a) Let $\Pi_0$ and $\Pi_1$ be two semi-Euclidean planes represented as full subplanes of Cartesian planes over Pythagorean ordered fields $F_1, F_2$ by convex subgroups $M_1 \subseteq (F_1, +)$ and $M_2 \subseteq (F_2, +)$. Show that $\Pi_0$ and $\Pi_1$ are isomorphic (as abstract Hilbert planes) if and only if there is an isomorphism $\varphi : F_1 \xrightarrow{\sim} F_2$ and a nonzero element $\lambda \in F_2$ such that $M_2 = \lambda \cdot \varphi(M_1)$.

(b) Similarly, let $\Pi_0$ and $\Pi_1$ be semi-hyperbolic planes represented in the Poincaré models in the unit circles over Euclidean ordered fields $F_1$ and $F_2$ by convex subgroups $M_1$ and $M_2$ of the multiplicative groups of positive elements. Show that $\Pi_0$ and $\Pi_1$ are isomorphic if and only if there is an isomorphism $\varphi : F_1 \xrightarrow{\sim} F_2$ such that $\varphi(M_1) = M_2$.

43.12 This and the following two exercises take place in an arbitrary Hilbert plane, using the notation of the calculus of reflections.

(a) If $A$ is a point and $b$ a line, show that $A \in b \Leftrightarrow Ab = bA$.

(b) If $a, b$ are two lines that meet, show that $c$ is an angle bisector of one of the angles between $a, b \Leftrightarrow ac = cb$.

(c) If $A, B$ are two distinct points, then a line $h$ is the perpendicular bisector of the segment $\overline{AB} \Leftrightarrow Ah = hB$.

43.13 Using the calculus of reflections, prove that the three angle bisectors of a triangle meet, as follows. Given the triangle $ABC$, let two angle bisectors $d, e$ meet at a point $P$. Drop a perpendicular $x$ from $P$ to $a$. Let $f = xed$ by the theorem of three reflections (Exercise 17.14). Then prove that $f$ is the angle bisector at $C$.

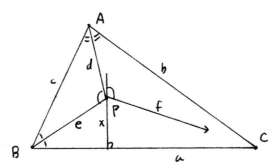

43.14 Using the calculus of reflections, prove that if two of the perpendicular bisectors of the sides of a triangle meet, then all three meet in the same point, as follows: Let two perpendicular bisectors $d, e$ meet at $P$. Let $x$ be the line $\overline{AP}$. Let $f = xed$. Then show that $f$ is the perpendicular bisector of $\overline{AC}$.

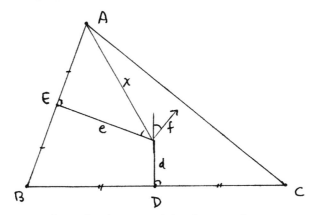

43.15 If you are curious about the theorem of the three medians in neutral geometry, look in Bachmann (1959) p. 74 and see whether you can understand the proof there.

43.16 Given an angle $BAC$, two rays $AD, AE$ making equal angles with the angle bisector of $BAC$ are called *isogonal conjugates*.

Let $ABC$ be a triangle, and let $AD, BE, CF$ be three concurrent lines in the triangle. Prove that the isogonal conjugates of $AD, BE, CF$, with respect to the angles of the original triangle, are concurrent.

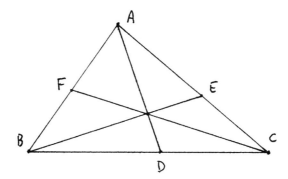

43.17 In a non-Archimedean hyperbolic plane, let $\Gamma$ be a circle of infinite radius $r$ (i.e., $\mu(r)$ is an infinite element of the field of ends $F$).

(a) Show that $\Gamma$ is not contained inside any polygon (cf. Exercises 36.3, 42.20).

(b) Show that the exterior of $\Gamma$ is not a segment-connected set (cf. Exercise 11.1).

43.18 Show that de Zolt's axiom $(Z)$ holds in any Hilbert plane.

# 8

**CHAPTER**

# Polyhedra

olyhedra are solid figures bounded by plane poly-gons. Most famous among these are the five regular, or Platonic, solids, identified classically with the four elements, earth, air, fire, water, and the whole uni-verse. Euclid begins his *Elements* with the construc-tion of an equilateral triangle (I.1) and ends in Book XIII with the construction of these regular solids. It has been suggested that Euclid's purpose in writing the *Elements* was to fully elucidate the geometry be-hind these five figures.

Euclid defines the tetrahedron, cube, octahedron, icosahedron, and dodeca-hedron by the number and type of faces they have. He then constructs each one inscribed in a sphere, and claims that only these five are possible. To make this exact, we need to supply the hypothesis of convexity, not stated explicitly by Euclid, and we need to prove that the figures so obtained are unique. In fact, it is not immediate what the definition of a regular polyhedron should be. We clarify this at the end of Section 44 by defining a *regular* polyhedron to be convex, with all of its faces congruent regular polygons, and with the same number of of faces meeting at each vertex. Then we can prove there are only five of these, and that in addition they have the further properties that all dihedral angles equal, they can be inscribed in a sphere, and the group of symmetries is transitive on the vertices.

To complete the proof we need Cauchy's rigidity theorem (Section 45), which says that a convex polyhedron is determined up to congruence by its faces and their combinatorial arrangement.

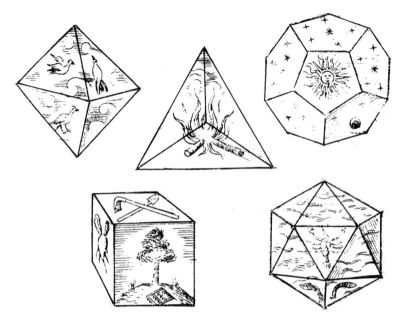

The theory developed to study the regular polyhedra also suffices to classify the *semiregular*, or Archimedean, solids, which are those solids whose faces are all regular polygons and that have at each vertex the same kinds of faces in the same cyclical order. There are two infinite series, the prisms and the antiprisms, and thirteen others (Section 46).

We also begin the study of a more difficult problem, to classify all *face-regular* polyhedra, which are those convex polyhedra having only regular polygons as faces. We complete the analysis for those having only equilateral triangles as faces in Section 45: There are five more besides the three regular solids with triangular faces. For the general problem, we show that there are only finitely many nonuniform face-regular polyhedra (46.3), but refer to Johnson (1966) for the complete classification.

In Section 47 we explore another interaction of geometry and abstract algebra by identifying the rotation groups of the regular polyhedra and showing that they, along with the cyclic and dihedral groups, are the only finite subgroups of the group of rotations of the sphere.

# 44    The Five Regular Solids

Our concern in this section will be first, to see what Euclid has done, and then to formulate exact definitions and study the question of existence and uniqueness of the five regular solids.

Up to this point in this book we have considered almost exclusively plane geometry. Here we move into solid, or three-dimensional, geometry. We will not take the time to set up axiomatic foundations for this solid geometry; instead, we let the reader make the straightforward extensions from Euclidean plane geometry to Euclidean solid geometry. A brief review of Euclid's *Elements*, Book XI, may be helpful. Or we may simply appeal to the "Cartesian" three-space over the real numbers ℝ, and use the methods of analytic geometry.

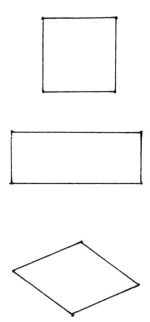

To start, let us review what we know about regular polygons in the plane. A triangle with three equal sides (equilateral) also has three equal angles (I.5). Conversely, a triangle with three equal angles (equiangular) has three equal sides (I.6). For polygons with four or more sides, however, having equal sides does not imply equal angles, nor conversely. A square is equilateral and equiangular. A rectangle is equiangular but not equilateral. A rhombus is equilateral but not equiangular.

We define a *regular* polygon in the plane to be a polygon that is both equilateral and equiangular.

## Proposition 44.1
*In the real Cartesian plane, for any n ≥ 3, there exists a regular polygon of n sides (n-gon) having a given segment as a side. Any two regular n-gons with a common side are congruent. The vertices of the regular n-gon lie on a circle. For any two vertices, there is a rotation of the n-gon to itself sending the first vertex to the second.*

*Proof*  For existence, just take a circle, and mark $n$ equidistant points on the circumference, subtending angles of $2\pi/n$ at the center. Expanding by a scale factor will make the side equal to any given segment.

To prove uniqueness, suppose we are given a regular $n$-gon with side $AB$. Bisect the equal angles at $A$, $B$, and let the angle bisectors meet at $O$. Then $O$ is equidistant from $A$ and $B$. Continuing this construction at the other vertices, one sees easily that $O$ is equidistant from all the vertices, so they lie on a circle with center $O$. Thus any two regular $n$-gons with a common side will be congruent.

A rotation through an angle of $2\pi/n$ at the center will send one vertex to the next. Multiples of this rotation will send the first vertex to any other desired vertex.

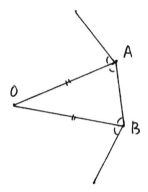

**Remark 44.1.1**

We make these elementary observations about regular polygons explicit to emphasize the analogy with the three-dimensional case to follow.

Note that we used the real numbers to be able to divide the angle $2\pi$ into $n$ equal pieces. If we work in the Cartesian plane over the field of constructible numbers, or if we ask for those regular polygons that are constructible using a compass and marked ruler, not all values of $n$ are possible. For detailed discussion of these questions, see Sections 29, 30, 31.

Now we come to the definition of the regular solids. Euclid does not give a general definition of a regular solid (we will do that later at the end of this section). Instead, in the definitions of Book XI of the *Elements*, he defines each one individually. A *pyramid* is a solid figure formed by joining a point to each of the vertices of a polygon in a plane not containing the point. Euclid does not use the word *tetrahedron*, but we will, defining it to be a triangular pyramid formed of four equilateral triangles. Euclid goes on to define a *cube* as a solid figure contained by six equal squares, the *octahedron* and *icosahedron* as solid figures bounded by 8 (resp. 20) equilateral triangles, and a *dodecahedron* as a figure bounded by 12 regular pentagons.

Let us also fix some terminology that we will use in discussing solid figures. A *polyhedron* is the surface of a solid figure bounded by plane polygons. When two polygons meet, they must have an entire edge in common. These plane polygons are the *faces* of the polyhedron. Their edges are the *edges* of the polyhedron, and their vertices are the *vertices* of the polyhedron. Where two faces meet along an edge, we have a *dihedral angle*. This is the angle between two rays, drawn in the two faces, from a point on the common edge, and both perpendicular to that edge. At a vertex, the angle in any face passing through that vertex is called a *face angle*. The collection of all the faces at a vertex makes a *solid angle*. The solid angle is not measured by a number, but we can speak of one solid angle being *congruent* to another if there is a rigid motion of the space (plus possibly a reflection) making one coincide with the other close to the vertex.

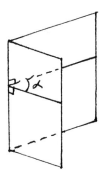

If all the vertices of a polyhedron lie on a sphere, we will say that the polyhedron is *inscribed* in the sphere. If all the faces of a polyhedron are tangent to a sphere, we will say that the polyhedron is *circumscribed* about the sphere.

Of course, defining something is no guarantee of its existence (think of a unicorn, for example). So our first job is to construct examples of these figures. Euclid does this in Book XIII, Propositions 13–17, by a very explicit and quite complicated method. For brevity we will use simpler methods, due to Legendre (1823), Appendix to Books VI, VII.

**Proposition 44.2**

*There exist tetrahedra, cubes, octahedra, icosahedra, and dodecahedra having the following properties*:

(a)  *In each figure, all the dihedral angles are equal.*
(b)  *The vertices of each figure lie on a sphere.*
(c)  *For any two vertices, there is a rigid motion of the figure onto itself sending the first vertex to the second.*

*Proof*   To make a tetrahedron, take an equilateral triangle of side 1, say. At its center, erect a line perpendicular to the plane of the triangle. On this line, find a point at distance 1 from one of the vertices of the triangle. This point will then be at distance 1 from all three vertices, so the pyramid from this point will be a tetrahedron.

The dihedral angles between the three new faces are obviously equal. But we observe that the construction could also have been made starting from one of these new faces, giving the same figure. Hence all the dihedral angles are equal.

Any four points not in a plane lie on a sphere, so the tetrahedron is inscribed in a sphere. A rotation about an axis passing through one vertex and the center of the opposite face will send any one vertex to another.

The cube we leave to the reader (Exercise 44.1).

To construct an octahedron, take a sphere of radius 1 and three mutually perpendicular diameters. Join the six points where these diameters meet the sphere, making 8 equilateral triangles. These form an octahedron. Any pair of adjacent vertices can be sent to any other by a succession of rotations about these axes, and this implies also that the dihedral angles are all equal.

The construction of the icosahedron is a little more complicated. We start with a regular pentagon (say of side 1) *BCDEF* in a plane. The pentagon lies on a circle (44.1), so from the center of the circle, erect a perpendicular to the plane of the pentagon, and on that line find a point *A* at distance 1 from *B*. Then *A* will also be at distance 1 from *C, D, E, F*. Joining *A* to the points *B, C, D, E, F*, we obtain a pentagonal pyramid with equilateral triangles as its upper faces. By symmetry, the dihedral angles between any two adjacent triangles are the same.

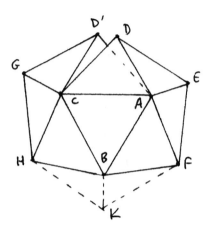

Make another such congruent pentagonal pyramid starting with a pentagon *B'A'D'GH* and label its top vertex *C'*. The dihedral angles in this new pyramid are equal to those in the other. Hence, if we glue the triangle *A'B'C'* onto the triangle *ABC*, the points *D* and *D'* will coincide. We get a figure made of eight equilateral triangles, with all dihedral angles equal.

Doing this once more, we get a convex figure of ten equilateral triangles with *ABC* in the center, and having all its dihedral angles equal. Furthermore, as we go around the six edges that form the outer boundary of this figure, the angle between any two successive edges is equal to the interior angle of a regular pentagon.

Now make another such figure of ten equilateral triangles. Because of the equality of edge angles and dihedral angles, the two will fit together perfectly to make an icosahedron with all dihedral angles equal.

The perpendiculars at the centers of any two adjacent faces will meet at a point *O* equidistant from the four vertices bounding these two faces. Because the dihedral angles are all equal, this construction propagates around the whole surface to show that the point *O* is equidistant from all the vertices. Hence the icosahedron is inscribed in a sphere with center *O*.

The construction we have given is clearly symmetric under a rotation of the initial triangle *ABC* into itself. On the other hand, since all the dihedral angles are equal, the construction could have been started anywhere. Thus there are rotations sending any vertex to an adjacent vertex, and a succession of these will be a rigid motion sending any vertex to any other vertex.

Lastly, to make a dodecahedron, take the icosahedron previously constructed. For each vertex of the icosahedron, join the five midpoints of the triangles meeting at that vertex. This makes a regular plane pentagon. The 12 pentagons thus constructed form a dodecahedron. Because of the symmetry of the icosahedron, the relation between any two adjacent pentagons is the same, so all the dihedral angles are equal.

The vertices of the dodecahedron are the midpoints of the faces of the icosahedron. These are all equidistant from the center of the sphere containing the icosahedron, so the vertices of the dodecahedron lie on a new smaller sphere, which is inscribed in the icosahedron.

A rotation of the icosahedron about an axis passing through two opposite vertices will send one triangle to an adjacent one, hence one vertex of the dodecahedron to an adjacent one. Successions of these will make rigid motions sending any vertex of the dodecahedron to any other.

## Remark 44.2.1

In this proof we have made use of some rotations and rigid motions of the five solids. We will study these rotations and rigid motions in more detail later, together with their group structure (Section 47).

## Remark 44.2.2

All of the steps in the above constructions could be carried out with ruler and compass in suitable planes. In particular, finding a point on a line at unit distance from a given point is just a matter of intersecting the line with a circle. Hence these polyhedra are all constructible with Euclidean tools. In particular, if we work in the real Cartesian three-space, their coordinates will lie in the field $K$ of constructible numbers. Put otherwise, these five solids exist in the Cartesian three-space over the field $K$.

Now we come to the question of uniqueness. Are the figures we have constructed the only polyhedra that satisfy Euclid's definitions?

Euclid himself gives us an answer, stated as an unnumbered proposition just after (XIII.18). He says that no other figure besides these five figures can be constructed that is contained by equilateral and equiangular (i.e., regular) polygons equal to each other.

His reasoning is as follows: If we use equilateral triangles, then we can put together 3, 4, or 5 of them at one vertex, but 6 would lie flat. If we use squares, we can put 3 at one vertex, but no more. If we use regular pentagons, again we can put 3 at a vertex. If we try to use hexagons, three of them would lie flat, so for a stronger reason we cannot use regular polygons of more sides.

These five cases, he says, correspond to the tetrahedron, octahedron, icosahedron, cube, and dodecahedron, respectively; hence there are no others.

Unfortunately, Euclid's conclusion is not correct as stated, because of some missing implicit hypotheses, nor is his proof of the corrected result complete.

To make a correct statement, we need first to require that the figures in question be *convex*. This means that for any two points on the surface of the polyhedron, the line segment between those points is entirely contained in the solid figure bounded by the polyhedron. We use the word convex in the strict sense, meaning also that no two faces lie in the same plane.

### Example 44.2.3
Otherwise we could have a figure such as the "punched-in icosahedron." Consider one vertex $A$ of an icosahedron, and let $BCDEF$ be the pentagon formed by the five adjacent vertices. Take off the pentagonal pyramid made by $ABCDEF$, and replace it by the pentagonal pyramid $A'BCDEF$, where $A'$ is the reflection of the point $A$ in the plane of $BCDEF$. The point $A'$ is then inside the original icosahedron, so the new figure is like an icosahedron elsewhere, but has a concavity at $A'$. Think of the shape of a soccer ball at the moment it is being kicked, so that the toe of the boot makes a concave spot in the ball. This is a figure bounded by 20 equal equilateral triangles, but it is not congruent to the one we constructed. So we must require convexity in order to have uniqueness.

Now Euclid's argument becomes correct insofar as it relates to what happens at a single vertex:

### Proposition 44.3
*In a convex polyhedron all of whose faces are equal regular polygons, the only possible configurations at a single vertex are 3, 4, or 5 triangles, 3 squares, or 3 pentagons.*

*Proof*  The argument given above now works. Because of the convexity at a vertex, the sum of the face angles at the vertex must be less than $2\pi$ (cf. Euclid (XI.21)), and the listed five cases are the only possibilities.

But even with the hypothesis of convexity, Euclid's original global statement is still not correct. Think of two equal tetrahedra, glued together along one face. This is a convex polyhedron (a *triangular dipyramid*) whose faces are 6 equilateral triangles, but it is not in our list.

What we need to assume (and this also was probably implicit in Euclid's thinking) is that the number of faces meeting at each vertex is the same. In the triangular dipyramid, we have three faces meeting at the two farthest points, and four faces meeting at each of the vertices along the glued face.

Now we can state a corrected version of Euclid's classification.

### Theorem 44.4
*Any polyhedron that is*

(a) *bounded by equal regular polygons,*
(b) *convex,*
(c) *has the same number of faces at each vertex,*

*is congruent (up to a scale factor) to one of the five constructed in (44.2). Furthermore, these five all have the additional properties*

(d) *all dihedral angles are equal,*
(e) *the vertices lie on a sphere,*
(f) *for any two vertices, there is a rigid motion of the figure taking one to the other.*

**Lemma 44.5**
*If a vertex $V$ of a polyhedron has three faces meeting it, with face angles $\alpha, \beta, \gamma$, then the three dihedral angles between these faces are uniquely determined by $\alpha, \beta, \gamma$. In particular, if $\alpha, \beta, \gamma$ are equal, the three dihedral angles will also be equal.*

*Proof* Take points $A, B, C$ equidistant from $V$ (say distance 1) on the three edges meeting at $V$. Holding the triangle $AVB$ fixed, let the triangle $AVC$ rotate around the line $AV$. As it does so, the point $C$ describes a circle on the unit sphere with center $V$. Similarly, as $BVC$ rotates around $BV$, the point $C$ describes another circle on the unit sphere centered at $V$. The point where these two circles meet is $C$ (or its reflection in the plane of $AVB$). Now it is clear that all three dihedral angles are uniquely determined by the three face angles of the triangles at $V$.

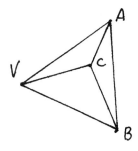

Another way to see this is to intersect the three faces with the unit sphere around $V$. We obtain a spherical triangle, which we call the *vertex figure* of the polyhedron at the vertex $V$. The sides of this spherical triangle are portions of great circles, subtending angles $\alpha, \beta, \gamma$ at the center of the sphere, so the lengths of the sides are just $\alpha, \beta, \gamma$ (in radians). Now use the fact that the three angles of a spherical triangle are determined by the three sides (and in fact can be calculated explicitly by the formulae of spherical trigonometry). These three angles, measured by their tangent lines, perpendicular to the radii of the sphere, are nothing but the dihedral angles of the original polyhedron.

**Remark 44.5.1**
Of course, the conclusion of the lemma is false for four or more faces. If you make an open figure of four equilateral triangles meeting at a single vertex, it is quite flexible: You can decrease two opposite dihedral angles while increasing the other two.

*Proof of Theorem*    Since we have assumed that the figure is convex and has the same number of faces at each vertex, we can apply (44.3), and so have to consider five cases.

*Case 1*    Three equilateral triangles at each vertex. Let $A$ be one of the vertices, and let $B, C, D$ be the adjacent vertices. Because the sides $AB, AC, AD, BC, BD, BC$ are all equal, $B, C, D$ form an equilateral triangle. The vertices $B, C, D$ each have two triangles already, so $BCD$ makes the third, and the whole figure has just these four equilateral triangles as faces. Because of (44.5), the dihedral angles are the same as those in the tetrahedron constructed in (44.2), so the two tetrahedra are congruent up to scale factor. Properties (d), (e), (f) follow from (44.2).

*Case 2*    Three squares at each vertex. By the lemma (44.5) the dihedral angles are uniquely determined. In this case they are right angles. Starting at one vertex, the three square faces fit on a cube. Continuing to adjacent vertices, the faces of our solid must coincide with those of the cube; hence it is a cube.

*Case 3*    Three regular pentagons at a vertex. This is similar to Case 2. Because the dihedral angles are uniquely determined, they must coincide with the dihedral angles of the dodecahedron constructed in (44.2). Staring at one vertex and working our way around, our figure must coincide with that dodecahedron, so the extra properties (d), (e), (f) follow from (44.2).

*Case 4*    Four equilateral triangles at a vertex. Here the lemma does not apply, so we must work harder. First we make a combinatorial argument to show that our figure is composed of eight equilateral triangles, in the same relative positions as the octahedron of (44.2).

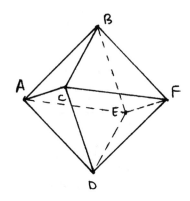

Let $A$ be one vertex. Since there are four faces meeting at $A$, there are four adjacent vertices $B, C, D, E$. Now, at $B$ we already have two of our faces and three of our edges, so there must be another vertex $F$ such that $BCF$ and $BEF$ are equilateral triangles. Now, at $C$ we have three of the required four faces: It follows that $CDF$ must be an equilateral triangle forming the fourth face. Similarly, $DEF$ is the fourth face at $E$, and now the figure is complete. Thus our figure is bounded by eight equilateral triangles in the same arrangement as the octahedron of (44.2).

However, this does not yet prove that our figure is congruent to the one constructed earlier. If you have built an octahedron or an icosahedron, you will have noticed that it is quite flexible in the intermediate stages of construction, and becomes rigid only when the last face is glued in position. So we are inclined to believe that the final shape is uniquely determined, but our model-building experience does not make a mathematical proof.

To prove that our figure is congruent to the earlier one, we argue as follows. Let $A, F$ be opposite vertices of our figure. Then $B, C, D, E$ are all equidistant from $A$ and $F$, so they must lie in the plane that bisects the segment $AF$. But they are also equidistant from $A$, so they lie on a sphere with center $A$ that intersects that plane in a circle. So $B, C, D, E$ lie on a circle. But they are also equidistant from each other, in order, so they form a square. Let $O$ be the center of the square. Then the segments $AF, BD, CE$ all pass through $O$, and are mutually perpendicular. If an edge of our figure has length 1, then $O$ is at distance $\sqrt{2}/2$ from $B, C, D, E$, and one then sees easily that $O$ is also at distance $\sqrt{2}/2$ from $A$ and $F$. Thus $A, B, C, D, E, F$ lie on a sphere with center $O$, and we have recovered the construction of (44.2). The properties (d), (e), (f) follow.

*Case 5*   Five equilateral triangles at each vertex. First we make a combinatorial argument to show that a figure made of triangles having five meeting at each vertex has 12 vertices and 20 faces, in the same arrangement as the icosahedron of (44.2). This step is left to reader (Exercise 44.5).

Now we have to show that a convex figure made of 20 equilateral triangles in the same arrangement as the icosahedron of (44.2) is congruent to that one. In that case the extra properties (d), (e), (f) will follow from (44.2). This is again the question of rigidity we encountered in Case 4, but for the icosahedron we do not know an elementary argument. We must refer to Cauchy's rigidity theorem (45.5) in the next section to complete the proof.

### Remark 44.5.2
Even when this theorem is completely proved, there still remains another question: Are the five figures constructed in (44.2) the only convex figures satisfying Euclid's definitions of tetrahedron, cube, octahedron, icosahedron, dodecahedron? This is a slightly different question, because instead of assuming (c) of (44.4) that each vertex has the same number of faces, we assume only that the total number of faces is given. For the tetrahedron it is obvious. For the cube and the dodecahedron, since it is possible only to have three faces at a vertex, the result follows from (44.4). But for the case of a convex figure made of 8 or 20 equilateral triangles, it is not obvious because there might be a different way of arranging the triangles with different numbers of them at different vertices. The result is nevertheless true, as we will see later when we classify convex polyhedra whose faces are all equilateral triangles (45.6.1).

**Definition**

We can now define a *regular polyhedron* to be a convex polyhedron whose faces are all equal regular polygons and having the same number of faces meeting at each vertex. It follows from (44.4) that the only regular polyhedra are the five Platonic solids constructed in (44.2), and that they all have the extra properties (d), (e), (f) of (44.4). Alternatively, one could define a regular polyhedron to be a convex polyhedron all of whose faces are equal regular polygons and that satisfies any one of the properties (d), (e), (f) of (44.4). See Exercise 44.6.

# Exercises

In these exercises, the words tetrahedron, cube, octahedron, dodecahedron, icosahedron refer to the *regular* polyhedra with 4, 6, 8, 12, 20 faces respectively.

44.1  Verify the existence of a cube having all the properties of Proposition 44.2.

44.2  If you join the centers of the faces of a cube by lines, show that this makes an octahedron. Conversely, show that joining the centers of the faces of an octahedron makes a cube. We say that the cube and the octahedron are *dual* solids.

44.3  Take a tetrahedron of side length 1, and around each vertex cut off a smaller tetrahedron of side length $\frac{1}{2}$. Show that what remains is an octahedron. Conclude that the dihedral angle of an octahedron and the dihedral angle of a tetrahedron are supplementary angles.

44.4  Make a model of each of the five regular solids. There are many ways to do this, and I would not want to limit your creative ingenuity, but I will tell you my favorite method. Lay out the faces on a flat piece of cardboard, with as many attached to each other as possible. Part of the fun is figuring out how to lay them out, but you can skip this step by looking in almost any geometry book (except this one) for a diagram. Then cut out the figure, and score with a knife the edges to be folded. Fold up and join edges of the solid figure by making a small double tab of cardboard to be glued inside the adjoining edges of the two faces. How to glue down the last face is another interesting problem I leave to you. When it is complete, you can paint the entire solid with different colors to emphasize the symmetries.

Another good medium for making quick models and experimenting is jelly beans (the small kind) and toothpicks. And when you are done, you can eat the jelly beans. One student of mine even made tetrahedra out of gingerbread triangles, glued together with cake frosting. That time the whole class enjoyed the models.

44.5  Show that a convex polyhedron whose faces are all triangles (not necessarily equilateral) having five faces meeting at each vertex must have 12 vertices and 20 faces.

44.6  In Theorem 44.4, if we assume (a) and (b), but instead of assuming (c), we assume any one of (d), (e), or (f), show that (c) and hence the rest of the theorem follows.

édris irregularibus, quibus tegitur Cubus intus. Huic fuccedit Icofaë-

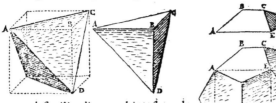

In priori fi-
gurâ appa-
ret Tetrae-
dron ACDF
latens in
Cubo; fic
ut quodli-
bet Tetrae-
dri planum
ut ACD, te-
gatur ab u-
no cubi an-
gulo ACDB
In fecunda
figura appa-
ret Cubus
AED latens
intus in Do-
decaedro;
fic ut quod-
libet Cubi
planum, ut
AED, tega-
tur a duob*
Dodecae-
dri angulis
feu Pétaedra
jà B C D E
quod eft fe-
ctile in tria
Tetraedra
diffimilia
per duo pla-
na, D C A,
& A B D.
Hic vides
Octaedron
infcriptum
Cubo; Ico-
fiedron Do-
decaedro,
Tetrae-
dron Te-
traedro.

dron 4. ob fimilitudinem, ultima fecundariarum, angulo folido plurilineari utenci-
um. Intimum eft Octoëdron 5. Cubi fimi-
le, & prima figura fecundariarum, cui ideò
primus locus interiorum debetur, quippe infcriptili; uti cubo circum-
fcriptili primus exteriorum.

Sunt autem notabilia duo veluti conjugia harum figurarum, ex

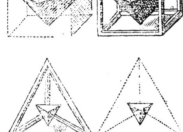

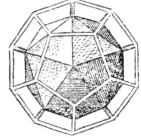

diverfis combinata claffibus: Ma-
res, Cubus & Dodecaëdron ex
primariis; feminæ, Octoëdron
& Icofiëdron ex fecundarijs; qui-
bus accedit una veluti cœlebs aut Androgynos, Tetraëdron; quia fibi
ipfi infcribitur, ut illæ fœmellæ maribus infcribuntur & veluti fubji-
ciuntur, & figna fexus fœminina mafculinis oppofita habent, angulos
fcilicet planiciebus.

Præterea ut Tetraëdron eft elementum, vifcera & veluti cofta Cubi Maris; fic Octaëdron fœmina, eft elementum & pars Tetraëdri, a-
liâ ratione: ita mediat Tetraëdron in hoc conjugio.

Præcipua connubiorum feu familiatum differentia in hoc confiftit
quòd Cubicæ quidem *Effabilis* eft proportio: nam Tetraëdron eft Tri-
ens de corpore Cubico, Octaëdron femiffis de Tetraëdrico, fexta pars
Cubi: Dodecaëdrici verò conjugij proportio eft *Ineffabilis* quidem,
fed *Divina*.

Harum duarum vocum copulatio jubet cavere Lectori, de earum
fig-ficatu. Vox enim *Ineffabilis* hic non denotat per fe nobilitatem
aliquam, ut aliàs in Theologia & rebus divinis; fed denotat conditio-

Plate XVI. A page from Kepler's *Harmonices Mundi* (1619) showing how one regular polyhedron may be inscribed in another. He calls the cube and the dodecahedron male, and
the octahedron and the icosahedron female, since the latter can be inscribed in the for-
mer, while the tetrahedron is androgynous, since it is inscribed in itself. Reprinted courtesy of the Bancroft Library of the University of California at Berkeley.

44.7 Take three $3 \times 5$ cards and arrange them so that they are in three mutually perpendicular planes, they all have the same center, and each one passes through a $3''$ slit in one of the others. Then the 12 corners of these three cards approximate the vertices of an icosahedron. Prove this as follows: Assume that the ratio of the sides of the cards is $\frac{1}{2}(1 + \sqrt{5}) \approx 1.618$ instead of its actual value $\frac{5}{3}$. Compute the coordinates of the twelve corners in a Cartesian 3-space whose axes run through the cards, and show that the distance between any two adjacent vertices is the same. (Use symmetry to limit your calculations to two, instead of 30.)

44.8 Let $d$ be the diameter of a sphere, and let $s$ be the side of an inscribed regular polyhedron. Show that $d$ and $s$ are related as follows:

(a) For a tetrahedron, $d^2 = \frac{3}{2}s^2$.

(b) For an octahedron, $d^2 = 2s^2$.

(c) For a cube, $d^2 = 3s^2$.

(d) For an icosahedron, $d^2 = \frac{1}{2}(5 + \sqrt{5})s^2$.

(e) For a dodecahedron, $d^2 = \frac{3}{2}(3 + \sqrt{5})s^2$.

44.9 Let $\alpha$ be the dihedral angle of a regular polyhedron. Verify that:

(a) For a tetrahedron, $\cos\alpha = \frac{1}{3}$.

(b) For an octahedron, $\cos\alpha = -\frac{1}{3}$.

(c) For a cube, $\cos\alpha = 0$.

(d) For an icosahedron, $\cos\alpha = -(\sqrt{5}/3)$, $\sin\alpha = \frac{2}{3}$.

(e) For a dodecahedron, $\cos\alpha = -(\sqrt{5}/5)$, $\tan\alpha = -2$.

# 45    Euler's and Cauchy's Theorems

To complete the classification of the regular solids, and for use in studying other classes of solids, we prove here two results of a more general nature about polyhedra. Euler's theorem gives a relation between the number of vertices, edges, and faces of a convex polyhedron. This is a special case of the so-called Euler characteristic of a surface studied in topology. Cauchy's rigidity theorem tells us that if two convex polyhedra have congruent faces, similarly arranged, then they are congruent as a whole. As an application we will classify all convex polyhedra made with only equilateral triangles.

**Theorem 45.1** (Euler)
*Given a convex polyhedron, let $v$ be the number of vertices, let $e$ be the number of edges, and let $f$ be the number of faces. Then*

$$v - e + f = 2.$$

*Proof* There are many proofs of this theorem, but here is one that is particularly easy to visualize. Since the polyhedron is convex, if you place your eye in the center of one face, you can see all the other faces with no overlap. If you pull your eye back just a little, you will also be able to see the edges of the face you are looking through. Then we project this image onto a plane. This distorts angles and distances, but the edges remain straight, so we obtain in the plane a figure made of various points and line segments connecting them,

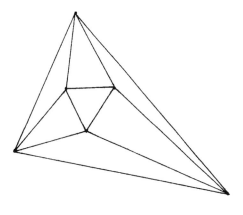

and the faces of the original polyhedron correspond to the plane polygons in this figure, plus one more face (the one you were looking through) that corresponds to the area of the plane outside the figure. Here, for example, is the plane figure you would get from an octahedron.

Now we perform two kinds of operations on this figure.

(a) Choose an edge that separates two faces, or that separates one face from the area outside the figure, and remove that edge. This decreases the number of edges by one, and also decreases the number of faces by one, since two faces are now joined together. So the expression $v - e + f$ is unchanged.

(b) If at some point in the procedure there is a vertex that has only one edge coming out of it, remove that vertex and that edge. This decreases both $v$ and $e$ by one, so again the expression $v - e + f$ is unchanged. For example, after removing edges $a, b, c$ in the diagram above, we are left with a vertex $A$ with just one edge $d$.

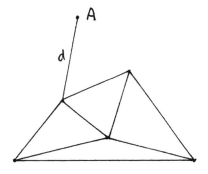

Let us think about what happens. We do step (a) as many times as possible. If step (a) is no longer possible, then there are no loops in the remaining graph of vertices and edges, so there must be some ends to the graph (since the graph is finite), and then we can do step (b). If there are no edges at all remaining, then the figure must be reduced to points only. But the original figure is connected, and it remains connected by performing step (a) or (b), so it is just one point. Then $v = 1$, $e = 0$, $f = 1$, so $v - e + f = 2$. Since the expression $v - e + f$

was unchanged by all the steps, the original $v - e + f$ also is equal to 2, as required.

**Remark 45.1.1**
The hypothesis "convex" in this theorem is stronger than necessary. For example, the result is still true for the punched-in icosahedron (44.2.3) even though it is not convex. The correct hypothesis for this theorem is that the polyhedron should be *simply connected*. However, to explain properly what this means, and to prove the theorem in this more general setting, we must refer the reader to a book on algebraic topology.

Euler's theorem has a very useful consequence relating to the face angles of a polyhedron. At any vertex of a convex polyhedron, the sum of the face angles of the faces at that vertex must be less than $2\pi$, as we have seen earlier (XI.21). So we define the *defect* $\delta_V$ at the vertex $V$ to be $2\pi$ minus the sum of the face angles at $V$. The defect $\delta_V$ is always positive.

**Corollary 45.2** (Descartes)
*In a convex polyhedron, the sum of the defects at all the vertices is equal to $4\pi$.*

*Proof*   We compute as follows:

$$\sum_V \delta_V = \sum_V \left(2\pi - \sum (\text{face angles at } V)\right)$$

$$= 2\pi v - \sum (\text{all face angles}),$$

where $v$ is the number of vertices. Now the sum of the face angles of an $n$-sided polygon is $(n - 2)\pi$. For each $n$, let $f_n$ be the number of faces having $n$ sides. Then the total number of faces is $f = \sum f_n$, and the number of edges $e$ is just $\frac{1}{2}\sum n f_n$. Combining these observations with the above, we have

$$\sum_V \delta_V = 2\pi v - \sum_n (n - 2)\pi f_n$$

$$= 2\pi v - \pi \sum_n n f_n + 2\pi \sum_n f_n$$

$$= 2\pi (v - e + f) = 4\pi,$$

as required.

**Remark 45.2.1**
This result is a powerful tool in that it limits the possible number of vertices with a particular configuration of faces in a convex polyhedron. We will see applications in several results below: (45.6), (46.3), (46.4.1).

Now we come to Cauchy's rigidity theorem. This was Cauchy's first mathematical accomplishment, before he went on to lay the foundations of a rigorous theory of convergence (Cauchy sequences) and develop the theory of functions of a complex variable (Cauchy–Riemann equations, Cauchy integral formula, etc.). A gap in the first lemma was discovered and repaired by Steinitz one hundred years later.

The problem is this. Suppose we have two convex polyhedra, made of congruent faces, similarly arranged. More precisely, this means we are given a one-to-one correspondence $\varphi$ from the set of faces $F_1, \ldots, F_f$ of the first polyhedron to the set of faces $F'_1, \ldots, F'_f$ of the second polyhedron, so that for each $i$, $F_i$ is congruent to $\varphi(F_i)$, and furthermore, $\varphi$ extends to one-to-one correspondences of vertices and edges preserving all incidence relations. Then we wish to conclude that the two polyhedra are congruent. Note that this formulation allows mirror images, so that congruence means rigid motion in space followed possibly by reflection in a plane.

Cauchy's idea is to study how the dihedral angles compare along corresponding edges. If all the dihedral angles are the same, then we can build the two polyhedra step by step into congruent figures. On the other hand, if the dihedral angles change, then we will track their increase or decrease around each vertex and eventually use Euler's theorem to make a contradiction. To study what happens at a single vertex, we intersect the faces of the polyhedron with a small sphere around the vertex. This produces a spherical polygon whose interior angles are precisely the dihedral angles of the original polyhedron. We call it the *vertex figure* at the vertex $V$. This leads us to the study of polygons with changing angles, which is the first lemma.

**Lemma 45.3** (Steinitz)
*Suppose given in the plane two convex polygons $A_1 A_2 \cdots A_n$ and $B_1 B_2 \cdots B_n$ with all sides equal except possibly the last: $A_i A_{i+1} = B_i B_{i+1}$ for all $i = 1, 2, \ldots, n-1$. Suppose also that the angles of the first polygon are less than or equal to the angles of the second, $\angle A_i \leq \angle B_i$, for $i = 2, \ldots, n-1$, with at least one strict inequality. Then $A_1 A_n < B_1 B_n$.*

*Proof* We proceed by induction on $n$.

*Case 1* For $n = 3$ it is elementary. In fact this is Euclid (I.24).

*Case 2* Suppose that $n \geq 4$ and for some $i$ that $\angle A_i = \angle B_i$. Then the triangle $A_{i-1} A_i A_{i+1}$ is congruent to the triangle $B_{i-1} B_i B_{i+1}$. So $A_{i-1} A_{i+1} = B_{i-1} B_{i+1}$, and the result follows by applying the induction hypothesis to the new polygons of $n-1$ vertices obtained by omitting $A_i$ and $B_i$.

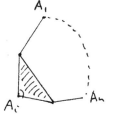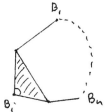

*Case 3* Suppose that $n \geq 4$ and all the $B$ angles are strictly bigger than the $A$ angles. Construct a point $A'_1$ such that $A_1A_2 = A'_1A_2$ and $\angle A'_1A_2A_3 = \angle B_2$. We compare $A_1 \cdots A_n$ to $A'_1A_2 \cdots A_n$ and then the latter to $B_1 \cdots B_n$. In the first comparison, $\angle A_3 = \angle A_3$, so $A_1A_2 < A'_1A_2$ by Case 2. In the second comparison, $\angle A_2 = \angle B_2$, so $A'_1A_2 < B_1B_2$ again by Case 2, and we are done.

But wait! There is a snag. If the new polygon $A'_1A_2 \cdots A_n$ is not convex, we cannot apply the earlier cases, and the proof fails.

*Case 4* If this happens, then there must be a point $A^*_1$ intermediate between $A_1$ and $A'_1$, for which $A^*_1A_2 = A_1A_2$, and $A^*_1, A_n, A_{n-1}$ are collinear.

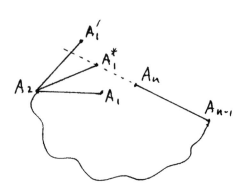

Then we first compare $A_1 \cdots A_n$ to $A^*_1A_2 \cdots A_n$ and obtain $A_1A_n < A^*_1A_n$ by Case 2. Next we compare the polygon $A^*_1A_2 \cdots A_{n-1}$ to $B_1B_2 \cdots B_{n-1}$, and obtain $A^*_1A_{n-1} < B_1B_{n-1}$ by the induction hypothesis. On the other hand, since $A^*_1, A_n, A_{n-1}$ are collinear, we have $A^*_1A_n = A^*_1A_{n-1} - A_{n-1}A_n$. Putting these together, we get

$$A_1A_n < A^*_1A_n$$
$$= A^*_1A_{n-1} - A_{n-1}A_n$$
$$< B_1B_{n-1} - B_{n-1}B_n$$
$$\leq B_1B_n,$$

where the last inequality is just the triangle inequality (I.20). Thus the proof is complete.

**Remark 45.3.1**

In fact, what we need for Cauchy's theorem is not this lemma for plane polygons, but the analogous result for polygons on the surface of a sphere. The extension to spherical polygons is not too difficult because the proof uses results from only the first part of Euclid, Book I, before the introduction of the parallel postulate. The verification that the needed results hold in spherical geometry is left to the reader (Exercises 45.3–45.8).

**Lemma 45.4**

*Let $A_1 \cdots A_n$ and $B_1 \cdots B_n$ be two convex polygons in the plane or on the sphere, with corresponding sides equal: $A_iA_{i+1} = B_iB_{i+1}$ for $i = 1, \ldots, n$ (interpreting $n + 1 = 1$). For each $i$, mark the vertex $A_i$ with $+$ if $\angle A_i < \angle B_i$, with $-$ if $\angle A_i > \angle B_i$, or with no* *mark if the angles are equal. Then either all corresponding angles are equal, or, as we make a circuit of the first polygon, ignoring unmarked vertices, the sign must change at least four times.*

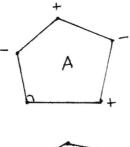

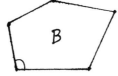

*Proof* Of course, the number of changes of sign is even, so if there were some, but less than four, there would be exactly two changes of sign. In that case one could draw a diagonal $A_iA_j$ cutting the polygon into two convex polygons, one of which contains only $-$ vertices, and the other only $+$ vertices. Applying the previous lemma to the $-$ side, we obtain $A_iA_j > B_iB_j$. Applying it to the $+$ side gives $A_iA_j < B_iB_j$, a contradiction.

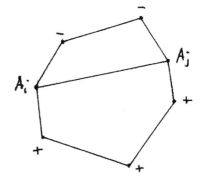

**Theorem 45.5** (Cauchy's Rigidity Theorem)
*Suppose we are given two convex polyhedra and a one-to-one map $\varphi$ from the set of faces of one to the other, so that corresponding faces are congruent, and are similarly arranged (as described above). Then the two polyhedra are congruent.*

*Proof* For each edge of the first polyhedron, we mark it $+, -$, or no mark, according as its dihedral angle is less than, greater than, or equal to the corresponding dihedral angle of the second polyhedron. At each vertex we intersect the polyhedron with a small sphere, and look at the resulting vertex figure. This is a convex spherical polygon, and its vertices inherit markings $+$ or $-$ from the edges, which by construction correspond to the increase or decrease of this polygon angle as compared to the vertex figure of the second polyhedron. We conclude from the lemma (45.4) that for each vertex, if we make a circuit of the edges meeting that vertex, either they are all unmarked, or there are at least 4 changes of sign. We will arrive at a contradiction by counting the total number of changes of sign in two different ways.

*Case 1*   Suppose that all edges are marked $+$ or $-$, in other words, all the dihedral angles are changing. Let $t$ denote the sum over all the vertices of the number of changes of sign of edges around that vertex. Then by the lemma (45.4) clearly $t \geq 4v$, where $v$ is the number of vertices.

On the other hand, let us count by faces. On a triangular face, two adjacent edges must have the same sign, so that face can contribute at most two changes of sign to its three vertices. Similarly, a face of $n$ sides can contribute at most $n$ changes of sign if $n$ is even, or $n - 1$ if $n$ is odd. In particular,

$$t \leq 2f_3 + \sum_{n \geq 4} n f_n,$$

where $f_n$ denotes the number of faces of $n$ sides. Putting together the two inequalities for $t$ gives

$$4v \leq 2f_3 + \sum_{n \geq 4} n f_n.$$

Now we use Euler's theorem, which can be written $v = e - f + 2$, and substitute $e = \frac{1}{2} \sum n f_n$ and $f = \sum f_n$. This gives

$$2 \sum (n - 2) f_n + 8 \leq 2f_3 + \sum_{n \geq 4} n f_n,$$

or

$$\sum_{n \geq 4} (n - 4) f_n + 8 \leq 0,$$

which is impossible, because the terms in the sum are all nonnegative.

*Case 2*   Now suppose there are some marked and some unmarked edges. Of course, if no edges are marked, all the dihedral angles are equal, and the two polyhedra are congruent. We will imitate the previous proof using only those vertices and edges that are marked. We call this a *net*. The *vertices* of the net are those vertices of the polyhedron that have a marked edge coming out of them. The *edges* of the net are the marked edges of the polyhedron. A *net-face* of the net is any maximal union of faces of the polyhedron that are not separated by edges of the net. A net-face is no longer a plane polygon, but it is a connected surface bounded by edges of the net.

Now we repeat the previous argument using only the vertices, edges, and net-faces of the net. Denote the numbers of these by $v', e', f'$. The argument is all the same, except for the application of Euler's theorem, which does not apply as stated, because a net may not be a polyhedron. Nevertheless, we can apply the proof of Euler's theorem (45.1) to the net, and the only difference is that the plane figure of points and lines may not be connected. At the end of the proof, there may be more than one point, so we find that $v' - e' + f' \geq 2$. With this inequality, the argument of Case 1 still works, so we have a contradiction.

The only remaining possibility is that all the dihedral angles are equal, so the two polyhedra are congruent.

### Remark 45.5.1
This result is false without the hypothesis of convexity. See Cromwell (1997), Chapter 6, for an interesting discussion of nonconvex flexible polyhedra.

As an application of the theorems of this section, we will classify all convex polyhedra that can be formed using only equilateral triangles (called *deltahedra* by some authors).

### Theorem 45.6
*There are exactly eight convex polyhedra all of whose faces are equilateral triangles. Each one is uniquely determined up to congruence, once the length of an edge is specified.*

*Proof*   Our strategy is this. Since the faces are all equilateral triangles, we know from (44.3) that at each vertex there must be 3, 4, or 5 triangles. Since the triangles are equilateral, the corresponding defect at such a vertex will be $\pi, 2\pi/3$, or $\pi/3$. Let $a$ be the number of vertices with 3 triangles, $b$ the number with 4 triangles, and $c$ the number with 5 triangles. Then $a, b, c$ are nonnegative integers, and according to Descartes's theorem (45.2) we have

$$\pi a + \frac{2}{3}\pi b + \frac{1}{3}\pi c = 4\pi,$$

or

$$a + \frac{2}{3}b + \frac{1}{3}c = 4.$$

This equation has only a finite number of solutions in nonnegative integers. So we will list all possible solutions, then discuss existence or nonexistence of the corresponding polyhedron until we have a complete classification. See Table 1 for the list of possible $a, b, c$.

As a first step, we can fill in the tetrahedron, octahedron, and icosahedron, which we know to exist. Next let us show that some combinations of $a, b, c$ are impossible. I claim that we cannot have a 3-face vertex adjacent to a 5-face vertex. Indeed, at a three-face vertex, the dihedral angles are uniquely determined (44.5) and are those of a tetrahedron. So imagine a tetrahedron sitting on one face of an octahedron. The joined vertices are then 5-face vertices, and two of

**Table 1.**   Convex polyhedra with equilateral triangular faces.

| $a$ | $b$ | $c$ | Name or Note | $v$ | $e$ | $f$ |
|---|---|---|---|---|---|---|
| 4 | 0 | 0 | tetrahedron | 4 | 6 | 4 |
| 3 | 1 | 1 | (1) | | | |
| 3 | 0 | 3 | (1) | | | |
| 2 | 3 | 0 | triangular dipyramid | 5 | 9 | 6 |
| 2 | 2 | 2 | (1) | | | |
| 2 | 1 | 4 | (1) | | | |
| 2 | 0 | 6 | (1) | | | |
| 1 | 4 | 1 | (2) | | | |
| 1 | 3 | 3 | (2) | | | |
| 1 | 2 | 5 | (2) | | | |
| 1 | 1 | 7 | (2) | | | |
| 1 | 0 | 9 | (2) | | | |
| 0 | 6 | 0 | octahedron | 6 | 12 | 8 |
| 0 | 5 | 2 | pentagonal dipyramid | 7 | 15 | 10 |
| 0 | 4 | 4 | snub disphenoid | 8 | 18 | 12 |
| 0 | 3 | 6 | tricapped triangular prism | 9 | 21 | 14 |
| 0 | 2 | 8 | bicapped square antiprism | 10 | 24 | 16 |
| 0 | 1 | 10 | (3) | | | |
| 0 | 0 | 12 | icosahedron | 12 | 30 | 20 |

the faces of the octahedron are in the same planes as faces of the tetrahedron (Exercise 44.3), so the figure is not convex in our strict sense. If we push one of the faces of the 5-vertex inward, to be no longer in the same plane as the face of the tetrahedron, the other one will be forced outward, so the figure will not be convex. Thus we cannot have a 3-vertex adjacent to a 5-vertex. Consequently, if $a > 0$, then $a + b \geq 4$, and if $c > 0$, then $b + c \geq 6$. This rules out all those cases indicated by note (1).

Next, suppose there is just one 3-vertex. The 3 adjacent vertices must be 4-vertices by the above argument, and then the figure closes at 6 faces to make a triangular dipyramid, that is, two tetrahedra glued together along one face. This gives the existence of the dipyramid $(a, b, c) = (2, 3, 0)$, and shows the impossibility of $a = 1$. This is note (2).

Now let us consider existence. For $(0, 5, 2)$ we have the *pentagonal dipyramid*, which is two pentagonal pyramids glued along their pentagonal face. For $(0, 3, 6)$ we have the *tricapped triangular prism*, which is formed as follows. Take a triangular prism—that is, two equilateral triangles in parallel planes, joined by three squares—and onto each square face glue a square pyramid. For $(0, 2, 8)$ we have the *bicapped square antiprism*. A *square antiprism* is made of two squares, in parallel planes, but with their axes tilted at 45° angles to each other, and joined by 8 equilateral triangles. On each square face, glue a square pyramid.

We can show that the case $(a,b,c) = (0,1,10)$ is impossible by an argument similar to note (2) above. If there were only one 4-vertex, and all the rest 5-vertices, the figure would grow from the 4-vertex the same as the bicapped square antiprism, and this would force another 4-vertex on the other side. This is note (3).

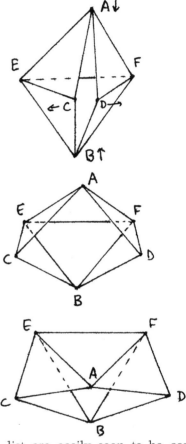

We have saved the case $(a,b,c) = (0,4,4)$ for last, since its existence is less elementary than the others. We call it the *snub disphenoid* after Johnson (1966). It has also been called a triangular dodecahedron, or Siamese dodecahedron. To show the existence of this figure, take a triangular dipyramid, and cut it open along two of its edges, from top to bottom. Now push the top and bottom vertices $A, B$ toward each other, forcing apart the split vertex $C, D$, and keeping $EF$ fixed. If you continue this process all the way, eventually the figure will lie flat, with $A$ on $B$ and $C, D$ at the outside corners.

Somewhere in between, there is a point where the distance of $AB$ and $CD$ are equal. Then the angles of the non-planar quadrilateral $ACBD$ are all equal. So you can take a second one of these figures and glue the two together along $ACBD$, with the roles of $AB$, $CD$ reversed, and obtain the desired figure.

Note that all the other polyhedra in this list are easily seen to be constructible by ruler and compass constructions. For the snub disphenoid, our existence proof used the intermediate value theorem in the real numbers to argue that as $AB$ decreases and $CD$ increases, there is a point where they become equal. In fact, this figure is not constructible—it requires the solution of a cubic equation to find its dimensions (Exercise 45.10).

Now we have ruled out the impossible cases, and have shown existence for the remaining cases. To show uniqueness, observe that for each total number of faces, there is only one triple $(a,b,c)$ possible, and this determines the arrangement of the faces. So by Cauchy's theorem (45.5) we conclude that the figures are unique up to congruence, after fixing a scale factor.

**Remark 45.6.1**

This also settles a question raised earlier (44.5.2), namely, the only convex figures that can be made with 8 or 20 equilateral triangles are the octahedron and the icosahedron, thus vindicating Euclid's definitions.

# Exercises

45.1  For each of the five regular solids:

(a)  Calculate the number of vertices, edges, and faces, and verify Euler's theorem.

(b)  Calculate the defect at a vertex and verify Descartes's theorem (Corollary 45.2).

45.2  Make an example to show that the result of Lemma 45.3 may fail if the polygons are not convex.

**Spherical Geometry**

In the following exercises we develop some elementary results of spherical geometry that are needed for the spherical form of Steinitz's lemma. We fix a sphere (of radius 1 for convenience) in Euclidean three-space $\mathbb{R}^3$. The *points* of the spherical geometry are points on the surface of the sphere. The *lines* are great circles on the sphere, that is, circles lying in a plane passing through the center of the sphere. A *circle* will be the set of points equidistant from a given point, or equivalently, the intersection of any plane with the sphere. We measure angles between lines and circles by the angle between their tangent lines in 3-space. We measure length of a line segment by the angle (in radians) that it subtends at the center of the sphere. So a complete great circle has length $2\pi$. A line from the north pole to the equator has length $\pi/2$.

We have seen earlier (Exercise 34.13) that this geometry does not satisfy Hilbert's axioms. However, we will see in these exercises that most of the results of the first part of Euclid's Book I still hold, with suitable modifications. When we speak of a *triangle* or a *polygon* we will always assume that it lies in a single hemisphere. In particular, the length of any side must be less than $\pi$. If we restrict our attention to one hemisphere, then the concepts of betweenness function well, and we can speak of the inside of a triangle, or of a convex polygon. To any of Euclid's propositions we will prefix "s" to denote the corresponding statement in spherical geometry. Thus for example, (sI.4) is the SAS theorem for spherical triangles.

45.3  Show that the construction of an equilateral triangle (sI.1) works for a line segment $AB$ of length less than $2\pi/3$, but fails if $2\pi/3 < AB < \pi$. What happens if the length of $AB$ is exactly $2\pi/3$?

45.4  Verify that (sI.2)–(sI.15) are all true, and note carefully when a different proof is necessary. Feel free to use the existence of rigid motions to prove congruences (cf. Exercise 34.13).

45.5 Give examples to show that (sI.16) and (sI.17) are false. What goes wrong with the proof of (I.16)?

45.6 Euclid uses (I.16) to prove (I.18), (I.19), and (I.20). This will not work in the spherical case. Instead, use (XI.20) to prove (sI.20). Then use (sI.20) to prove (sI.19), and finally use (sI.19) to prove (sI.18) by contradiction.

45.7 Coming to (sI.22), the construction of a triangle from three given sides, recall that in our definition of a spherical triangle, we assume that it is contained (properly) in a single hemisphere. Show then that the sum of the sides of any triangle must be less than $2\pi$, and this condition must be added to the condition of (I.22) to make the construction possible.

45.8 Show that (sI.23)–(sI.26) are all ok, using the results proved above. In particular, this gives us (sI.24) used in the Steinitz lemma.

45.9 Make models of each of the new polyhedra in Table 1.

45.10 In the construction of the snub disphenoid described in the proof of Theorem 45.6, let $H$ be the point where $AB$ meets the plane of $CEFD$. Let $\theta$ be the angle $EHF$, and let $x = \cos\theta$.

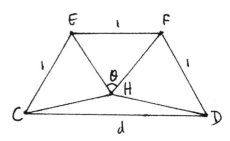

(a) Show that for $AB = CD$, $x$ satisfies an irreducible cubic equation with integer coefficients (taking the edge length $EF$ to be 1). Hence the disphenoid is not constructible by ruler and compass.

(b) Does this cubic equation require a real square root or an angle trisection for its solution (cf. Section 31)?

(c) Solve the equation above and use it to get an approximate value for $d = AB = CD$. *Answer: $d \approx 1.28917$.*

(d) Compute the dihedral angle along the edge $EF$. *Answer*: $96°11'54''$.

45.11 Imitating the constructions used to make the solids in Table 1,

(a) show that a square dipyramid is the same as an octahedron, and

(b) a bicapped pentagonal antiprism is the same as an icosahedron.

# 46   Semiregular and Face-Regular Polyhedra

After discussing the regular polyhedra and the polyhedra made from equilateral triangles, it is natural to ask what other convex polyhedra can be made using only regular polygons as faces. We call these *face-regular* polyhedra. Among these the most symmetric are the *semiregular,* or *Archimedean,* solids. We define these to be convex polyhedra having only regular polygons as faces, and *uniform* in the sense that each vertex has the same number of the same kinds of faces, in the same cyclic order at each vertex. (This terminology is not universal. Some authors use the word "uniform" to denote the stronger condition that there is a rigid motion sending any vertex to any other—cf. (46.2) below.)

We can describe a semiregular solid by giving the configuration of faces at a single vertex. The symbol $(a_1, \ldots, a_k)$ will describe a vertex having an $a_1$-sided regular polygon, an $a_2$-sided polygon, ..., and an $a_k$-sided polygon, in that order. So the symbol $(3, 3, 3)$ would mean three equilateral triangles at each vertex— this is the regular tetrahedron. The symbol $(3, 4, 3, 4)$ describes a solid in which each vertex has two equilateral triangles and two squares, alternating with each

other as you make a circuit around the vertex. An example of this is the *cuboctahedron.* Take a cube and mark the middle of each edge. Join the markings on each face to make a smaller square. Then cut off each corner of the cube along the lines. This leaves a solid with 6 squares and 8 equilateral triangles having the configuration $(3, 4, 3, 4)$ at each vertex.

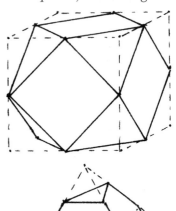

Another way to cut down a regular solid is illustrated by the *truncated tetrahedron.* Take a tetrahedron and mark the thirds of each edge. Join the markings on each face to make a regular hexagon. Cut off the vertices of the tetrahedron along the lines. This leaves a figure with four regular hexagons and four equilateral triangles as faces, and the arrangement $(3, 6, 6)$ at each vertex.

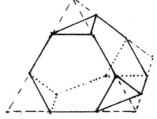

By these methods you can make many more semiregular solids. Try it.

Another construction is to take two congruent regular $n$-gons, for any $n \geq 3$, in parallel planes, lined up with each other, and join corresponding edges with squares. This makes an $n$-sided *prism.* Its symbol is $(4, 4, n)$. When $n = 4$ it is a cube.

If you again take two regular $n$-gons in parallel planes, but rotate one by $\pi/n$, so that the vertices of one are lined up with the edges of the other, and join them by equilateral triangles (adjusting the height as needed), you get an $n$-sided *antiprism.* For $n = 3$ this is an octahedron.

Now we will see that aside from the infinite families of prisms and antiprisms, there is only a finite number of semiregular solids. They are often called

**Plate XVII.** The snub cube (left), made of 6 squares and 32 triangles, is one of the Archimedean semi-regular polyhedra (46.1), (Exercise 46.4). The truncated tetrahedron (middle), made of 4 hexagons and 4 triangles, is another of the Archimedean solids (46.1). The disphenocingulum (right) is a face-regular solid made of 4 squares and 20 triangles (Exercise 46.7).

**Plate XVIII.** The stretched cube (left), made of 6 squares and 12 triangles, does not exist: cf. (46.4.1), (Exercise 46.8). The snub disphenoid (middle), made of 12 triangles, is one of the face-regular solids (45.6), (Exercise 45.10). The tricapped triangular prism (right), made of 14 triangles, is another of the face-regular solids (45.6).

**Plate XIX.** The bilunabirotunda (left) is a face-regular solid made of 4 pentagons, 2 squares, and 8 triangles (Exercise 46.9). The dodecahedron (middle), made of 12 pentagons, is one of the five regular solids (44.2). This model is made from 6 sheets of origami paper by folding only (no glue). On the right is an icosahedron (20 triangular faces) made of jellybeans and toothpicks, with a tetrahedron inside it, to help visualize how the tetrahedron group is a subgroup of the icosahedral group (Exercise 47.9).

**Plate XX.** The snub dodecahedron (left) is made of 12 pentagons and 80 triangles. The truncated icosahedron, also called a soccer ball (right), is made of 12 pentagons and 20 hexagons. Both are Archimedean semi-regular solids (46.1). These models were made as a group project by my freshman seminar class. Each person made one pentagon with attached triangles or hexagons, then we assembled the pieces with tape.

*Archimedean* solids, because they were studied in a lost book of Archimedes (cf. Pappus (1876), Book V, Sections 19 ff). They were rediscovered and classified by Kepler.

### Theorem 46.1
*Aside from the five regular polyhedra and the two infinite families of prisms and antiprisms, there are just thirteen (and one variant) other semiregular polyhedra.*

*Proof*  Our strategy is similar to that used in classifying the convex figures made of triangles. First we will use numerical criteria to limit the possible behavior at a vertex, and then we will discuss existence and uniqueness of the corresponding solids.

Since the solids are convex, the sum of the face angles at a vertex must be less than $2\pi$ (cf. proof of (44.3)). The face angle of a regular $n$-gon is $(n-2)/n\pi$, so at a vertex $(a_1, \ldots, a_k)$ we must have

$$\sum \frac{a_i - 2}{a_i} \pi < 2\pi,$$

which gives

$$\sum_{i=1}^{k} \frac{2}{a_i} > k - 2.$$

This is our main numerical restriction on possible vertex configurations.

*Case 1*  Each vertex has three faces, say $(a, b, c)$. Then our inequality is

$$\frac{1}{a} + \frac{1}{b} + \frac{1}{c} > \frac{1}{2}.$$

If $a, b, c$ are all equal, then $a = 3, 4$, or $5$. These correspond to regular solids, as we have seen before.

If at least two of $a, b, c$ are different, say $a \neq b$, then as we go around the edges of a $c$-face, the adjacent faces must alternate between $a$ and $b$. It follows that $c$ is an even number. This restriction, together with our inequality, limits the possible triples $(a, b, c)$ to those shown in Table 2 (Exercise 46.1).

*Case 2*  Each vertex has four faces, say $(a, b, c, d)$. In this case our inequality is

$$\frac{1}{a} + \frac{1}{b} + \frac{1}{c} + \frac{1}{d} > 1.$$

If $a = 3$, there is another limitation. Look at what happens around a triangle. At each vertex of the triangle, $b$ and $d$ are adjacent faces, sharing an edge with the

**Table 2.**   The semiregular polyhedra.

| Vertex Figure | Name | $v$ | $f_3$ | $f_4$ | $f_5$ | Other $f_n$ |
|---|---|---|---|---|---|---|
| $(3,3,3)$ | tetrahedron | 4 | 4 | | | |
| $(4,4,4)$ | cube | 8 | | 6 | | |
| $(5,5,5)$ | dodecahedron | 20 | | | 12 | |
| $(4,4,n)$ | $n$-sided prism, $n \geq 3$, $n \neq 4$ | $2n$ | | $n$ | | $f_n = 2$ |
| $(3,6,6)$ | truncated tetrahedron | 12 | 4 | | | $f_6 = 4$ |
| $(4,6,6)$ | truncated octahedron | 24 | | 6 | | $f_6 = 8$ |
| $(5,6,6)$ | truncated icosahedron | 60 | | | 12 | $f_6 = 20$ |
| $(3,8,8)$ | truncated cube | 24 | 8 | | | $f_8 = 6$ |
| $(3,10,10)$ | truncated dodecahedron | 60 | 20 | | | $f_{10} = 12$ |
| $(4,6,8)$ | truncated cuboctahedron | 48 | | 12 | | $f_6 = 8$, $f_8 = 6$ |
| $(4,6,10)$ | truncated icosidodecahedron | 120 | | 30 | | $f_6 = 20$, $f_{10} = 12$ |
| $(3,3,3,3)$ | octahedron | 6 | 8 | | | |
| $(3,3,3,n)$ | $n$-faced antiprism $n \geq 4$ | $2n$ | $2n$ | | | $f_n = 2$ |
| $(3,4,3,4)$ | cuboctahedron | 12 | 8 | 6 | | |
| $(3,5,3,5)$ | icosidodecahedron | 30 | 20 | | 12 | |
| $(3,4,4,4)$ | $\left\{\begin{array}{l}\text{rhombicuboctahedron}\\\text{pseudorhombicuboctahedron}\end{array}\right\}$ | 24 | 8 | 18 | | |
| $(3,4,5,4)$ | rhombicosidodecahedron | 60 | 20 | 30 | 12 | |
| $(3,3,3,3,3)$ | icosahedron | 12 | 20 | | | |
| $(3,3,3,3,4)$ | snub cube | 24 | 32 | 6 | | |
| $(3,3,3,3,5)$ | snub dodecahedron | 60 | 80 | | 12 | |

triangle, while $c$ is opposite the triangle at each vertex. It follows that $b = d$. This restriction, together with the inequality, limits possible vertex types to those in Table 2 (Exercise 46.2).

*Case 3*   Each vertex has five faces. Our inequality is

$$\sum_{i=1}^{5} \frac{2}{a_i} > 3,$$

and this already limits us to the three cases shown. As we cannot have six or more faces at a vertex, these are all the possibilities (Exercise 46.3).

Next we come to the questions of existence and uniqueness. Once we have fixed a vertex arrangement, one can check easily that the global arrangement of the polyhedron is uniquely determined (with one exception noted below), so uniqueness will follow from Cauchy's theorem (45.5).

As for existence, all except the last two can be constructed using Euclidean tools, by methods similar to those suggested above, so I will leave to you the pleasure of figuring out the details and building models of as many as you like.

One word about the rhombicuboctahedron, which is a special case. This
solid can be made as follows. Take a
cube, and in the middle of each face
draw a smaller square with edges parallel to the edges of the cube. Now remove the edges of the cube, and join
squares in adjacent faces by squares
(after adjusting the size of the smaller
squares appropriately). The corners of
the original cube are then replaced by
equilateral triangles, and we obtain the
*rhombicuboctahedron*. It is a sort of

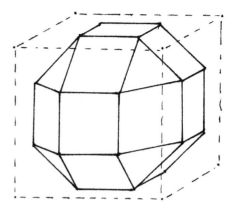

sphere wrapped up around the equator with a ribbon of 8 squares, and two other
similar ribbons in mutually perpendicular north–south axes. Or you can think of
it as an octagonal prism containing the equator, with caps on the top and bottom. If you rotate the top cap by 45°, you get a new figure, which is still semiregular with vertex $(3, 4, 4, 4)$, but not congruent to the original figure. This is
the *pseudorhombicuboctahedron*, discovered only in 1930 (see Ball (1940), p. 137).

The last two in the list, the snub cube and the snub dodecahedron, are different in that they cannot be constructed by any simple operation applied to the
regular solids. They are also different in that each one comes in a left-handed
and a right-handed version, congruent to each other by reflection, but not by
any orientation-preserving motion of 3-space. We give in Exercise 46.4 a method
of constructing the snub cube by solving a cubic equation, and I leave to you to
find a construction of the snub dodecahedron.

### Corollary 46.2
*The semiregular polyhedra have the following additional properties (cf. (44.4)):*

(d) *The dihedral angles at one vertex are equal to the dihedral angles at every
other vertex.*

(e) *The vertices lie on a sphere.*

(f) *Except for the pseudorhombicuboctahedron, there is a congruence of the solid
taking any vertex to any other vertex.*

*Proof*   This is a consequence of the classification, because we note in each case
that the vertex figure determines the global arrangement of faces, and that
arrangement is the same starting from any vertex. Thus Cauchy's theorem
(45.5) gives a congruence. This proves (f), which implies (d) and (e).

In the case of the pseudorhombicuboctahedron, its construction from the
normal rhombicuboctahedron shows that it satisfies (d) and (e). But the rigid
motions of this figure are not transitive on the vertices (Exercise 46.5).

Next we consider the nonuniform convex face-regular polyhedra. We have seen five of these in the classification of solids bounded by equilateral triangles (45.6). Others can be made by cutting or gluing polyhedra we already know. For example, start with an icosahedron. The five faces at a single vertex are bounded by a regular pentagon. Cutting the figure by the plane of that pentagon gives a pentagonal pyramid on one side, and leaves a *diminished icosahedron*, with one pentagonal face and 15 triangle faces, on the other side. If we cut off a second pentagonal face parallel to the first one, we get a pentagonal antiprism. But if we cut off a second pentagonal face adjoining the first, we get a *bidiminished icosahedron*. Then we can cut off a third pentagonal face, making a *tridiminished icosahedron* with 3 pentagonal and 5 trianuglar faces. This last figure is minimal in the sense that it cannot be separated further into the union of regular-faced polyhedra, so it is called an *elementary* face-regular polyhedron. The tetrahedron is already elementary as it stands. But the octahedron can be separated into two square pyramids.

On can also make face-regular polyhedra by gluing others together. For example, one can glue a square pyramid onto one face of a cube, or onto two opposite faces of a cube. Or one could cut out the middle section of the rhombicuboctahedron and glue the top and bottom caps together (in two different ways) to make new figures.

According to Johnson's classification (1966) there are 91 nonuniform convex face-regular polyhedra. The complete classification of these is not a simple matter, so we will confine ourselves to proving that their number is finite.

### Theorem 46.3
*There is only a finite number of nonuniform convex face-regular polyhedra (up to congruence, after fixing the length of an edge).*

*Proof*  The key point (see lemma below) is to show that for $n$ sufficiently large, any convex face-regular polyhedron with an $n$-face must be a prism or an antiprism, which is uniform. It follows that for nonuniform solids, there is only a finite number of possible face types, and hence only a finite number of possible vertex configurations. Each of these has a positive defect, and there is only a finite number of ways of choosing these to add up to $4\pi$ (45.2). Then there is only a finite number of ways of arranging these vertex types into a global figure, and by Cauchy's theorem any two with the same arrangement are congruent. So it remains only to prove the following lemma.

## Lemma 46.4

*There is an $n_0$ such that if a convex face-regular polyhedron has a face with $n \geq n_0$ sides, then it is either a prism or an antiprism.*

*Proof*   This lemma is actually true for $n_0 = 11$, but to make the proof simpler, we will prove it for $n_0 = 42$. Let us consider what happens at a vertex of the $n$-gon, and to begin with we consider a 3-face vertex $(a, b, n)$. We know from the proof of (46.1) that

$$\frac{1}{a} + \frac{1}{b} + \frac{1}{n} > \frac{1}{2}.$$

On the other hand, in order to make a 3-dimensional figure, the sum of the face angles of the $a$ and $b$ faces must be greater than the face angle of the $n$-gon. This gives a second inequality

$$\frac{1}{a} + \frac{1}{b} - \frac{1}{n} < \frac{1}{2}.$$

From these two it follows that

$$\left| \frac{1}{2} - \frac{1}{a} - \frac{1}{b} \right| < \frac{1}{n}.$$

Now, the minimum nonzero value of the expression on the left, for $a, b \geq 3$, is $1/42$ (Exercise 46.6). So if we take $n \geq 42$, this inequality implies

$$\frac{1}{a} + \frac{1}{b} = \frac{1}{2},$$

so $(a, b) = (3, 6)$ or $(4, 4)$. Thus we have shown that for $n \geq 42$, the only possible 3-faced vertex configurations at a vertex of the $n$-gon are $(3, 6, n)$ and $(4, 4, n)$.

A 5-faced vertex including an $n$-face for $n \geq 6$ is impossible, so let us consider a 4-face vertex $(a, b, c, n)$. In this case, the same argument as above shows that

$$\left| 1 - \frac{1}{a} - \frac{1}{b} - \frac{1}{c} \right| < \frac{1}{n}.$$

The minimum nonzero value of this expression for $a, b, c \geq 3$ is $\frac{1}{12}$, so for $n \geq 12$, we obtain

$$\frac{1}{a} + \frac{1}{b} + \frac{1}{c} = 1,$$

in which case $a = b = c = 3$. So if we have a 4-face vertex along the $n$-gon, it must be $(3, 3, 3, n)$.

Now we will show that the resulting
figure must be a prism or an antiprism.
First suppose $A$ is a $(3,6,n)$ vertex, and
let $B$ be the vertex at the other end of
the 3-6 edge. Because of the dihedral
angle along the 3-6 edge, the angle between the two other edges at $B$ will be
the same as the face angle at $A$—cf.
(44.5). Therefore, the remaining faces at
$B$ must be another $n$-gon, or $(3,6), (4,4)$,
or $(3,3,3)$ by the argument above. But
the latter three cases lead to face angles
at $B$ totaling $2\pi$, which is impossible.
Therefore, the third face at $B$ is another
$n$-gon. Then the third vertex $C$ of the
triangle $ABC$ becomes a $(3,n,n)$ vertex,
which is impossible. Thus $(3,6,n)$ cannot occur.

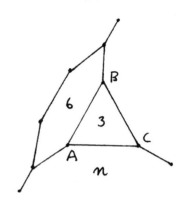

Next suppose there is a $(4,4,n)$ vertex $A$. Let $B$ be the vertex at the other
end of the 4-4 edge. Then as above, the
angle between the edges at $B$ is the
same as at $A$. We cannot add $(3,6)$ or
$(4,4)$ or $(3,3,3)$, as before, so we must
have another $n$-face at $B$. This forces the
whole figure to be an $n$-prism.

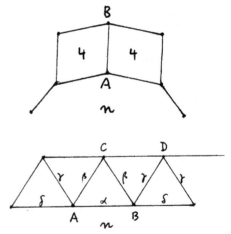

Now suppose there is a $(3,3,3,n)$
vertex at $A$. At the next vertex of the $n$-
gon we then have at least one triangle.
We have seen that $(3,6,n)$ is impossible,
so it is also a $(3,3,3,n)$ vertex. Thus all
the vertices of the $n$-gon are $(3,3,3,n)$
vertices. These vertices are not rigid,
but if we fix one dihedral angle $\alpha$ along the edge $AB$, the other dihedral angles
between the triangles are all determined, as are the angles between edges at the
new vertices $C, D$, etc. Furthermore, we see that the possible values of these angles repeat after moving over two vertices. If the angles $\alpha, \delta$ along the $n$-gon are
all equal, then the angles at $C, D$ will be equal to those at $A, B$, and by the same
reasoning as in the two previous cases, the remaining face at $C$ must be another
$n$-gon. This forces the figure to be the $n$-antiprism.

If, on the other hand, $\alpha$ and $\delta$ are different, then the angles at $C, D$ will be one
greater and one lesser than those at $A, B$. We may assume that the angle at $C$ is
greater than the one at $A$. Then the remaining face at $C$ will be an $n'$-gon for

some $n' > n$. But the $n'$-gon also passes through $D$, so the angle at $D$ must be equal to the one at $C$, a contradiction.

Thus we see that the only convex face-regular polyhedra containing an $n$-gon for $n \geq 42$ are the prisms and the antiprisms, as required.

**Example 46.4.1**

To illustrate the process of finding face-regular solids, let me give an example. One day I was searching for solids made of only squares and triangles (Exercise 46.7) and discovered the following one, which I call a *stretched cube*. Take a cube, choose two opposite vertices $A, B$, and cut it in two pieces so that one piece has the three faces at $A$ and the other piece has the three faces at $B$. Now pull the two pieces apart just far enough so as to fill in the gap with equilateral triangles. First I imagine the construction. Then I make a sketch, to see how the faces will fit together. Then I list the vertex types and check that the sum of the deficiencies is $4\pi$.

| Number | Vertex Type | Deficiency | Total Deficiency |
|--------|-------------|------------|------------------|
| 2 | $(4, 4, 4)$ | $\pi/2$ | $\pi$ |
| 6 | $(3, 3, 4, 4)$ | $\pi/3$ | $2\pi$ |
| 6 | $(3, 3, 3, 3, 4)$ | $\pi/6$ | $\pi$ |
| | | | $4\pi$ |

Then I make a cardboard model. I had to squeeze a bit to make the last couple of faces fit in place, but that inaccuracy did not seem more than the usual margin of error in my models. It was only later that I realized that this polyhedron is impossible (Exercise 46.8). So now I like to hold the model up in front of my class and say, "This polyhedron does not exist."

# Exercises

46.1 Show that the only possible triples of integers $(a, b, c)$ all greater than or equal to 3, satisfying the inequality

$$\frac{1}{a} + \frac{1}{b} + \frac{1}{c} > \frac{1}{2}$$

and the additional property that whenever two are distinct, the third is even, are those listed in Table 2.

46.2 Show that the only possible quadruples $(a, b, c, d)$ of integers greater than or equal to 3 (up to cyclic permutation) satisfying

$$\frac{1}{a} + \frac{1}{b} + \frac{1}{c} + \frac{1}{d} > 1$$

and if $a = 3$, then $b = d$, are those listed in Table 2.

46.3 If $a_1, \ldots, a_k$ is a collection of $k \geq 5$ integers, all greater than or equal to 3, satisfying the inequality

$$\sum_{i=1}^{k} \frac{2}{a_i} > k - 2,$$

show that $k = 5$, and the only possibilities are those listed in Table 2.

46.4 (The snub cube.) In a plane, consider a tilted square with vertices $(\pm a, \pm b)$, $(\pm b, \mp a)$. Now consider a cube of side 2 in three-space, centered at the origin, and put one of these tilted squares on each face, always with the same orientation as seen from the outside of the cube. Thus the corners of the tilted square in the top face will have coordinates $(a, b, 1)$, etc. Join the vertices of these tilted squares to nearby vertices in the other faces, to get a figure of 6 squares and 32 triangles. Now write equations in $a, b$ to express that the sides of these triangles are all the same length (including the sides of the tilted squares). Show that two of these equations imply the rest, and then eliminate $a$ or $b$ to obtain one irreducible cubic polynomial with integer coefficients. Solving this equation will construct the snub cube, which is therefore not constructible with ruler and compass.

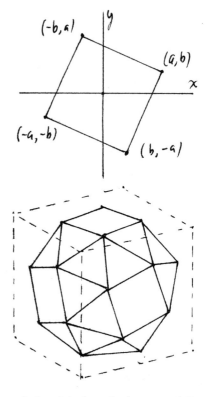

46.5 Explain in what way two vertices of the pseudorhombicuboctahedron are different so as to prevent the existence of a rigid motion or congruence sending one to the other.

46.6 (a) Show that the minimum nonzero value of the expression $|1/2 - (1/a) - (1/b)|$ as $a, b$ range over all integers greater than or equal to 3 is $\frac{1}{42}$.

(b) Similarly, the minimum nonzero value of $|1 - (1/a) - (1/b) - (1/c)|$ for $a, b, c \geq 3$ is $\frac{1}{12}$.

46.7 See how many convex face-regular polyhedra you can discover using only equilateral triangles and squares.

46.8 Give a convincing reason why the stretched cube of (46.4.1) does not exist.

46.9 Some of the most interesting of the nonuniform face-regular polyhedra are those elementary ones that do not arise from cutting up uniform solids. For example, see whether you can make a model (and prove the existence) of a *bilunabirotunda*, having four pentagonal faces, two squares, and 8 equilateral triangles.

# 47  Symmetry Groups of Polyhedra

We will make use of some group theory and a little linear algebra to study the symmetry of polyhedra. The purpose of this section is to elucidate the geometry of polyhedra by finding their symmetry groups, and at the same time to illustrate some concepts of group theory by their applications to geometry.

We measure the symmetry of a figure by looking for ways in which the figure is congruent to itself. So we define a *symmetry* of a figure to be a one-to-one mapping of the figure onto itself that is a congruence, i.e., that preserves all distances and angles. One symmetry followed by another is again a symmetry; the inverse mapping of a symmetry is a symmetry; and composition of symmetries is an associative operation. Hence the set of all symmetries of a figure (including the identity map) is a group, which we call the *symmetry group* of the figure.

For a simple example, look at an equilateral triangle in the plane, and label its vertices $1, 2, 3$. A clockwise rotation through an angle of $2\pi/3$ maps the triangle onto itself by a congruence, so that is a symmetry. This symmetry induces a permutation of the vertices,

$1 \to 2 \to 3 \to 1$, which we represent by the symbol $(123)$. If we perform this rotation twice, we get another symmetry, which permutes the vertices by $(132)$. A third application of the same rotation brings us to the identity, which we denote by $e$.

Another kind of symmetry is obtained by dropping the altitude from the vertex 1 to the midpoint of the side 23, and reflecting the figure in that line. This induces the permutation $(23)$ on the vertices. Reflections in two other axes give the permutations $(12)$ and $(13)$.

A symmetry of the triangle is completely determined if we know what it does to the vertices. So we can list the symmetries so far mentioned by giving the corresponding permutations:

$$e,\ (123),\ (132),\ (12),\ (13),\ (23).$$

It happens that these are all possible permutations of the three symbols $1, 2, 3$.

We conclude (a) that we have now listed all possible symmetries of the triangle, and (b) that the group of symmetries of the triangle is isomorphic to the group of permutations of three symbols, called the *symmetric group on three letters*, and denoted by the symbol $S_3$. We notice in this example that there are three different kinds of symmetries: the identity, the rotations, and the reflections.

An isosceles triangle has a group of only two symmetries: the identity and one reflection. A triangle of three unequal sides has a symmetry group consisting of only the identity. This is a *trivial* symmetry group.

A regular polygon of $n$ sides has a symmetry group consisting of $2n$ elements. The rotation through an angle of $2\pi/n$ generates a cyclic subgroup of $n$ elements, consisting of the identity and $n-1$ rotations through angles $2\pi i/n$ for $i = 1, 2, \ldots, n-1$. Then there are $n$ reflections in lines passing through the vertices and the midpoints of the sides of the polygon, making $2n$ symmetries in all. This group is called the *dihedral group* $D_n$ of order $2n$. (Note that terminology in the literature is not consistent. Some authors call this group $D_{2n}$.)

Naturally, the situation in three dimensions is more complicated. To fix our ideas and to illustrate what can happen, let us consider the symmetries of a tetrahedron. We label its vertices $1, 2, 3, 4$.

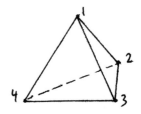

One way to make a symmetry is to rotate the figure around a line. Take, for example, the line through the vertex 1 and the midpoint of the opposite side. Rotating around this axis induces a symmetry of the equilateral triangle and a permutation of the vertices of the tetrahedron (234) that leaves 1 fixed. Twice this will give (243).

On the other hand, if we take an axis through the vertex 2 and the midpoint of its opposite side, we obtain symmetries that induce the permutations (134) and (143).

In the case of a plane figure, it is obvious that one rotation followed by another rotation is again a rotation (or the identity), since all the rotations are around the same point. But in three dimensions, if we rotate first around one axis and then around another axis, we will certainly obtain a symmetry of some kind (i.e., a congruence), but it is not obvious in general whether this will be another rotation. Try it with a model of a tetrahedron and see what happens! Since a symmetry of the tetrahedron is completely determined by what it does to the vertices, we can at least compute the permutation induced by the composition of two rotations. Let us do (234) first followed by (134). Then, reading from left to right

$$(234)(134) = (13)(24).$$

So there is a symmetry that interchanges 1 and 3 and also interchanges 2 and 4. If we take an axis through the midpoint of the side 13 and the midpoint of the side 24, a rotation of $\pi$ around this axis will induce this symmetry. So in this case, the composition of rotations around two different axes is equal to a rotation around a third axis. We will see later that this is true in general (47.2).

Besides rotating the figure around an axis, there are several other ways we can contemplate making symmetries.

One is reflection in a plane. For example, if we consider the plane containing the edge 12 and the midpoint of the edge 34, we can reflect the figure in this plane and obtain a symmetry (34) that leaves 1 and 2 fixed.

Another method is to pick up the figure, turn it around any way, and then replace it in the same spot. Call this a *rigid motion* in 3-space.

Among the abstract symmetries, defined simply as congruences of the figure with itself, we can consider those that *preserve orientation*. Imagine a creature sitting on one vertex of the figure who numbers the faces at that vertex in a clockwise order. If the ordering is still clockwise after the congruence, we say that it preserves orientation. Otherwise, it *reverses orientation*. For example, rotations preserve orientation, and reflections reverse orientation.

How are all these kinds of symmetries related to each other? Let $R$ denote the set of rotations, plus the identity. It is clear that rotations are rigid motions, but it is not obvious that they form a group—we have to show that the product of two rotations is again a rotation. Let $G_0$ denote the group of rigid motions. This group is clearly contained in $G_1$, the group of all orientation-preserving symmetries. And $G_1$ is contained in $G$, the group of all symmetries:

$$R \subseteq G_0 \subseteq G_1 \subseteq G.$$

We will now show, by a counting argument, that $R = G_0 = G_1$ and $G_1 < G$. First we count elements in $R$. For each vertex, there are two rotations leaving that vertex fixed. For each pair of opposite edges, there is one rotation sending those edges to themselves. Adding in the identity makes at least twelve elements in $R$. On the other hand, $G_1 \neq G$ because there are reflections that do not preserve orientation. And since a symmetry is determined by its effect on the vertices, $G$ is isomorphic to a subgroup of the symmetric group $S_4$, which has 24 elements:

$$R \subseteq G_1 < G \subseteq S_4.$$

Now, $R$ has at least 12 elements, and $S_4$ has 24 elements, and the order of a subgroup divides the order of a group, so we conclude that $R = G_0 = G_1$ and $G \cong S_4$. This shows that $R$ is a group, the *group of rotations* of the tetrahedron, and the whole group of symmetries is isomorphic to $S_4$. The permutations in $R$ are of the type (123) or (12)(34), which are *even* permutations, so in fact $R$ is isomorphic to the *alternating group* $A_4$ of even permutations of four letters. Thus we have proved the following result.

### Proposition 47.1

*The group of all symmetries of the tetrahedron is isomorphic to the symmetric group $S_4$. The rotations of the tetrahedron form a subgroup of order* 12, *isomorphic to the alternating group $A_4$.*

A similar analysis of the rotations and symmetries of the octahedron is in Exercises 47.3–47.6. Before discussing the icosahedron, we prove a general result.

### Proposition 47.2

*The composition of two rotations of a polyhedron is again a rotation (or the identity). More generally, any orientation-preserving symmetry of a polyhedron is a rotation about some line (or the identity).*

*Proof*  Since rotations preserve orientation, it will be sufficient to prove the second statement. So let $\varphi$ be an orientation-preserving symmetry of a polyhedron. First we invoke the notion of *centroid* (or center of gravity) of a solid figure (see, e.g., Lines (1965), Chapter IX). It is clear that $\varphi$ sends the centroid to itself. Taking the centroid to be the origin of a coordinate system for $\mathbb{R}^3$, we can extend $\varphi$ to an *isometry* of $\mathbb{R}^3$ leaving the origin fixed. In other words, $\varphi$ preserves distances and angles.

Now we think of $\mathbb{R}^3$ as a three-dimensional vector space over $\mathbb{R}$. Since vector addition is defined by the parallelogram law, $\varphi(v_1 + v_2) = \varphi(v_1) + \varphi(v_2)$. It is also clear that $\varphi(\lambda v) = \lambda \varphi(v)$ for any $\lambda \in \mathbb{R}$. In other words, $\varphi$ is a *linear* map of $\mathbb{R}^3$ into itself. The scalar product $\langle v, w \rangle$ can be defined as $|v| \cdot |w| \cdot \cos \theta$, where $\theta$ is the angle between the two vectors. Since $\varphi$ is an isometry, it preserves this expression, and so $\varphi$ preserves scalar product. In other words, $\varphi$ is an *orthogonal* linear transformation. Its determinant will be $\pm 1$, with $+1$ preserving orientation and $-1$ reversing orientation.

The characteristic polynomial of $\varphi$ has degree 3, so it will have a real root. In other words, there is a real eigenvector $e$ with $\varphi(e) = \lambda(e)$. Because $\varphi$ is orthogonal, $\lambda = \pm 1$.

If there is an eigenvector $e$ with $\varphi(e) = e$, then $\varphi$ leaves the line containing $e$ fixed, and induces an orientation-preserving orthogonal map of a plane perpendicular to $e$. This will be a rotation in the plane, so $\varphi$ is the rotation around the axis of $e$ as required.

If there is no eigenvector $e$ with $\varphi(e) = e$, but only an $e$ with $\varphi(e) = -e$, then the line of $e$ is sent into itself, and $\varphi$ induces an orientation-reversing map on the perpendicular plane. This is the reflection in a line of that plane that has a fixed vector, contradicting the hypothesis that there was none.

So we conclude that $\varphi$ is the rotation around a line, as required.

## Remark 47.2.1

Thus it makes sense to speak of the *rotation group* of any polyhedron, and this group is identical with the group of rigid motions of the figure into itself and the group of all orientation-preserving symmetries.

Now let us discuss the rotation group $G$ of the icosahedron. Knowing that all the rotations form a group, we can use some group theory to find the order of the group, without listing all the individual elements. Let $A$ be one vertex of the icosahedron, and let $H_A$ be the subgroup of $G$ consisting of those rotations that leave $A$ fixed. This is the *stabilizer* subgroup of $A$. If a rotation leaves $A$ fixed, its axis must be the line through $A$ and its opposite, or *antipodal*, point. A rotation through $2\pi/5$ around this axis sends the icosahedron into itself, and generates the group $H_A$, which has order 5.

Next, we look for the *orbit* of $A$ under the action of the whole group $G$, that is, the set of points to which $A$ can be sent by elements of $G$. A rotation of order 3 around an axis through the center of a face adjoining $A$ sends $A$ to one of its neighboring vertices $B$. In the same way, any vertex can be sent to any of its neighboring vertices, and thus the orbit of $A$ under $G$ is the entire set of 12 vertices of the icosahedron. For any vertex $C$, the set of elements of $G$ that send $A$ to $C$ is a left coset of $H_A$, of the form $gH_A$, where $g \in G$ and $g(A) = C$. The number of cosets is called the *index* of the subgroup, and one knows that the order of $G$ is the product of the order of the subgroup $H_A$, which is 5, and the index, which is equal to the number of elements in the orbit of $A$, which is 12. So the order of $G$ is 60.

In fact, it is not too hard to count all the elements of $G$ directly. Each vertex, together with its antipode, corresponds to a subgroup of order 5, which contains the identity and four elements of order 5. So there are $4 \times \frac{12}{2} = 24$ elements of order 5.

Rotation around an axis through the center of opposite faces has order 3. There are two of these for each pair of opposite faces, hence $2 \times \frac{20}{2} = 20$ elements of order 3.

Rotation around an axis through the midpoints of opposite edges has order 2. So there are $1 \times \frac{30}{2} = 15$ elements of order 2. Summing up, we have

| | |
|---|---|
| identity | 1 |
| elements of order 5 | 24 |
| elements of order 3 | 20 |
| elements of order 2 | 15 |
| | 60 |

Next, let us look at subgroups of $G$ and their relation to the geometry of the icosahedron. We have already seen that the stabilizer of a vertex is a subgroup of order 5. Similarly, the stabilizer of the midpoint of a face is of order 3, and the stabilizer of the midpoint of an edge is of order 2.

If we consider an axis through two opposite vertices, the stabilizer of this line includes the stabilizer of one vertex, but allows also rotations that send the vertex to its antipode. This is a dihedral group $D_5$. Similarly, the stabilizer of an axis through the middle of two opposite faces is a dihedral group $D_3$ (isomorphic to the symmetric group $S_3$). The stabilizer of an axis through the middles of two opposite edges is $D_2$ (isomorphic to the Klein four-group $V$).

These subgroups have orders $2, 3, 5, 4, 6, 10$. There are also subgroups of order 12 (Exercise 47.9). On the other hand, we can show that certain other orders of subgroups are impossible. Let us show, for example, that there is no subgroup $H$ of order 15. If there were, then by Cayley's theorem it would contain an element of order 5, which would be a rotation around a vertex $A$. Then $H$ would contain the group generated by that element, which is $H_A$. The orbits of the set of vertices under the action of $H_A$ are $A$, its antipode, and two orbits of 5 vertices each. The group $H$ would also contain an element of order 3, so the orbit of $A$ under $H$ would contain at least 6 elements, and so the order of $H$ would be at least $5 \times 6 = 30$, a contradiction. (See also Exercise 47.10.)

Now let us discuss conjugation and normal subgroups. Two elements $a, b \in G$ are *conjugate* if there exists a $g \in G$ with $b = gag^{-1}$. If $a$ is a rotation around a vertex $A$, then $b$ is the same kind of rotation around the vertex $B = g(A)$. Indeed, $g^{-1}$ takes the vertex $B$ back to $A$, $a$ performs the rotation, and $g$ takes $A$ back to $B$. So geometrically, two rotations are conjugate if they are rotations through the same angle around two different axes. In particular, since any point of the icosahedron can be moved to its antipode, every rotation is conjugate to its own inverse. Thus all elements of order 2 are conjugate, and all elements of order 3 are conjugate. But the elements of order 5 fall into two conjugacy classes: The rotations of $\pm 2\pi/5$ form one class, and the rotations of $\pm 4\pi/5$ form the other class. They are distinguished by the property that the first kind map some faces to an adjacent face, while the second kind map no face to an adjacent face. Thus the whole group is divided into conjugacy classes with $1, 12, 12, 20$, and 15 elements, respectively.

A *normal subgroup* of a group $G$ is a subgroup $N$ that is stable under conjugation: $gNg^{-1} = N$ for any $g \in G$. If a normal subgroup contains an element $a$, it must also contain all the conjugates of $a$. In this way we can verify that the rotation group of the icosahedron contains no normal subgroups except the identity $\{e\}$ and the whole group $G$. Indeed, a normal subgroup $N$ contains 1, its order divides 60, and $N$ must be a union of 1 together with some subset of the conjugacy classes of orders $12, 12, 20, 15$. There is no sum of these numbers that divides 60 except 1 and 60. Thus $G$ has no nontrivial normal subgroups, and we say that $G$ is a *simple* group.

Finally, we show that the icosahedral group $G$ is isomorphic to the alternating group $A_5$ on 5 letters. Define a *frame* of the icosahedron to be a set of three mutually perpendicular axes through the middles of opposite edges. There are five such frames. Any rotation of the icosahedron induces a permutation of these five frames. An element of order 5 gives a permutation $(abcde)$. An element of order 3 gives a permutation of the form $(abc)$. An element of order 2 gives a permutation of the form $(ab)(cd)$. These are all even permutations, so we obtain a homomorphism from $G$ to $A_5$. Clearly the map is injective, and the two groups have the same order, so $G \cong A_5$. Thus we have proved the following.

### Proposition 47.3

*The group of rotations of the icosahedron is a simple group of order* 60, *isomorphic to the alternating group* $A_5$.

The exercises contain more examples of rotation groups, symmetry groups, and their properties.

The rotations of a polyhedron induce rotations of the sphere, and so determine certain finite subgroups of the *special orthogonal* group SO(3) of orthogonal linear transformations with determinant 1. The next theorem shows that the rotation groups of the regular polyhedra (plus the cyclic and dihedral groups) are in fact the only possible finite subgroups of SO(3).

### Theorem 47.4

*Any finite subgroup of* SO(3) *(the rotation group of a sphere) is isomorphic to one of the following*:

$$C_n, \text{ cyclic, for } n \geq 1,$$
$$D_n, \text{ dihedral, for } n \geq 2,$$
$$T, \text{ the tetrahedral group, } \cong A_4,$$
$$O, \text{ the octahedral group, } \cong S_4,$$
$$I, \text{ the icosahedral group, } \cong A_5.$$

*Furthermore, if two finite subgroups of* SO(3) *are isomorphic as abstract groups, then they are conjugate as subgroups of* SO(3).

*Proof*  Fix a sphere $S$ centered at the origin, and regard SO(3) as the group of rotations of the sphere. Let $G$ be a finite subgroup of SO(3), i.e., a finite group of rotations of the sphere $S$, and let $N$ be the order of $G$.

Each nonidentity element of $G$ is a rotation about some axis passing through the center of the sphere (47.2). The points where this axis meets the sphere are the two *poles* of the rotation. Since the group $G$ is finite, the set $P$ of all the poles of all the rotations will be a finite set. Furthermore, $G$ acts as a group of permutations on this set, because if $x \in P$ is a pole of a rotation $a \in G$, and if $g$ is any element of $G$, then $g(x)$ is a pole of the conjugate rotation $gag^{-1}$ (cf. discussion of the icosahedron above).

For each $x \in P$, let $H_x \subseteq G$ be the stabilizer subgroup of $x$, and let $P_x \subseteq P$ be the orbit of $x$ under the action of $G$. Then $r_x n_x = N$, where $r_x$ is the order of the subgroup $H_x$, and $n_x$ is the number of points in the orbit $P_x$ (cf. discussion of stabilizers and orbits in the case of the icosahedron above).

On the other hand, we can count the nonidentity elements of $G$ as follows. Each one has two poles, and for each pole $x$ there are $r_x - 1$ elements in the subgroup $H_x$. If we sum over the poles, then

$$\frac{1}{2}\sum_{x \in P}(r_x - 1) = N - 1.$$

Now write $P$ as the union of its orbits $P_i$, $i = 1, \ldots, t$, under $G$. Within an orbit, the numbers $r_x$ and $n_x$ are the same, so we can rewrite this sum as a sum over the $t$ orbits of $P$,

$$\frac{1}{2}\sum_{i=1}^{t} n_i(r_i - 1) = N - 1.$$

Now recall $r_i n_i = N$ for each $i$, and divide by $N$. This gives the fundamental equation

$$\sum_{i=1}^{t}\left(1 - \frac{1}{r_i}\right) = 2 - \frac{2}{N}.$$

We shall carry out the classification of possible subgroups $G$ by examining possible solutions of this equation for integers $r_i \geq 2$ and $N \geq 1$, remembering that $r_i$ divides $N$ for each $i$.

Since each $r_i \geq 2$, the left-hand side is at least $\frac{1}{2}t$, while the right-hand side is less than 2. We conclude that $t \leq 3$.

*Case 1*   $t = 1$. The only solution of the equation is $r = N = 1$, which does not satisfy our restriction $r \geq 2$, but we can associate it with the trivial subgroup $G = \{e\}$.

*Case 2*   $t = 2$. In this case the equation reduces to

$$\frac{2}{N} = \frac{1}{r_1} + \frac{1}{r_2}.$$

Remembering that $r_i n_i = N$ and multiplying through by $N$, we get

$$n_1 + n_2 = 2.$$

The only solution here is $n_1 = n_2 = 1$, so $r_1 = r_2 = N$. Thus there is just one axis with its two poles, and $G$ is isomorphic to a cyclic group of order $N$, for any $N \geq 1$.

*Case 3* $t = 3$. In this case our equation gives

$$\frac{1}{r_1} + \frac{1}{r_2} + \frac{1}{r_3} = 1 + \frac{2}{N}.$$

In particular, the left-hand side must be greater than 1, and the only triples of $r_1, r_2, r_3 \geq 2$ that achieve this are $(2, 2, n), (2, 3, 3), (2, 3, 4), (2, 3, 5)$.

If $(r_1, r_2, r_3) = (2, 2, n)$, then $N = 2n$, and $n_3 = 2$. Thus there is one axis having a cyclic group of rotations of order $n$, and there are two other orbits of $n$ axes each of twofold rotations. Thus $G$ is a dihedral group $D_n$.

Now suppose $(r_1, r_2, r_3) = (2, 3, 3)$. Then we obtain $N = 12$ and $(n_1, n_2, n_3) = (6, 4, 4)$. Choose one of the orbits of four points and call them $A, B, C, D$. The stabilizer $H_A$ is a group of order 3, leaving $A$ fixed and permuting $B, C, D$ cyclically. Hence $B, C, D$ are equidistant from $A$. The same argument applies to $B, C, D$, so all four points are equidistant from each other. Joining them by lines makes a tetrahedron inscribed in the sphere. Now, $G$ permutes $A, B, C, D$, so it induces rotations of the tetrahedron, and we get a group homomorphism $G \rightarrow T$, where $T$ is the rotation group of the tetrahedron. The image of $G$ clearly generates $T$, and both groups have the same order, so $G$ is isomorphic to $T$.

We leave the remaining two cases to the reader (Exercise 47.16).

For the last statement of the theorem, suppose that $G_1$ and $G_2$ are two finite groups of rotations of the sphere that are isomorphic as abstract groups. Then they are isomorphic to the same one in the list. But the proof of the theorem shows more. In the cases $C_n, D_n$, there is a principal axis around which there are rotations through $2\pi/n$, and this axis determines the group. For $C_n$, we can move the principal axis of $G_1$ to that of $G_2$ by a rotation $g$ of the sphere, and this same element creates a conjugacy $G_2 \cong gG_1g^{-1}$. For $D_n$, we can require in addition that $g$ take a secondary axis of $G_1$ to $G_2$, and then again $G_2 = gG_1g^{-1}$.

In the case of $T, O, I$, we showed that there is an inscribed tetrahedron, octahedron, or icosahedron, and $G$ is its group of rotations. We need only find a rotation $g$ of the sphere that takes two neighboring vertices of the first figure to the second. The rest will follow, and then $G_2 \cong gG_1g^{-1}$ will be the required conjugacy.

# Exercises

47.1 Label the vertices of a regular pentagon $1, 2, 3, 4, 5$, and list all the symmetries of the pentagon as permutations of these five symbols. Show that the resulting group, isomorphic to $D_5$, is actually a subgroup of the alternating group $A_5$ inside of $S_5$.

47.2 (a) List all the rotations of a tetrahedron as permutations of the four vertices.

(b) List the other 12 symmetries of the tetrahedron. Which of these are given by reflection in a plane? Show that those that are not reflections can be described as *screw reflections*, namely, reflection in a plane followed by a rotation about an axis perpendicular to the plane.

47.3  Mark the vertices of an octahedron $1, 2, \ldots, 6$. List all the rotations of the octahedron by the permutations they induce on the vertices. How many elements of each kind are there? What are their orders? How many in all?

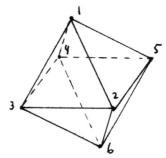

47.4  The octahedron has four axes $a, b, c, d$ running through the centers of opposite faces. Any rotation induces a permutation of $a, b, c, d$. Thus we get a map $\varphi : R \to S_4$ from the set of rotations to the symmetric group on the four letters $a, b, c, d$. Show that $R$ has at least 24 elements, show that the map $\varphi$ is injective, and conclude that $R$ is a group isomorphic to $S_4$.

47.5  Find subgroups of the group of rotations of the octahedron isomorphic to $C_2, C_3, C_4, D_2 = V, D_3 = S_3, D_4$, and describe them in terms of the geometry of the octahedron.

47.6  Show that the group of all symmetries of the octahedron is a group of order 48.

47.7  Give a geometric proof that the composition of two rotations of the sphere about arbitrary axes is equal to another rotation of the sphere, by using spherical geometry. Show that it must have a fixed point (cf. Exercise 17.11 and note that translations do not exist on the sphere).

47.8  Extend the proof of Proposition 47.2 to show that any symmetry of a polyhedron is either the identity, a rotation, a reflection in a plane, or a screw-reflection (cf. Exercise 47.2).

47.9  Show that it is possible to select 4 faces of an icosahedron in such a way that the subgroup of rotations that preserve that set of four faces is isomorphic to the group $T$ of rotations of the tetrahedron.

47.10  Show that the group of rotations of the icosahedron contains no subgroup of order 20 or 30.

47.11  Show that the full group of symmetries of the icosahedron is a group of order 120 that is not isomorphic to the symmetric group $S_5$.

47.12  In the group $G$ of rotations of the icosahedron, show that one can find elements $a, b$ with the properties

(a) $a$ and $b$ together generate the whole of $G$;

(b) $a^2 = e, b^3 = e, (ab)^5 = e$.

Now show that $G$ is the largest group with the properties (a), (b), in the following sense: If $G'$ is another group, generated by elements $x, y$ satisfying $x^2 = y^3 = (xy)^5 = e$, and if $\varphi : G' \to G$ is a homomorphism that sends $x, y$ to $a, b$, then $\varphi$ is an isomorphism. We say that $G$ is the group given by *generators* $a, b$ and *relations* (b) above.

47.13 (a) Give a criterion in terms of the geometry of the axes and angles of rotation for two rotations of a polyhedron to commute with each other (i.e., $ab = ba$).

(b) Give a similar criterion for a rotation to commute with reflection in a plane.

(c) Show that the antipodal map commutes with all other symmetries.

47.14 (a) Show that the group of rotations of an $n$-sided prism and an $n$-sided antiprism are both isomorphic to the dihedral group $D_n$.

(b) Show, however, that the full group of symmetries of the $n$-prism is not isomorphic (as an abstract group) to the symmetries of the $n$-antiprism. *Hint*: To show that two groups are not isomorphic, you must find some group-theoretic property true of one but not the other, such as having an element of a certain order, or having a certain number of elements of a given order, or having an element of order 2 in the *center*, i.e., that commutes with every other elements.

47.15 (a) Show that the rotation group and the full symmetry group of a cube are the same as for an octahedron.

(b) Ditto for the dodecahedron and icosahedron.

(c) Examine the list of semiregular polyhedra (Theorem 46.1) and find in each case the rotation group and the full symmetry group. Pay special attention to the snub cube and the pseudorhombicuboctahedron.

47.16 Complete the proof of Theorem 47.4 by showing that in the cases $(r_1, r_2, r_3) = (2, 3, 4)$ or $(2, 3, 5)$ there is an octahedron or icosahedron inscribed in the sphere in such a way that $G$ is identified with its group of rotations.

47.17 If $H$ is any subgroup of one of the groups $C_n, D_n, T, O, I$, then of course $H$ is isomorphic to a subgroup of SO(3) also. Verify that this does not contradict Theorem 47.4 by showing directly that any subgroup of any of the groups in this list is also isomorphic to one of the groups in the list.

47.18 Show that the full symmetry group of the tetrahedron and the rotation group of the octahedron are isomorphic as abstract groups, but they are not conjugate as subgroups of $O(3)$, the orthogonal group of all symmetries of the sphere.

47.19 Consider the group of transformations of the set $\mathbb{C} \cup \{\infty\}$ generated by the operations

$$\alpha : z' = -z,$$

$$\beta : z' = z^{-1},$$

$$\gamma : z' = \frac{z+i}{z-i}.$$

(a) Show that $\alpha, \beta, \gamma$ generate a group $G$ of order 12.

(b) Show that $G$ permutes the set $\{0, \pm 1, \pm i, \infty\}$.
(c) Lift the operation of $G$ on $\mathbb{C}$, considered as a plane, to the unit sphere by the stereographic projection, so that $\pm 1, \pm i$ land on the equator. Then show that $G$ is identified with the group of rotations of the octahedron with vertices $\{0, \pm 1, \pm i, \infty\}$.

It was on that night that he dreamed his dream of titanic basalt towers—dripping with slime and ocean ooze and fringed with great sea mats—their wierdly proportioned bases buried in gray-green muck and their non-Euclidean-angled parapets fading into the watery distances of that unquiet submarine realm.

– from *Rising with Surtsey*
by Brian Lumley
in Tales of the Cthulhu Mythos
by H. P. Lovecraft
*&* divers hands
Arkham, Sauk City (1990) p. 315
Reprinted by permission of
Arkham House Publishers, Inc.
Sauk City, WI, USA

# Appendix: Brief Euclid

For reference we include abbreviated statements of the most frequently quoted results from Euclid's *Elements*.

**Book I. Definitions**
1. A *point* is that which has no part.
2. A *line* is length without breadth.
4. A *straight line* lies evenly with its points.
8. A *plane angle* is the inclination of two lines.
10. When the two adjacent angles are equal it is a *right angle*.
15. A *circle* is a line all of whose points are equidistant from one point.
20. A triangle with two equal sides is *isosceles*.
23. *Parallel* straight lines are lines in the same plane that do not meet, no matter how far extended in either direction.

**Postulates**
1. To draw a line through two points.
2. To extend a given line.
3. To draw a circle with given center through a given point.
4. All right angles are equal.
5. If a line crossing two other lines makes the interior angles on the same side less than two right angles, then these two lines will meet on that side when extended far enough.

**Common Notions**
1. Things equal to the same thing are equal.
2. Equals added to equals are equal.

3. Equals subtracted from equals are equal.
4. Things which coincide are equal.
5. The whole is greater than the part.

## Propositions
1. To construct an equilateral triangle on a given segment.
2. To draw a segment equal to a given segment at a given point.
3. To cut off a smaller segment from a larger segment.
4. Side–angle–side (SAS) congruence for triangles.
5. The base angles of an isosceles triangle are equal.
6. If the base angles are equal, the triangle is isosceles.
7. It is not possible to put two triangles with equal sides on the same side of a segment.
8. Side–side–side (SSS) congruence for triangles.
9. To bisect an angle.
10. To bisect a segment.
11. To construct a perpendicular to a line at a given point on the line.
12. To drop a perpendicular from a point to a line not containing the point.
13. A line standing on another line makes angles equal to two right angles.
15. Vertical angles are equal.
16. The exterior angle of a triangle is greater than either opposite interior angle.
17. Any two angles of a triangle are less than two right angles.
18. If one side of a triangle is greater than another, then the angle opposite it is greater than the other.
19. If one angle of a triangle is greater than another, then the side opposite it is greater than the other.
20. Any two sides of any triangle are greater than the third.
22. To construct a triangle, given three sides, provided any two are greater than the third.
23. To reproduce a given angle at a given point and side.
24. Two sides equal but included angle greater of two triangles implies base greater.
25. Two sides equal and greater base implies greater angle.
26. Angle–side–angle (ASA) and angle–angle–side (AAS) congruence for triangles.
27. Alternate interior angles equal implies parallel lines.
28. Exterior angle equal to opposite interior, or two interior angles equal to two right angles, implies parallel lines.
29. A line crossing two parallel lines makes alternate interior angles equal.
30. Lines parallel to the same line are parallel.
31. To draw a line parallel to a given line through a given point.
32. Sum of angles of a triangle is two right angles, and exterior angle equals the sum of opposite interior angles.

33. Lines joining endpoints of equal parallel lines are equal and parallel.
34. The opposite sides and angles of a parallelogram are equal.
35. Parallelograms on the same base and in the same parallels are equal.
36. Parallelograms on equal bases in the same parallels are equal.
37. Triangles on the same base in the same parallels are equal.
38. Triangles on equal bases in the same parallels are equal.
39. Equal triangles on the same base on the same side are in the same parallels.
40. Equal triangles on equal bases on the same side are in the same parallels.
41. A parallelogram is twice the triangle on the same base in the same parallels.
42. To construct a parallelogram with a given angle equal to a given triangle.
43. Parallelograms on opposite sides of the diagonal of a parallelogram are equal.
44. To construct a parallelogram with given side and angle equal to a given triangle.
45. To construct a parallelogram with a given angle equal to a given figure.
46. To construct a square on a given segment.
47. (Theorem of Pythagoras) The square on the hypotenuse is equal to the sum of the squares on the sides of a right triangle.
48. If the sum of the squares on two sides equals the square on the third side, the triangle is right.

## Book II. Propositions
1. The rectangle contained by two lines is the sum of the rectangles contained by one and the segments of the other.
4. The square on the whole line is equal to the squares on its two segments plus twice the rectangle on the two segments.
5. The square on half a line is equal to the rectangle on the unequal segments plus the square of the difference.
6. The rectangle on a line plus an added piece with the added piece, plus the square of half the segment, is equal to the square of the half plus the added piece.
11. To cut a line so that the rectangle on the whole and one segment is equal to the square on the other segment (extreme and mean ratio).
14. To construct a square equal to a given figure.

## Book III. Propositions
1. To find the center of a circle.
2. The segment joining two points of a circle lies inside the circle.
5. If two circles intersect, they do not have the same center.
6. If two circles are tangent, they do not have the same center.
10. Two circles can intersect in at most two points.
11, 12. If two circles are tangent, their centers lie in a line with the point of tangency.
16. The line perpendicular to a diameter at its end is tangent to the circle, and

the angle between the tangent line and the circle is less than any rectilineal angle.

17. To draw a tangent to a circle from a point outside the circle.
18. A tangent line to a circle is perpendicular to the radius at the point of tangency.
19. The perpendicular to a tangent line at the point of tangency will pass through the center of the circle.
20. The angle at the center is twice the angle at a point of the circumference subtending a given arc of a circle.
21. Two angles from points of a circle subtending the same arc are equal.
22. The opposite angles of a quadrilateral in a circle are equal to two right angles.
31. The angle in a semicircle is a right angle.
32. The angle between a tangent line and a chord of a circle is equal to the angle on the arc cut off.
35. If two chords cut each other, the rectangle on the segments of one chord is equal to the rectangle on the segments of the other chord.
36. From a point outside a circle, let a tangent and a secant line be drawn. Then the square of the tangent line is equal to the rectangle formed by the two segments from the point to the circle on the secant line.
37. From a point outside a circle, if two lines cut the circle, so that the square of one is equal to the rectangle formed by the segments of the other, then the first is a tangent line.

## Book IV. Propositions

1. To inscribe a given segment in a circle.
2. To inscribe a triangle, equiangular to a given triangle, in a circle.
3. To circumscribe a triangle, equiangular to a given triangle, around a circle.
4. To inscribe a circle in a triangle.
5. To circumscribe a circle around a triangle.
10. To construct an isosceles triangle whose base angles are twice the vertex angle.
11. To inscribe a regular pentagon in a circle.
12. To circumscribe a regular pentagon around a circle.
15. To inscribe a regular hexagon in a circle.
16. To inscribe a regular 15-sided polygon in a circle.

## Book V. Definitions

4. Magnitudes are said to *have a ratio* if either one, being multiplied, can exceed the other.
5. Four magnitudes $a$, $b$; $c$, $d$ are *in the same ratio* if for any whole numbers $m$, $n$, we have $ma > nb$ or $ma = nb$ or $ma < nb$ if and only if $mc > nd$ or $mc = nd$ or $mc < nd$ respectively.

## Book VI. Propositions
1. Triangles of the same height are in the same ratio as their bases.
2. A line is parallel to the base of a triangle if and only if it cuts the sides proportionately.
3. A line from a vertex of a triangle to the opposite side bisects the angle if and only if it cuts the opposite side in proportion to the remaining sides of the triangle.
4. The sides of equiangular triangles are proportional.
5. If the sides of two triangles are proportional, their angles are equal.
6. If two triangles have one angle equal and the sides containing the angle proportional, the triangles will be similar.
8. The altitude from the right angle of a right triangle divides the triangle into two triangles similar to each other and to the whole.
12. To find a fourth proportional to three given lines.
13. To find a mean proportional between two given lines.
16. Four lines are proportional if and only if the rectangle on the extremes is equal to the rectangle on the means.
30. To cut a line in extreme and mean ratio.
31. Any figure on the hypotenuse of a right triangle is equal to the sum of similar figures on the sides of the triangle.

## Book X. Propositions
1. Given two unequal quantities, if one subtracts from the greater a quantity greater than its half, and repeats this process enough times, there will remain a quantity lesser than the smaller of the two original quantities.
117. (not in Heath, but in Commandino). The diagonal of a square is incommensurable with its side.

## Book XI. Definitions
25. A *cube* is a polyhedron made of six equal squares.
26. An *octahedron* is a polyhedron made of eight equal equilateral triangles.
27. An *icosahedron* is a polyhedron made by twenty equal equilateral triangles.
28. A *dodecahedron* is a polyhedron made by twelve equal regular pentagons.

## Propositions
21. The plane angles in a solid angle make less than four right angles.
28. A parallelepiped is bisected by its diagonal plane.
29, 30. Parallelepipeds on the same base and of the same height are equal.
31. Parallelepipeds on equal bases, of the same height, are equal.

## Book XII. Propositions
2. Circles are in the same ratio as the squares of their diameters.
3. A pyramid is divided into two pyramids and two prisms.

5. Pyramids of the same height on triangular bases are in the same ratio as their bases.

7. A prism with a triangular base is divided into three equal triangular pyramids.

## Book XIII. Propositions

7. If at least three angles of an equilateral pentagon are equal, the pentagon will be regular.

10. In a circle, the square on the side of the inscribed pentagon is equal to the square on the side of the inscribed hexagon plus the square on the side of the inscribed decagon.

13. To inscribe a tetrahedron in a sphere.

14. To inscribe an octahedron in a sphere.

15. To inscribe a cube in a sphere.

16. To inscribe an icosahedron in a sphere.

17. To inscribe a dodecahedron in a sphere.

18. (Postscript). Besides these five figures there is no other contained by equal regular polygons.

# Notes

**Section 1.** To appreciate this text you should have a copy of Euclid's *Elements* handy. The most natural choice for an English-speaking reader is Heath's authoritative translation and commentary (1926) available in an inexpensive Dover reprint.

The "hard problem" in this section is discussed in Coxeter and Greitzer (1967).

**Section 5.** For the history of the theorems of the three medians and the three altitudes of a triangle, see Tropfke (1923) vol. 4, pp. 163, 164. Both are implicit in the work of Archimedes, but were not included in the repertoire of elementary geometry until much later.

The proof given for (5.6) is due to Gauss, Werke (1870–77), vol. 4, p. 396.

We give several different proofs that the altitudes of a triangle meet in a point:

(a)  Gauss's proof (5.6)
(b)  using the Euler line (5.7)
(c)  using cyclic quadrilaterals (Exercise 5.7)
(d)  using the angle bisectors of the orthic triangle (5.10)
(e)  by analytic geometry (13.1)
(f)  in the Poincaré model (Exercise 39.17)
(g)  using angle bisectors (Exercise 40.14)
(h)  by non-Euclidean analytic geometry (41.13)
(i)  using the calculus of reflections (43.15).

The Euler line (5.7) was discovered by Euler in 1765. The nine-point circle appeared in a paper of Brianchon and Poncelet in 1821, and independently in a paper of Feuerbach in 1822. Feuerbach found only six of the nine points, but he also proved the remarkable result that this circle is tangent to the inscribed circle and the three exscribed circles of the triangle.

**Section 6.** The axioms presented in this and subsequent Sections 7, 8, 9 are essentially the same as the axioms proposed by Hilbert in his *Foundations of Geometry* (1971). We have made a few small changes.

First of all, Hilbert postulates a set of points and a set of lines, together with a relation of incidence "a point lies on a line." We, however, postulate a set of points, and take lines to be subsets of the set of points, so that the incidence of a point and line simply becomes membership in the set.

Second, Hilbert formulates his axioms for 3-dimensional space. We have taken only the plane axioms because they exhibit all the essential features of the geometry we need.

Third, Hilbert is a minimalist. For example, in the (SAS) axiom (C6), having assumed $AB \cong DE$, $AC \cong DF$, and $\angle BAC = \angle EDF$, he postulates only that $\angle ABC = \angle DEF$. He then proves as a theorem that $\angle ACB \cong \angle DFE$ and $BC \cong EF$. This degree of minimalism in the axioms seems unnecessary for an elementary text such as ours, so we have simplified in a couple of places by making an axiom slightly stronger than necessary.

Exercise 6.3. Unfortunately, we do not have the space to develop the ideas of projective geometry further. But see, e.g., Hartshorne (1967) for an introduction.

Exercise 6.9. Kirkman's schoolgirl problem was first published in the *Lady's and Gentlement's Diary* for (1850). See Ball (1940), Chapter X, for an extensive discussion. One solution can be found by taking the fifteen points of the projective 3-space over the field of 2 elements to be the girls, and the 35 lines of 3 points each to be the rows. Then with a little care you can find five lines that fill the space and an automorphism $\sigma$ of order 7 that cycles those five lines through the set of all the lines in such a way as to solve the problem.

**Section 7.** It is possible to take (7.1) as an axiom and then prove (B4), as is done in Greenberg (1993).

**Section 12.** Here is what Dedekind says about continuity (from *Stetigkeit und Irrationale Zahlen* (1872)):

> I find the essence of continuity in the following principle: "If all the points of a line fall into two classes in such a way that each point of the first class lies to the left of each point of the second class, then there exists one and only one point that gives rise to this division of all the points into two classes, this cutting of the line into two pieces."

As mentioned before, I believe I am not wrong if I assume that everyone will immediately admit the truth of this assertion; most of my readers will be very disappointed to realize that by this triviality the mystery of continuity will be revealed. I am very glad if everyone finds the above principle so clear and so much in agreement with his own conception of a line; for I am not in a position to give any kind of proof of its correctness; nor is anyone else. The assumption of this property of the line is nothing else than an axiom by which we first recognize the continuity of the line, through which we think continuity into the line [die Stetigkeit in die Linie hineindenken]. If space has any real existence at all, it does not necessarily need to be continuous; countless properties would remain the same if it was discontinuous. And if we knew for certain that space was discontinuous, still nothing could hinder us, if we so desired, from making it continuous in our thought by filling up its gaps; this filling up would consist in the creation of new point-individuals, and would have to be carried out in accord with the above principle.

**Section 14.** The theorem of Pappus occurs in a different form in Pappus (1876), Book VII, Proposition 139.

I got the idea for Exercises 14.4–14.13 from a paper of Sturmfels.

**Section 18.** While the possibility of a non-Archimedean geometry was perhaps fore-shadowed by the controversy about the angle between a circle and its tangent line, started by Peletier and Clavius in the sixteenth century, the first serious study of a non-Archimedean geometry is due to Veronese at the end of the nineteenth century. Our treatment follows Hilbert's approach via fields. See Enriques (1907), Chapter VII.

**Section 19.** For the segment arithmetic, we follow Hilbert's *Foundations* (1971), Chapter III, with simplifications by Enriques described in Supplement II. The idea of making an arithmetic of line segments goes back to Descartes in *La Géométrie* (1637), except that he made no effort to justify the usual rules of arithmetic as applied to line segments.

**Section 22.** Awareness of the problem of area grew in the nineteenth century (see Simon (1906), Section 15). We follow Hilbert's treatment in his *Foundations* (1971), Chapter IV. Hilbert was the first to recognize the importance of Archimedes' axiom in the theory. Terminology is not uniform. What we call "equal content" is sometimes called "equicomplementable" (German: ergänzungsgleich). What we call "equidecomposable" or "equivalent by dissection" is sometimes called "equivalent by finite decomposition" (German: zerlegungsgleich, or teilungsgleich). The axiom (Z) is named after A. de Zolt, who attempted in 1881 to prove this statement geometrically (cf. Simon (1906), Section 15).

**Section 24.** For the history of the Bolyai–Gerwien theorem, see Simon (1906), Section 15. For practical notes on efficient dissections, see Lindgren (1964). For

Exercise 24.8, see Dudeney (1929). A new book by Frederickson (1997) promises to become the standard reference for dissectors.

**Section 25.** There is a vast literature on the problem of squaring the circle and attempts to understand the analytic as well as the geometrical significance of the number $\pi$. In fact, this single problem has been the catalyst for significant advances in many branches of mathematics over its 4000-year history. A full discussion of the subject would lead way beyond the confines of this book. For an approach to the literature, see, for example, Beckmann (1971), Hobson (1953), Simon (1906), Section 6, and Rudio (1892).

**Section 26.** The correspondence between Gauss and Gerling can be found in Gauss, Werke (1870–77), vol. VIII, pp. 241 ff.

**Section 27.** This treatment of the Dehn invariant is based on Cartier's Bourbaki seminar talk (1985). See also Boltianskii (1978) for a detailed treatment of the problem.

**Section 28.** There is an extensive literature on these classical problems and attempts at their solution. See, for example, Klein (1895), Enriques (1907), or Lebesgue (1950). Descartes (1637) was already aware that the first two required cubic equations and could not be solved by quadratic equations. The first proof of the impossibility was given by Wantzel (1837), but his proof has a gap. The first complete proof is due to Petersen (1871), cf. also Petersen (1878). Our proof using the notion of the degree of a field extension and the characterization (28.7) using the Galois group are apparently due to van der Waerden (1930).

For the squaring of the circle, see notes to Section 25.

**Section 29.** Gauss's proof of the constructibility of the regular 17-gon can be found in his *Disquisitiones Arithmeticae*, reprinted in his Werke (1870–77), vol. I.

The construction given in the text is due to Maywald: See the book of Goldenring (1915), p. 16, who collected more than twenty different constructions.

**Section 30.** For references on the use of the marked ruler for solving the classical problems of trisecting the angle and doubling the cube, see Pappus (1876) Book III, Section 7, Enriques (1907) II, pp. 204 ff and pp. 233 ff and Knorr (1986) pp. 341 ff.

In his history of the conic sections in antiquity, Zeuthen (1886) expresses the opinion that the early geometers accepted the use of the marked ruler (German: Einschiebung; French: règle à glissière; Italian: riga segnata; Greek: neusis) along with the ruler and compass as a legitimate tool of construction, and that it was only from the time of Plato and Euclid that a strict distinction was made of those problems that could be solved with ruler and compass only. Once the theory of conics was well developed, the use of conics to solve the "solid" problems was preferred to the marked ruler.

The construction of (30.1) is ascribed by Pappus to "the ancients," but he says that the construction of (30.2) is due to Nicomedes (see Pappus (1878), Book IV, Proposition 32).

Viète's construction of the regular heptagon was published in his *Supplementum Geometriae* of 1593 (reprinted in Viète (1970)). Similar constructions have been rediscovered periodically: See Collins (1866), Plemelj (1912), Bieberbach (1952), Gleason (1988), among others. Archimedes' work on the heptagon (Exercise 30.6) was lost until 1926, when it was found in an Arabic manuscript (see Knorr (1986) pp. 178 ff).

**Section 31.** The story of the discovery of the solutions to the cubic and quartic equations in the early sixteenth century by Ferro, Tartaglia, Cardano, and Ferrari is one of the most colorful chapters of the history of mathematics (see, for example, Eves (1953), Section 8–8). The solution of the *casus irreducibilis* of the cubic equation by trisecting an angle is due to Viète.

**Section 32.** Consult your favorite algebra book. One I like is Stewart (1989).

**Section 33.** References to the work of Proclus, Tacquet, Clairaut, Clavius, Simson, and Playfair can be found under their names in the References.

For more details on the history of the theory of parallels and the discovery of non-Euclidean geometry, see the book of Bonola (1955), which includes the texts of Bolyai and Lobachevsky, or the book of Engel and Stäckel (1895), which includes selections from the work of Wallis, Saccheri, Lambert, Gauss, Schweikart, and Taurinus.

Among more recent texts, the books of Wolfe (1945) and Greenberg (1993) have very readable accounts of the development of non-Euclidean geometry.

**Section 34.** The results on Saccheri quadrilaterals are due to Saccheri (1733), except that he used continuity arguments in the proof of dividing all geometries into three cases (34.7). A proof of this theorem without using continuity was first given by Lambert (see his work in Engel and Stäckel (1895), esp. Section 57, p. 187). The present proof of the key proposition (34.4) is due to Bonola (1955), Section 14.

The theory of limiting parallel rays is originally due to Gauss (Werke (1870–77), vol. 8, pp. 202–209), but there was a gap in his proof of transitivity (34.11), because you cannot assume of three nonintersecting lines that one is in between the other two (cf. Moise (1963) Section 24.2). This gap is filled by (34.12).

**Section 35.** Theorem (35.2), that the semielliptic case (Saccheri's hypothesis of the obtuse angle) is impossible, was first proved by Saccheri (1733). The present proof is the correct part of the proof given by Legendre (1823) for his Proposition 19, Book I, in which he claimed to show that the angle sum in any triangle is equal to 2RA. The last part of his proof used an untenable limit argument.

The example of a semielliptic plane (Exercise 34.14) is essentially the same as an example given by Dehn (1900) of what he called a non-Legendrean geometry, since it did not satisfy the conclusion of (35.2).

**Section 36.** The main ideas for the proof of the non-Euclidean Bolyai–Gerwien theorem (36.6) are already present in Gerwien's second paper of (1833), in which he treated the case of spherical polygons. The full treatment without the use of Archimedes' axiom is due to Finzel (1912). The unwound circle group is my invention. While Finzel must have had something like this in mind, he did not make explicit in what group his area function took its values.

**Section 37.** According to Max Simon (1906), p. 93, the notion of circular inversion first appears in the work of Poncelet, followed by Steiner, Quetelet, Magnus, and Plücker, all in the first half of the nineteenth century. Since then it has become a useful standard technique.

Constructions with compass alone were studied by Mascheroni (1797), who proved that compass alone suffices to carry out any construction possible with ruler and compass (Exercise 37.26).

The cross-ratio occurs in Pappus (1876), Book VII, Proposition 129, except that instead of a ratio of ratios, which could not be expressed in the language of that time, it was a proportion between the rectangle $AP \cdot BQ$ and the rectangle $AQ \cdot BP$. This proposition easily implies the invariance of cross-ratio under projection (Exercise 37.14).

**Section 39.** A note on the consistency of non-Euclidean geometry. The discoverers of these geometries, Gauss, Bolyai, and Lobachevsky, seem to have been convinced of the existence of these geometries by the extensive theory they developed and its internal coherence. Lobachevsky also noted that the formulas of hyperbolic trigonometry could be obtained by taking the formulas of spherical trigonometry for a sphere whose radius is imaginary.

Still, a rigorous proof of consistency (or at least consistency relative to Euclidean geometry) is best made by producing a model. Credit for the first such model goes to Beltrami (1868), who found that the geometry on surfaces of constant negative curvature in 3-space behaves like a hyperbolic plane. Beltrami seems also to have been aware of the several representations of hyperbolic geometry in the Euclidean plane, used later by Klein and Poincaré. Klein preferred the "projective" model, where the plane is represented by the points inside a fixed circle, and the lines by chords of Euclidean lines in that circle. This gives the most direct connection with the projective metric of Cayley, and allows one to use the powerful methods of projective geometry.

Poincaré (1882), on the other hand, was led through his investigations of automorphic functions to a conformal model using the points of the Euclidean upper half-plane, in which the lines become half-circles orthogonal to the *x*-axis. This model can be easily transformed into the model we use, whose points are the interior of a circle, and whose lines are segments of circles orthogonal to the fixed circle, which was developed extensively by Carslaw (1916).

For references, see Bonola (1955), Poincaré's *Science and Hypothesis* (1905), and Klein's *Nichteuklidische Geometrie* (1927).

We chose the Poincaré model for this book because it can be developed directly from Euclidean geometry using inversion in circles, and because it seems the most direct and elementary approach. We could not have developed the Klein model fully without a considerable excursion into projective geometry, which would extend beyond the scope of the present book.

Exercise 39.22. The theorem that the medians of a triangle meet in a point is still true without the hypothesis of a circumscribed circle, but the proof, using projective geometry, is more difficult. See Greenberg (1993), p. 277, also Baldus–Löbell (1953), p. 102, Liebmann (1923), p. 22. There is a proof in neutral geometry using the calculus of reflections (Exercise 43.15).

Some authors call the example of Exercise 39.28 a *half-elliptic* plane, because any two lines have either a common point or a common orthogonal. However, I find this terminology misleading, because this plane is semihyperbolic, not semielliptic.

The example (Exercise 39.31) of a plane in which not every segment is the side of an equilateral triangle is new, although Pambuccian (1998) has independently found a similar example.

**Section 40.** The axiom (L) was proposed by Hilbert in his article "A new development of Bolyai–Lobachevskian geometry," reprinted as Appendix III to his *Foundations of Geometry* (1971). This article also contains the proofs of (40.5), existence of a common perpendicular, and (40.6) existence of the enclosing line.

The proof suggested in Exercise 40.14 is due to Liebmann (1923), p. 31.

**Section 41.** The construction of the field of ends (41.2), (41.3), and (41.4), together with the equation of a line (41.6) is taken from Hilbert's article "A new development ..." cited above. After he has verified the field axioms, he says that "the construction of the geometry poses no further difficulties" (p. 147).

In the remainder of the section we have worked out various applications of Hilbert's field. The proofs of (41.9)–(41.13), using the multiplicative distance function (41.7) to avoid the real numbers, are new.

Bolyai's parallel construction first appears in Bolyai (1832), Section 34. There are several different proofs in the literature:

(a) proofs using the relations between right triangles and quadrilaterals with three right angles (called Spitzecke by Liebmann), and using hyperbolic trigonometry. See Greenberg (1993) p. 413, Bonola (1955), App. III, and Liebmann (1923) p. 35.

(b) a proof using a prism in hyperbolic 3-space — Bonola (1955), App. III.

(c) a purely geometric proof using the Hjelmslev midline theorem, due to Liebmann: See Carslaw (1916) p. 73 or Wolfe (1945) p. 95.

(d) proofs in the Klein model, using the projective geometry of the ambient plane: Greenberg (1993) p. 269; Baldus–Löbell (1953) p. 93.

**Section 42.** The formulae of hyperbolic trigonometry (42.2) and (42.3) are in Lobachevsky (1914), pp. 38, 41. The hyperbolic law of sines and law of cosines (Exercises 42.6, 42.7c) are in (loc. cit.) p. 44.

Our derivation of these formulae directly from Hilbert's field of ends, and independent of the real numbers, is, as far as I know, new, except that Szász (1953) has derived equivalent formulae using the isomorphism of the hyperbolic plane with the Poincaré model.

Bolyai, in his *Science of Absolute Space* (1832; English translation in Bonola (1955)), Section 43, studied the area of a circle, and recognized that it could be "squared" or not, depending on the arithmetic properties of the number $A/\pi$.

**Section 43.** The characterization of hyperbolic planes (43.3) is a natural consequence of the ideas set forth in Hilbert's "Neue Begründung ..." article (Appendix III of *Foundations* (1971)). It gives a direct and elementary proof of a result that otherwise could be derived only as a consequence of the much deeper theorem of Pejas (1961). We have reformulated Pejas's theorem somewhat for clarity. To understand the development of the ideas leading to Pejas's theorem is tantamount to reviewing the entire history of the role of projective geometry in the foundations of elementary geometry. Some useful references are Greenberg (1979), the comments of Bachmann (1959) pp. 25–26, running comments in Hessenberg–Diller (1967), Dehn's appendix to Pasch–Dehn (1976), and the encyclopedia article of Enriques (1907–10).

**Section 44.** The picture of the five regular solids at the beginning of this section is from Kepler, *Harmonices Mundi* (1619). Reprinted courtesy of the Bancroft Library of the University of California at Berkeley.

# References

[1] Altshiller-Court, N., *College Geometry*, Johnson Publishing Company, Richmond, VA (1925); 2nd ed. Barnes & Noble, New York (1952).

[2] Apollonii, P., *Conicorum Libri Quattuor*. (Trans. F. Commandino) Bononiae (1566).

[3] Archibald, R.C., *Euclid's book on divisions of figures, with a restoration based on Woepke's text and on the Practica Geometriae of Leonardo Pisano*, University Press, Cambridge (1915).

[4] Archimedes, *Quae supersunt omnia cum Eutocii Ascalonitae commentariis*. Ex recensione Josephi Torelli, Clarendon, Oxford (1792).

[5] Artin, E., *Geometric Algebra*, Interscience, New York (1957).

[6] Bachmann, F., *Aufbau der Geometrie aus dem Spiegelungsbegriff*, Springer, New York (1973) (first published 1959).

[7] Bachmann, F., Zur Parallelenfrage, *Abh. Math. Sem. Univ. Hamburg* **27** (1964) 173–192.

[8] Baldus, R. and Löbell, F., *Nichteuklidische Geometrie, hyperbolische Geometrie der Ebene*, 3rd Ed., de Gruyter, Berlin (1953).

[9] Ball, W.W.R., *Mathematical Recreations & Essays*, revised by H.S.M. Coxeter, 11th edition, Macmillan, London (1940) (first published 1892).

[10] Beckmann, P., *A history of $\pi$* (PI), 2nd ed., Golem, Boulder (1971).

[11] Beltrami, E., Saggio di interpretazione della geometria non euclidea, *Giornale di Mat.* **6** (1868) 284–312.

[12] Bieberbach, L., *Theorie der geometrischen Konstruktionen*, Birkhäuser, Basel (1952).

**495**

[13] Birkhoff, G.D., A set of postulates for plane geometry (based on scale and protractor), *Annals of Math.* **33** (1932) 329–345.

[14] Blanchet, M.A., *Eléments de Géométrie par A.M. Legendre, avec additions et modifications*, 7 ème édition, Firmin Didot, Paris (1859).

[15] Blumenthal, L.M., *A Modern View of Geometry*, W.H. Freeman, San Francisco (1961).

[16] Boltianskii, V.G., *Hilbert's Third Problem*, Winston, Washington (1978).

[17] Bolyai, J., *Appendix scientiam spatii absolute veram exhibens: a veritate aut falsitate Axiomatis XI Euclidei (a priori haud unquam decidenda) independentem* (1832) (English translation in Bonola (1955)).

[18] Bonola, R., *Non-Euclidean Geometry, a critical and historical study of its development.* English translation by H.S. Carslaw. With a supplement containing the Dr. George Bruce Halsted translations of the Science of Absolute Space by John Bolyai, and The Theory of Parallels, by Nicholas Lobachevski, Dover, New York (1955) (English translation of Bonola, first published 1912).

[19] Borsuk, K. and Szmielew, W., *Foundations of Geometry, Euclidean and Bolyai–Lobachevskian Geometry, Projective Geometry*, North–Holland, Amsterdam (1960).

[20] Callahan, J.J., *Euclid or Einstein. A proof of the parallel theory and a critique of metageometry.* Devin–Adair, New York (1931).

[21] Camerer, J.G., *Apollonii de Tactionibus, quae supersunt, ac maxime lemmata Pappi, in hos libros Graece nunc primum edita, e codicibus mscptis, cum Vietae librorum Apollonii restitutione, adjectis observationibus, computationibus, ac problematis Apolloniani historia*, Ettinger, Gothae (1795).

[22] Carslaw, H.S., *The Elements of Non-Euclidean Plane Geometry and Trigonometry*, Longmans, Green, London (1916).

[23] Cartier, P., Décomposition des polyèdres: le point sur le troisième problème de Hilbert, *Sém. Bourbaki* **646** (1985).

[24] Cederberg, J.N., *A Course in Modern Geometries*, Springer, New York (1989).

[25] Clairaut, M., *Elémens de Géométrie*, Cellot & Jombert, Paris (1775) (first published 1741).

[26] Clavius, C., *Euclidis Elementorum, libri XV, accessit XVI de solidorum regularium cuiuslibet intra quodlibet comparatione. Omnes perspicuis Demonstrationibus, accuratisque Scholiis illustrati, ac multarum rerum accessione locupletati*, 3rd ed., Ciotti, Coloniae (1591) (first published 1574).

[27] Collins, M., Solution de question 656, *Nouvelles Annales de Math.* **5** (1866) 226.

[28] Commandino, F., *De gli Elementi d'Euclide libri quindici, con gli scholii antichi, tradotti prima in lingua latina, & con commentarij illustrati, et hora d'ordine dell'istesso transportati nella nostra vulgare, & da lui riveduti.* In Casa di F. Commandino, Urbino (1575) (translation of his Latin ed. of 1572).

[29] Coxeter, H.S.M., *Non-Euclidean Geometry*, University Press, Toronto (1942).

[30] Coxeter, H.S.M., *Introduction to Geometry*, Wiley, New York (1961).

[31] Coxeter, H.S.M. and Greitzer, S.L., *Geometry Revisited*, Random House, New York (1967).

[32] Cromwell, P.R., *Polyhedra*, Cambridge University Press (1997).

[33] Dedekind, J.W.R. *Stetigkeit und irrationale Zahlen*, Braunschweig (1872).

[34] Dehn, M., Die Legendre'schen Sätze über die Winkelsumme im Dreieck, *Math. Annalen* **53** (1900) 404–439.

[35] Descartes, R., *La Géométrie*, first published as an appendix to the *Discours de la Métode* (1637). First separate French edition (1664). English translation: *The Geometry of René Descartes*, tr. D.E. Smith and M.L. Latham, Chicago, Open Court (1925).

[36] Dodgson, C.L., *Euclid and his Modern Rivals*, Macmillan, London (1879).

[37] Dudeney, H.E., *The Canterbury Puzzles, and other curious problems*, Nelson, London (1929) (first published 1908).

[38] Dudley, U., *A Budget of Trisections*, Springer, New York (1987).

[39] Engel, F. and Stäckel, P., *Die Theorie der Parallellinien von Euklid bis auf Gauss, eine Urkundensammlung zur Vorgeschichte der nichteuklidischen Geometrie*, Teubner, Leipzig (1895).

[40] Enriques, F., Prinzipien der Geometrie, *Enz. Math. Wiss.* III.I.1 (1907–1910) 1–129.

[41] Enriques, F., *Fragen der Elementargeometrie, II Teil: Die Geometrischen Aufgaben, ihre Lösung und Lösbarkeit*, Teubner, Leipzig (1907).

[42] Euclid, *Elements* (see different editions under the names of Clavius, Commandino, Heath, Peletier, Peyrard, Playfair, Simson, Tacquet).

[43] Eves, H., *An Introduction to the History of Mathematics*, Rinehart, New York (1953).

[44] Eves, H., *College Geometry*, Jones and Bartlett, Boston (1995).

[45] Finzel, A., Die Lehre vom Flächeninhalt in der allgemeinen Geometrie, *Math. Annalen* **72** (1912) 262–284.

[46] Frederickson, G.N., *Dissections, Plane and Fancy*, Cambridge University Press (1997).

[47] Freudenthal, H., Zur Geschichte der Grundlagen der Geometrie, *Nieuw Archief voor Wiskunde*, **4** (1957) 105–142.

[48] Gauss, C.F., *Werke*, Göttingen (1870–1877).

[49] Gerwien, P., Zerschneidung jeder beliebigen Anzahl von gleichen geradlinigen Figuren in dieselben Stücke, *Journal für die riene und angewandte Mathematik*, ed. Crelle **10** (1833) 228–234.

[50] Gerwien, P., Zerschneidung jeder beliebigen Menge verschieden gestalteter Figuren von gleichem Inhalt auf der Kugelfläche in dieselben Stücke, *Journal für die reine und angewandte Mathematik* (Crelle) **10** (1833) 235–240.

[51] Gleason, A.M., Angle trisection, the heptagon, and the triskaidecagon, *Amer. Math. Monthly* (1988) 185–194.

[52] Goldenring, R., *Die elementargeometrischen Konstruktionen des regelmässigen Siebzehnecks, eine historisch-kritische Darstellung*, Teubner, Leipzig (1915).

[53] Greenberg, M., Euclidean and non-Euclidean geometries without continuity, *Amer. Math. Monthly* **86** (1979) 757–764.

[54] Greenberg, M.J., Aristotle's axiom in the foundations of geometry, *J. Geometry* **33** (1988) 53–57.

[55] Greenberg, M.J., *Euclidean and non-Euclidean Geometries, Development and History*, 3rd ed., Freeman, New York (1993).

[56] Hadamard, J., *Leçons de géométrie élémentaire*, Paris (1901–1906).

[57] Hall, M., *The Theory of Groups*, Macmillan, New York (1959).

[58] Hartshorne, R., *Foundations of Projective Geometry*, Benjamin, New York (1967).

[59] Heath, T.L., *The thirteen books of Euclid's Elements, translated from the text of Heiberg, with introduction and commentary*, 2nd ed., 3 vols, University Press, Cambridge (1926) (Dover reprint 1956).

[60] Herstein, I.N., *Topics in Algebra*, 2nd ed., Xerox, Lexington (1975).

[61] Hessenberg, G. and Diller, J., *Grundlagen der Geometrie*, 2nd ed., de Gruyter, Berlin (1967).

[62] Hilbert, D., *Foundations of Geometry* (Grundlagen der Geometrie), 2nd English ed., translated by Leo Unger from the tenth German edition, revised and enlarged by Dr. Paul Bernays, Open Court, La Salle (1971) (first published 1899).

[63] Hjelmslev, J., Neue Begründung der ebenen Geometrie, *Math. Ann.* **64** (1907) 449–474.

[64] Hobson, E.W., *Squaring the Circle*, Chelsea, New York (1953) (first published 1913).

[65] Johnson, N.W., Convex polyhedra with regular faces, *Canadian J. Math.* **18** (1966) 169–200.

[66] Klein, F., *Vorträge über ausgewählte Fragen der Elementargeometrie*, ausgearbeitet von F. Tägert, Teubner, Leipzig (1895).

[67] Klein, F., *Vorlesungen über Nicht-Euklidische Geometrie*, neu bearbeitet von W. Rosemann, Chelsea, New York, n.d. (first published 1927).

[68] Kline, M., *Mathematical Thought from Ancient to Modern Times*, Oxford University Press, New York (1972).

[69] Knorr, W.R., *The Ancient Tradition of Geometric Problems*, Birkhäuser, Boston (1986).

[70] Lebesgue, H., *Leçons sur les constructions géométriques, professées au Collège de France en* 1940–1941, Gauthier–Villars, Paris (1950).

[71] Legendre, A.M., *Eléments de Géométrie*, avec des notes, 12 ème éd., Firmin Didot, Paris (1823).

[72] Leistner, J.I.C., *Unwiderrufflicher, Wohlgegründter und Ohnendlicher Beweis der wahren Quadratur des Circuls, oder des Durchmessers zu seinem Umcreysz, wie* 1225 *zu* 3844 *oder* 3844 *zu* 1225, Heyinger, Wienn (1737).

[73] Liebmann, H., *Nichteuklidische Geometrie*, 3rd ed., de Gruyter, Berlin (1923) (first published 1905).

[74] Lindgren, H., *Geometric Dissections*, van Nostrand, Princeton (1964).

[75] Lines, L., *Solid Geometry*, Dover Publications (1965) (first published in 1935).

[76] Lobachevsky, N., *Geometrical researches on the theory of parallels*, tr. G.B. Halsted, Open Court, La Salle IL (1914) (first published in 1840).

[77] Lobachevsky, N.I., *Collection complète des oeuvres géométriques de N.J. Lobatcheffsky*, Kazan (1866).

[78] Loomis, E.S., *The Pythagorean Proposition, its demonstrations analyzed and classified, and bibliography of sources for data of the four kinds of "proof,"* Nat. Council of Teachers of Math., Washington (1968) (first published 1940).

[79] Lucas, E., *Récréations Mathématiques*, Gauthier–Villars, Paris (1882), 4 vols.

[80] Lyusternik, L.A., *Convex Figures and Polyhedra*, Dover, New York (1963).

[81] Martin, G.E., *The Foundations of Geometry and the Non-Euclidean Plane*, Springer, New York (1975).

[82] Martin, G.E., *Geometric Constructions*, Springer, New York (1998).

[83] Mascheroni, L., *Geometria del Compasso*, Pavia (1797).

[84] Millman, R.S. and Parker, G.D., *Geometry*, Springer, New York (1991).

[85] Moise, E.E., *Elementary Geometry from an Advanced Standpoint*, Addison–Wesley, Reading (1963).

[86] Montucla, J.F., *Histoire des Mathématiques*, Blanchard, Paris (1968) (first published 1758).

[87] Pambuccian, V., Zum Stufenaufbau des Parallelenaxioms, *J. Geometry* **51** (1994) 79–88.

[88] Pambuccian, V., Zur Existenz gleichseitiger Dreiecke in *H*-Ebenen, *Journal of Geometry* **63** (1998) 147–153.

[89] Pappus, *Pappi Alexandrini collectionis quae supersunt*, ed. F. Hultsch, Berlin (1876), 3 vols.

[90] Pasch, M. and Dehn, M., *Vorlesungen über neuere Geometrie*, 2nd ed. mit einem Anhang: Die Grundlegung der Geometrie in historischer Entwicklung, Springer, Berlin (1976) (first published 1882).

[91] Pejas, W., Die Modelle des Hilbertschen Axiomensystems der absoluten Geometrie, *Math. Annalen* **143** (1961) 212–235.

[92] Peletier, J., *In Euclidis Elementa Geometrica Demonstrationum Libri sex*, Tornaesium, Lugduni (1557).

[93] Petersen, J., Om Ligninger, der løses ved Kvadratrod, med Anvendelse paa Problemers Løsning ved Passer og Lineal, Kopenhagen (1871).

[94] Petersen, J., *Theorie der algebraischen Gleichungen*, Kopenhagen (1878).

[95] Peyrard, F., *Les Elémens de Géométrie d'Euclide, traduits littéralement, et suivis d'un Traité du Cercle, du Cylindre, du Cône et de la Sphère; de la Mesure des Surfaces et des Solides; avec des Notes*, 2nd ed., Louis, Paris (1809) (first published 1804).

[96] Playfair, J., *Elements of Geometry, containing the first six books of Euclid*, Edinburgh (1795).

[97] Plemelj, J., Die Siebenteilung des Kreises, *Monatshefte für Math. u. Physik* **23** (1912) 309–311.

[98] Poincaré, H., Théorie des groups fuchsiens, *Acta Math.* **1** (1882) 1–62 and *Oeuvres*, Tome II (1952) 108–168.

[99] Poincaré, H., *Science and Hypothesis*, trans. W.J. Greenstreet, London (1905).

[100] Proclus, *A Commentary on the first book of Euclid's Elements*, translated with introduction and notes by Glenn R. Morrow, University Press, Princeton (1970).

[101] Rudio, F., *Archimedes, Huygens, Lambert, Legendre. Vier Abhandlungen über die Kreismessung, Deutsch herausgegeben und mit einer Uebersicht über die Geschichte des Problems von der Quadratur des Zirkels von den ältesten Zeiten bis auf unsere Tage*, Teubner, Leipzig (1892).

[102] Saccheri, G., *Euclides ab omni naevo vindicatus: sive conatus Geometricus quo stabiliuntur prima ipsa universae Geometriae Principia*, edited and translated by George Bruce Halsted, Open Court, Chicago (1920) (first published 1733).

[103] Simon, M., *Ueber die Entwicklung der Elementar-Geometrie im XIX Jahrhundert*, Teubner, Leipzig (1906).

[104] Simson, R., *The Elements of Euclid, viz. the first six books, together with the eleventh and twelfth. The errors, by which Theon, or others, have long ago vitiated these books, are corrected, and some of Euclid's demonstrations are restored.* Glasgow (1756); later edition (1803).

[105] Smith, D.E., *A Source Book in Mathematics*, Dover, New York (1959) (first published 1929).

[106] Sommerville, M.A., *Bibliography of Non-Euclidean Geometry*, Harrison, London (1911).

[107] Steck, M., *Bibliographia Euclideana*, Gerstenberg, Hildesheim (1981).

[108] Steiner, J., *Gesammelte Werke*, ed. K. Weierstrass, Berlin, G. Reimer (1881, 1882), 2 vols.

[109] Steinitz, E., *Algebraische Theorie der Körper*, Chelsea, New York (1950) (first published 1910).

[110] Stewart, I., *Galois Theory*, 2nd ed., Chapman & Hall, London (1989).

[111] Stillwell, J., *Mathematics and Its History*, Springer, New York (1989).

[112] Szász, P., Ueber die Hilbertsche Begründung der hyperbolischen Geometrie, *Acta Math. Acad. Sci. Hungaricae* **4** (1953) 243–250.

[113] Tacquet, A., *Elementa Geometriae planae ac solidae, quibus accedunt selecta ex Archimede theoremata*, Manfrè, Patavii (1738).

[114] Thomas-Stanford, C., *Early Editions of Euclid's Elements*, Wofsy, San Francisco (1977).

[115] Tropfke, J., *Geschichte der Elementarmathematik, vol. IV, Ebene Geometrie*, de Gruyter, Berlin (1923).

[116] van der Waerden, B.L., *Moderne Algebra*, Berlin, Springer (1930), 2 vols.

[117] Viète, F., *Opera Mathematica*, Olms, Hildesheim (1970) (reprint of 1646 edition).

[118] von Staudt, K.G.C., *Geometrie der Lage*, Bauer u. Raspe, Nürnberg (1847).

[119] Wantzel, P.L., Recherches sur les moyens de reconnaître si un problème de géométrie peut se résoudre avec la règle et le compas, *J. Math.* **2** (1837) 366–372.

[120] Wolfe, H.E., *Introduction to Non-Euclidean Geometry*, Holt, Rinehart, Winston, New York (1945).

[121] Zeuthen, H.G., *Die Lehre von den Kegelschnitten im Altertum*, A.F. Höst, Kopenhagen (1886).

# List of Axioms

I1–I3　　incidence axioms (Section 6)
P　　　　Playfair's axiom (Section 6)
B1–B4　betweenness axioms (Section 7)
C1–C3　congruence for line segments (Section 8)
C4–C6　congruence for angles (Section 9)
E　　　　circle–circle intersection (Section 11)
A　　　　Archimedes' axiom (Section 12)
D　　　　Dedekind's axiom (Section 12)
Z　　　　de Zolt's axiom (Section 22)
L　　　　existence of limiting parallel rays (Section 40)

A *Hilbert plane* (Section 10) is a geometry satisfying (I1)–(I3), (B1)–(B4), and (C1)–(C6).

A *Euclidean plane* (Section 12) is a Hilbert plane satisfying (P) and (E).

A *hyperbolic plane* (Section 40) is a Hilbert plane satisfying (L).

A Hilbert plane is *semi-Euclidean, semielliptic,* or *semihyperbolic* (Section 34) according as the sum of the angles in a triangle is $= 2\text{RA}, > 2\text{RA}$, or $< 2\text{RA}$.

The *Cartesian plane* (Section 14) over a field $F$ is the usual analytic geometry on the set $F^2$.

The *Poincaré model* (Section 39) over a field $F$ is the non-Euclidean geometry on the set of points inside a fixed circle.

For other axioms, acronyms, and definitions, see the Index.

# Index of Euclid's Propositions

# Index

# Undergraduate Texts in Mathematics

*(continued from page ii)*

**Franklin:** Methods of Mathematical Economics.

**Frazier:** An Introduction to Wavelets Through Linear Algebra

**Gamelin:** Complex Analysis.

**Gordon:** Discrete Probability.

**Hairer/Wanner:** Analysis by Its History. *Readings in Mathematics.*

**Halmos:** Finite-Dimensional Vector Spaces. Second edition.

**Halmos:** Naive Set Theory.

**Hämmerlin/Hoffmann:** Numerical Mathematics. *Readings in Mathematics.*

**Harris/Hirst/Mossinghoff:** Combinatorics and Graph Theory.

**Hartshorne:** Geometry: Euclid and Beyond.

**Hijab:** Introduction to Calculus and Classical Analysis.

**Hilton/Holton/Pedersen:** Mathematical Reflections: In a Room with Many Mirrors.

**Hilton/Holton/Pedersen:** Mathematical Vistas: From a Room with Many Windows.

**Iooss/Joseph:** Elementary Stability and Bifurcation Theory. Second edition.

**Isaac:** The Pleasures of Probability. *Readings in Mathematics.*

**James:** Topological and Uniform Spaces.

**Jänich:** Linear Algebra.

**Jänich:** Topology.

**Jänich:** Vector Analysis.

**Kemeny/Snell:** Finite Markov Chains.

**Kinsey:** Topology of Surfaces.

**Klambauer:** Aspects of Calculus.

**Lang:** A First Course in Calculus. Fifth edition.

**Lang:** Calculus of Several Variables. Third edition.

**Lang:** Introduction to Linear Algebra. Second edition.

**Lang:** Linear Algebra. Third edition.

**Lang:** Short Calculus: The Original Edition of "A First Course in Calculus."

**Lang:** Undergraduate Algebra. Second edition.

**Lang:** Undergraduate Analysis.

**Laubenbacher/Pengelley:** Mathematical Expeditions.

**Lax/Burstein/Lax:** Calculus with Applications and Computing. Volume 1.

**LeCuyer:** College Mathematics with APL.

**Lidl/Pilz:** Applied Abstract Algebra. Second edition.

**Logan:** Applied Partial Differential Equations.

**Lovász/Pelikán/Vesztergombi:** Discrete Mathematics.

**Macki-Strauss:** Introduction to Optimal Control Theory.

**Malitz:** Introduction to Mathematical Logic.

**Marsden/Weinstein:** Calculus I, II, III. Second edition.

**Martin:** Counting: The Art of Enumerative Combinatorics.

**Martin:** The Foundations of Geometry and the Non-Euclidean Plane.

**Martin:** Geometric Constructions.

**Martin:** Transformation Geometry: An Introduction to Symmetry.

**Millman/Parker:** Geometry: A Metric Approach with Models. Second edition.

**Moschovakis:** Notes on Set Theory.

**Owen:** A First Course in the Mathematical Foundations of Thermodynamics.

**Palka:** An Introduction to Complex Function Theory.

**Pedrick:** A First Course in Analysis.

**Peressini/Sullivan/Uhl:** The Mathematics of Nonlinear Programming.

**Prenowitz/Jantosciak:** Join Geometries.

## Undergraduate Texts in Mathematics